A course in mathematics for students of physics: 1

A COURSE IN

# mathematics

FOR STUDENTS OF
PHYSICS: 1

PAUL BAMBERG
SHLOMO STERNBERG

The right of the University of Cambridge to print and sell all manner of books was granted by Henry VIII in 1534. The University has printed and published continuously since 1584.

CAMBRIDGE UNIVERSITY PRESS

*Cambridge*

*New York Port Chester*

*Melbourne Sydney*

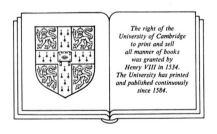

Published by the Press Syndicate of the University of Cambridge
The Pitt Building, Trumpington Street, Cambridge CB2 1RP
40 West 20th Street, New York, NY 10011, USA
10 Stamford Road, Oakleigh, Melbourne 3166, Australia



First published 1988
Reprinted 1990

Printed in Great Britain at the University Press, Cambridge

*British Library cataloguing in publication data*

Bamberg, Paul
A course in mathematics for students of
physics: 1.
1. Mathematical physics.
I. Title. II. Sternberg, Shlomo
510′.2453 QC20

*Library of Congress cataloguing in publication data*

Bamberg, Paul G.
A course in mathematics for students of physics: 1.
Bibliography
Includes index.
1. Mathematics–1961–. I. Sternberg, Shlomo.
II. Title.
QA37.2.B36 1988 510 86–2230

ISBN 0 521 25017 X

TM

# CONTENTS OF VOLUME 1

# CONTENTS OF VOLUME 2

# PREFACE

This book, with apologies for the pretentious title, represents the text of a course we have been teaching at Harvard for the past eight years. The course is aimed at students with an interest in physics who have a good grounding in one-variable calculus. Some prior acquaintance with linear algebra is helpful but not necessary. Most of the students simultaneously take an intensive course in physics and so are able to integrate the material learned here with their physics education. This also is helpful but not necessary. The main topics of the course are the theory and physical application of linear algebra, and of the calculus of several variables, particularly the exterior calculus. Our pedagogical approach follows the 'spiral method' wherein we cover the same topic several times at increasing levels of sophistication and range of application, rather than the 'rectilinear approach' of strict logical order. There are, we hope, no vicious circles of logical error, but we will frequently develop a special case of a subject, and then return to it for a more general definition and setting only after a broader perspective can be achieved through the introduction of related topics. This makes some demands of patience and faith on the part of the student. But we hope that, at the end, the student is rewarded by a deeper intuitive understanding of the subject as a whole.

Here is an outline of the contents of the book in some detail. The goal of the first four chapters is to develop a familiarity with the algebra and analysis of square matrices. Thus, by the end of these chapters, the student should be thinking of a matrix as an object in its own right, and not as a square array of numbers. We deal in these chapters almost exclusively with $2 \times 2$ matrices, where the most complicated of the computations can be reduced to solving quadratic equations. But we always formulate the results with the higher-dimensional case in mind. We begin Chapter 1 by explaining the relation between the multiplication law of $2 \times 2$ matrices and the geometry of straight lines in the plane. We develop the algebra of $2 \times 2$ matrices and discuss the determinant and its relation to area and orientation. We define the notion of an abstract vector space, in general, and explain the concepts of basis and change of basis for one- and two-dimensional vector spaces.

In Chapter 2 we discuss conformal linear geometry in the plane, that is, the geometry of lines and angles, and its relation to certain kinds of $2 \times 2$ matrices. We also discuss the notion of eigenvalues and eigenvectors, so important in quantum mechanics. We use these notions to give an algorithm for computing the powers of a matrix. As an application we study the basic properties of Markov chains.

The principal goal of Chapter 3 is to explain that a system of homogeneous linear differential equations with constant coefficients can be written as $\mathrm{d}\mathbf{u}/\mathrm{d}t = A\mathbf{u}$ where $A$ is a matrix and $\mathbf{u}$ is a vector, and that the solution can be written as $\mathrm{e}^{At}\mathbf{u}_0$ where $\mathbf{u}_0$ gives the initial conditions. This of course requires us to explain what is meant by the exponential of a matrix. We also describe the qualitative behavior of solutions and the inhomogeneous case, including a discussion of resonance.

Chapter 4 is devoted to the study of scalar products and quadratic forms. It is rich in physical applications, including a discussion of normal modes and a detailed treatment of special relativity.

Chapters 5 and 6 present the basic facts of the differential calculus. In Chapter 5 we define the differential of a map from one vector space to another, and discuss its basic properties, in particular the chain rule. We give some physical applications such as Kepler motion and the Born approximation. We define the concepts of directional and partial derivatives, and linear differential forms.

In Chapter 6 we continue the study of the differential calculus. We present the vector versions of the mean-value theorem, of Taylor's formula and of the inverse function theorem. We discuss critical point behavior and Lagrange multipliers.

Chapters 7 and 8 are meant as a first introduction to the integral calculus. Chapter 7 is devoted to the study of linear differential forms and their line integrals. Particular attention is paid to the behavior under change of variables. Other one-dimensional integrals such as arc length are also discussed.

Chapter 8 is devoted to the study of exterior two-forms and their corresponding two-dimensional integrals. The exterior derivative is introduced and invariance under pullback is stressed. The two-dimensional version of Stokes' theorem, i.e. Green's theorem, is proved. Surface integrals in three-space are studied.

Chapter 9 presents an example of how the results of the first eight chapters can be applied to a physical theory – optics. It is all in the nature of applications, and can be omitted without any effect on the understanding of what follows.

In Chapter 10 we go back and prove the basic facts about finite-dimensional vector spaces and their linear transformations. The treatment here is a straightforward generalization, in the main, of the results obtained in the first four chapters in the two-dimensional case. The one new algorithm is that of row reduction. Two important new concepts (somewhat hard to get used to at first) are introduced: those of the dual space and the quotient space. These concepts will prove crucial in what follows.

Chapter 11 is devoted to proving the central facts about determinants of $n \times n$

matrices. The subject is developed axiomatically, and the basic computational algorithms are presented.

Chapters 12–14 are meant as a gentle introduction to the mathematics of shape, that is, algebraic topology. In Chapter 12 we begin the study of electrical networks. This involves two aspects. One is the study of the 'wiring' of the network, that is, how the various branches are interconnected. In mathematical language this is known as the topology of one-dimensional complexes. The other is the study of how the network as a whole responds when we know the behavior of the individual branches, in particular, power and energy response. We give some applications to physically interesting networks.

In Chapter 13 we continue the study of electrical networks. We examine the boundary-value problems associated with capacitive networks and use these methods to solve some classical problems in electrostatics involving conductors.

In Chapter 14 we give a sketch of how the one-dimensional results of Chapters 12 and 13 generalize to higher dimensions.

Chapters 15–18 develop the exterior differential calculus as a continuous version of the discrete theory of complexes. In Chapter 15 the basic facts of the exterior calculus are presented: exterior algebra, $k$-forms, pullback, exterior derivative and Stokes' theorem.

Chapter 16 is devoted to electrostatics. We suggest that the dielectric properties of the vacuum give the continuous analog of the capacitance of a network, and that these dielectric properties are what determine Euclidean geometry in three-dimensional space. The basic facts of potential theory are presented.

Chapter 17 continues the study of the exterior differential calculus. The main topics are vector fields and flows, interior products and Lie derivatives. These are applied to magnetostatics.

Chapter 18 concludes the study of the exterior calculus with an in-depth discussion of the star operator in a general context.

Chapter 19 can be thought of as the culmination of the course. It applies the results of the preceding chapters to the study of Maxwell's equations and the associated wave equations.

Chapters 20 and 21 are essentially independent of Chapters 9–19 and can be read independently of them. They are not usually included in our one-year course. But Chapters 1–9, 20 and 21 would form a self-contained unit for a shorter course.

The material in Chapter 20 is a relatively standard treatment of the theory of functions of a complex variable, suitable for students at the level of this book.

Chapter 21 discusses some of the more elementary aspects of asymptotics.

Chapter 22 shows how the exterior calculus can be used in classical thermodynamics, following the ideas of Born and Carathéodory.

The book is divided into two volumes, with Chapters 1–11 in volume 1.

Most of the mathematics and all of the physics presented in this book were developed by the first decade of the twentieth century. The material is thus at least seventy-five years old. Yet much of the material is not yet standard in the elementary courses (although most of it with the possible exception of network theory must be learned for a grasp of modern physics, and is studied at some stage of the physicist's career). The reasons are largely historical. It was apparent to Hamilton that the real and complex numbers were insufficient for the deeper study of geometrical analysis, that one wants to treat the number pairs or triplets of the Cartesian geometry in two and three dimensions as objects in their own right with their own algebraic properties. To this end he developed the algebra of quaternions, a theory which had a good deal of popularity in England in the middle of the nineteenth century. Quaternions had several drawbacks: they more naturally pertained to four, rather than to three dimensions – the geometry of three dimensions appeared as a piece of a larger theory rather than having a natural existence of its own; also, they have *too much* algebraic structure, the relation between quaternion multiplication, for example, and geometric constructions in three dimensions being somewhat complicated. (The first of these objections would, of course be regarded far less seriously today. But it would be replaced by an objection to a theory that is *limited* to four dimensions.) Eventually, the three-dimensional *vector algebra* with its scalar and vector products was distilled from the theory of quaternions. It was conjoined with the necessary differential operations, and give rise to the *vector analysis* as finally developed by Gibbs and promulgated by him in a famous and very influential text.

So vector analysis, with its grad, div, curl etc. became the standard language in which the geometric laws of physics were taught. Now while vector analysis is well suited to the geometry of three-dimensional Euclidean space, it has a number of serious drawbacks. First, and least serious, is that the essential unity of the subject is obscured. Thus the fundamental theorem of the calculus, Green's theorem, Gauss' theorem and Stokes' theorem are all aspects of the same theorem (now called Stokes' theorem). But this is not at all clear in the vector analysis treatment. More serious is that the fundamental operators involve the Euclidean structure (for example grad and div) or the three-dimensional structure and orientation as well (for example curl). Thus the theory is wedded to a three-dimensional orientated Euclidean space. A related problem is that the operators do not behave nicely under general changes of coordinates – their expression in non-rectangular coordinates being unwieldy. Already Poincaré, in his fundamental scientific and philosophical writings which led to the theory of relativity, stressed the need to distinguish between those laws of geometry and physics which are 'topological', i.e. depend only on the differential structure of space and so are invariant under smooth deformations, and those which depend on more geometrical structure such as the notion of distance. One of the major impacts of the theory of relativity on mathematics was to encourage the study of higher-dimensional spaces, a study which had existed in the previous mathematical literature, but was not regarded as central to the study of geometry. Another was to emphasize general coordinate changes. The vector analysis was not up to these two tasks and so was supplemented in the more advanced literature by *tensor analysis*. But tensor analysis with its jumble of indices has a number of serious drawbacks, the most serious of which being that it is extraordinarily difficult to tell which operations have any geometric significance and which are artifacts of the coordinate system. Thus, while it is reasonably well-suited for computation, it is hard to assess exactly what it is that one is computing. The whole purpose of the development initiated by Hamilton – to have a calculus whose objects have a perceived geometrical significance – was vitiated. In order to make the theory work one had to introduce a relatively sophisticated geometrical construct, such as an affine connection. Even with such constructs the geometric meanings of the operations are obscure. In fact tensor analysis never displaced the intuitively clear vector analysis from the elementary curriculum.

It is generally accepted in the mathematics community, and gradually being accepted in the physics community, that the most suitable framework for geometrical analysis is the exterior differential calculus of Grassmann and Cartan. This calculus has the advantage that its computational rules are simple and concise, that its objects have a transparent geometrical significance, that it works in all

**Maxwell's equations in the course of history**
The constants $c, \mu_0$, and $\varepsilon_0$ are set to 1.

| **The homogeneous equation** | **The inhomogeneous equation** |
|---|---|
| Earliest form | |
| $\frac{\partial B_x}{\partial x}+\frac{\partial B_y}{\partial y}+\frac{\partial B_z}{\partial z}=0$ | $\frac{\partial E_x}{\partial x}+\frac{\partial E_y}{\partial y}+\frac{\partial E_z}{\partial z}=\rho$ |
| $\frac{\partial E_z}{\partial y}-\frac{\partial E_y}{\partial z}=-\dot{B}_x$ | $\frac{\partial B_z}{\partial y}-\frac{\partial B_y}{\partial z}=j_x+\dot{E}_x$ |
| $\frac{\partial E_x}{\partial z}-\frac{\partial E_z}{\partial x}=-\dot{B}_y$ | $\frac{\partial B_x}{\partial z}-\frac{\partial B_z}{\partial x}=j_y+\dot{E}_y$ |
| $\frac{\partial E_y}{\partial x}-\frac{\partial E_x}{\partial y}=-\dot{B}_z$ | $\frac{\partial B_y}{\partial x}-\frac{\partial B_x}{\partial y}=j_z+\dot{E}_z$ |
| At the end of the last century | |
| $\nabla\cdot\mathbf{B}=0$ <br> $\nabla\times\mathbf{E}=-\dot{\mathbf{B}}$ | $\nabla\cdot\mathbf{E}=\rho$ <br> $\nabla\times\mathbf{B}=\mathbf{j}+\dot{\mathbf{E}}$ |
| At the beginning of this century | |
| ${*F^{\beta\alpha}}_{,\alpha}=0$ | ${F^{\beta\alpha}}_{,\alpha}=j^\beta$ |
| Mid-twentieth-century | |
| $\mathrm{d}F=0$ | $\delta F=J$ |

dimensions, that it behaves well under maps and changes of coordinates, that it has an essential unity to its principal theorems and that it clearly distinguishes between the 'topological' and 'metrical' properties. The geometrical laws of physics take on a simple and elegant form in terms of the exterior calculus. To emphasize this point, it might be useful to reproduce the above table, taken from Thirring's *Course on Mathematical Physics.*

Hermann Grassmann (1809–77) published his *Ausdehnungslehre* in 1844. It was not appreciated by the mathematical community and was dismissed by the leading German mathematicians of his time. In fact, Grassmann was never able to get a university position in mathematics. He remained a high-school teacher throughout his career. (Nevertheless, he seemed to have a happy and productive life. He raised a large family and was recognized as an expert on Sanskrit literature.) Towards the end of his life he tried again, with another edition of his *Ausdehnungslehre*, but this fared no better than the first. Only one or two mathematicians of his time, such as Möbius, appreciated his work. Nevertheless, the *Ausdehnungslehre* (or calculus of extension) contains for the first time many of the notions central to modern mathematics and most of the algebraic structures used in this book. Thus vector spaces, exterior algebra, exterior and interior products and a form of the generalized Stokes' theorem all make their appearance.

Elie Cartan (1869–1951) is now universally recognized as the leading geometer of our century. His early work, of such overwhelming importance for modern mathematics, on Lie groups and on systems of partial differential equations was done in relative obscurity. But, by the 1920s, his work became known to the broad mathematical community, due, in part, to the writings of Hermann Weyl who presented novel expositions of his work at a time when the theory of Lie groups began to play a central role in mathematics and in physics. Cartan's work on the theory of principal bundles and connections is now basic to the theory of elementary particles (where it goes under the generic name of 'gauge theories'). In 1922 Cartan published his book *Leçons sur les invariants intégraux* in which he showed how the exterior differential calculus, which he had invented, was a flexible tool, not only for geometry but also for the variational calculus and a wide variety of physical applications. It has taken a while, but, as we have mentioned above, it is now recognized by mathematicians and physicists that this calculus is the appropriate vehicle for the formulation of the geometrical laws of physics. Accordingly, we feel that it should displace the 'vector calculus' in the elementary curriculum and have proceeded accordingly.

Some explanation is in order for the time and effort devoted to the theory of electrical networks, a subject not usually considered as part of the elementary curriculum. First of all there is a purely pedagogical justification. The subject always goes over well with the students. It provides a down-to-earth illustration of such concepts as dual space and quotient space, concepts which frequently seem overly abstract and not readily accepted by the student. Also, in the discrete, algebraic setting of network theory, Stokes' theorem appears as essentially a definition, and a natural one at that. This serves to motivate the d operator and Stokes' theorem in the setting of the exterior calculus. There are deeper, more philosophical reasons for our decision to emphasize network theory. It has been recognized for about a century that the forces that hold macroscopic bodies together are essentially electrical in character. Thus (in the approximation where the notion of rigid body and Euclidean geometry makes sense, that is, in the non-relativistic realm) the concept of a rigid body, and hence of Euclidean geometry, derives from electrostatics. The frontiers of physics, both in the very small (the study of elementary particles) and the very large (the study of cosmology) have already begun to reopen fundamental questions as to the geometry of space and time. We thought it wise to bring some of the issues relating geometry to physics before the student even at this early stage of the curriculum. The advent of the computer, and also some of the recent theories of physics will, no doubt, call into question the discrete versus the continuous character of space and time (an issue raised by Riemann in his dissertation on the foundations of geometry). It is to be hoped that our discussion may be of some use to those who will have to deal with this problem in the future.

Of course, we have had to omit several important topics due to the limitation of a one-year course. We do not discuss infinite-dimensional vector spaces, in particular Hilbert spaces, nor do we define or study abstract differentiable manifolds and their properties. It has been our experience that these topics make too heavy a demand on the sophistication of the student, and the effort involved in explaining them is best expended elsewhere. Of course, at various places in the text we have to pay the price for not having these concepts at our disposal. More serious is the omission of a serious discussion of Fourier analysis, classical mechanics and probability theory. These topics are touched upon but not presented as a coherent subject of study. Our only excuse is that a thorough study of each would probably require a semester's course, and substantive treatments from the modern viewpoint are available elsewhere. A suggested guide to further reading is given at the end of the book.

We would like to thank Prof. Daniel Goroff for a careful reading of the manuscript and for making many corrections and fruitful suggestions for improvement. We would also like to thank Jeane Morris for her excellent typing and her devoted handling of the production of the manuscript from the inception of the project to its final form, over a period of eight years.

# 1

# Linear transformations of the plane



In Chapter 1 we explain the relation between the multiplication law of $2 \times 2$ matrices and the geometry of straight lines in the plane. We develop the algebra of $2 \times 2$ matrices and discuss the determinant and its relation to area and orientation. We define the notion of an abstract vector space, in general, and explain the concepts of basis and change of basis for one- and two-dimensional vector spaces.

## 1.1. Affine planes and vector spaces

The familiar Euclidean plane of high-school plane geometry arose early in the history of mathematics because its properties are readily discovered by physical experiments with a tabletop or blackboard. Through our experience in using rulers and protractors, we are inclined to accept 'length' and 'angle' as concepts which are as fundamental as 'point' and 'line'. We frequently have occasion, though, both in pure mathematics and in its applications to physics and other disciplines, to consider planes for which straight lines are defined but in which no general notion of length is defined, or in which the usual Euclidean notion of length is not appropriate. Such a plane may be represented on a sheet of paper, but the physical distance between two points on the paper, as measured by a ruler, or the angle between two lines, as measured by a protractor, need have no significance.

An example of such a plane is the one used to describe graphically the motion of particles along a line (the $x$-axis). A point $P$ or $Q$ in this plane represents the physical concept of *event*, something which has a time and place. A line $l$ also

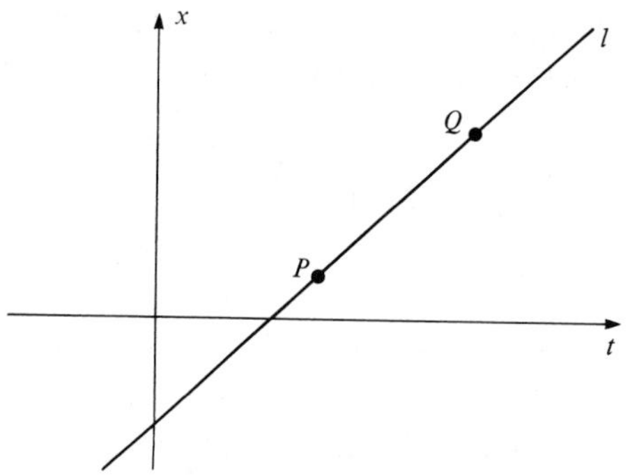


**Figure 1.1**

has physical significance; it corresponds to the motion of a particle which is subject to no force. We can compare the lengths of segments along the $t$-axis (time intervals) or along the $x$-axis (distances). Yet the distance between $P$ and $Q$, as measured with a ruler, is devoid of physical significance. Furthermore, the origin in such a plane, where the axes cross, is of no fundamental physical significance.

The mathematical concept of *real affine plane* is the appropriate one to represent this and many other 'two-dimensional' situations. An affine plane contains *points*, which we shall represent by upper case letters $P$, $Q$, etc., and straight lines, which we shall call simply *lines* and represent by lower-case letters $l$, $m$, etc. As our model for the affine plane, we shall follow Descartes and consider the set of all pairs of real numbers as our plane. A typical point is then an ordered pair of real numbers denoted by $\begin{bmatrix} x \\ y \end{bmatrix}$. This plane is called $A\mathbb{R}^2$. The A stands for affine, and is to remind us that we have no preferred origin. The $\mathbb{R}$ stands for the collection of real numbers, and the superscript 2 indicates that we are considering pairs of real numbers. (When we plot the plane on paper, the usual convention is to plot $x$ horizontally and $y$ vertically along perpendicular axes. However, the notion of 'perpendicular' or the size of any angle is undefined for us at the moment. We could just as well plot $x$ and $y$ along any axes.) A line is a particular kind of set of points. We assume that you are familiar with (straight) lines from your previous studies of geometry, and, in particular, that you are acquainted with the description of lines in analytic geometry.

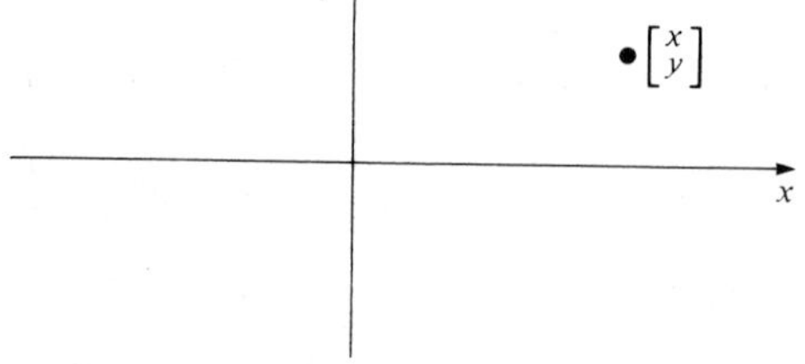


**Figure 1.2**

The lines of the affine plane $A\mathbb{R}^2$ can be described in various ways. One way is to give an equation satisfied by the points of the line, for example

$$l=\left\{\begin{bmatrix}x\\y\end{bmatrix}\middle|\, ax+by=c\right\}.$$

This is to be read as '$l$ is the set of points $\begin{bmatrix}x\\y\end{bmatrix}$ such that the equation $ax+by=c$ is satisfied'. Here it is assumed that $a$ and $b$ are not both zero.

This method of characterizing a line is a little inconvenient because the parameters $a, b, c$ which characterize the line are not unique. For example

$$\left\{\begin{bmatrix}x\\y\end{bmatrix}\middle|\, ax+by=c\right\}$$

and

$$\left\{\begin{bmatrix}x\\y\end{bmatrix}\middle|\, 3ax+3by=3c\right\}$$

are the same line. More generally the parameters $ra, rb, rc$, for $r\neq 0$, describe the same straight line as $a, b, c$.

A second method of characterizing a line in $A\mathbb{R}^2$ is in terms of two points lying on the line. Given two distinct points $P_0=\begin{bmatrix}x_0\\y_0\end{bmatrix}$ and $P_1=\begin{bmatrix}x_1\\y_1\end{bmatrix}$, we construct the line through $P_1$ and $P_2$ as the set of all points

$$\begin{bmatrix}x\\y\end{bmatrix}=\begin{bmatrix}x_0+t(x_1-x_0)\\y_0+t(y_1-y_0)\end{bmatrix}$$

where the parameter $t$ ranges over the real numbers. This description of a line is even more redundant than the previous one: we can replace our points $P_1$ and $P_0$ by any other pair of distinct points on the same line.

Another convenient way of describing a straight line (a more 'dynamic' as opposed to a 'static' way) is to give a point on the line and the 'direction vector of the line': thus the set of all points of the form

$$\left\{\begin{bmatrix}x_0\\y_0\end{bmatrix}+t\begin{pmatrix}u\\v\end{pmatrix}\middle|\, t\in\mathbb{R}\right\}\quad\text{where}\quad\begin{pmatrix}u\\v\end{pmatrix}\neq\begin{pmatrix}0\\0\end{pmatrix}\text{ is a fixed vector}$$

is a line. $\Bigg($Here we think of the line as being traversed by a particle moving with 'velocity vector' $\begin{pmatrix}u\\v\end{pmatrix}$ and situated at $\begin{bmatrix}x_0\\y_0\end{bmatrix}$ at time zero.$\Bigg)$ Here we have used four parameters to describe the line. But we can multiply $\begin{pmatrix}u\\v\end{pmatrix}$ by any non-zero scalar and get the same line (just traversed with different velocity) and we can displace $\begin{bmatrix}x_0\\y_0\end{bmatrix}$ along the line, showing that we have two redundant parameters.

Of course, this ties in with our second description if

$$u = x_1 - x_0,$$
$$v = y_1 - y_1.$$

There is a fourth, familiar description of a line which is not redundant, but has the awkward feature that it does not describe absolutely all lines in the same way. If $a$ and $b$ are any real numbers, the set

$$l = \left\{ \begin{bmatrix} x \\ y \end{bmatrix} \middle| y = ax + b \right\}$$

is a straight line which intersects the $y$-axis at the point $\begin{bmatrix} 0 \\ b \end{bmatrix}$ and which has 'slope' $a$; i.e., for points on the line, an increase in one unit of $x$ implies an increase in $a$ units of $y$. This set is a line, and the description is not redundant, for we have described $a$ and $b$ in terms of geometric properties of the line. But not all lines are of this form. We must add the lines which are parallel to the $y$-axis, and which have the description

$$l = \left\{ \begin{bmatrix} x \\ y \end{bmatrix} \middle| x = d \right\}$$

From a strictly logical point of view, we should take one of the four descriptions given above as our *definition* of a straight line; for example, we should say that, by definition, a line is a subset, $l$, of $A\mathbb{R}^2$ such that there are three real numbers $a, b$, and $c$ with $a$ and $b$ not both zero such that

$$l = \left\{ \begin{bmatrix} x \\ y \end{bmatrix} \middle| ax + by = c \right\}.$$

We should then *prove* that such a subset can be given by either of the other three descriptions. We shall not go into such logical niceties here, since you have seen, or can construct, such arguments from elementary analytic geometry.

It is important to remember that an affine plane has no origin and that it makes no sense to add points of an affine plane. We attach no special significance to the point $\begin{bmatrix} 0 \\ 0 \end{bmatrix}$, and we resist the temptation to add points like $\begin{bmatrix} 2 \\ 1 \end{bmatrix}$ and $\begin{bmatrix} 3 \\ 6 \end{bmatrix}$ 'coordinate by coordinate'. There is, however, a closely related mathematical structure, called a two-dimensional *vector space*, in which an operation of addition *is* defined. We construct a vector space from an affine plane by associating with any pair of points the 'displacement vector' $\overrightarrow{PQ}$ whose 'tail' is at $P$ and whose 'head' is at $Q$. We denote vectors by lowercase bold letters: $\mathbf{v}, \mathbf{w}$, etc. A vector $\mathbf{v}$ is also given as a pair of real numbers, for example $\mathbf{v} = \begin{pmatrix} 5 \\ 2 \end{pmatrix}$. (Notice that we use ( ) for vectors and not [ ] as for points.) The vector $\mathbf{v} = \begin{pmatrix} 5 \\ 2 \end{pmatrix}$ is to be thought

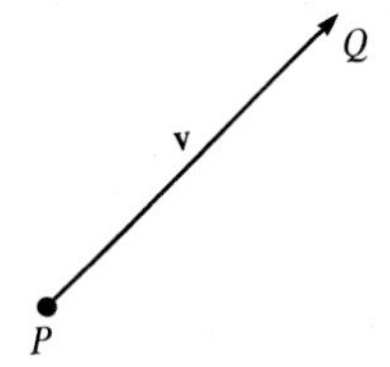


**Figure 1.3**

of as that displacement which carries the point $\begin{bmatrix}1\\3\end{bmatrix}$ into $\begin{bmatrix}6\\5\end{bmatrix}$, carries the point $\begin{bmatrix}-3\\2\end{bmatrix}$ into $\begin{bmatrix}2\\4\end{bmatrix}$ and, in general, carries any point $P=\begin{bmatrix}x\\y\end{bmatrix}$ into $Q=\begin{bmatrix}x+5\\y+2\end{bmatrix}$. Thus each vector **v** determines a (particular kind of) transformation of the affine plane into itself, a rigid translation of the whole plane. If $P$ is any point in the plane, we will denote the displaced point $Q$ by $P$"+"**v**: the "+" is a symbol for this operation of vectors on points. Thus **v** sends $P$ into $Q=P$"+"**v**. Explicitly, if $P=\begin{bmatrix}x\\y\end{bmatrix}$ and $\mathbf{v}=\begin{pmatrix}a\\b\end{pmatrix}$, then $P$"+"$\mathbf{v}=\begin{bmatrix}x+a\\y+b\end{bmatrix}$.

We put quotation marks about the + sign because the operation is between two different kinds of object, points and vectors, and so differs from the usual notion of addition. Similarly, given any pair of points $P$ and $Q$, there is a unique vector $\mathbf{v}=Q$"−"$P$ such that

$$P\text{“}+\text{”}\mathbf{v}=Q.$$

We put quotation marks around the − because it relates different kinds of objects, it gives a vector from a pair of points. You should convince yourself, by working out some examples on graph paper, that two pairs of points, $P$, $Q$ and $R$, $S$ determine the same vector, i.e., $Q$"−"$P=S$"−"$R$, if and only if $\overrightarrow{PQ}$ and $\overrightarrow{RS}$ are opposite sides of a parallelogram. For this reason, one frequently finds it said that a vector is determined by 'magnitude and direction'. But we want to refrain from introducing either magnitude or direction as they are not invariant concepts for us.

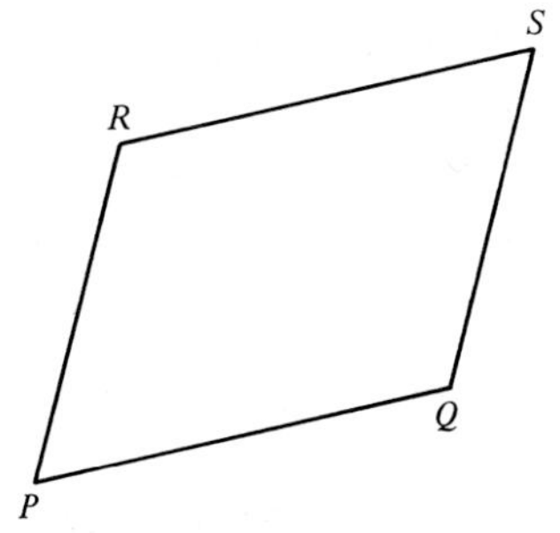


**Figure 1.4**

We can define the sum of two vectors: if $\mathbf{u} = \begin{pmatrix} a \\ b \end{pmatrix}$ and $\mathbf{v} = \begin{pmatrix} c \\ d \end{pmatrix}$, define their sum by

$$\mathbf{u} + \mathbf{v} = \begin{pmatrix} a + c \\ b + d \end{pmatrix}.$$

Notice that

$$P \text{``}+\text{''} (\mathbf{u} + \mathbf{v}) = (P \text{``}+\text{''} \mathbf{u}) \text{``}+\text{''} \mathbf{v} \tag{1.1}$$

since, if $\mathbf{u} = \begin{pmatrix} a \\ b \end{pmatrix}$, $\mathbf{v} = \begin{pmatrix} c \\ d \end{pmatrix}$ and $P = \begin{bmatrix} x \\ y \end{bmatrix}$, then both the left and the right hand side of the above equation equal $\begin{bmatrix} a + c + x \\ b + d + y \end{bmatrix}$. The equation (1.1) says that the displacement corresponding to $\mathbf{u} + \mathbf{v}$ can be obtained by successively applying the displacement $\mathbf{v}$ and then the displacement $\mathbf{u}$. Notice that $\mathbf{u} + \mathbf{v} = \mathbf{v} + \mathbf{u}$. We can visualize the addition of vectors by the familiar parallelogram law: if we start with a point $P$ and write $R = P \text{``}+\text{''} \mathbf{u}$, $Q = P \text{``}+\text{''} \mathbf{v}$ and $S = P \text{``}+\text{''} (\mathbf{u} + \mathbf{v})$, then the four points $P, Q, S, R$ lie at the four vertices of a parallelogram. You should convince yourself of this fact by working out some examples on graph paper. The proof of this fact goes as follows. For any vector $\mathbf{v} = \begin{pmatrix} a \\ b \end{pmatrix}$ and any real number $t$, define their product, $t\mathbf{v}$, by $t\mathbf{v} = \begin{pmatrix} ta \\ tb \end{pmatrix}$. If $\mathbf{v} \neq \begin{pmatrix} 0 \\ 0 \end{pmatrix}$ and $P$ is any point, the set

$$l = \{P \text{``}+\text{''} t\mathbf{v}\} \quad (\text{as } t \text{ varies over } \mathbb{R})$$

is a straight line passing through $P$ (just look at the third of our four descriptions of straight lines). If $R$ is some other point, then the line

$$m = \{R \text{``}+\text{''} s\mathbf{v}\} \quad (\text{as } s \text{ varies over } \mathbb{R})$$

and $l$ will intersect, i.e., have some point in common, if and only if there are some $s_1$ and $t_1$ such that

$$R \text{``}+\text{''} s_1\mathbf{v} = P \text{``}+\text{''} t_1\mathbf{v}$$

which means that

$$R = P \text{``}+\text{''} (t_1 - s_1)\mathbf{v}$$

and hence, for every $s$, that

$$R \text{``}+\text{''} s\mathbf{v} = P \text{``}+\text{''} (s + t_1 - s_1)\mathbf{v}.$$

This means that the lines $m$ and $l$ coincide. In other words, either the lines $l$ and $m$ coincide, or they do not intersect, i.e., either they are the same or they are parallel. Now let us go back to our diagram for vector addition. If $\mathbf{v} \neq \begin{pmatrix} 0 \\ 0 \end{pmatrix}$, then the point $Q = P \text{``}+\text{''} \mathbf{v}$ lies on the line $l$ through $P$ and the point $S = R \text{``}+\text{''} \mathbf{v}$ lies on the line $m$ through $R$. There are now two possibilities: if the point $R$ does *not* lie on the line $l$, so that $\mathbf{u} \neq t\mathbf{v}$ for any $t$, the lines $l$ and $m$ are parallel. A similar argument applies to the other two sides and we conclude that the figure is a parallelogram. If $\mathbf{u} = t\mathbf{v}$, then all four points lie on the line $l$. We can still view this picture as a sort of 'degenerate' parallelogram:

**Figure 1.5**

If either $\mathbf{u}$ or $\mathbf{v} = \begin{pmatrix} 0 \\ 0 \end{pmatrix}$, the picture degenerates further:

**Figure 1.6**

We say that the vectors $\mathbf{u}$ and $\mathbf{v}$ are *linearly dependent* if there are numbers $r$ and $s$, not both zero, such that

$$r\mathbf{u} + s\mathbf{v} = 0.$$

If $r \neq 0$ we can solve this equation for $u$ to obtain $\mathbf{u} = -(s/r)\mathbf{v}$ and if $s \neq 0$ we can solve this equation for $\mathbf{v} = -(r/s)\mathbf{u}$. In either case, the 'addition parallelogram' degenerates into segments on a line $\left(\text{or if } \mathbf{u} = \mathbf{v} = \begin{pmatrix} 0 \\ 0 \end{pmatrix}, \text{ into a single point}\right)$. This is the reason for the term linearly dependent. If two vectors are not linearly dependent, we say that they are *linearly independent*.

The zero vector $\begin{pmatrix} 0 \\ 0 \end{pmatrix}$, denoted by $\mathbf{0}$, has the same point for its head and tail. It is called an *additive identity* because

$$\mathbf{0} + \mathbf{v} = \mathbf{v} + \mathbf{0} = \mathbf{v} \quad \text{for all } \mathbf{v}.$$

The set of all vectors $\mathbf{v} = \begin{pmatrix} x \\ y \end{pmatrix}$ where $x$ and $y$ are arbitrary real numbers is called $\mathbb{R}^2$. The space $\mathbb{R}^2$ is an example of a *vector space*, to be defined in the next section. The notational distinction between $\mathbb{R}^2$ and $A\mathbb{R}^2$ lies in the fact that in $\mathbb{R}^2$ the point $\begin{pmatrix} 0 \\ 0 \end{pmatrix}$ has a special significance (it is the additive identity) and the addition of two vectors in $\mathbb{R}^2$ makes sense. These do not hold for $A\mathbb{R}^2$.

## 1.2. Vector spaces and their affine spaces

It is easy to check that the operations of addition of vectors in $\mathbb{R}^2$ and for multiplying vectors by real numbers satisfy the following collection of axioms:

**Laws for addition of vectors**

| | |
|---|---|
| Associative law of addition: | $(\mathbf{u}+\mathbf{v})+\mathbf{w}=\mathbf{u}+(\mathbf{v}+\mathbf{w})$. |
| Commutative law of addition: | $\mathbf{u}+\mathbf{v}=\mathbf{v}+\mathbf{u}$. |
| Existence of additive identity: | there is a vector $\mathbf{0}$ such that $\mathbf{0}+\mathbf{v}=\mathbf{v}$ for all $\mathbf{v}$. |
| Existence of additive inverse: | for every $\mathbf{v}$ there is a $-\mathbf{v}$ such that $\mathbf{v}+(-\mathbf{v})=\mathbf{0}$. |

**Laws involving the multiplication of vectors by real numbers**

| | |
|---|---|
| 'One' acts as multiplicative identity: | $1\mathbf{v}=\mathbf{v}$ for every $\mathbf{v}$. |
| Associative and distributive laws: | for any real numbers $r$ and $s$ and any vectors $\mathbf{u}$ and $\mathbf{v}$ |

$$(rs)\mathbf{v}=r(s\mathbf{v})$$
$$(r+s)\mathbf{v}=r\mathbf{v}+s\mathbf{v}$$
$$r(\mathbf{u}+\mathbf{v})=r\mathbf{u}+r\mathbf{v}.$$

The above axioms are known as the axioms for a *vector space*. By definition, a vector space is a collection, $V$, of objects, $\mathbf{u}$, $\mathbf{v}$, etc., called vectors, such that we are given a binary operation, $+$, which assigns to every pair of vectors $\mathbf{u}$ and $\mathbf{v}$ a third vector $\mathbf{u}+\mathbf{v}$ and a multiplication which assigns to every real number $t$ and every vector $\mathbf{v}$ another vector $t\mathbf{v}$ such that the above axioms hold.

We have verified that $\mathbb{R}^2$ is an example of a vector space. As a second example, we could take $\mathbb{R}^3$ where a vector now consists of a triplet

$$\mathbf{v}=\begin{pmatrix}a\\b\\c\end{pmatrix}$$

of real numbers. Addition of vectors is done componentwise as in $\mathbb{R}^2$:

$$\text{if } \mathbf{v}_1=\begin{pmatrix}a_1\\b_1\\c_1\end{pmatrix} \text{ and } \mathbf{v}_2=\begin{pmatrix}a_2\\b_2\\c_2\end{pmatrix}, \text{ then } \mathbf{v}_1+\mathbf{v}_2=\begin{pmatrix}a_1+a_2\\b_1+b_2\\c_1+c_2\end{pmatrix}.$$

The space $\mathbb{R}^3$ is just the space of vectors in our familiar three-dimensional space. We shall study the concept of dimension later on. We could also consider the space $\mathbb{R}=\mathbb{R}^1$ of the real numbers themselves as a vector space. Here addition is just the ordinary addition and multiplication ordinary multiplication. When we introduce the notion of dimension, this will be an example of a one-dimensional vector space.

As a different looking example of a vector space, consider the collection of all polynomials. We can add two polynomials:

$$(1+3x+7x^2)+(2-x^2+x^4-x^6)=3+3x+6x^2+x^4-x^6,$$

just add the coefficients. We can also multiply a polynomial by a real number:

$$7(1+3x+3x^2)=7+21x+21x^2.$$

You should check that the axioms for a vector space are satisfied. We can also consider the space of polynomials of at most a given degree. For example, the most general polynomial of degree at most two is of the form

$$P = ax^2 + bx + c.$$

The sum of two such polynomials

$$P_1 = a_1x^2 + b_1x + c_1 \quad \text{and} \quad P_2 = a_2x^2 + b_2x + c_2$$

is

$$P_1 + P_2 = (a_1 + a_2)x^2 + (b_1 + b_2)x + c_1 + c_2.$$

For example, if

$$P_1 = 3x^2 + 2x + 1, \quad P_2 = 7x^2 - 10x + 2$$

then

$$P_1 + P_2 = 10x^2 - 8x + 3.$$

The set of polynomials of degree at most two is also a vector space. Notice that it 'looks like' $\mathbb{R}^3$ in the sense that the preceding equations look like

$$\begin{pmatrix} 3 \\ 2 \\ 1 \end{pmatrix} + \begin{pmatrix} 7 \\ -10 \\ 2 \end{pmatrix} = \begin{pmatrix} 10 \\ -8 \\ 3 \end{pmatrix}.$$

We will return to this point later.

Suppose that we are *given* a vector space $V$; for example, $V$ could be $\mathbb{R}^1$, $\mathbb{R}^2$ or $\mathbb{R}^3$. By an *affine space* associated to $V$, we mean a set $A$ consisting of points $P, Q$, etc., and an operation "+" which assigns to each $P \in A$ and each $\mathbf{v} \in V$ another point in $A$ which is denoted by $P$"+"$\mathbf{v}$. This rule is subject to the following axioms:

| | |
|---|---|
| Associative law: | $(P$"+"$\mathbf{u})$"+"$\mathbf{v} = P$"+"$(\mathbf{u} + \mathbf{v})$ for any $P \in A$ and $\mathbf{u}, \mathbf{v} \in V$. |
| 'Zero' acts as identity: | $P$"+"$\mathbf{0} = P$ for any $P \in A$. |
| Transitivity: | given any two points $P$ and $Q \in A$, there is a $\mathbf{v} \in V$ such that $P$"+"$\mathbf{v} = Q$. |
| Faithfulness: | if, for any $P$, the equality $P$"+"$\mathbf{u} = P$"+"$\mathbf{v}$ holds, then $\mathbf{u} = \mathbf{v}$. |

Combining the last two axioms, we can say that, given any two points $P$ and $Q$, there is a *unique* vector $\mathbf{v}$ such that $P$"+"$\mathbf{v} = Q$. It is then sometimes convenient to write $\mathbf{v} = Q$"−"$P$.

The notion of a vector space and associated affine space lies at the basis of three centuries of physical thought, from Newtonian mechanics through special relativity and quantum mechanics. The purpose of the present chapter is to develop most of the key ideas in the study of these structures by examining the intuitively simple case of the two-dimensional* vector space $\mathbb{R}^2$. Let us begin, however, with some

* We will give a precise definition of the term 'two-dimensional' in §1.12, of 'one-dimensional' in a few lines, and of the general concept of the dimension of a vector space in Chapter 10.

comments about the one-dimensional case. Here the concepts are so 'obvious' that a detailed discussion of them may appear so pedantic as to be non-intuitive. Yet it is worth the effort.

A vector space $V$ is called *one-dimensional* if it satisfies the following two conditions: (i) it possesses some vector $\mathbf{v} \neq \mathbf{0}$; and (ii) if $\mathbf{v} \neq \mathbf{0}$, then any $\mathbf{u} \in V$ can be written as $\mathbf{u} = r\mathbf{v}$ for some real number $r$. Notice that the $r$ in this equation is unique: if

$$r_1\mathbf{v} = r_2\mathbf{v},$$

then we claim that $r_1 = r_2$. Indeed, from $r_1\mathbf{v} = r_2\mathbf{v}$ we can write

$$(r_1 - r_2)\mathbf{v} = \mathbf{0}.$$

If $r_1 - r_2 \neq 0$, then setting $s = (r_1 - r_2)^{-1}$, we have

$$\begin{aligned} \mathbf{0} = s[(r_1 - r_2)\mathbf{v}] &= (s(r_1 - r_2))\mathbf{v} \\ &= 1\mathbf{v} \\ &= \mathbf{v}, \end{aligned}$$

so $\mathbf{v} = \mathbf{0}$, contradicting our original assumption that $\mathbf{v} \neq \mathbf{0}$. (You should check exactly which of the vector space axioms we used at each stage of the preceding argument.) Once we have chosen a $\mathbf{v} \neq \mathbf{0}$ in a one-dimensional vector space, then to each vector $\mathbf{u}$ there is assigned a real number, $r$,

$$\mathbf{u} \to r \quad \text{where} \quad \mathbf{u} = r\mathbf{v}.$$

If $\mathbf{u}_1 = r_1\mathbf{v}$ and $\mathbf{u}_2 = r_2\mathbf{v}$, then $\mathbf{u}_1 + \mathbf{u}_2 = (r_1 + r_2)\mathbf{v}$. Thus $\mathbf{u}_1 + \mathbf{u}_2$ corresponds to $r_1 + r_2$. Similarly, if $\mathbf{u} = r\mathbf{v}$ and $t$ is any real number, then $t\mathbf{u} = (tr)\mathbf{v}$ so that $t\mathbf{u}$ corresponds to $tr$. In short, every vector corresponds to a real number, and the vector operations correspond to the operations on $\mathbb{R}^1$. We say that we have an *isomorphism* of the one-dimensional vector space $V$ with $\mathbb{R}^1$. This identification of $V$ with $\mathbb{R}^1$ depends on the choice of $\mathbf{v}$. A choice of $\mathbf{v}$ is called a choice of *basis* of $V$, and the number $r$ associated to $u$ via $\mathbf{u} = r\mathbf{v}$ is called the coordinate of $\mathbf{u}$ relative to the basis $\mathbf{v}$. Suppose we choose a different basis, $\mathbf{v}'$. Here $\mathbf{v}' = a\mathbf{v}$ where $a$ is some non-zero real number. If $\mathbf{u} = r\mathbf{v}$, then

$$\mathbf{u} = (ra^{-1})a\vec{v}$$

so

$$\mathbf{u} = r'\mathbf{v}' \quad \text{where} \quad r' = a^{-1}r.$$

Thus, changing the basis, by replacing $\mathbf{v}$ by $a\mathbf{v}$, has the effect of changing the coordinate of any vector by replacing the coordinate $r$ of any vector by $a^{-1}r$. The choice of a basis in a one-dimensional vector space is much like the choice of a unit for some physical quantity. If we change our units of mass from kilograms to grams, an object that weighs 1.3 kilograms now weighs 1300 grams. The difference is that, for many familiar physical quantities, the measurement of any object is given by positive numbers (or zero) only. It usually makes no sense to say that something has negative volume or mass, etc. An exception is in the theory of electricity, where electric charge can be positive or negative. For instance, we might imagine situations in which we might want to choose the charge of the electron as our unit. In terms of this basis, the electron would have charge $+1$ instead of $-1.602\,191 \times 10^{-19}$ coulombs, where the coulomb is a 'standard unit', i.e., a basis that has been agreed upon by international convention.

Let $A$ be an affine space associated to the one-dimensional vector space $V$. If we pick some point $O$ in $A$, then every other point, $P$, determines a vector $\mathbf{u} = P\text{“}-\text{”}O$. If we also choose a basis, $\mathbf{v}$ of $V$, then each $P$ gets assigned a number, $x(P)$, where

$$P = O\text{“}+\text{”}x(P)\mathbf{v}.$$

We call $x(P)$ the coordinate of $P$, but here we had to make two choices: we had to choose an 'origin' $O$, which allowed us to identify points with vectors, and then we had to choose a basis of $V$, which allowed us to identify vectors with numbers. If we change our basis, by replacing $\mathbf{v}$ by $\mathbf{v}' = a\mathbf{v}$, then $x$ is replaced by $x'$ where

$$x'(P) = a^{-1}x(P).$$

If, in addition, we replace $O$ by $O'$, where $O' = O\text{“}+\text{”}\mathbf{w}$, then

$$P\text{“}-\text{”}O' = (P\text{“}-\text{”}O) - \mathbf{w}.$$

If $\mathbf{w} = b\mathbf{v}'$, then this has the effect of replacing $x'$ by $x''$, where now

$$x''(P) = a^{-1}x(P) - b.$$

We should compare the above discussion with Newton's introduction of the concept of absolute time. Newton wrote:

> Absolute, true and mathematical time or duration flows evenly and equably from its own nature and independent of anything external; relative, apparent and common time is some measure of duration by means of motion (as by the motion of a clock) which is commonly used instead of true time.

In our terminology, what Newton said is that there exists a concept of absolute time, and the set of all absolute times has the structure of a one-dimensional affine space. The idea of 'flowing evenly and equably' is made mathematically more precise by the assertion that there is the action, given by "+", of a one-dimensional vector space $V$ on the set of all times. It is this postulated action which allows us to compare different intervals of time. Newton's distinction between 'true' and 'common' time corresponds to our discussion of the degree of arbitrariness involved in introducing coordinates on the affine line.

We should pause for a moment and ponder over this abstract postulate of Newton, which lay at the cornerstone of physics for over two centuries. We have, each of us, our own psychological perception of time. Our psychological time differs in many important respects from Newton's absolute time. The first striking difference is that for us time has a definite direction. The future is to some extent unknown and subject to our volition and intervention. (In many European languages, for example, the future tense is indicated by volition (in English 'I will go' – 'I wish to go') or compulsion (in French 'j'irai' – 'I have to go').)

The past is, to some extent, known or remembered. Yet Newton's laws of motion are insensitive to the change of direction of time. If we were to run a motion picture of Newton's (and to all extents and purposes the actual) planetary system backwards, we would discover no discrepancy with Newton's laws. The second difference is that our psychological time does not 'flow evenly and equably', at least in comparison with Newton's absolute time. We have certain bodily functions which are recurrent, and so suggest to us a notion of a time interval: we get hungry a 'certain amount of time' after having had our last meal. But this is very variable, being determined by the level of our blood sugar, which in turn depends on what exactly we ate, what we have been doing in the interim, our overall physiological profile, etc. Also, our psychological perception of these intervals of time varies greatly. Time passes quickly when we are interested and excited by what we are doing, and slowly when we are bored. Nevertheless, our internal rhythms appear to be somewhat correlated to periodicities in the world about us; from the earliest records of civilization, the measurement of 'external time', whether for civil or for scientific purposes, has always been based on the revolution of the celestial bodies. The period of apparent revolution of the sun, i.e., the interval between successive crossings of a meridian, has been the usual standard for a day. The Egyptians divided the day into 24 hours of equal length, while the Greeks divided the period from sunrise to sunset into twelve equal hours, and similarly the night. These subdivisions were marked off by various devices such as sundials during the day or water clocks. (Those who adopted the Greek system had to furnish their water clocks with some compensating device so that the hours could be modified according to the needs of the season.) All of these devices have in common that they move in one direction with psychological time – the shadow of the sundial moves in the same direction every day, the water always runs downhill. (The civil day itself is irregular, due to the varying motion of the sun on the celestial sphere. The simplest relatively accurate measure of time is the sidereal day. This is the revolution of the earth about its axis, and is measured by observing some fixed star: the period between two successive transits of some fixed star across some meridian line is a sidereal day. A civil day is, on the average, about four minutes longer than a sidereal day.)

The earliest clocks seem to have come into use in Europe during the thirteenth century, but were highly inaccurate. The first major step in the improvement of the clock came in the seventeenth century when Galileo discovered that the time intervals between swings of a pendulum were constant (as measured against a normal pulse beat, for instance). He seems to have made little practical use of this information, except for the invention of a little instrument for doctors to use in measuring the pulse of their patients. His son, however, is said to have applied the pendulum to clocks. From then on, the development of mechanical clocks was fairly rapid. Thus, it was just around the time of Newton that one finally had a method, which, in principle, could divide time into arbitrarily small equal intervals. It is also worth noting that the direction of rotation of the hands of a mechanical clock is entirely conventional. It is also easily reversible. By a simple change of the gearing, we can make the hands rotate counterclockwise instead of clockwise. It is interesting to speculate how much the development of mechanical clocks had to do with Newton's conception of time.

## 1.3. Functions and affine functions

In the next few sections we will study those transformations of $A\mathbb{R}^2$ into itself which carry straight lines into straight lines. We must begin with some general discussion of the notion of 'transformation' or 'function'.

Let $W$ and $X$ be sets. A rule $f: W \to X$ which assigns one element $f(w)$ of $X$ to each $w \in W$ is called a *function* (or *map*, or *mapping*, or *operator*) from $W$ to $X$. The set $W$ is called the *domain* of $f$. If $A$ is a subset of $W$, we let $f(A)$ denote the subset of $X$ consisting of the element $f(w)$ where $w \in A$:

$$f(A) = \{f(w) | w \in A\}.$$

The set $f(W)$ is called the *image* of $f$: in general, it is a subset of $X$.

For example, suppose $f$ is the map of $\mathbb{R}^2$ into itself given by $f(P) = P$"+"$\mathbf{v}$ where $\mathbf{v}$ is a fixed vector. Then $f(A)$ is obtained from $A$ by 'translating $A$ through $\mathbf{v}$'. If $A = l = \{P + t\mathbf{u}\}$ is a line, then $f(l) = \{P + \mathbf{v} + t\mathbf{u}\}$ is another line. Thus the image of a line under a translation is another line.

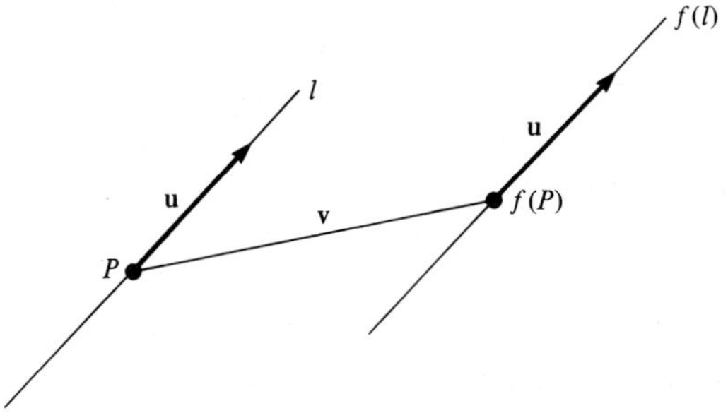


**Figure 1.7**

This notion of function is very general and powerful. The only restriction, really, is that the 'output' of the function must be well-defined. It is not acceptable, for example, to have a function $f: \mathbb{R} \to \mathbb{R}$ with the property that $f(1) = 2$ and $f(1) = 3$. There would be nothing wrong, however, with a function $f: \mathbb{R} \to \mathbb{R}^2$ for which $f(1) = \begin{pmatrix} 2 \\ 3 \end{pmatrix}$.

Certain standard terminology concerning the domain and range of $f$ is worth learning.

1. If two distinct elements $w_1, w_2 \in W$ are always mapped into distinct points $x_1$, $x_2 \in X$, then $f$ is called *injective* (or one-to-one). Equivalently, $f$ is injective if $f(w_1) = f(w_2)$ implies $w_1 = w_2$.

2. If the image of $f$, $f(W)$, is the entire set $X$, then $f$ is called *surjective* (or onto). Equivalently, if the equation $f(w)=x$ has *at least one* solution for each $x \in X$, then $f$ is surjective.
3. If $f$ is both injective and surjective, it is called *bijective* (or one-to-one onto). Equivalently, $f$ is bijective if the equation $f(w)=x$ has a *unique* solution $w$ for each $x \in X$. In this case there exists a function $f^{-1}: X \to W$, called the *inverse* of $f$, which maps each $x \in X$ into the unique $w$ for which $f(w)=x$.

Figure 1.8 may help you visualize why a function must be both injective and surjective in order to be invertible.

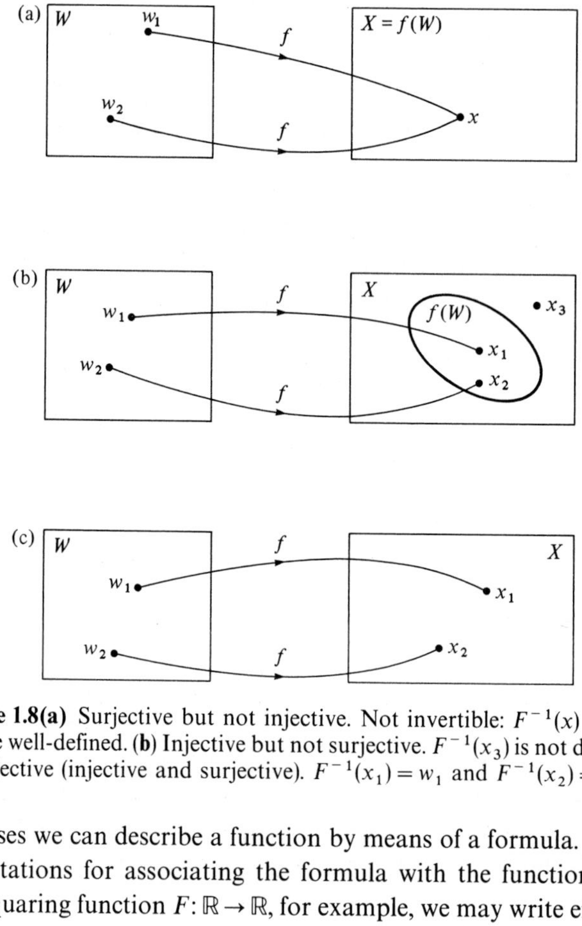


**Figure 1.8(a)** Surjective but not injective. Not invertible: $F^{-1}(x)$ would not be well-defined. (**b**) Injective but not surjective. $F^{-1}(x_3)$ is not defined. (**c**) Bijective (injective and surjective). $F^{-1}(x_1)=w_1$ and $F^{-1}(x_2)=w_2$.

In many cases we can describe a function by means of a formula. There are two equivalent notations for associating the formula with the function. To describe the familiar squaring function $F: \mathbb{R} \to \mathbb{R}$, for example, we may write either $F(x)=x^2$ or

$$F: x \mapsto x^2.$$

whichever notation we use, the symbol $x$ is a 'dummy' having nothing to do with $F$: the same function is described by

$$F(t)=t^2$$

or by

$$F: \alpha \mapsto \alpha^2.$$

A function described by a formula can involve more than one numerical argument, for example

$$G(x, y) = 2x + 3y$$

or

$$G: (x, y) \mapsto 2x + 3y.$$

This function $G$ takes the ordered pair of numbers $(x, y)$ and produces the number $2x + 3y$.

One further notion that applies to functions is that of *composition*. Let $W, X, Y, Z$ all denote sets, and suppose we have functions

$$f: W \to X,$$
$$g: X \to Y,$$
$$h: Y \to Z.$$

We denote the function which takes $w \in W$, operates on it with $f$ to obtain an element of $X$, then operates on that element with $g$ to produce an element of $Y$, by $g \circ f$, called the *composition* of $g$ with $f$. More succinctly, $(g \circ f)(w) = g(f(w))$. Notice that the composition

$$h \circ g \circ f: W \to Z$$

($f$ followed by $g$ followed by $h$) is the same as $h \circ (g \circ f)$ or as $(h \circ g) \circ f$. Thus the operation of composition obeys an 'associative law' just as does multiplication of real numbers.

We turn now to functions on affine lines and planes and on vector spaces, beginning with one-dimensional examples which, although important, are so subtle that they can easily be overlooked.

Let $\mathbb{A}$ be an affine line, illustrated in Figure 1.9. Given any ruler, we can choose

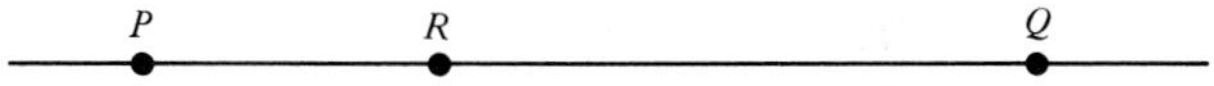


**Figure 1.9**

an origin and orientation for this line and assign a coordinate to each point on the line. Mathematically speaking, we have chosen an origin, $O$, of $\mathbb{A}$ and a basis, $\mathbf{v}$, of $V$ as described in the last section. Thus we construct an affine coordinate function

$$x: \mathbb{A} \to \mathbb{R}.$$

Of course, there are many possible affine coordinate functions on a line, and which one we construct depends on our origin and unit of measurement. We call $x$ a *coordinate* function because it is invertible: knowing $x(P)$, we can reconstruct $P$. Notice that $x$ preserves the 'interpolation property' of a real affine line: if

$$R = (1 - t)P + tQ,$$

then

$$x(R) = (1 - t)x(P) + tx(Q).$$

In particular, if $R$ is the midpoint of the segment $PQ$, then

$$x(R) = \tfrac{1}{2}x(P) + \tfrac{1}{2}x(Q).$$

You have probably never thought of this $x$ as a *function* before. You cannot write a formula for it. Yet you can hardly do elementary physics without it, because it is what lets you express other functions on a line in terms of formulas. If, for example, the force which acts on a particle on a line is a function of position

$$f: \mathbb{A} \to \mathbb{R}$$

you cannot write a formula for $f$, but you can introduce an affine coordinate

$$x: \mathbb{A} \to \mathbb{R}$$

and a function $F: \mathbb{R} \to \mathbb{R}$ and write $f(P) = F(x(P)) = (F \circ x)(P)$. This is what a formula like Force $= \sin x$, used to represent a function on a line, really means.

Time is an affine line whose points are 'instants'. The affine coordinate function $t: \mathbb{A} \to \mathbb{R}$ assigns a number to each instant. To define $t$ we use a clock. Clocks which run at different rates lead to different functions $t$, but any 'good' clock yields an affine function. A defective clock, for example a pendulum clock whose pendulum varies in length because of temperature change, would yield a non-affine coordinate function.

The motion of a particle along a straight line determines a function from one real affine line $\mathbb{A}_t$ (time) to another real affine line $\mathbb{A}_x$ (space). This function $f: \mathbb{A}_t \to \mathbb{A}_x$ acts on an instant of time $E$ to yield a point $P$ on the line, so that $P = f(E)$. We cannot write a formula for $f$ because $E$ and $P$ are not numbers. If we want a formula to describe the particle's motion, we have to introduce affine coordinate functions $t$ and $x$. Then we can write

$$x(P) = F(t(E))$$

where $F: \mathbb{R} \to \mathbb{R}$ can be represented by a formula like $F(\alpha) = x_0 + v_0\alpha + \tfrac{1}{2}a\alpha^2$.

## 1.4. Euclidean and affine transformations

A map $f: \mathbb{R}^2 \to \mathbb{R}^2$ is called a *Euclidean* transformation if $f$ preserves distance. This means that for any two points $P_1 = \begin{bmatrix} x_1 \\ y_1 \end{bmatrix}$ and $P_2 = \begin{bmatrix} x_2 \\ y_2 \end{bmatrix}$, the distance from $f(P_1)$ to $f(P_2)$ is the same as the distance from $P_1$ to $P_2$. If we express $f$ in terms of two functions $\phi: \mathbb{R}^2 \to \mathbb{R}$ and $\psi: \mathbb{R}^2 \to \mathbb{R}$ so that

$$f\begin{bmatrix} x \\ y \end{bmatrix} = \begin{bmatrix} \phi(x, y) \\ \psi(x, y) \end{bmatrix}$$

this condition amounts to the requirement

$$[\phi(x_2, y_2) - \phi(x_1, y_1)]^2 + [\psi(x_2, y_2) - \psi(x_1, y_1)]^2 = (x_2 - x_1)^2 + (y_2 - y_1)^2$$

for all values of $x_1, y_1, x_2, y_2$.

Euclidean geometry can be thought of as the study of those properties of subsets of the plane which are invariant under the application of any Euclidean transformation. For instance, if $A$ is a circle and $f$ is a Euclidean transformation, then $f(A)$ is again a circle. If $l$ is a straight line, then $f(l)$ is again a straight line. It is clear from the definition that, if $f$ and $g$ are Euclidean transformations, then $g \circ f$ is again Euclidean.

A map $f: \mathbb{R}^2 \to \mathbb{R}^2$ is called an *affine* transformation if it carries straight lines into straight lines. Thus $f(l)$ must be a straight line for any straight line $l$. For example, suppose $f$ is the transformation defined by

$$f\begin{bmatrix} x \\ y \end{bmatrix} = \begin{bmatrix} 2x + y + 1 \\ y - x + 5 \end{bmatrix}.$$

The most general straight line in the plane is given by an equation of the form

$$ax + by + c = 0.$$

That is,

$$l = \left\{ \begin{bmatrix} x \\ y \end{bmatrix} \middle| ax + by + c = 0 \right\}.$$

So

$$f(l) = \left\{ \begin{bmatrix} w \\ z \end{bmatrix} \middle| w = 2x + y + 1, z = y - x + 5 \quad \text{and} \quad ax + by + c = 0 \right\}.$$

But we can solve the equations

$$w = 2x + y + 1$$
$$z = y - x + 5$$

for $x$ and $y$ in terms of $w$ and $z$.

The solution is

$$x = \tfrac{1}{3}(w - 1) - \tfrac{1}{3}(z - 5),$$
$$y = \tfrac{1}{3}(w - 1) + \tfrac{2}{3}(z - 5)$$

so the condition

$$ax + by + c = 0$$

can be written as

$$a[\tfrac{1}{3}(w - 1) - \tfrac{1}{3}(z - 5)] + b[\tfrac{1}{3}(w - 1) + \tfrac{2}{3}(z - 5)] + c = 0$$

or as

$$\tfrac{1}{3}(a + b)w + (\tfrac{2}{3}b - \tfrac{1}{3}a)z + c + \tfrac{4}{3}a - \tfrac{11}{3}b = 0.$$

In other words

$$f(A) = \left\{ \begin{bmatrix} w \\ z \end{bmatrix} \middle| ew + gz + h = 0 \right\}$$

where

$$e = \tfrac{1}{3}(a + b),$$
$$g = \tfrac{2}{3}b - \tfrac{1}{3}a,$$

and

$$h = c + \tfrac{4}{3}a - \tfrac{11}{3}b.$$

This is again a straight line.

Notice that $f$ is *not* a Euclidean transformation. Affine geometry consists of the study of those properties of subsets of the plane which are invariant under all one-to-one affine transformations. Thus, if $A$ is a circle, $f(A)$ need not be a circle (but will be an ellipse as we shall see later on).

Suppose that $f$ is such an affine transformation. Then $f$ carries straight lines into straight lines (by definition) and parallel straight lines into parallel straight lines (since distinct points go into distinct points). Thus $f$ carries parallelograms into parallelograms. Thus the concept of a parallelogram makes sense in affine geometry (figure 1.10) (while the concept of rectangle or square does not (figure 1.11)).

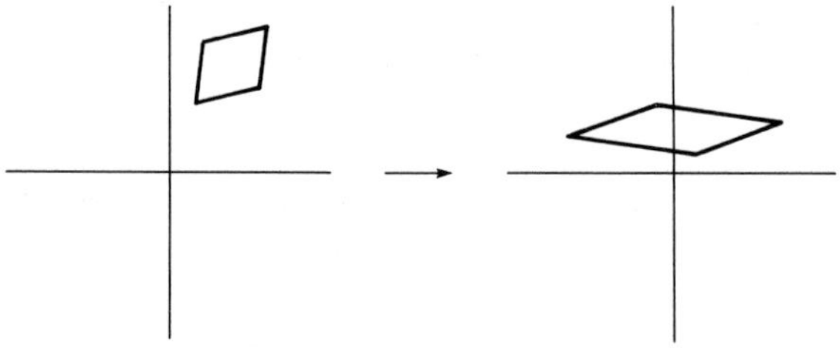

**Figure 1.10**

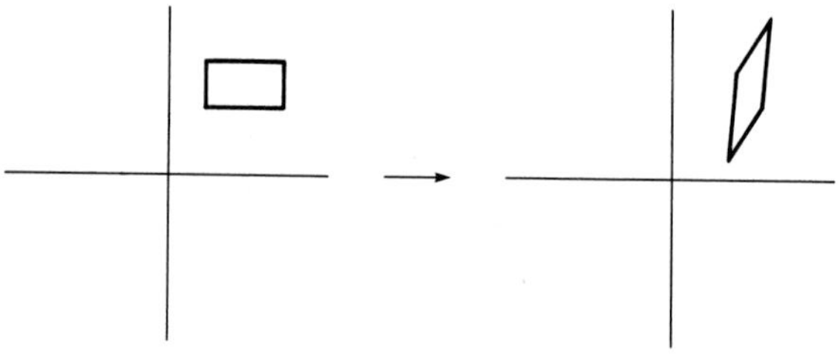

**Figure 1.11**

## 1.5. Linear transformations

The simplest kind of affine (and Euclidean) transformations are the *translations*

$$\begin{bmatrix} x \\ y \end{bmatrix} \to \begin{bmatrix} x+a \\ y+b \end{bmatrix}.$$

By a translation we can move any point of the plane into any other point. Before proceeding further it is convenient to restrict attention to affine transformations that keep one point, say $\begin{bmatrix} 0 \\ 0 \end{bmatrix}$, fixed. We can then get to any other point by applying a translation.

Let $f$ be a one-to-one affine transformation which keeps the origin fixed. Choose $O = \begin{bmatrix} 0 \\ 0 \end{bmatrix}$ as the origin. We can now identify a point $P = \begin{bmatrix} x \\ y \end{bmatrix}$ with its *position*

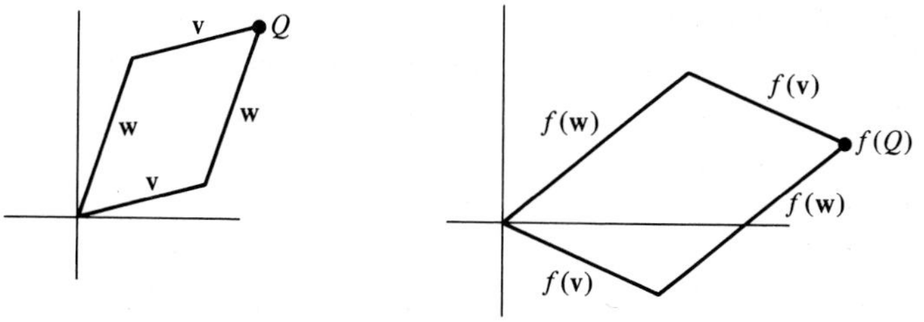


**Figure 1.12**

*vector* $\mathbf{v}=\begin{pmatrix}x\\y\end{pmatrix}$, so $P=O$ "+" $\mathbf{v}$. We shall, accordingly, drop the [ ] notation and the distinction between $A\mathbb{R}^2$ and $\mathbb{R}^2$. Since $f$ carries parallelograms into parallelograms, it follows immediately that, if the position vector of $Q$ is $\mathbf{v}+\mathbf{w}$, the position vector of $f(Q)$ is $f(\mathbf{v})+f(\mathbf{w})$. Therefore

$$f(\mathbf{v}+\mathbf{w})=f(\mathbf{v})+f(\mathbf{w}), \tag{1.2}$$

if $\mathbf{v}\neq\mathbf{w}$. We can now show that $f$ preserves ratios of segments along any line. From the parallelogram spanned by $\mathbf{w}, \mathbf{v}, \mathbf{v}+\mathbf{w}$ and $2\mathbf{v}$, we see that

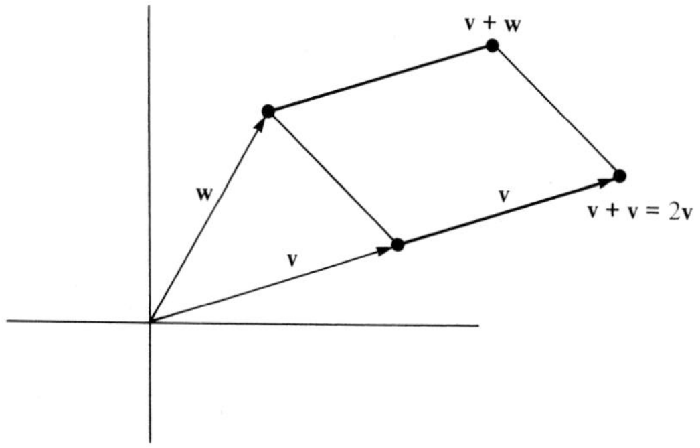


**Figure 1.13**

$$f(2\mathbf{v})=2f(\mathbf{v})$$

so (1.2) holds also when $\mathbf{v}=\mathbf{w}$. By repeating the argument,

$$f(n\mathbf{v})=nf(\mathbf{v})$$

for any integer $n\geqslant 0$. Applied to $(1/m)\mathbf{v}$, this implies

$$f(a\mathbf{v})=af(\mathbf{v})$$

for any rational number $a\geqslant 0$.

From the parallelogram with vertices $O, \mathbf{w}, -\mathbf{v}+\mathbf{w}, -\mathbf{v}$ we see that

$$f(-\mathbf{v})=-f(\mathbf{v})$$

so that

$$f(a\mathbf{v})=af(\mathbf{v})$$

for any rational number, $a$, positive or negative, and all $\mathbf{v}$.

If we *assume* that $f$ is continuous, it would follow that

$$f(a\mathbf{v}) = af(\mathbf{v})$$

for all real numbers, $a$. It turns out that it is not necessary to make this assumption. That is, it follows from properties of the real number system that knowing that $f$ carries lines into lines in the plane implies that $f$ is continuous, and hence that $f(av) = af(v)$ for all real numbers $a$ and all vectors $v$. The proof of this fact is a little tricky, and we shall present it in an appendix at the end of this chapter. For the moment we shall restrict attention to those affine transformations which do satisfy $f(a\mathbf{v}) = af(\mathbf{v})$ for all real $a$, although, as we said, this turns out not to be a restriction at all. For such $f$, we have the identity

$$f(a\mathbf{v} + b\mathbf{w}) = af(\mathbf{v}) + bf(\mathbf{w}) \tag{1.3}$$

for any real numbers $a$ and $b$ and for any vectors $\mathbf{v}$ and $\mathbf{w}$ in $\mathbb{R}^2$.

A map $f: \mathbb{R}^2 \to \mathbb{R}^2$ satisfying (1.3) is called a *linear* transformation of the plane. We have converted the study of *affine* transformations of $\mathbb{R}^2$ which hold the origin fixed into the study of linear transformations of the vector space $\mathbb{R}^2$.

Any map of $\mathbb{R}^2 \to \mathbb{R}^2$ satisfying (1.3) is linear, by definition. Not every linear transformation is one-to-one. For example, the transformation which maps every vector in $\mathbb{R}^2$ into the zero vector,

$$f\begin{pmatrix} x \\ y \end{pmatrix} = \begin{pmatrix} 0 \\ 0 \end{pmatrix} \quad \text{for all} \begin{pmatrix} x \\ y \end{pmatrix},$$

is linear, but not one-to-one.

If $f$ is a linear transformation,

$$f(\mathbf{v} + t\mathbf{w}) = f(\mathbf{v}) + tf(\mathbf{w}).$$

If $f$ is also one-to-one, then $\mathbf{w} \neq 0$ implies $f(\mathbf{w}) \neq 0$. Thus $f$ carries the line $\{\mathbf{v} + t\mathbf{w} | t \in \mathbb{R}\}$ into the line $\{f(\mathbf{v}) + tf(\mathbf{w}) | t \in \mathbb{R}\}$, so $f$ carries lines into lines. Hence every one-to-one linear transformation is affine. A *one-to-one* linear transformation is called *regular* or *non-singular.* A linear transformation which is *not* one-to-one is called *singular.* We have seen that every regular linear transformation is affine. We shall see that the singular ones collapse the whole plane either into the origin or into a line.

It is clear that if $f$ and $g$ are linear transformations (regular or not) then $g \circ f$ is again a linear transformation. Indeed, $(g \circ f)(a\mathbf{v} + b\mathbf{w}) = g(af(\mathbf{v}) + bf(\mathbf{w}))$ since $f$ is linear. Since $g$ is linear this equals $ag \circ f(\mathbf{v}) + bg \circ f(\mathbf{w})$ which shows that $g \circ f$ is linear.

To summarize: Linear transformations are, by definition, those $f$ which satisfy (1.3) for all pairs of vectors $\mathbf{v}$ and $\mathbf{w}$ and all real numbers $a$ and $b$. An affine transformation is a one-to-one map of $\mathbb{R}^2$ into itself which carries lines into lines Any affine transformation can be written as a (non-singular) linear transformation followed by a translation; that is, any affine transformation $f$ satisfies

$$f(\mathbf{w}) = l(\mathbf{w}) + \mathbf{v}$$

where $l$ is a regular linear transformation. Conversely, every $f$ of this form is affine.

## 1.6. The matrix of a linear transformation

Let $f$ be a linear transformation. We can write any $\begin{pmatrix} x \\ y \end{pmatrix}$ in the plane as

$$\begin{pmatrix} x \\ y \end{pmatrix} = x\begin{pmatrix} 1 \\ 0 \end{pmatrix} + y\begin{pmatrix} 0 \\ 1 \end{pmatrix}$$

so that

$$f\begin{pmatrix} x \\ y \end{pmatrix} = xf\begin{pmatrix} 1 \\ 0 \end{pmatrix} + yf\begin{pmatrix} 0 \\ 1 \end{pmatrix}.$$

This formula shows how $f$ is completely determined by what it does to the two basis vectors $\begin{pmatrix} 1 \\ 0 \end{pmatrix}$ and $\begin{pmatrix} 0 \\ 1 \end{pmatrix}$. Suppose that $f\begin{pmatrix} 1 \\ 0 \end{pmatrix} = \begin{pmatrix} a \\ c \end{pmatrix}$ and $f\begin{pmatrix} 0 \\ 1 \end{pmatrix} = \begin{pmatrix} b \\ d \end{pmatrix}$. Then $f$ is completely determined by the four numbers $a, b, c$, and $d$,

$$f\begin{pmatrix} x \\ y \end{pmatrix} = \begin{pmatrix} ax + by \\ cx + dy \end{pmatrix}.$$

We write these four numbers as a square *matrix*

$$\mathrm{Mat}(f) = \begin{pmatrix} a & b \\ c & d \end{pmatrix}$$

where the first column is the image of $\begin{pmatrix} 1 \\ 0 \end{pmatrix}$ and the second column is the image of $\begin{pmatrix} 0 \\ 1 \end{pmatrix}$ The image of any point $\begin{pmatrix} x \\ y \end{pmatrix}$ is then given by

$$\begin{pmatrix} a & b \\ c & d \end{pmatrix}\begin{pmatrix} x \\ y \end{pmatrix} = \begin{pmatrix} ax + by \\ cx + dy \end{pmatrix} \tag{1.4}$$

We regard (1.4) as a multiplication rule, telling us how to multiply the vector $\begin{pmatrix} x \\ y \end{pmatrix}$ by the matrix $\begin{pmatrix} a & b \\ c & d \end{pmatrix}$, to give another vector. It says to take the row $\times$ column for each of the two components. Thus the top component is $ax + by$ which is obtained from the top row $(a, b)$ of the matrix and the column $\begin{pmatrix} x \\ y \end{pmatrix}$. Similarly for the bottom component.

For example, suppose that $R_\theta$ is counterclockwise rotation of the plane through angle $\theta$. Then

$$R_\theta\begin{pmatrix} 1 \\ 0 \end{pmatrix} = \begin{pmatrix} \cos\theta \\ \sin\theta \end{pmatrix}$$

and

$$R_\theta\begin{pmatrix} 0 \\ 1 \end{pmatrix} = \begin{pmatrix} -\sin\theta \\ \cos\theta \end{pmatrix}$$

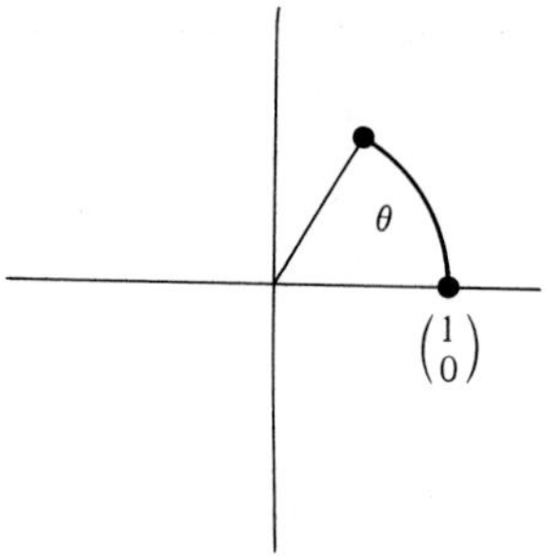


**Figure 1.14**

so that $R_\theta$ has the matrix

$$\begin{pmatrix} \cos\theta & -\sin\theta \\ \sin\theta & \cos\theta \end{pmatrix}.$$

The image of any point $\begin{pmatrix} x \\ y \end{pmatrix}$ is given by

$$\begin{pmatrix} \cos\theta & -\sin\theta \\ \sin\theta & \cos\theta \end{pmatrix}\begin{pmatrix} x \\ y \end{pmatrix} = \begin{pmatrix} (\cos\theta)x - (\sin\theta)y \\ (\sin\theta)x + (\cos\theta)y \end{pmatrix}.$$

The formula (1.4) shows how to assign a linear transformation to each matrix. We can thus identify $2 \times 2$ matrices with linear transformations of $\mathbb{R}^2$.

## 1.7. Matrix multiplication

Suppose that $F$ is a linear transformation whose matrix is $\begin{pmatrix} a & b \\ c & d \end{pmatrix}$ and $G$ is a linear transformation whose matrix is $\begin{pmatrix} e & f \\ g & h \end{pmatrix}$. Then $F \circ G$ is again a linear transformation. It has a matrix whose first column is

$$(F \circ G)\begin{pmatrix} 1 \\ 0 \end{pmatrix} = F\begin{pmatrix} e \\ g \end{pmatrix} = \begin{pmatrix} a & b \\ c & d \end{pmatrix}\begin{pmatrix} e \\ g \end{pmatrix} = \begin{pmatrix} ae + bg \\ ce + dg \end{pmatrix}.$$

The second column is

$$(F \circ G)\begin{pmatrix} 0 \\ 1 \end{pmatrix} = F\begin{pmatrix} f \\ h \end{pmatrix} = \begin{pmatrix} a & b \\ c & d \end{pmatrix}\begin{pmatrix} f \\ h \end{pmatrix} = \begin{pmatrix} af + bh \\ cf + dh \end{pmatrix}.$$

Thus we define the 'multiplication' of matrices to correspond to composition of linear transformations, $(\text{Mat}\, F) \times (\text{Mat}\, G) = \text{Mat}\,(F \circ G)$. The rule for multiplication is

$$\begin{pmatrix} a & b \\ c & d \end{pmatrix}\begin{pmatrix} e & f \\ g & h \end{pmatrix} = \begin{pmatrix} ae + bg & af + bh \\ ce + dg & cf + dh \end{pmatrix}.$$

For any position in the product matrix we take the same row from the first matrix and the same column from the second matrix and multiply row by column.

For example, if $R_\theta$ is (counterclockwise) rotation through angle $\theta$ and $R_\phi$ is rotation through angle $\phi$, then $R_\theta \cdot R_\phi = R_{\theta+\phi}$ and

$$\begin{pmatrix} \cos\theta & -\sin\theta \\ \sin\theta & \cos\theta \end{pmatrix} \times \begin{pmatrix} \cos\phi & -\sin\phi \\ \sin\phi & \cos\phi \end{pmatrix}$$

$$= \begin{pmatrix} \cos\theta\cos\phi - \sin\theta\sin\phi & -\cos\theta\sin\phi - \sin\theta\cos\phi \\ \sin\theta\cos\phi + \cos\theta\sin\phi & \cos\theta\cos\phi - \sin\theta\sin\phi \end{pmatrix}$$

Comparing this with the matrix of $R_{\theta+\phi}$

$$\begin{pmatrix} \cos(\theta+\phi) & -\sin(\theta+\phi) \\ \sin(\theta+\phi) & \cos(\theta+\phi) \end{pmatrix}$$

gives the standard trigonometric formulae for $\cos(\theta+\phi)$ and $\sin(\theta+\phi)$. Thus you need no longer remember the identities for the sine and cosine of the sum of two angles. You can derive them from the more general rule of matrix multiplication.

Notice that matrix multiplication, in general, is *not* commutative: for example,

$$\begin{pmatrix} 1 & 2 \\ 0 & 3 \end{pmatrix}\begin{pmatrix} 4 & 0 \\ 0 & 5 \end{pmatrix} = \begin{pmatrix} 4 & 10 \\ 0 & 15 \end{pmatrix}$$

while

$$\begin{pmatrix} 4 & 0 \\ 0 & 5 \end{pmatrix}\begin{pmatrix} 1 & 2 \\ 0 & 3 \end{pmatrix} = \begin{pmatrix} 4 & 8 \\ 0 & 15 \end{pmatrix}.$$

(Two *rotations* of $\mathbb{R}^2$ do commute with one another since it does not matter through which angle we rotate first. But, *in general*, two matrices need not commute.)

As an illustration of matrix multiplication, we prove a 'triple product decomposition' which will be used later on. This decomposition states that any matrix $\begin{pmatrix} a & b \\ c & d \end{pmatrix}$ with $a \neq 0$ can be written as a triple product of the form

$$\begin{pmatrix} a & b \\ c & d \end{pmatrix} = \begin{pmatrix} 1 & 0 \\ y & 1 \end{pmatrix}\begin{pmatrix} r & 0 \\ 0 & s \end{pmatrix}\begin{pmatrix} 1 & x \\ 0 & 1 \end{pmatrix}. \tag{1.5}$$

To prove this result we simply devise a procedure for determining $y, r, s$, and $x$. We first multiply the matrices on the right. Since

$$\begin{pmatrix} r & 0 \\ 0 & s \end{pmatrix}\begin{pmatrix} 1 & x \\ 0 & 1 \end{pmatrix} = \begin{pmatrix} r & rx \\ 0 & s \end{pmatrix},$$

we want

$$\begin{pmatrix} a & b \\ c & d \end{pmatrix} = \begin{pmatrix} 1 & 0 \\ y & 1 \end{pmatrix}\begin{pmatrix} r & rx \\ 0 & s \end{pmatrix} = \begin{pmatrix} r & rx \\ ry & ryx + s \end{pmatrix}.$$

Now we can equate corresponding entries in the left-hand and right-hand matrices. First, $a = r$, and since by assumption $a \neq 0$, $r \neq 0$. Next, $b = rx$ and so $x = b/r = b/a$ (remember that $a \neq 0$). Similarly, $c = ry$ and so $y = c/r = c/a$. Finally, $d = rxy + s$ and so

$$s = d - rxy = d - r(b/r)(c/r) = d - (bc)/a.$$

A similar decomposition, important in the analysis of lens systems, is

$$\begin{pmatrix} a & b \\ c & d \end{pmatrix} = \begin{pmatrix} 1 & y \\ 0 & 1 \end{pmatrix} \begin{pmatrix} 0 & f \\ e & 0 \end{pmatrix} \begin{pmatrix} 1 & x \\ 0 & 1 \end{pmatrix}$$

valid for any matrix with $c \neq 0$. The proof of this decomposition is simple: again, just multiply out the triple product and equate corresponding matrix entries on both sides of the equation.

## 1.8. Matrix algebra

Let $F$ and $G$ be two linear transformations of $\mathbb{R}^2$. We define their sum by

$$(F+G)(\mathbf{v}) = F(\mathbf{v}) + G(\mathbf{v}).$$

Notice that

$$\begin{aligned}(F+G)(a\mathbf{v}+b\mathbf{w}) &= F(a\mathbf{v}+b\mathbf{w}) + G(a\mathbf{v}+b\mathbf{w}) \\ &= aF(\mathbf{v}) + bF(\mathbf{w}) + aG(\mathbf{v}) + bG(\mathbf{w}) \\ &= a(F(\mathbf{v}) + G(\mathbf{v})) + b(F(\mathbf{w}) + G(\mathbf{w})) \\ &= a(F+G)(\mathbf{v}) + b(F+G)(\mathbf{w}).\end{aligned}$$

Thus $F+G$ is again a linear transformation. It is clear that this addition is associative and commutative, that the zero transformation, $0(\mathbf{v}) = 0$, for all $\mathbf{v}$, is the zero for this addition and that $(-F)(\mathbf{v}) = -F(\mathbf{v})$ defines the negative of $F$, i.e., $(-F)+F=0$, where 0 in this equation stands for the zero linear transformation.

If $H$ is a third linear transformation, then composition, represented by matrix multiplication, has the following property:

$$\begin{aligned}H\circ(F+G)(v) &= H[(F+G)(v)] \\ &= H[F(v)+G(v)] \\ &= H[F(v)] + H[G(v)] \\ &= (H\circ F)(v) + (H\circ G)(v)\end{aligned}$$

for all $\mathbf{v}$, or, in short,

$$H\circ(F+G) = H\circ F + H\circ G,$$

and, similarly,

$$(F+G)\circ H = F\circ H + G\circ H.$$

Thus multiplication is distributive relative to this addition.

It follows directly from the definition of the sum of linear transformations that if the matrices of $F$ and $G$ are

$$\operatorname{Mat}(F) = \begin{pmatrix} a & b \\ c & d \end{pmatrix} \quad \text{and} \quad \begin{pmatrix} e & f \\ g & h \end{pmatrix} = \operatorname{Mat}(G)$$

then the matrix of $F+G$ is

$$\operatorname{Mat}(F+G) = \begin{pmatrix} a+e & b+f \\ c+g & d+h \end{pmatrix} = \operatorname{Mat}(F) + \operatorname{Mat}(G).$$

In other words, we add matrices by adding the entries at each position. We can also multiply a matrix by a number: $2\begin{pmatrix} a & b \\ c & d \end{pmatrix} = \begin{pmatrix} 2a & 2b \\ 2c & 2d \end{pmatrix}$. Notice that

$$2\begin{pmatrix} a & b \\ c & d \end{pmatrix} = \begin{pmatrix} 2a & 2b \\ 2c & 2d \end{pmatrix} = \begin{pmatrix} 2 & 0 \\ 0 & 2 \end{pmatrix}\begin{pmatrix} a & b \\ c & d \end{pmatrix}.$$

We have now defined addition and multiplication for $2 \times 2$ matrices, and the rules for addition and multiplication satisfy most of the familiar rules for adding and multiplying numbers. Thus:

Addition is commutative and associative with the existence of a zero and a negative;

Multiplication is associative with the existence of an identity, $\begin{pmatrix} 1 & 0 \\ 0 & 1 \end{pmatrix}$, and is distributive over addition.

There are, however, two important differences:

(1) multiplication is *not* commutative;

(2) the product of two non-zero matrices can be zero, so the cancellation law for multiplication need not hold:

$$\begin{pmatrix} 0 & 1 \\ 0 & 0 \end{pmatrix} \times \begin{pmatrix} 0 & 1 \\ 0 & 0 \end{pmatrix} = \begin{pmatrix} 0 & 0 \\ 0 & 0 \end{pmatrix}.$$

Nevertheless, we shall see that in many respects we can deal with linear transformations as if they were numbers.

Instead of linear transformations of the vector space $\mathbb{R}^2$, we could consider linear transformations of the vector space $\mathbb{R}^3$. The vectors of $\mathbb{R}^3$ are described as $\begin{pmatrix} x \\ y \\ z \end{pmatrix}$ and there are now three basis vectors, $\begin{pmatrix} 1 \\ 0 \\ 0 \end{pmatrix}$, $\begin{pmatrix} 0 \\ 1 \\ 0 \end{pmatrix}$ and $\begin{pmatrix} 0 \\ 0 \\ 1 \end{pmatrix}$. The most general linear transformation of $\mathbb{R}^3$ is now described by a $3 \times 3$ matrix of the form

$$\begin{pmatrix} a_{11} & a_{12} & a_{13} \\ a_{21} & a_{22} & a_{23} \\ a_{31} & a_{32} & a_{33} \end{pmatrix}$$

where the first column is the image of the first basis vector $\begin{pmatrix} 1 \\ 0 \\ 0 \end{pmatrix}$, etc. The formula for multiplication is

$$\begin{pmatrix} a_{11} & a_{12} & a_{13} \\ a_{21} & a_{22} & a_{23} \\ a_{31} & a_{32} & a_{33} \end{pmatrix} \times \begin{pmatrix} b_{11} & b_{12} & b_{13} \\ b_{21} & b_{22} & b_{23} \\ b_{31} & b_{32} & b_{33} \end{pmatrix} = \begin{pmatrix} c_{11} & c_{12} & c_{13} \\ c_{21} & c_{22} & c_{23} \\ c_{31} & c_{32} & c_{33} \end{pmatrix}$$

where, for any $i$ and $j$ ranging over $1, 2, 3$,

$$c_{ij} = a_{i1}b_{1j} + a_{i2}b_{2j} + a_{i3}b_{3j}.$$

Again, for any position, the row from the first matrix multiplies the column from the second. Thus, for example, taking $i = 2$ and $j = 3$ in the above formula corresponds to the diagram

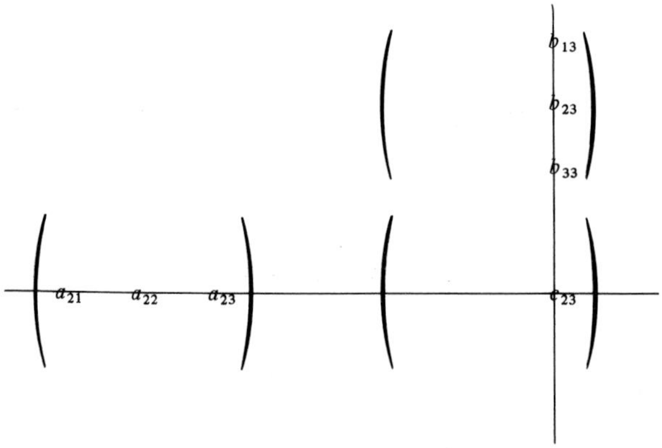


**Figure 1.15**

The law for addition is again positionwise addition. The various associative, distributive laws apply as before, as does the commutative law for addition.

Equally well, we could consider $4 \times 4$, $5 \times 5$, or in general $n \times n$ matrices. Also we can multiply a vector in $\mathbb{R}^3$ by a $3 \times 3$ matrix:

$$\begin{pmatrix} a & b & c \\ d & e & f \\ g & h & i \end{pmatrix}\begin{pmatrix} x \\ y \\ z \end{pmatrix} = \begin{pmatrix} ax + by + cz \\ dx + ey + fz \\ gx + hy + iz \end{pmatrix}$$

and, more generally, vectors in $\mathbb{R}^n$ by $n \times n$ matrices.

## 1.9. Areas and determinants

We return to the plane. Let $f$ be a non-singular linear transformation of the plane. Since $f(\mathbf{v} + t\mathbf{w}) = f(\mathbf{v}) + tf(\mathbf{w})$, we know that $f$ carries lines into lines and hence is an affine transformation. Thus $f$ carries squares into parallelograms. Furthermore, let $\square_{\mathbf{v}}$ be the unit square whose left-hand lower corner is at $\mathbf{v}$,

$$\begin{aligned} \square_{\mathbf{v}} &= \left\{ \mathbf{v} + s\begin{pmatrix} 1 \\ 0 \end{pmatrix} + t\begin{pmatrix} 0 \\ 1 \end{pmatrix} \middle| 0 \leqslant s < 1, \quad 0 \leqslant t < 1 \right\} \\ &= \{\mathbf{v} + \mathbf{w} \mid \mathbf{w} \in \square_0\}. \end{aligned}$$

Then the image of $\square_{\mathbf{v}}$ under $f$, which we denote by $f(\square_{\mathbf{v}})$, is just a translate of the image of $\square_0$ under $f$:

$$\begin{aligned} f(\square_{\mathbf{v}}) &= \{f(\mathbf{v}) + f(\mathbf{w}) \mid \mathbf{w} \in \square_0\} \\ &= \{f(\mathbf{v}) + \mathbf{u} \mid \mathbf{u} \in f(\square_0)\} \end{aligned}$$

and thus $f(\square_{\mathbf{v}})$ has the same area as $f(\square_0)$. The same clearly holds if we consider a square of any size, not necessarily the unit square. On the other hand, we can

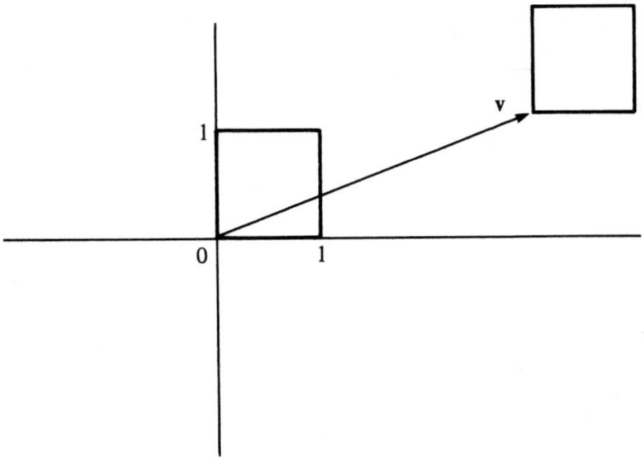


**Figure 1.16**

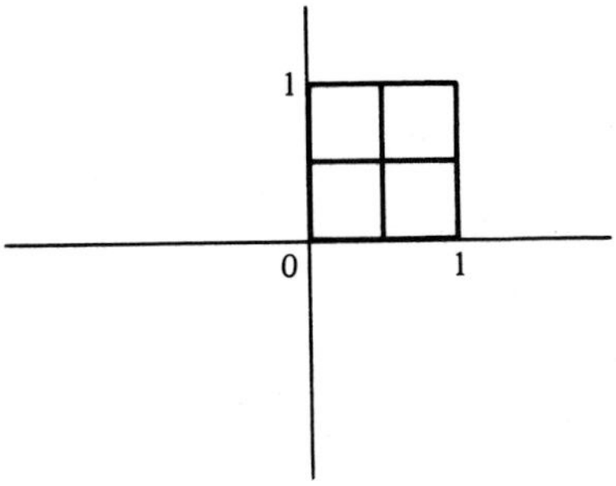


**Figure 1.17**

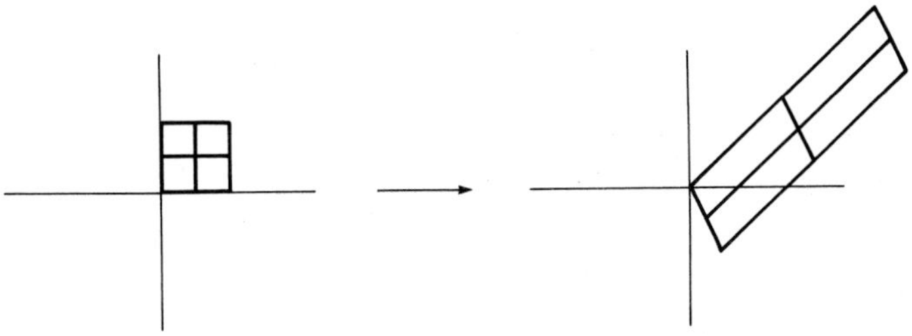

**Figure 1.18**

subdivide the unit square into four congruent squares and their images are all congruent and fit together to form the image of $\square_0$; thus each of these images has area equal to $\frac{1}{4} \times$ (the area of $f(\square_0)$).

By repeated subdivision we conclude that, if $\square$ is any square whose side length is $1/2^k$, then

$$\frac{\text{area } f(\square)}{\text{area } \square}$$

is a number which is independent of $\square$ (and of the size $2^k$). Let us denote this number by Ar ($f$), so that

$$\mathrm{Ar}(f) = \frac{\mathrm{area}\cdot(f(\square))}{\mathrm{area}\ \square}.$$

If $D$ is any region in the plane

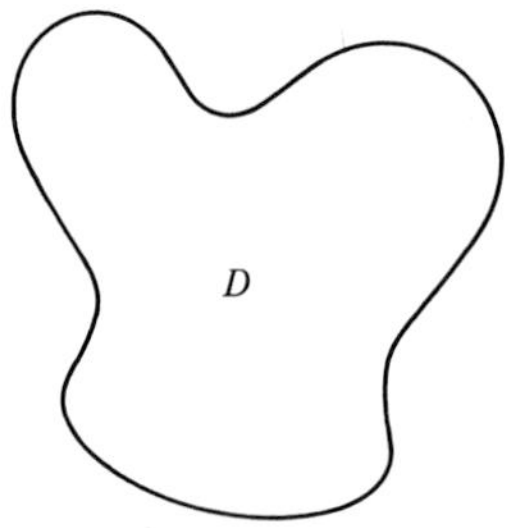


**Figure 1.19**

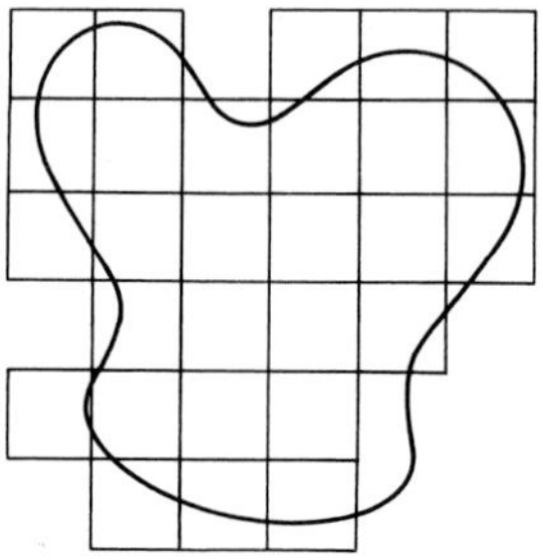

**Figure 1.20**

we can approximate it by a union of squares (and its image by the image parallelograms) so that

$$\mathrm{Ar}(f) = \frac{\mathrm{area}\, f(D)}{\mathrm{area}\ D}$$

for *any* (nice) region. (Strictly speaking, we should approximate it from the inside and the outside. If we assume that we can cover the boundary by a finite union of small squares whose total area can be made as small as we like, then the total area of the parallelograms covering the image of the boundary will also be as small as we like. Hence the approximation is legitimate. This is the meaning of our qualification that the region be 'nice'.)

Thus $\mathrm{Ar}(f)$ gives the factor which tells us how area changes when we apply $f$. If $f$ and $g$ are two non-singular linear transformations,

$$\frac{\mathrm{area}\,(f\circ g)(\square)}{\mathrm{area}\ \square} = \frac{\mathrm{area}\, f\circ g(\square)}{\mathrm{area}\, g(\square)} \times \frac{\mathrm{area}\, g(\square)}{\mathrm{area}\ \square}$$

so

$$\mathrm{Ar}(f \circ g) = (\mathrm{Ar}(f)) \times (\mathrm{Ar}(g)).$$

We now compute $\mathrm{Ar}(f)$ for some special cases by inspection then compose them to get the general case. Notice that $\mathrm{Ar}(f) = \text{area } [f(\square_0)] = \text{area of the}$ image of the unit square under $f$.

**Case 1a**: $f$ is represented by

$$F = \mathrm{Mat}(f) = \begin{pmatrix} r & 0 \\ 0 & s \end{pmatrix} \quad r > 0 \quad s > 0$$

$\mathrm{Ar}(f) = rs.$

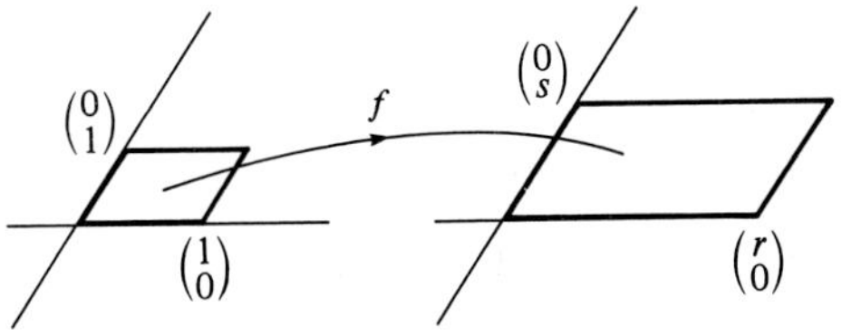


**Figure 1.21(a)**

**Case 1b**: $f$ is represented by

$$F = \begin{pmatrix} r & 0 \\ 0 & s \end{pmatrix} \quad r < 0 \quad s > 0$$

$\mathrm{Ar}(f) = |rs|.$

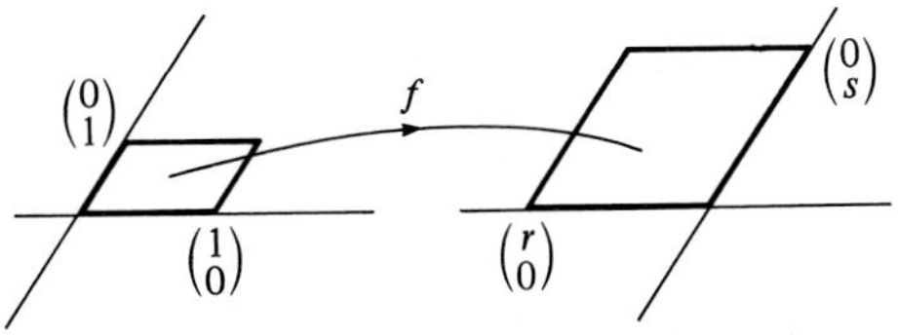


**Figure 1.21(b)**

In general it is clear that $\mathrm{Ar}(f) = |rs|$ in any case where $F$ is a diagonal matrix.

**Case 2a**: $f$ is represented by

$$F = \begin{pmatrix} 1 & x \\ 0 & 1 \end{pmatrix}$$

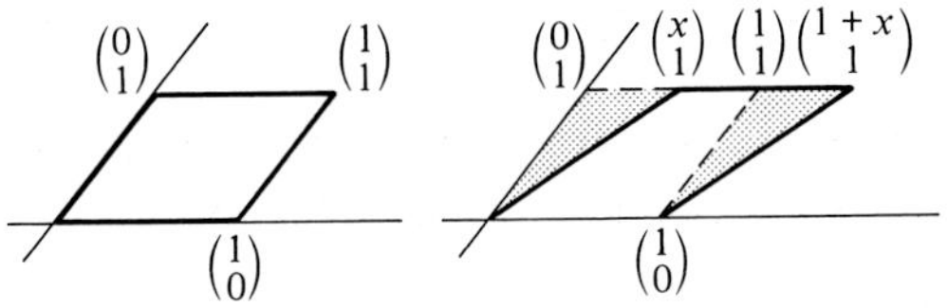

**Figure 1.22(a)**

Since the shaded triangle with vertices $\begin{pmatrix}1\\0\end{pmatrix}, \begin{pmatrix}1\\1\end{pmatrix}, \begin{pmatrix}1+x\\1\end{pmatrix}$ can be obtained from the shaded triangle with vertices $\begin{pmatrix}0\\0\end{pmatrix}, \begin{pmatrix}0\\1\end{pmatrix}, \begin{pmatrix}x\\1\end{pmatrix}$ by a translation $\left(\text{adding } \begin{pmatrix}1\\0\end{pmatrix} \text{ to each vertex}\right)$, the image of the unit parallelogram has the same area as the unit parallelogram. That is, area is unchanged by this *shear transformation.* $\mathrm{Ar}(f) = 1$ in this case.

**Case 2b**: $f$ is represented by

$$F = \begin{pmatrix}1 & 0\\ y & 1\end{pmatrix}$$

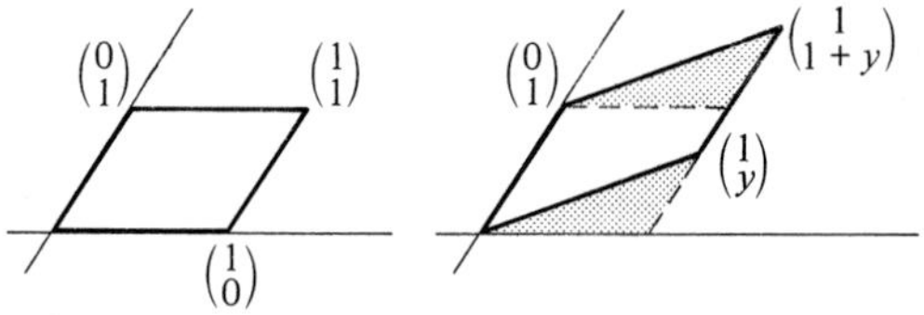


**Figure 1.22(b)**

Again the two shaded triangles have the same area, and so $\mathrm{Ar}(f) = 1$ in this case also.

For any matrix $F = \begin{pmatrix}a & b\\ c & d\end{pmatrix}$ we define its *determinant* by

$$\mathrm{Det}\, F = ad - bc. \tag{1.6}$$

We wish to prove the basic formula

$$\mathrm{Ar}(f) = |\mathrm{Det}\, F|, \tag{1.7}$$

We have proved this formula for each of the three kinds of matrices listed above. To prove it in general we make use of the following important property of determinants:

$$\mathrm{Det}\,(F \circ G) = (\mathrm{Det}\, F) \times (\mathrm{Det}\, G)$$

which can be verified by direct multiplication: if $G = \begin{pmatrix}e & f\\ g & h\end{pmatrix}$ then

$$\begin{aligned}\mathrm{Det}\,(F \circ G) &= \mathrm{Det}\begin{pmatrix}ae+bg & af+bh\\ ce+dg & cf+dh\end{pmatrix}\\ &= (ae+bg)(cf+dh) - (af+bh)(ce+dg)\\ &= (ad-bc)(eh-fg) = (\mathrm{Det}\, F) \times (\mathrm{Det}\, G).\end{aligned}$$

From the two rules $\mathrm{Ar}(f \circ g) = (\mathrm{Ar}\, f) \times (\mathrm{Ar}\, g)$ and $\mathrm{Det}(F \circ G) = (\mathrm{Det}\, F) \times (\mathrm{Det}\, G)$ we conclude that the formula

$$\boxed{\mathrm{Ar}\, f = |\mathrm{Det}\, F|}$$

is true for any matrix that can be written as a product of matrices for which we already know the formula to be true. We proved in section 7 (equation (1.5)) that if $a \neq 0$ we can write

$$\begin{pmatrix} a & b \\ c & d \end{pmatrix} = \begin{pmatrix} 1 & 0 \\ y & 1 \end{pmatrix}\begin{pmatrix} r & 0 \\ 0 & s \end{pmatrix}\begin{pmatrix} 1 & x \\ 0 & 1 \end{pmatrix}.$$

We have thus proved the formula for all matrices with $a \neq 0$. To deal with the case $a = 0$ we can proceed in either of two ways:

(i) Direct verification:

$$\mathrm{Ar}\begin{pmatrix} 0 & b \\ c & d \end{pmatrix} = |bc| = \left|\mathrm{Det}\begin{pmatrix} 0 & b \\ c & d \end{pmatrix}\right|.$$

(Details of the proof are left to the reader.)

(ii) Continuity argument: We can notice that both $\mathrm{Ar}\, f$ and $\mathrm{Det}\, F$ are continuous functions of the entries of $F$ (i.e., if we change the entries slightly, the values of $\mathrm{Ar} f$ and of $\mathrm{Det}\, F$ change only slightly). Now, if $\begin{pmatrix} 0 & b \\ c & d \end{pmatrix}$ is non-singular, so is $\begin{pmatrix} e & b \\ c & d \end{pmatrix}$ for sufficiently small $e$ $\left(\text{indeed } \mathrm{Ar}\begin{pmatrix} e & b \\ c & d \end{pmatrix} \text{ is non-zero}\right)$. Thus, since we know that the equality

$$\mathrm{Ar}\begin{pmatrix} e & b \\ c & d \end{pmatrix} = \left|\mathrm{Det}\begin{pmatrix} e & b \\ c & d \end{pmatrix}\right|$$

is true for all $e$ close to zero, we conclude that it is true for $e = 0$ as well.

We should point out the significance of the sign of $\mathrm{Det}\, F$ when $\mathrm{Det}\, F \neq 0$. (We have given a meaning to its absolute value.) The meaning, at present, is best illustrated by example. The transformation $\begin{pmatrix} 1 & 0 \\ 0 & -1 \end{pmatrix}$ is a reflection about the $x$-axis.

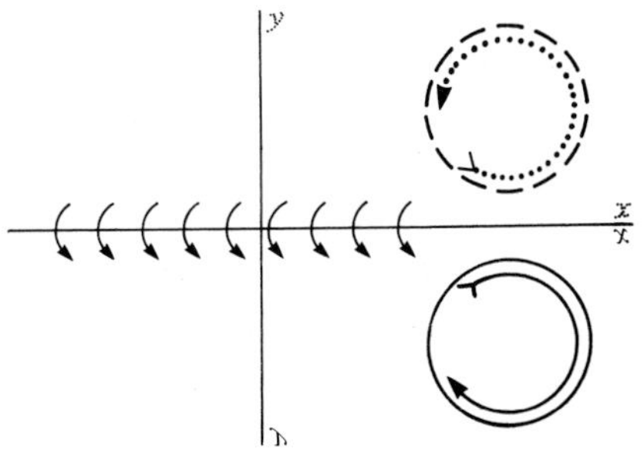


**Figure 1.23**

It has the effect of switching counterclockwise rotation into clockwise rotation. Thus the fact that the determinant is negative has to do with the fact that the orientation of the plane is reversed.

As an illustration, let us consider a Euclidean linear transformation of the plane, represented by $F$. Since $F\begin{pmatrix}1\\0\end{pmatrix}$ has length 1, we can write

$$F\begin{pmatrix}1\\0\end{pmatrix}=\begin{pmatrix}\cos\theta\\\sin\theta\end{pmatrix}$$

for some $\theta$. Since $F\begin{pmatrix}0\\1\end{pmatrix}$ has length 1 and is perpendicular to $\begin{pmatrix}\cos\theta\\\sin\theta\end{pmatrix}$ we have the two possibilities

$$F\begin{pmatrix}0\\1\end{pmatrix}=\begin{pmatrix}-\sin\theta\\\cos\theta\end{pmatrix}\quad\text{or}\quad F\begin{pmatrix}0\\1\end{pmatrix}=\begin{pmatrix}\sin\theta\\-\cos\theta\end{pmatrix}.$$

In the first case we have a rotation through angle $\theta$.

$$F=\begin{pmatrix}\cos\theta & -\sin\theta\\\sin\theta & \cos\theta\end{pmatrix}\quad\text{and}\quad \operatorname{Det}F=1.$$

In the second case the transformation $F=\begin{pmatrix}\cos\theta & \sin\theta\\\sin\theta & -\cos\theta\end{pmatrix}$, for which $\det F=-1$ is a *reflection* about the line through $\begin{pmatrix}\cos\frac{1}{2}\theta\\\sin\frac{1}{2}\theta\end{pmatrix}$. It follows from the addition formulas of trigonometry that

$$\begin{pmatrix}\cos\theta & \sin\theta\\\sin\theta & -\cos\theta\end{pmatrix}\begin{pmatrix}\cos\frac{1}{2}\theta\\\sin\frac{1}{2}\theta\end{pmatrix}=\begin{pmatrix}\cos\frac{1}{2}\theta\\\sin\frac{1}{2}\theta\end{pmatrix}$$

so that the vector $\begin{pmatrix}\cos\frac{1}{2}\theta\\\sin\frac{1}{2}\theta\end{pmatrix}$ is fixed and the line through the origin containing this vector is mapped into self. Furthermore, $F\circ F=\begin{pmatrix}1 & 0\\0 & 1\end{pmatrix}$, so that $F$ is indeed a reflection.

## 1.10. Inverses

For any matrix $F=\begin{pmatrix}a & b\\c & d\end{pmatrix}$, we define $F^{\mathrm{a}}$ by

$$F^{\mathrm{a}}=\begin{pmatrix}d & -b\\-c & a\end{pmatrix}.$$

Direct multiplication shows that

$$F^{\mathrm{a}}F=FF^{\mathrm{a}}=\begin{pmatrix}\operatorname{Det}F & 0\\0 & \operatorname{Det}F\end{pmatrix}=\operatorname{Det}F\begin{pmatrix}1 & 0\\0 & 1\end{pmatrix}.$$

$G$ is called the (multiplicative) inverse of $F$ if $GF=FG=\begin{pmatrix}1 & 0\\0 & 1\end{pmatrix}$. The above equation shows that $F$ cannot have an inverse if $\operatorname{Det} F = 0$: if $FG = \begin{pmatrix} 1 & 0 \\ 0 & 1 \end{pmatrix}$, we see from the above that

$$F^{\mathrm{a}} = F^{\mathrm{a}} F G = \begin{pmatrix} \operatorname{Det} F & 0 \\ 0 & \operatorname{Det} F \end{pmatrix} G = (\operatorname{Det} F) G$$

which is impossible if $\operatorname{Det} F = 0$ and $F \neq \begin{pmatrix} 0 & 0 \\ 0 & 0 \end{pmatrix}$; if $F = \begin{pmatrix} 0 & 0 \\ 0 & 0 \end{pmatrix}$ then it is certainly impossible to find a $G$ such that $FG = \begin{pmatrix} 1 & 0 \\ 0 & 1 \end{pmatrix}$. The same equation, $F^{\mathrm{a}} = (\operatorname{Det} F)G$, shows that if $\operatorname{Det} F \neq 0$ then

$$G = (\operatorname{Det} F)^{-1} F^{\mathrm{a}}$$

is the inverse of $F$.

A direct check, by multiplication, shows that the above formula does give the inverse of $F$. We have thus proved the following theorem:

> A matrix $F$ has an inverse if and only if $\operatorname{Det} F \neq 0$. If $\operatorname{Det} F \neq 0$ then the inverse matrix, $F^{-1}$ has the formula
>
> $$F^{-1} = \begin{pmatrix} \dfrac{d}{\operatorname{Det} F} & \dfrac{-b}{\operatorname{Det} F} \\ \dfrac{-c}{\operatorname{Det} F} & \dfrac{a}{\operatorname{Det} F} \end{pmatrix}. \tag{1.8}$$

We should understand the geometric meaning of $F^{-1}$; it 'undoes' the effect of $F$. If we apply first $F$ and then $F^{-1}$ then we are back to the identity transformation. We see also that $FF^{-1} = \begin{pmatrix} 1 & 0 \\ 0 & 1 \end{pmatrix}$ (by direct multiplication if you like).

Notice that, if $F$ is singular, it cannot have an inverse, since $F\mathbf{v} = F\mathbf{w}$ implies $F^{-1}F\mathbf{v} = F^{-1}F\mathbf{w}$ or $\mathbf{v} = \mathbf{w}$ for any $F$ having an inverse.

It is reasonable that the condition $\operatorname{Det} F = 0$ corresponds to the singularity of $F$ in view of the interpretation of $\operatorname{Det} F$ in terms of area. Indeed, suppose that the parallelogram spanned by the origin and $\begin{pmatrix} a \\ c \end{pmatrix}$ and $\begin{pmatrix} b \\ d \end{pmatrix}$ has non-zero area, meaning that $\begin{pmatrix} a \\ c \end{pmatrix}$ and $\begin{pmatrix} b \\ d \end{pmatrix}$ do not lie on same straight line through the origin. This means that $\operatorname{Det} \begin{pmatrix} a & b \\ c & d \end{pmatrix} \neq 0$, since $\left| \operatorname{Det} \begin{pmatrix} a & b \\ c & d \end{pmatrix} \right|$ is the area of this parallelogram. In this case the inverse matrix exists, so we can write

$$\begin{pmatrix} 1 & 0 \\ 0 & 1 \end{pmatrix} = \begin{pmatrix} a & b \\ c & d \end{pmatrix} \begin{pmatrix} e & f \\ g & h \end{pmatrix} \quad \text{where} \quad F^{-1} = \begin{pmatrix} e & f \\ g & h \end{pmatrix}.$$

This is the same as saying that $\begin{pmatrix} 1 \\ 0 \end{pmatrix} = e\begin{pmatrix} a \\ c \end{pmatrix} + g\begin{pmatrix} b \\ d \end{pmatrix}$ and $\begin{pmatrix} 0 \\ 1 \end{pmatrix} = f\begin{pmatrix} a \\ c \end{pmatrix} + h\begin{pmatrix} b \\ d \end{pmatrix}$.

This, of course implies that we can express *any* vector in the plane as a linear combination of $\begin{pmatrix} a \\ c \end{pmatrix}$ and $\begin{pmatrix} b \\ d \end{pmatrix}$. ($e, f, g$ and $h$ are just numbers.)

Recall that the vectors $\mathbf{u}$ and $\mathbf{v}$ are said to be *linearly dependent* if there are numbers $r$ and $s$, not both zero, such that

$$r\mathbf{u} + s\mathbf{v} = 0.$$

If $r \neq 0$, we can solve this equation for $\mathbf{u}$ to obtain $\mathbf{u} = -(s/r)\mathbf{v}$ and, if $s \neq 0$, we can solve this equation for $\mathbf{v} = -(r/s)\mathbf{u}$. In either case, the 'addition parallelogram' degenerates into segments on a line $\left(\text{or, if } \mathbf{u} = \mathbf{v} = \begin{pmatrix} 0 \\ 0 \end{pmatrix}, \text{ into a single point}\right)$. This is the reason for the term linearly dependent. If two vectors are not linearly dependent, we say that they are *linearly independent.*

If $\mathbf{u} = \begin{pmatrix} x \\ y \end{pmatrix}$ and $\mathbf{v} = \begin{pmatrix} z \\ t \end{pmatrix}$, then $\mathbf{u}$ and $\mathbf{v}$ are linearly independent if and only if the matrix

$$M = \begin{pmatrix} x & z \\ y & t \end{pmatrix}$$

is non-singular. Indeed, the matrix $M$ carries the unit vectors $\begin{pmatrix} 1 \\ 0 \end{pmatrix}$ and $\begin{pmatrix} 0 \\ 1 \end{pmatrix}$ into $\mathbf{u}$ and $\mathbf{v}$ respectively. Thus $\mathbf{u}$ and $\mathbf{v}$ will lie on the same line if and only if $M$ carries the unit square into a degenerate 'parallelogram' of zero area, that is, if and only if $\operatorname{Det} M = 0$.

Suppose that $\mathbf{u}$ and $\mathbf{v}$ *are* linearly independent and introduce the matrix $M$ as above. Let $\mathbf{w}$ be any vector in the plane $\mathbb{R}^2$. Since $M$ is non-singular, we can form the matrix $M^{-1}$ and consider the vector $M^{-1}\mathbf{w}$. This is a well-defined vector in $\mathbb{R}^2$ and hence we can write

$$M^{-1}\mathbf{w} = \begin{pmatrix} a \\ b \end{pmatrix} = a\begin{pmatrix} 1 \\ 0 \end{pmatrix} + b\begin{pmatrix} 0 \\ 1 \end{pmatrix}.$$

If we apply $M$ to both sides of this equation, we get

$$MM^{-1}\mathbf{w} = \mathbf{w} = aM\begin{pmatrix} 1 \\ 0 \end{pmatrix} + bM\begin{pmatrix} 0 \\ 1 \end{pmatrix} = a\mathbf{u} + b\mathbf{v}.$$

Thus if $\mathbf{u}$ and $\mathbf{v}$ are linearly independent, every vector in the plane can be written as a linear combination of $\mathbf{u}$ and $\mathbf{v}$. Conversely, suppose that $\mathbf{u}$ and $\mathbf{v}$ are vectors such that every vector in the plane can be written as a linear combination of $\mathbf{u}$ and $\mathbf{v}$. Then $\mathbf{u}$ and $\mathbf{v}$ clearly cannot lie on the same line through the origin, since this would imply that every vector in the plane would have to lie on this line. Thus

> Vectors $\mathbf{u}$ and $\mathbf{v}$ are linearly independent if and only if every vector in the plane can be written as a linear combination of $\mathbf{u}$ and $\mathbf{v}$.

Suppose that $F$ is a matrix and $\mathbf{u}_1$ and $\mathbf{u}_2$ are any pair of linearly independent vectors. The parallelogram spanned by $\mathbf{u}_1$ and $\mathbf{u}_2$ has non-zero area, hence $F(\mathbf{u}_1)$ and $F(\mathbf{u}_2)$ will be linearly independent if and only if $\operatorname{Det} F \neq 0$.

Thus the following assertions are all equivalent:

(1) $F = \begin{pmatrix} a & b \\ c & d \end{pmatrix}$ has an inverse;

(2) $\operatorname{Det} F \neq 0$;

(3) the vectors $\begin{pmatrix} a \\ c \end{pmatrix}$ and $\begin{pmatrix} b \\ d \end{pmatrix}$ do not lie on the same line through the origin;

(4) every vector in the plane is a linear combination of $\begin{pmatrix} a \\ c \end{pmatrix}$ and $\begin{pmatrix} b \\ d \end{pmatrix}$;

(5) for some pair $\mathbf{u}_1, \mathbf{u}_2$ of vectors, the vectors $F(\mathbf{u}_1)$ and $F(\mathbf{u}_2)$ are linearly independent;

(6) For *any* pair of linearly independent vectors, $\mathbf{v}_1, \mathbf{v}_2$, the vectors $F(\mathbf{v}_1)$ and $F(\mathbf{v}_2)$ are linearly independent;

(7) $F$ is not singular.

(1.9)

Let us use the preceding considerations to illustrate some reasoning in *affine geometry*. We first remark that, in affine geometry, not only does the length of a segment make no sense, but also the comparative lengths of two segments which do not lie on the same line make no sense. Indeed, if $\mathbf{u}$ and $\mathbf{v}$ are two independent vectors, there will be a unique linear transformation, $f$, which sends $\mathbf{u} \to r\mathbf{u}$ and $\mathbf{v} \to s\mathbf{v}$ for any non-zero numbers $r$ and $s$. Thus, by adjusting $s/r$, we can make the ratio of the lengths of $f(\mathbf{u})$ and $f(\mathbf{v})$ anything we please. On the other hand, the ratio of lengths of two segments lying on the same line *does* make sense. Indeed, since translations preserve length, we may assume that the line $l$ and its image $f(l)$ both pass through the origin. Since rotations preserve length, we may apply a rotation and assume that $f(l) = l$. But then if $0 \neq \mathbf{u} \subset l$, the image $f(\mathbf{u})$ also lies in $l$ so $f(\mathbf{u}) = c\mathbf{u}$ for some constant $c$ and hence $f(\mathbf{v}) = c\mathbf{v}$ for any $\mathbf{v} \subset l$. Thus $f$ changes the length of all segments on $l$ by the same factor $|c|$.

We should also point out that given any two triangles $\Delta_1$ and $\Delta_2$ there is an affine transformation, $f$, with $f(\Delta_1) = \Delta_2$. Indeed, by translating, we may assume that one of the vertices of $\Delta_1$ is the origin. Let $\mathbf{u}_1$ and $\mathbf{v}_1$ be the two remaining vertices. The vectors $\mathbf{u}_1$ and $\mathbf{v}_1$ are linearly independent since $\mathbf{0}, \mathbf{u}_1, \mathbf{v}_1$ are vertices of a triangle so do not lie on a line. Similarly we may assume that the vertices of $\Delta_2$ are $\mathbf{0}, \mathbf{u}_2, \mathbf{v}_2$. But then there is a unique linear $f$ with $f(\mathbf{u}_1) = \mathbf{u}_2$ and $f(\mathbf{v}_1) = \mathbf{v}_2$.

Now consider the following proposition: *for any triangle, the three lines joining the vertices to the midpoints of the opposite sides intersect at a single point.*

This is an assertion in affine geometry – the notion of midpoint makes sense, as does the assertion that three lines meet at a common point. To *prove* this theorem, it is enough to verify it for a single triangle, since we can find an affine transformation carrying any triangle into any other, and, if the theorem is true for one, it must be true for the other. But the theorem is clearly true for equilateral triangles. So we have proved the theorem in general.

## 1.11. Singular matrices

Let us examine what can happen when $\operatorname{Det} F = 0$. There are two alternatives:

$$\text{either } F = \begin{pmatrix} 0 & 0 \\ 0 & 0 \end{pmatrix} \quad \text{or} \quad F \neq \begin{pmatrix} 0 & 0 \\ 0 & 0 \end{pmatrix}$$

If $F$ is the zero matrix, then $F$ maps every vector into $\mathbf{0}$; it collapses the whole plane into the origin. In the alternative case where $F$ is not the zero matrix, but $\operatorname{Det} F = 0$, we claim that the following two assertions hold:

(i) there is a line, $l$, such that $F(\mathbf{u}) \in l$ for every $\mathbf{u}$ in $\mathbb{R}^2$. Furthermore, every $\mathbf{v} \in l$ is of the form $\mathbf{v} = F(\mathbf{u})$. In other words $F$ collapses the plane onto the line $l$.

(ii) there is a line, $k$, such that $F(\mathbf{w}) = \mathbf{0}$ if and only if $\mathbf{w} \in k$. In other words, $F$ collapses $k$ into the origin, and does not send any vector not in $k$ into $\mathbf{0}$.

Let us prove assertion (i). Let

$$\mathbf{c}_1 = \begin{pmatrix} a \\ c \end{pmatrix} = F\left(\begin{pmatrix} 1 \\ 0 \end{pmatrix}\right) \quad \text{and} \quad \mathbf{c}_2 = \begin{pmatrix} b \\ d \end{pmatrix} = F\left(\begin{pmatrix} 0 \\ 1 \end{pmatrix}\right)$$

be the two columns of $F = \begin{pmatrix} a & b \\ c & d \end{pmatrix}$.

Since $F$ is not the zero matrix, $\mathbf{c}_1$ and $\mathbf{c}_2$ can not both be equal to $\mathbf{0}$. On the other hand, if $\mathbf{c}_1$ and $\mathbf{c}_2$ did not lie on the same line, then by the equivalence of assertion (3) and assertion (1) of (1.9) (on the preceding page) we would conclude that $\operatorname{Det} F \neq 0$, contrary to our assumption. Thus $\mathbf{c}_1$ and $\mathbf{c}_2$ lie on a line. Call this line $l$. Every vector $\mathbf{u}$ can be written as

$$\mathbf{u} = \begin{pmatrix} x \\ y \end{pmatrix} = x\begin{pmatrix} 1 \\ 0 \end{pmatrix} + y\begin{pmatrix} 0 \\ 1 \end{pmatrix}$$

so $F(\mathbf{u}) = x\mathbf{c}_1 + y\mathbf{c}_2$ lies on $l$. If $\mathbf{c}_1 \neq \mathbf{0}$, every $\mathbf{v} \in l$ can be written as $\mathbf{v} = x\mathbf{c}_1$ for some number $x$, hence $\mathbf{v} = F\left(\begin{pmatrix} x \\ 0 \end{pmatrix}\right)$. Otherwise, we must have $\mathbf{c}_2 \neq \mathbf{0}$ and thus $\mathbf{v} = y\mathbf{c}_2$ for some number $y$ and hence $\mathbf{v} = F\left(\begin{pmatrix} 0 \\ y \end{pmatrix}\right)$. This completes the proof of (i).

Let us now prove (ii). Let $\mathbf{b}_1$ and $\mathbf{b}_2$ be the first and second columns of the matrix $F^a$ so that

$$\mathbf{b}_1 = \begin{pmatrix} d \\ -c \end{pmatrix} \quad \text{and} \quad \mathbf{b}_2 = \begin{pmatrix} -b \\ a \end{pmatrix}.$$

Since $\operatorname{Det} F^a = \operatorname{Det} F = 0$, we know from (1.9) that $\mathbf{b}_1$ and $\mathbf{b}_2$ must be linearly dependent, and they can't both be equal to $\mathbf{0}$ since some entry of $F$ is $\neq 0$ by hypothesis. So they span a line. Call it $k$. Direct computation shows that $F\mathbf{b}_1 = \mathbf{0} = F\mathbf{b}_2$ so every $\mathbf{w} \in k$ satisfies $F\mathbf{w} = \mathbf{0}$. If $\mathbf{w} = \begin{pmatrix} x \\ y \end{pmatrix}$ satisfies $F\mathbf{w} = \mathbf{0}$ then $\mathbf{w}$ must satisfy the equation

$$ax - by = 0.$$

This is the equation for a line, unless $a = b = 0$, i.e. $\mathbf{b}_2 = 0$, and this line must then be $k$. If $\mathbf{b}_2 = \mathbf{0}$, then $c$ and $d$ can not both vanish. But $\mathbf{w}$ must also satisfy

$$cx - dy = 0,$$

and this is the equation of a line, and the line must be $k$. This proves (ii).

For any $F$ whatsoever, let $\operatorname{im}(F)$ denote the subset of $\mathbb{R}^2$ consisting of all elements of the form $F(\mathbf{u})$. In symbols,

$$\operatorname{im}(F) = \{\mathbf{v} \mid \mathbf{v} = F(\mathbf{u}) \quad \text{for some } \mathbf{u}\}.$$

The set $\operatorname{im}(F)$ is pronounced as 'the image of $F$'. Similarly, we define the 'kernel of $F$' written as $\ker(F)$ to be the set of vectors which are sent into $\mathbf{0}$ by $F$. In symbols,

$$\ker(F) = \{\mathbf{w} \mid f(\mathbf{w}) = \mathbf{0}\}.$$

There are thus three possibilities:

(a) $\operatorname{Det} F \neq 0$. Then $\operatorname{im}(F) = \mathbb{R}^2$ and $\ker(F) = \{\mathbf{0}\}$.

(b) $\operatorname{Det} F = 0$ but $F$ is not the zero matrix. Then $\operatorname{im}(F) = l$ is a line and $\ker(F) = k$ is a line. In other words both $\operatorname{im}(F)$ and $\ker(F)$ are one-dimensional vector spaces.

(c) $F$ is the zero matrix. Then $\operatorname{im}(F) = \{\mathbf{0}\}$ and $\ker(F) = \mathbb{R}^2$.

If we think of $\{\mathbf{0}\}$ as being a 'zero-dimensional' vector space we see that in all cases we have

$$\text{dimension of } \operatorname{im}(F) + \text{dimension of } \ker(F) = 2.$$

**Special kinds of singular transformations.** We now examine some special kinds of singular transformations.

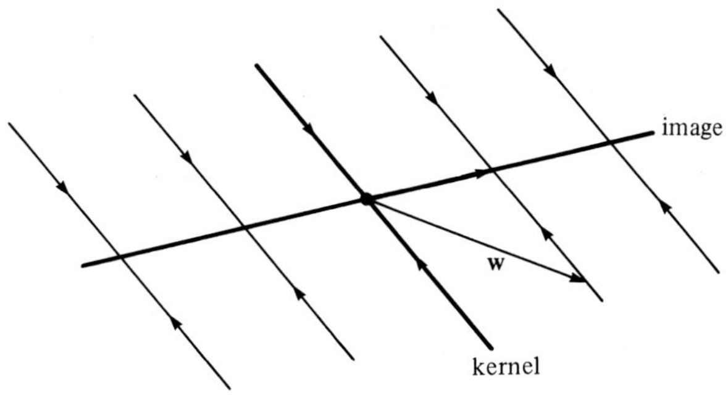


**Figure 1.24**

**1. Projections**

We call $p$ a *projection* if it has the following property:

*if* $\mathbf{v}$ *is a vector in the image of* $p, p(\mathbf{v}) = \mathbf{v}$.

This is rather special, since all we can expect in general for a singular transformation $f$ is that $f(\mathbf{v})$ lies on the same line as $\mathbf{v}$; i.e., $f(\mathbf{v}) = \alpha\mathbf{v}$ for some number $\alpha$. Now, if $\mathbf{w}$ is an arbitrary vector, $\mathbf{v} = p(\mathbf{w})$ is in the image of $p$, and

$$(p \circ p)(\mathbf{w}) = p(\mathbf{v}) = \mathbf{v} = p(\mathbf{w}).$$

It follows that $p \circ p = p$, and so the matrix $P$ which represents a projection satisfies $P^2 = P$. Thus

$$\begin{pmatrix} a & b \\ c & d \end{pmatrix}\begin{pmatrix} a & b \\ c & d \end{pmatrix} = \begin{pmatrix} a^2 + bc & ab + bd \\ ac + cd & bc + d^2 \end{pmatrix} = \begin{pmatrix} a & b \\ c & d \end{pmatrix}$$

So $ab + bd = b$, and if $b \neq 0$, $a + d = 1$. Furthermore, $ac + cd = c$, so if $c \neq 0$, $a + d = 1$. Even if $b = 0$ and $c = 0$, we have $a^2 = a$, $d^2 = d$, and $\operatorname{Det} P = ad = 0$, so either $a = 1$, $d = 0$ or $a = 0$, $d = 1$, or $P = 0$. Therefore, unless $P = 0$, the *trace* of $P$ defined as $\operatorname{tr} P = a + d$ must equal 1.

To summarize: a non-zero (singular) projection $p$ satisfies $p \circ p = p$; its matrix $P$ satisfies $P^2 = P$ and has zero determinant $(ad - bc = 0)$ and unit trace $(a + d = 1)$.

Conversely, suppose that $ad - bc = 0$ and $a + d = 1$. Then $ab + bd = (a + d)b = b$ and $ac + cd = c$, while $a^2 + bc = a^2 + ad = a$, and $bc + d^2 = d$. Thus $P^2 = P$ and $p$ is a projection onto a line.

More generally, let us call an operator $p$ a *projection* if $p^2 = p$. Then there are three possibilities.

(1) $P$ is non-singular. In this case, we can multiply the equation $P^2 = P$ on both sides by $P^{-1}$ to obtain

$$P = \begin{pmatrix} 1 & 0 \\ 0 & 1 \end{pmatrix}.$$

In this case, $\operatorname{tr} P = 2 = \dim(\operatorname{image}(P))$, where we write dim for dimension.

(2) $P$ is singular but not zero. This is the case considered earlier. Here $P$ maps $\mathbb{R}^2$ onto a line and is the identity when restricted to this line. Here $\operatorname{tr} P = 1 = \dim(\operatorname{im}(p))$, where we write im for image.

(3) $P = \begin{pmatrix} 0 & 0 \\ 0 & 0 \end{pmatrix}$ so $P$ maps the whole plane to the origin and $\operatorname{tr} P = 0 = \dim(\operatorname{im}(p))$.

In all cases, we have $\operatorname{tr} P = \dim(\operatorname{im}(p))$.

**2. Nilpotents**

For a general singular matrix, the two lines $\operatorname{im} F$ and $\ker F$ will be different. Let us consider a special kind of transformation, $n$, with the property that its image and its kernel are the same. Applying $n$ to any vector $\mathbf{w}$ yields a vector $\mathbf{v} = n(\mathbf{w})$ which is the image of $n$ and hence also in the kernel. It follows that $n(\mathbf{v}) = n \circ n(\mathbf{w}) = \mathbf{0}$. Thus $n \circ n$ collapses the entire plane into the origin, and the matrix $N$ representing $n$ must satisfy $N^2 = 0$. So

$$\begin{pmatrix} a & b \\ c & d \end{pmatrix}\begin{pmatrix} a & b \\ c & d \end{pmatrix} = \begin{pmatrix} a^2 + bc & ab + bd \\ ac + cd & bc + d^2 \end{pmatrix} = \begin{pmatrix} 0 & 0 \\ 0 & 0 \end{pmatrix}.$$

In particular, $ab + bd = 0$ and $ac + cd = 0$, so if $b \neq 0$ or $c \neq 0$, then $a + d = 0$. If $bc = 0$, then $a^2 = d^2 = 0$, so $N = \begin{pmatrix} 0 & 0 \\ 0 & 0 \end{pmatrix}$. In every case, then, $N$ has zero trace. Conversely, if $\operatorname{tr} N = a + d = 0$ and $\det N = ad - bc = 0$, then $N^2 = 0$.

Let us call a matrix *nilpotent* if some power of it vanishes. Thus $N$ is nilpotent if $N^k = 0$ for some $k$. For such a matrix, we must have $\det N = 0$, for otherwise we could keep multiplying the equation $N^k = 0$ by $N^{-1}$ until we get $N = 0$. Thus $ad - bc = 0$ and hence

$$\begin{pmatrix} a & b \\ c & d \end{pmatrix}^2 = \begin{pmatrix} (a+d)a & (a+d)b \\ (a+d)c & (a+d)d \end{pmatrix}.$$

Then

$$N^k = \begin{pmatrix} (a+d)^{k-1}a & (a+d)^{k-1}b \\ (a+d)^{k-1}c & (a+d)^{k-1}d \end{pmatrix}.$$

This can only vanish if $(a + d) = 0$. But then, we already know that $N^2 = 0$. Thus, in the plane, a matrix $N$ is nilpotent if and only if $N^2 = 0$ and this holds if and only if

$$\det N = 0 \quad \text{and} \quad \operatorname{tr} N = 0.$$

## 1.12. Two-dimensional vector spaces

A vector space $V$ is called *two-dimensional* if we can find two vectors, $\mathbf{u}_1$ and $\mathbf{u}_2$ in $V$, such that every $\mathbf{v} \in V$ can be written uniquely as

$$\mathbf{v} = a_1\mathbf{u}_1 + a_2\mathbf{u}_2$$

The word 'uniquely' means that if

$$\mathbf{v} = a_1\mathbf{u}_1 + a_2\mathbf{u}_2 \quad \text{and} \quad \mathbf{v} = b_1\mathbf{u}_1 + b_2\mathbf{u}_2$$

then we must have

$$a_1 = b_1 \quad \text{and} \quad a_2 = b_2.$$

An ordered pair, $\mathbf{u}_1, \mathbf{u}_2$ of vectors with the above property is called a *basis* of the vector space. Such a choice of basis determines a map $L = L_{\mathbf{u}_1,\mathbf{u}_2}$ of $V$ onto $\mathbb{R}^2$

$$V \xrightarrow{L} \mathbb{R}^2$$

by

$$L(\mathbf{v}) = \begin{pmatrix} a_1 \\ a_2 \end{pmatrix} \quad \text{if} \quad \mathbf{v} = a_1\mathbf{u}_1 + a_2\mathbf{u}_2.$$

The 'uniqueness' part of our assumption above guarantees that the map $L$ is well-defined; the components $a_1$ and $a_2$ are completely determined by $\mathbf{v}$. The map is also onto: given $\begin{pmatrix} a_1 \\ a_2 \end{pmatrix}$ the vector $\mathbf{v} = a_1\mathbf{u}_1 + a_2\mathbf{u}_2$ clearly satisfies $L(v) = \begin{pmatrix} a_1 \\ a_2 \end{pmatrix}$.

If $\mathbf{v} = a_1\mathbf{u}_1 + a_2\mathbf{u}_2$ and $\mathbf{w} = b_1\mathbf{u}_1 + b_2\mathbf{u}_2$ then

$$\begin{aligned} \mathbf{v} + \mathbf{w} &= a_1\mathbf{u}_1 + a_2\mathbf{u}_2 + b_1\mathbf{u}_1 + b_2\mathbf{u}_2 \\ &= (a_1 + b_1)\mathbf{u}_1 + (a_2 + b_2)\mathbf{u}_2 \end{aligned}$$

so

$$L(\mathbf{v} + \mathbf{w}) = L(\mathbf{v}) + L(\mathbf{w})$$

and similarly

$$L(r\mathbf{v}) = rL(\mathbf{v})$$

for any real number $r$ and any vector $\mathbf{v}$ of $V$. We say that $L$ is an *isomorphism** of $V$ with $\mathbb{R}^2$. It allows us to identify all operations on and properties of the vector space $V$ with operations on and properties of $\mathbb{R}^2$, just as in the one-dimensional case, a choice of basis allowed us to translate properties of a one-dimensional vector space into those of $\mathbb{R}^1$. Of course, just as in the one-dimensional case, the isomorphism, $L$, depends on the choice of basis. Thus, the choice of basis, $\{\mathbf{u}_1, \mathbf{u}_2\}$ is the two-dimensional analog of a 'choice of units'. Only those properties which are independent of the choice of basis will be interesting to us and of true geometrical character. We shall shortly study how $L$ changes with a change of basis. For the moment, let us observe that the basis $\{\mathbf{u}_1, \mathbf{u}_2\}$ can be recovered from $L$. Indeed

$$\mathbf{u}_1 = L^{-1}\begin{pmatrix} 1 \\ 0 \end{pmatrix} \quad \text{and} \quad \mathbf{u}_2 = L^{-1}\begin{pmatrix} 0 \\ 1 \end{pmatrix}.$$

So *giving a basis is the same as giving an isomorphism*, $L: V \to \mathbb{R}^2$. Given $L$, simply define $\mathbf{u}_1$ and $\mathbf{u}_2$ by the preceding equation. Since every vector in $\mathbb{R}^2$ can be written uniquely as a linear combination of $\begin{pmatrix} 1 \\ 0 \end{pmatrix}$ and $\begin{pmatrix} 0 \\ 1 \end{pmatrix}$ and since $L$ is an isomorphism, it follows that every vector of $V$ can be written uniquely as a linear combination of $\mathbf{u}_1$ and $\mathbf{u}_2$, and the isomorphism associated with $\{\mathbf{u}_1, \mathbf{u}_2\}$ is clearly $L$.

A *linear transformation* $F: V \to V$ is a map of $V$ into $V$ which satisfies our usual identity:

$$F(a\mathbf{u} + b\mathbf{v}) = aF(\mathbf{u}) + bF(\mathbf{v}).$$

A choice of basis gives an identification $L: V \to \mathbb{R}^2$ and we can define a linear transformation of $\mathbb{R}^2$ by

$$LFL^{-1}.$$

Here $L^{-1}: \mathbb{R}^2 \to V$, then $F: V \to V$ and $L: V \to \mathbb{R}^2$. It is best to visualize the situation by a diagram:

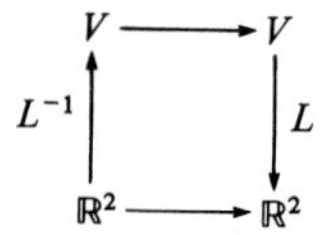


* In mathematics, the word isomorphism means a one-to-one mapping which preserves all the relevant structure. For vector spaces, $V$ and $W$, we say that a map $L$ from $V$ to $W$ is an isomorphism if it is linear, is one-to-one and onto.

The transformation $LFL^{-1}$ going from $\mathbb{R}^2 \to \mathbb{R}^2$ along the bottom is obtained by going up, across and down. Now any linear transformation of $\mathbb{R}^2 \to \mathbb{R}^2$ is given by a matrix. Thus, once we have chosen the basis $L$, we have associated a matrix

$$\mathrm{Mat}_L(F) = \mathrm{Mat}(LFL^{-1})$$

to any linear operator $F: V \to V$. If $G: V \to V$ is a second linear transformation, then

$$LGFL^{-1} = LGL^{-1}LFL^{-1}$$

so

$$\mathrm{Mat}_L(G \circ F) = \mathrm{Mat}_L(G)\mathrm{Mat}_L(F).$$

In other words, composition of linear transformations goes over into matrix multiplication. Similarly for addition of linear transformations. Thus the algebra of linear transformations on $V$ gets translated into the algebra of $2 \times 2$ matrices.

The space $\mathbb{R}^2$ is itself a vector space. It has a 'natural' basis consisting of $\begin{pmatrix} 1 \\ 0 \end{pmatrix}$ and $\begin{pmatrix} 0 \\ 1 \end{pmatrix}$. If $f: \mathbb{R}^2 \to \mathbb{R}^2$ is a linear transformation, its matrix relative to this natural basis is the matrix $F$, in the language of the preceding few sections. The map $L: \mathbb{R}^2 \to \mathbb{R}^2 = V$ corresponding to this basis is just the identity $I$. Thus the relation between $f$ and $F$ should be written as

$$F = \mathrm{Mat}_I(f).$$

From a strictly logical point of view we should have used the notation $\mathrm{Mat}_I(f)$ instead of $F$ from the very beginning, but it would have been too cumbersome. From now on, once we have the idea of a linear transformation on a general vector space, we shall drop the distinction between lower case letters and upper case letters.

The assignment of $\mathrm{Mat}_L(F)$ to $F$ does depend on an artifact, namely on the choice of basis. We now must examine what happens when we change the basis. So suppose that we are given two bases. This means that we are given two isomorphisms, $L: V \to \mathbb{R}^2$ and $M: V \to \mathbb{R}^2$. Then we can consider the matrix $B = ML^{-1}$: $\mathbb{R}^2 \to \mathbb{R}^2$, so

$$M = BL.$$

We can visualize the situation by the diagram:

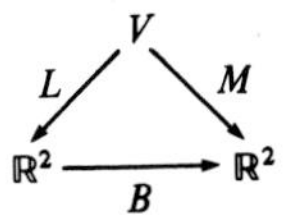


The matrix $B$ is called the 'change of basis matrix' (relative to the bases $L$ and $M$). It is the two-dimensional analog of the factor 1000 by which we have to multiply all numerical values of masses when we pass from kilograms to grams in our choice of unit. To repeat: $L(\mathbf{v})$ and $M(\mathbf{v})$ are two points in $\mathbb{R}^2$ corresponding to the *same* point $\mathbf{v}$ in $V$ by the two choices, $L$ and $M$, of bases. These two points in $\mathbb{R}^2$ are related to one another by the change of basis matrix:

$$M(\mathbf{v}) = BL(\mathbf{v}).$$

The matrix $B$ gives an isomorphism of $\mathbb{R}^2 \to \mathbb{R}^2$, i.e. it is non-singular. (It is clear that, if we are given $L$ and also given a non-singular matrix $B$, then we can define $M = BL$, and this $M$ is an isomorphism of $V$ with $\mathbb{R}^2$. Thus, once we have fixed some basis, $L$, of $V$, the set of all other bases is parameterized by the set of all invertible $2 \times 2$ matrices, $B$.)

Now suppose that $F: V \to V$ is a linear transformation. Then

$$\mathrm{Mat}_L(F) = LFL^{-1}$$

and

$$\mathrm{Mat}_M(F) = MFM^{-1}.$$

But $MFM^{-1} = (BL)F(BL)^{-1} = BLFL^{-1}B^{-1} = B(LFL^{-1})B^{-1}$, so

$$\mathrm{Mat}_M(F) = B\,\mathrm{Mat}_L(F)B^{-1}.$$

This important formula tells us how the two matrices of the *same* linear transformation are related to one another when we know the change of basis matrix, $B$.

For a given linear transformation, $F: V \to V$, it may be possible to *choose* a basis, $L$, so that $\mathrm{Mat}_L(F)$ has a particularly convenient or instructive form. For example, suppose that $F: V \to V$ sends all of $V$ onto a line and sends this line into $\mathbf{0}$, in other words suppose that $\mathrm{im}\,F = \ker F$. Let us choose $\mathbf{u}_2$ to be some vector that does *not* belong to $\ker F$ and set $\mathbf{u}_1 = F(\mathbf{u}_2)$, so $\mathbf{u}_1 \neq \mathbf{0}$ and $F(\mathbf{u}_1) = \mathbf{0}$. We take $\mathbf{u}_1, \mathbf{u}_2$ as our basis. Then $LFL^{-1}\begin{pmatrix}1\\0\end{pmatrix} = LF(\mathbf{u}_1) = L(\mathbf{0}) = \begin{pmatrix}0\\0\end{pmatrix}$ and $LFL^{-1}\begin{pmatrix}0\\1\end{pmatrix} = LF(\mathbf{u}_2) = L(\mathbf{u}_1) = \begin{pmatrix}1\\0\end{pmatrix}$. So, for this choice of basis we have

$$\mathrm{Mat}_L(F) = \begin{pmatrix}0 & 1\\0 & 0\end{pmatrix}. \tag{1.10}$$

Now in this entire discussion, there is nothing to prevent us from considering the case where our vector space, $V$, happens to be $\mathbb{R}^2$ itself. When we identified a linear transformation with a matrix, it was with respect to the standard basis. In other words, when we wrote $F$ in sections 1.5 and 1.6, it should have been written as $\mathrm{Mat}_I(F)$. So, for example, let $N$ be a non-zero nilpotent matrix. Thus $N = \mathrm{Mat}_I(F)$, where $F$ is a linear transformation of $\mathbb{R}^2$ with $\ker F = \mathrm{im}\,F$. (In words we would say that $N$ is the matrix of the linear transformation $F$ relative to the standard basis.) From the preceding considerations we know that we can find some other basis, $L$, relative to which (1.10) holds. By the change of basis formula (the change of basis from $L$ to $I$) we know that

$$N = B\begin{pmatrix}0 & 1\\0 & 0\end{pmatrix}B^{-1}. \tag{1.11}$$

We have thus proved: given any non-zero nilpotent matrix $N$, we can find an invertible matrix, $B$ such that (1.11) holds. We shall return to these kinds of considerations (and, in particular, how to find $B$) in the next chapter.

For an important application to physics of the results of this chapter please turn to Chapter 9. There we show how Gaussian optics is really the study of $2 \times 2$ matrices. Most of Chapter 9 can be read with only a knowledge of Chapter 1.

## Appendix: the fundamental theorem of affine geometry

We wish to prove the following:

*Let $f$ be an affine transformation of $\mathbb{R}^2$ satisfying $f\begin{pmatrix}0\\0\end{pmatrix}=\begin{pmatrix}0\\0\end{pmatrix}$. Then $f$ is linear.*

In proving this theorem, we can make a number of simplifying reductions. Notice that, if $g$ is an invertible linear transformation, then $g\circ f$ is linear if and only if $f$ is linear. Now $f\begin{pmatrix}1\\0\end{pmatrix}$ and $f\begin{pmatrix}0\\1\end{pmatrix}$ cannot lie on the same line through the origin. They are thus linearly independent and hence we can find a linear transformation $g$ with $g\circ f\begin{pmatrix}1\\0\end{pmatrix}=\begin{pmatrix}1\\0\end{pmatrix}$ and $g\circ f\left(\begin{pmatrix}0\\1\end{pmatrix}\right)=\begin{pmatrix}0\\1\end{pmatrix}$. Thus, replacing $f$ by $g\circ f$, it is enough to prove the following:

*Let $f$ be an affine transformation satisfying $f\left(\begin{pmatrix}0\\0\end{pmatrix}\right)=\begin{pmatrix}0\\0\end{pmatrix}$, $f\left(\begin{pmatrix}1\\0\end{pmatrix}\right)=\begin{pmatrix}1\\0\end{pmatrix}$ and $f\left(\begin{pmatrix}0\\1\end{pmatrix}\right)=\begin{pmatrix}0\\1\end{pmatrix}$. Then $f$ is the identity transformation.*

*Proof.* From section 1.2 we know that $f\left(\begin{pmatrix}1\\1\end{pmatrix}\right)=f\left(\begin{pmatrix}1\\0\end{pmatrix}\right)+f\left(\begin{pmatrix}0\\1\end{pmatrix}\right)=\begin{pmatrix}1\\0\end{pmatrix}+\begin{pmatrix}0\\1\end{pmatrix}=\begin{pmatrix}1\\1\end{pmatrix}$. $\left(\text{In fact, we proved that } f\left(\begin{pmatrix}r\\s\end{pmatrix}\right)=\begin{pmatrix}r\\s\end{pmatrix} \text{ whenever } r \text{ and } s \text{ are rational.}\right)$ Thus $f$ carries the $x$-axis, the $y$-axis and the line $x=y$ $\left(\text{which is the line through } \begin{pmatrix}0\\0\end{pmatrix} \text{ and } \begin{pmatrix}1\\1\end{pmatrix}\right)$ into themselves.

Thus, for any real number $a$

$$f\left(\begin{pmatrix}a\\0\end{pmatrix}\right)=\begin{pmatrix}\phi(a)\\0\end{pmatrix}$$

where $\phi$ is some function. (We want to prove $\phi(a)=a$ for all $a$.) Similarly $f\left(\begin{pmatrix}0\\b\end{pmatrix}\right)=\begin{pmatrix}0\\\psi(b)\end{pmatrix}$ for some function $\psi$, and since

$$f\left(\begin{pmatrix}a\\b\end{pmatrix}\right)=f\left(\begin{pmatrix}a\\0\end{pmatrix}\right)+f\left(\begin{pmatrix}0\\b\end{pmatrix}\right)$$

we have

$$f\left(\begin{pmatrix} a \\ b \end{pmatrix}\right) = \begin{pmatrix} \phi(a) \\ \psi(b) \end{pmatrix}.$$

We claim that the functions $\phi$ and $\psi$ are the same. Indeed, consider the line $x = a$. It is parallel to the $y$-axis, and hence its image under $f$ must be parallel to the $y$-axis and hence its image must be the line $x = \phi(a)$. Now the line $x = a$ intersects the line $x = y$ at the point $\begin{pmatrix} a \\ a \end{pmatrix}$, and the line $x = \phi(a)$ intersects the line $x = y$ at $\begin{pmatrix} \phi(a) \\ \phi(a) \end{pmatrix}$. Hence

$$f\left(\begin{pmatrix} a \\ a \end{pmatrix}\right) = \begin{pmatrix} \phi(a) \\ \phi(a) \end{pmatrix}$$

and so $\phi(a) = \psi(a)$ for all $a$.

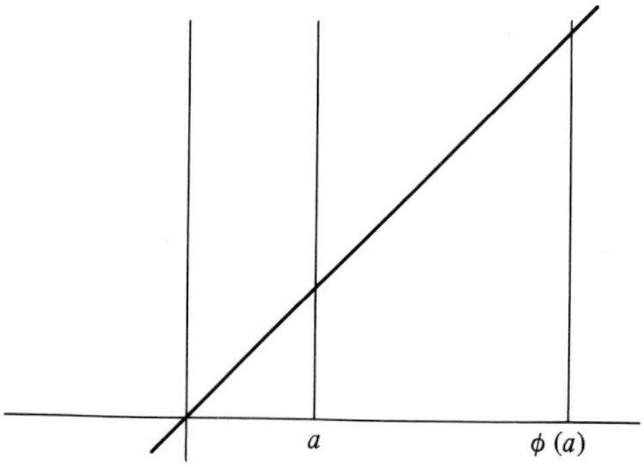


**Figure 1.25**

We know that $f\left(\begin{pmatrix} a+b \\ 0 \end{pmatrix}\right) = f\left(\begin{pmatrix} a \\ 0 \end{pmatrix}\right) + f\left(\begin{pmatrix} b \\ 0 \end{pmatrix}\right)$ so

$$\phi(a+b) = \phi(a) + \phi(b).$$

All of this is essentially the same level of argument as in section 1.2. We now establish the surprising fact that

$$\phi(ab) = \phi(a)\phi(b).$$

Indeed, consider figure 1.26:

The line joining $\begin{pmatrix} 1 \\ 1 \end{pmatrix}$ to $\begin{pmatrix} b \\ 0 \end{pmatrix}$ is parallel to the line joining $\begin{pmatrix} a \\ a \end{pmatrix}$ to $\begin{pmatrix} ab \\ 0 \end{pmatrix}$. Thus the value $ab$ can be obtained by parallels and intersections. Therefore, drawing the same diagram for $\begin{pmatrix} \phi(a) \\ 0 \end{pmatrix}$ and $\begin{pmatrix} \phi(b) \\ 0 \end{pmatrix}$ we see that

$$\begin{pmatrix} \phi(ab) \\ 0 \end{pmatrix} = f\begin{pmatrix} ab \\ 0 \end{pmatrix} = \begin{pmatrix} \phi(a)\phi(b) \\ 0 \end{pmatrix}$$

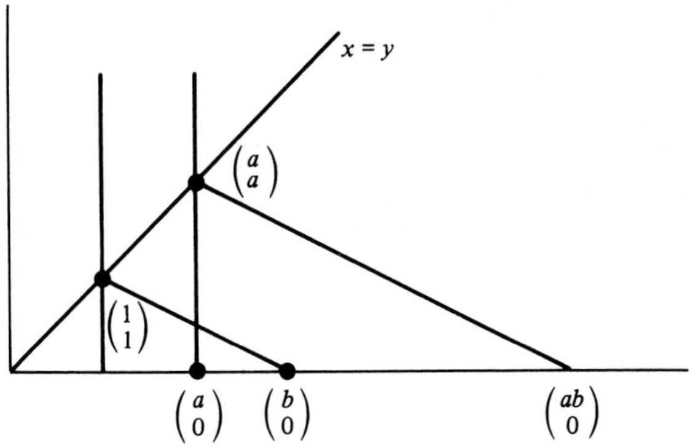


**Figure 1.26**

so

$$\phi(ab) = \phi(a)\phi(b).$$

Now a real number $x$ is positive if and only if $x = y^2$ for some other number $y$. Then

$$\phi(x) = \phi(y^2) = \phi(y)^2$$

so

$$x > 0 \quad \text{implies} \quad \phi(x) > 0.$$

Thus $a - b > 0$ implies $\phi(a) - \phi(b) > 0$. Thus if

$$r < a < s$$

then

$$\phi(r) < \phi(a) < \phi(s).$$

Now for any real number $a$ we can find rational numbers $r$ and $s$ with $r < a < s$ and $s - r$ as small as we please. But, for rational numbers, $\phi(r) = r$ and $\phi(s) = s$. Thus

$$r < \phi(a) < s.$$

Hence $|a - \phi(a)| < s - r$. Since $s - r$ can be chosen arbitrarily small, this implies that $\phi(a) = a$ for all real numbers. QED

---

## Summary

A Transformations of the plane

You should be able to define the terms affine transformation, linear transformation, and Euclidean transformation.

You should be able to identify geometric properties that are preserved by affine transformations and properties that are preserved by Euclidean transformation.

B Matrix algebra

You should know how to add and multiply two square matrices of the same size.

You should be able to calculate the determinant of a $2 \times 2$ matrix and to write down the inverse of an invertible $2 \times 2$ matrix.

C Matrices and linear transformations

Given sufficient information about a linear transformation of the plane, you should be able to write down the $2 \times 2$ matrix that represents the transformation.

You should understand the significance of matrix multiplication in terms of composition of linear transformations and be able to apply this relationship.

You should be able to determine the image and kernel of the transformation represented by a given $2 \times 2$ matrix.

You should be able to identify $2 \times 2$ matrices that represent transformations with special properties (rotations, reflections, projections, nilpotent transformations).

---

## Exercises

1.1 Here are some theorems of Euclidean plane geometry. Decide whether each is a valid statement in *affine* plane geometry.

(a) The medians of a triangle meet at a point which is 2/3 of the way from each vertex to the midpoint of the opposite side.

(b) The angle bisectors of an isosceles triangle are equal in length.

(c) The diagonals of a rhombus are perpendicular.

(d) The diagonals of a parallelogram bisect each other.

(e) Let $PQR$ and $P'Q'R'$ be two triangles such that the lines $PQ$ and $P'Q'$ are parallel, $QR$ and $Q'R'$ are parallel, and $PR$ and $P'R'$ are parallel. Then the three lines $PP'$, $QQ'$, and $RR'$ are either parallel or concurrent.

1.2(a) Let $\mathbb{A}_1$ and $\mathbb{A}_2$ be affine lines. Let $x$ be an affine coordinate function on $\mathbb{A}_1$; let $y$ be an affine coordinate function on $\mathbb{A}_2$. Let $f: \mathbb{A}_1 \to \mathbb{A}_2$ be an *affine* mapping. Associated with $f$ is a function $F: \mathbb{R} \to \mathbb{R}$ such that if $Q = f(P)$, then $y(Q) = F \circ x(P)$. Show that the most general formula for $F$ is $F(\alpha) = r\alpha + s$.

(b) Let $x' = ax + b$, $y' = cy + d$, so that $x'$ and $y'$ are new affine coordinate functions on $\mathbb{A}_1$ and $\mathbb{A}_2$ respectively. If $y(Q) = F \circ x(P)$ where $F(\alpha) = r\alpha + s$, find the formula for the function $F'(\beta)$ such that $y'(Q) = F' \circ x'(P)$.

1.3 A function $u: \mathbb{R}^2 \to \mathbb{R}$ is *affine* if it is an affine function on each line of the plane and if, for any parallelogram, $u(P) + u(R) = u(Q) + u(S)$ where the vertices are labeled as in figure 1.27. Suppose that $u: \mathbb{R}^2 \to \mathbb{R}$ is affine and that $u\begin{bmatrix}1\\2\end{bmatrix} = 3$, $u\begin{bmatrix}3\\3\end{bmatrix} = 8$, $u\begin{bmatrix}2\\-1\end{bmatrix} = 9$.

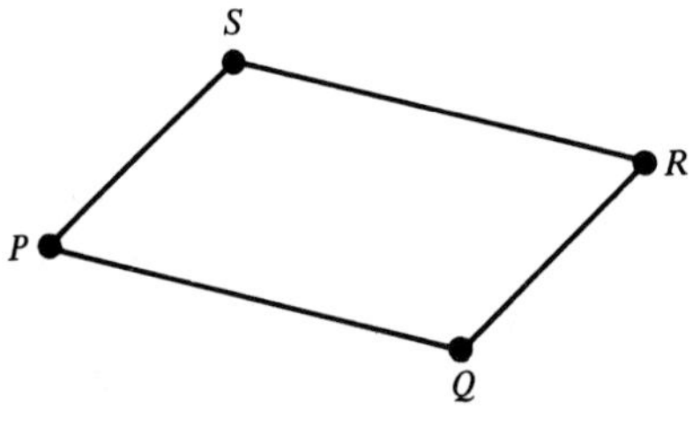


**Figure 1.27**

(a) Find a formula for $u\begin{bmatrix} x \\ y \end{bmatrix}$.

(b) Sketch $\mathbb{R}^2$, showing the lines $u = \text{constant}$.

1.4 Find the image of the rectangle $ABCDE$ shown in figure 1.28 under the linear transformation represented by each of the following matrices. In each case calculate the determinant of the transformation and verify that the area and orientation of the image of the rectangle are correctly predicted by this determinant.

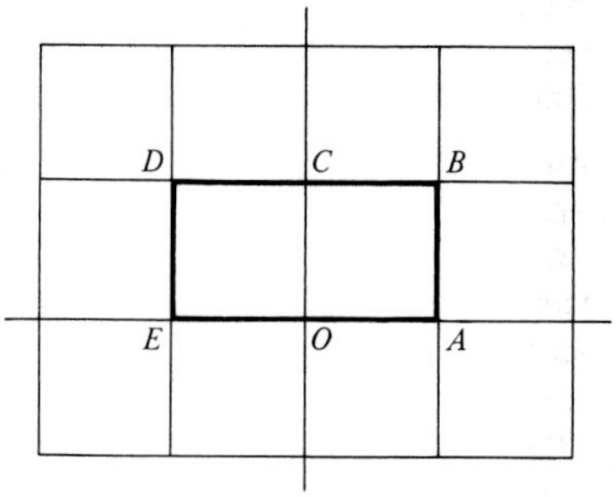


**Figure 1.28**

(a) The rotation $R_{\pi/2} = \begin{pmatrix} 0 & -1 \\ 1 & 0 \end{pmatrix}$.

(b) The rotation $R_{\pi/4} = (1/\sqrt{2})\begin{pmatrix} 1 & -1 \\ 1 & 1 \end{pmatrix}$.

(c) The 'distortion' $D_2 = \begin{pmatrix} 2 & 0 \\ 0 & 1/2 \end{pmatrix}$.

(d) The 'Lorentz transformation' $L_2 = \begin{pmatrix} 5/4 & 3/4 \\ 3/4 & 5/4 \end{pmatrix}$.

(e) The shear transformation $S_1 = \begin{pmatrix} 1 & 1 \\ 0 & 1 \end{pmatrix}$.

(f) The shear transformation $S_1' = \begin{pmatrix} 1/2 & 1/2 \\ -1/2 & 3/2 \end{pmatrix}$.

(g) The reflection $M_0 = \begin{pmatrix} 1 & 0 \\ 0 & -1 \end{pmatrix}$.

(h) The reflection $M_{\pi/4} = \begin{pmatrix} 0 & 1 \\ 1 & 0 \end{pmatrix}$.

(i) The projection $P_{\pi/4} = \begin{pmatrix} 1/2 & 1/2 \\ 1/2 & 1/2 \end{pmatrix}$.

(j) The nilpotent transformation $N_0 = \begin{pmatrix} 0 & 1 \\ 0 & 0 \end{pmatrix}$.

(k) The nilpotent transformation $N_{\pi/4} = \begin{pmatrix} -1/2 & 1/2 \\ -1/2 & 1/2 \end{pmatrix}$.

1.5. Calculate algebraically each of the following products of matrices defined in Exercise 1.4, and interpret geometrically the transformation defined by the product matrix

(a) $R_{\pi/2}R_{\pi/4}$
(b) $S_1^2$
(c) $R_{\pi/2}M_0$
(d) $P_{\pi/4}^2$
(e) $N_{\pi/4}^2$

1.6. Calculate the inverse of each of the following matrices, and interpret the result geometrically.

(a) $R_{\pi/4} = (1/\sqrt{2})\begin{pmatrix} 1 & -1 \\ 1 & 1 \end{pmatrix}$.

(b) $L_2 = \begin{pmatrix} \frac{5}{4} & \frac{3}{4} \\ \frac{3}{4} & \frac{5}{4} \end{pmatrix}$.

(c) $S_1 = \begin{pmatrix} 1 & 1 \\ 0 & 1 \end{pmatrix}$.

(d) $M_{\pi/4} = \begin{pmatrix} 0 & 1 \\ 1 & 0 \end{pmatrix}$.

1.7. For each of the following matrices of determinant zero, determine the image and the kernel.

(a) $P_{\pi/4} = \begin{pmatrix} \frac{1}{2} & \frac{1}{2} \\ \frac{1}{2} & \frac{1}{2} \end{pmatrix}$.

(b) $N_{\pi/4} = \begin{pmatrix} -\frac{1}{2} & \frac{1}{2} \\ -\frac{1}{2} & \frac{1}{2} \end{pmatrix}$.

(c) $A = \begin{pmatrix} 1 & 2 \\ -1 & -2 \end{pmatrix}$.

1.8. Apply the triple product decomposition proved in section 1.7 to express the matrix

$$\begin{pmatrix} 2 & 4 \\ 6 & 8 \end{pmatrix}$$

in the form

$$\begin{pmatrix} 1 & 0 \\ y & 1 \end{pmatrix}\begin{pmatrix} r & 0 \\ 0 & s \end{pmatrix}\begin{pmatrix} 1 & x \\ 0 & 1 \end{pmatrix}.$$

1.9. Devise a procedure for writing any matrix $\begin{pmatrix} a & b \\ c & d \end{pmatrix}$ with $c \neq 0$ as a triple product

$$\begin{pmatrix} 1 & y \\ 0 & 1 \end{pmatrix}\begin{pmatrix} 0 & f \\ e & 0 \end{pmatrix}\begin{pmatrix} 1 & x \\ 0 & 1 \end{pmatrix}$$

and apply this procedure to the matrix $\begin{pmatrix} -12 & 26 \\ -4 & -8 \end{pmatrix}$.

1.10. Prove that $\operatorname{Ar}\begin{pmatrix} 0 & b \\ c & d \end{pmatrix} = \left|\operatorname{Det}\begin{pmatrix} 0 & b \\ c & d \end{pmatrix}\right|$ by

(a) direct verification (find the image of the unit square), and by

(b) using the decomposition in Exercise 1.9, which works even when $a = 0$.

1.11. Construct a $2 \times 2$ matrix which represents each of the following transformations of the plane:

(a) A transformation $P$, satisfying $P^2 = P$, which maps the entire plane onto the line $y = 2x$ and which maps the line $y = -2x$ into the origin.

(b) A shear transformation $S$ which carries every point on the line $y = 2x$ into itself, which transforms the $y$-axis into the line $y = -x$, and which satisfies the condition $(S - I)^2 = 0$.

(c) A transformation which carries $\begin{pmatrix} 1 \\ 2 \end{pmatrix}$ into $\begin{pmatrix} 2 \\ 4 \end{pmatrix}$ and which carries $\begin{pmatrix} 1 \\ -2 \end{pmatrix}$ into $\begin{pmatrix} \frac{1}{2} \\ -1 \end{pmatrix}$.

(d) A nilpotent transformation $N$, satisfying $N^2 = 0$, whose image and kernel are both the line $y = 3x$.

1.12. For practice in multiplying $3 \times 3$ matrices, consider the two matrices

$$U = \begin{pmatrix} 0 & 1 & 0 \\ 0 & 0 & 1 \\ 0 & 0 & 0 \end{pmatrix} \quad L = \begin{pmatrix} 0 & 0 & 0 \\ 1 & 0 & 0 \\ 0 & 1 & 0 \end{pmatrix}.$$

Calculate $UL$, $LU$ and $U^3$.

1.13. Define the determinant of a $3 \times 3$ matrix by

$$\begin{pmatrix} a_{11} & a_{12} & a_{13} \\ a_{21} & a_{22} & a_{23} \\ a_{31} & a_{32} & a_{33} \end{pmatrix} = a_{11}\operatorname{Det}\begin{pmatrix} a_{22} & a_{23} \\ a_{32} & a_{33} \end{pmatrix} - a_{12}\operatorname{Det}\begin{pmatrix} a_{21} & a_{23} \\ a_{31} & a_{33} \end{pmatrix}$$

$$+ a_{13}\operatorname{Det}\begin{pmatrix} a_{21} & a_{22} \\ a_{31} & a_{32} \end{pmatrix}$$

$$= a_{11}a_{22}a_{33} - a_{11}a_{23}a_{32} - a_{12}a_{21}a_{33} + a_{12}a_{23}a_{31}$$
$$+ a_{13}a_{21}a_{32} - a_{13}a_{22}a_{31}.$$

Prove that

$$\operatorname{Det}(F \circ G) = \operatorname{Det} F \times \operatorname{Det} G.$$

1.14. Show that, if the matrix

$$\begin{pmatrix} a_{11} & a_{12} & a_{13} \\ a_{21} & a_{22} & a_{23} \\ a_{31} & a_{32} & a_{33} \end{pmatrix} = F$$

satisfies the conditions $a_{11} \neq 0$ and $a_{11}a_{22} - a_{12}a_{21} \neq 0$, then we can write

$F$ as a triple product

$$F = \begin{pmatrix} 1 & 0 & 0 \\ y_{21} & 1 & 0 \\ y_{31} & y_{32} & 1 \end{pmatrix} \begin{pmatrix} e & 0 & 0 \\ 0 & f & 0 \\ 0 & 0 & g \end{pmatrix} \begin{pmatrix} 1 & x_{12} & x_{13} \\ 0 & 1 & x_{23} \\ 0 & 0 & 1 \end{pmatrix}.$$

1.15. In $\mathbb{R}^3$ we can define Vol $F$ for any non-singular linear transformation $F$ much as we defined Ar in the plane. Thus

$$\text{Vol } F = \frac{\text{volume } F(D)}{\text{volume } D}$$

for any region $D$ and, in particular,

$$\text{Vol}(F \circ G) = \text{Vol} F \times \text{Vol} G$$

and Vol $F = \text{volume } F(\square)$, where $\square$ is the unit cube. Prove that

$$\text{Vol } F = |\text{Det } F|.$$

1.16. Consider an affine transformation of the plane which does *not* leave the origin fixed:

$$F\begin{pmatrix} x \\ y \end{pmatrix} = A\begin{pmatrix} x \\ y \end{pmatrix} + \begin{pmatrix} a \\ b \end{pmatrix}$$

where $A$ represents an affine transformation which leaves the origin fixed. Show that such a transformation can be represented by a $3 \times 3$ matrix:

$$\left(\begin{array}{cc|c} \multicolumn{2}{c|}{A} & a \\ & & b \\ \hline 0 & 0 & 1 \end{array}\right)$$

where $A$ is a $2 \times 2$ matrix, provided the vector $\begin{pmatrix} x \\ y \end{pmatrix}$ in the plane is represented by the three-component vector $\begin{pmatrix} x \\ y \\ 1 \end{pmatrix}$. You should verify the following:

(a) Such a $3 \times 3$ matrix has determinant equal to Det $A$.

(b) When such a $3 \times 3$ matrix acts on $\begin{pmatrix} x \\ y \\ 1 \end{pmatrix}$, the third component of the resulting vector is 1.

(c) The matrices $T(a, b) \equiv \begin{pmatrix} 1 & 0 & a \\ 0 & 1 & b \\ 0 & 0 & 1 \end{pmatrix}$ represent pure translations, and they obey the composition law

$$T(a, b)T(c, d) = T(a + c, b + d).$$

From a geometric point of view we can give the following interpretation to Exercise 1.16: We are considering the affine plane as the plane $z = 1$ in $\mathbb{R}^3$. We have identified the group of affine motions as a group of linear transformations in $\mathbb{R}^3$. Now we can identify the point $\begin{pmatrix} x \\ y \\ 1 \end{pmatrix}$ in the plane $z = 1$ with the line

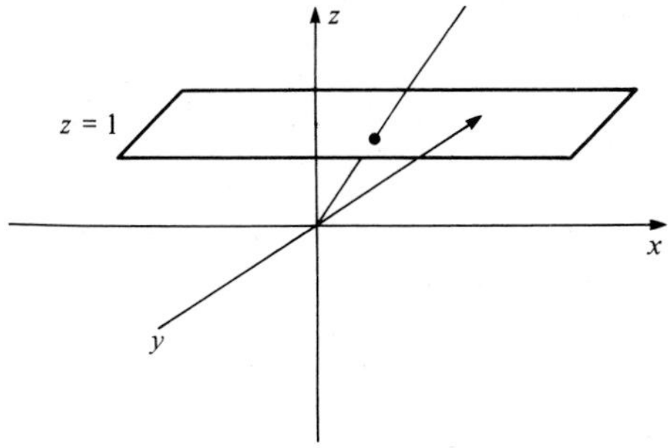


Figure 1.29

joining that point to the origin: any point on our plane determines a unique line through the origin; a line through the origin that intersects the plane $z = 1$ intersects it at a unique point. We can thus identify 'points' in our affine plane with certain kinds of lines through the origin in $\mathbb{R}^3$: those that intersect the plane $z = 1$. The advantage to this interpretation is that it gives us a grip on the notion of (artistic) perspective: two plane figures in $\mathbb{R}^3$ (not containing the origin) are 'in perspective' from the origin if they determine the same family of lines through the origin. This suggests

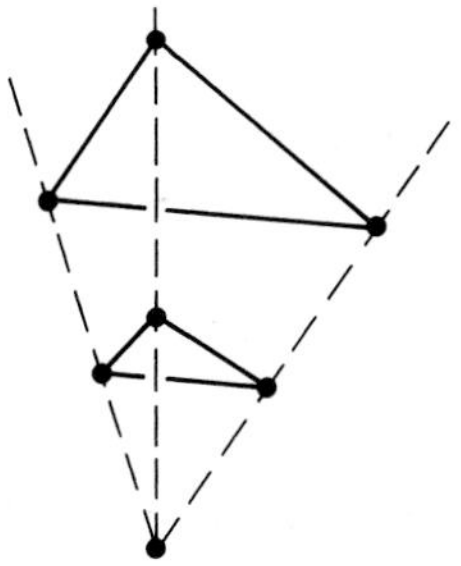

Figure 1.30

that we consider a new geometry in which 'points' are lines through the origin in $\mathbb{R}^3$; in other words, we drop the requirement that the line must intersect the $z = 1$ plane. The new 'points' that we have added are those lines through the origin in $\mathbb{R}^3$ which lie in the $z = 0$ plane, as these are the only lines through the origin which do not meet the plane $z = 1$. Let $\begin{pmatrix} a \\ b \\ 0 \end{pmatrix} \neq \begin{pmatrix} 0 \\ 0 \\ 0 \end{pmatrix}$ be a point of $\mathbb{R}^3$ in the $z = 0$ plane and let $P$ denote the line through the origin and $\begin{pmatrix} a \\ b \\ 0 \end{pmatrix}$ so $P$ is one of our new 'points'. Thus $P = \left\{ \begin{pmatrix} at \\ bt \\ 0 \end{pmatrix} \right\}_{t \in \mathbb{R}}$. From the point of view of $\mathbb{R}^3$, where $P$ is a line, we can approximate

$P$ by the family of lines through the origin $P_\varepsilon$ where

$$P_\varepsilon = \left\{ \begin{pmatrix} at \\ bt \\ \varepsilon t \end{pmatrix}_{t \in \mathbb{R}} \right\}.$$

As $\varepsilon \to 0$, $P_\varepsilon \to P$. But $P_\varepsilon$ intersects the $z = 1$ plane, when $t = 1/\varepsilon$, at the point $\begin{pmatrix} a/\varepsilon \\ b/\varepsilon \\ 1 \end{pmatrix}$.

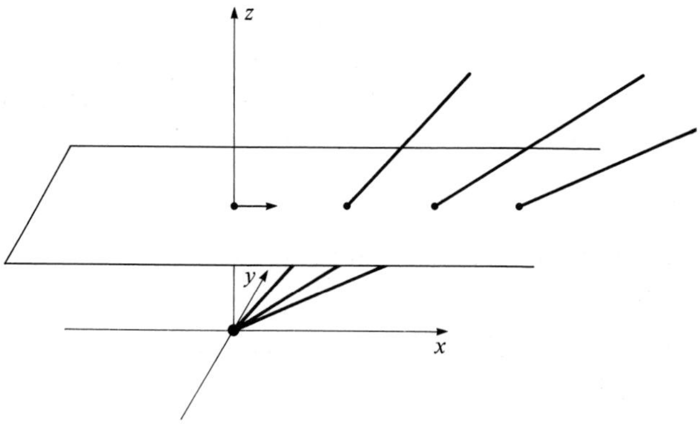


**Figure 1.31**

The points $P_\varepsilon$ in the affine plane tend to infinity as $\varepsilon \to 0$ in a definite direction given by the vector $\begin{pmatrix} a \\ b \end{pmatrix}$. We can thus think of the new 'point' $P$ as a 'point at infinity' of the affine plane. These new 'points at infinity' were first introduced in the theoretical study of perspective by artists and geometers of the fifteenth and sixteenth centuries.

We have thus introduced a new space, called $\mathbb{P}^2$, the *projective plane*. A 'point' of $\mathbb{P}^2$ is just a line through the origin in $\mathbb{R}^3$. Let us now see how to define a 'line' in $\mathbb{P}^2$. From the point of view of $\mathbb{R}^3$, a 'line' in the affine plane $z = 1$ consist of a family of lines through the origin which intersect the plane $z = 1$ along a straight line.

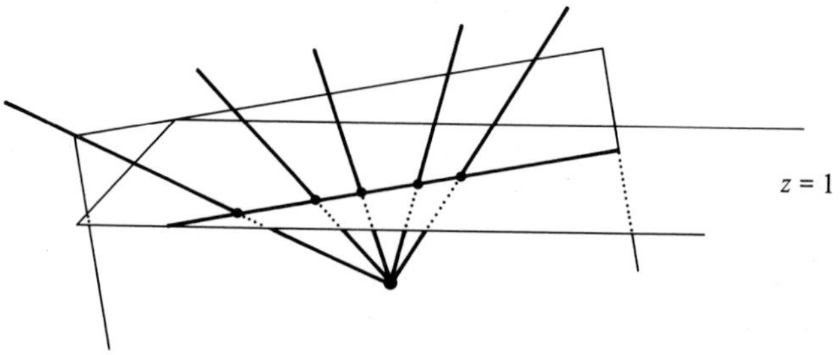


**Figure 1.32**

But this means that the family of lines through the origin sweep out a plane in $\mathbb{R}^3$. In other words, from the $\mathbb{R}^3$ point of view, a line in the affine plane is just a plane through the origin in $\mathbb{R}^3$ which intersects the $z = 1$ plane. It is now clear what to do: we drop this last intersection condition and define a 'line' in $\mathbb{P}^2$ to be a plane through the origin in $\mathbb{R}^3$. There is only one plane through the origin in $\mathbb{R}^3$ which does not intersect the plane $z = 1$ and that is the plane $z = 0$. We have thus to add just one 'line at infinity'. A 'point' $P$ lies on the 'line' $l$ if the line through the origin lies in the plane through the origin, $l$. Two distinct 'points', $P$ and $Q$ (that is, two distinct lines through the origin) determine a unique plane through the origin, i.e., two distinct 'points' determine a unique 'line'. Any two distinct planes through the origin in $\mathbb{R}^3$ intersect in a line through the origin. Thus any two 'lines' in $\mathbb{P}^2$ intersect in a 'point'. (Notice that this is different from affine geometry where two lines can be parallel. Two parallel lines in the affine plane intersect 'at infinity' in the projective plane.)

To summarize:

A 'point' in $\mathbb{P}^2$ is a line through the origin in $\mathbb{R}^3$;

A 'line' in $\mathbb{P}^2$ is a plane through the origin in $\mathbb{R}^3$;

Any two distinct 'points' lie on a unique 'line';

Any two distinct 'lines' intersect at a unique 'point'.

Any invertible $3 \times 3$ matrix acts on $\mathbb{R}^3$ so as to carry lines through the origin into lines through the origin and planes through the origin into planes through the origin.

1.17. (a) Show that any invertible $3 \times 3$ matrix determines a one-to-one transformation of the projective plane, $\mathbb{P}^2$, which carries 'lines' into 'lines'.

(b) Show that two invertible $3 \times 3$ matrices $A$ and $B$ determine the same transformation of $\mathbb{P}^2$ if and only if $A = cB$ for some non-zero real number, $c$.

1.18. Three vectors $\mathbf{u}, \mathbf{v}$ and $\mathbf{w}$ in $\mathbb{R}^3$ are called linearly independent if no equation of the form

$$a\mathbf{u} + b\mathbf{v} + c\mathbf{w} = \mathbf{0}$$

can hold unless $a, b$ and $c$ are all zero. Show that if $\mathbf{u}, \mathbf{v}$ and $\mathbf{w}$ are linearly independent, then there exists a unique $3 \times 3$ matrix $A$ such that

$$A\mathbf{u} = \begin{pmatrix}1\\0\\0\end{pmatrix}, \quad A\mathbf{v} = \begin{pmatrix}0\\1\\0\end{pmatrix} \quad \text{and} \quad A\mathbf{w} = \begin{pmatrix}0\\0\\1\end{pmatrix}$$

and that $A$ is invertible. (The general version of this theorem for any finite-dimensional vector space will be proved later.)

1.19. (a) Let $P_1, P_2, P_3$ be the 'points' in $\mathbb{P}^2$ given by the lines through the origin and $\begin{pmatrix}1\\0\\0\end{pmatrix}$, $\begin{pmatrix}0\\1\\0\end{pmatrix}$ and $\begin{pmatrix}0\\0\\1\end{pmatrix}$ respectively. Let $Q_1, Q_2, Q_3$ be any three 'points' of $\mathbb{P}^2$ which do not lie on the same 'line'. Show that there is an invertible $3 \times 3$ matrix which carries $Q_1$ into $P_1$, $Q_2$ into $P_2$ and $Q_3$ into $P_3$.

(b) Let $Q_1, Q_2, Q_3, Q_4$ be the four 'points' in $\mathbb{P}^2$, no three of which lie on the same line. Let $R_1, R_2, R_3, R_4$ be another set of four 'points', no three of which lie on a 'line'. Show that there exists a $3 \times 3$ matrix which carries $Q_1$ to $R_1$, $Q_2$ to $R_2$, $Q_3$ to $R_3$ and $Q_4$ to $R_4$.

(c) Prove the 'fundamental theorem of projective geometry' which asserts that any one-to-one transformation of $\mathbb{P}^2$ which carries 'lines' into 'lines' comes from a $3 \times 3$ matrix. (Hint: Reduce to the fundamental theorem of affine geometry proved in the appendix to this chapter.)

1.20. As an illustration of the use of 1.19(b), prove Fano's theorem which says. Let $A, B, C, D$ be four points, no three of which lie on a line

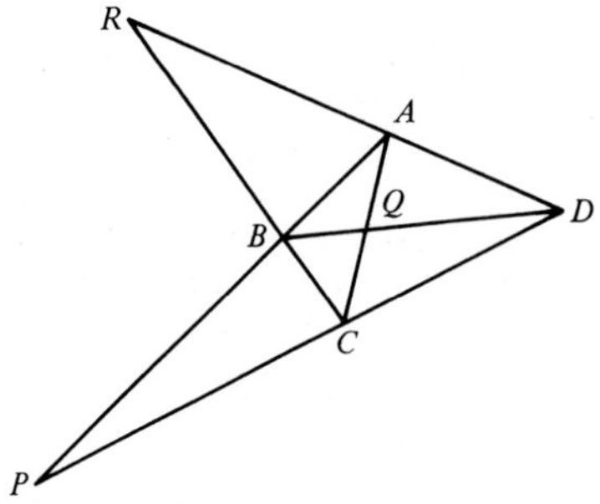


**Figure 1.33**

Let $P$ be the point of intersection of $AB$ and $CD$.
Let $Q$ be the point of intersection of $AC$ and $BD$.
Let $R$ be the point of intersection of $AD$ and $BC$.
Then $P, Q$ and $R$ do *not* lie on a line.
(Hint: Reduce to a special case; for example, $A, B, C$, the three vertices of an equilateral triangle and $D$ its center.)

# 2

# Eigenvectors and eigenvalues



In Chapter 2 we discuss conformal linear geometry in the plane, that is, the geometry of lines and angles, and its relation to certain kinds of $2 \times 2$ matrices. We also discuss the notion of eigenvalues and eigenvectors, so important in quantum mechanics. We use these notions to give an algorithm for computing the powers of a matrix. As an application we study the basic properties of Markov chains.

## 2.1. Conformal linear transformations

We wish to consider those linear transformations $f$ of $\mathbb{R}^2$ that

(1) preserve angle,

(2) preserve orientation, i.e., $\operatorname{Det} F > 0$,

where $F$ is a matrix representing $f$. Notice that, if $f$ and $g$ are two such linear transformations, so is their composition $g \circ f$.

Suppose that $f$ preserves angle and orientation. We can find some rotation $r_{-\theta}$ such that $r_{-\theta} \circ f$ takes $\begin{pmatrix} 1 \\ 0 \end{pmatrix}$ into a point on the positive $x$-axis. Then $r_{-\theta} \circ f\left(\begin{pmatrix} 0 \\ 1 \end{pmatrix}\right)$ lies on the positive $y$-axis since $r_{-\theta} \circ f$ preserves angles and $\operatorname{Det}(r_{-\theta} \circ f) > 0$. Thus the

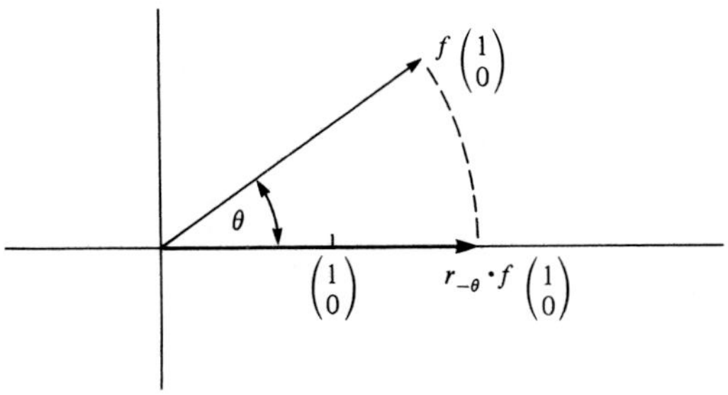


**Figure 2.1**

matrix representing $r_{-\theta}\circ f$ is of the form

$$r_{-\theta}\circ f=\begin{pmatrix} r & 0 \\ 0 & s \end{pmatrix}$$

with $r>0$, $s>0$. Since $r_{-\theta}\circ f$ preserves angles, it must carry the line through $\begin{pmatrix} 1 \\ 1 \end{pmatrix}$ into itself. To say that $\begin{pmatrix} r & 0 \\ 0 & s \end{pmatrix}\begin{pmatrix} 1 \\ 1 \end{pmatrix}$ lies on the line through $\begin{pmatrix} 1 \\ 1 \end{pmatrix}$ means that $r=s$. Thus

$$r_{-\theta}\circ f=\begin{pmatrix} r & 0 \\ 0 & r \end{pmatrix}.$$

The matrix representing $f$ is therefore of the form

$$\begin{aligned} F&=\begin{pmatrix} r & 0 \\ 0 & r \end{pmatrix}\begin{pmatrix} \cos\theta & -\sin\theta \\ \sin\theta & \cos\theta \end{pmatrix} \\ &=\begin{pmatrix} r\cos\theta & -r\sin\theta \\ r\sin\theta & r\cos\theta \end{pmatrix}=\begin{pmatrix} a & -b \\ b & a \end{pmatrix} \end{aligned}$$

where $a=r\cos\theta$, $b=r\sin\theta$. It is clear that any such matrix preserves angle and satisfies $\operatorname{Det} F=a^2+b^2=r^2>0$.

Conversely, any non-zero matrix of the form $\begin{pmatrix} a & -b \\ b & a \end{pmatrix}$ preserves angle and orientation since, starting with

$$\begin{pmatrix} a & -b \\ b & a \end{pmatrix}, \quad a^2+b^2\neq 0$$

we can set $r^2=a^2+b^2$ and then find $\theta$ such that

$$\cos\theta=ar^{-1}, \quad \sin\theta=br^{-1}$$

since $a\leqslant r$ and $\sin^2\theta+\cos^2\theta=(a^2+b^2)/r^2=1$. And therefore it follows that

$$\begin{pmatrix} a & -b \\ b & a \end{pmatrix}=\begin{pmatrix} r\cos\theta & -r\sin\theta \\ r\sin\theta & r\cos\theta \end{pmatrix}.$$

Thus the most general matrix of the form

$$\begin{pmatrix} a & -b \\ b & a \end{pmatrix}, \quad a^2+b^2\neq 0,$$

preserves angle and orientation, with

$$\operatorname{Det}\begin{pmatrix} a & -b \\ b & a \end{pmatrix}=a^2+b^2.$$

The product of any two such matrices is clearly such a matrix, but notice in addition that

$$\begin{pmatrix} a & -b \\ b & a \end{pmatrix}\begin{pmatrix} a' & -b' \\ b' & a' \end{pmatrix}=\begin{pmatrix} aa'-bb' & -(ba'+a'b) \\ ba'+ab' & aa-bb' \end{pmatrix}=\begin{pmatrix} a' & -b' \\ b' & a' \end{pmatrix}\begin{pmatrix} a & -b \\ b & a \end{pmatrix}$$

so that, in this case, multiplication is commutative. Furthermore, the inverse exists, unless $a=b=0$, since the determinant $= a^2+b^2$. Finally

$$\begin{pmatrix} a & -b \\ b & a \end{pmatrix} + \begin{pmatrix} a' & -b' \\ b' & a' \end{pmatrix} = \begin{pmatrix} a+a' & -(b+b') \\ b+b' & a+a' \end{pmatrix}$$

so that the *sum* of two matrices of the form $\begin{pmatrix} a & -b \\ b & a \end{pmatrix}$ is again of this type. This is somewhat remarkable and not to have been expected from the definition. Let us call a matrix of the form $\begin{pmatrix} a & -b \\ b & a \end{pmatrix}$ *conformal*. (We allow the possibility that $a=b=0$. Thus the non-zero conformal matrices are the ones that preserve angles and orientation.)

We have proved that the set of all conformal matrices is closed under addition and multiplication, that multiplication *is* commutative for such matrices, and each non-zero conformal matrix has an inverse. Thus conformal matrices behave very much like numbers.

We can write any conformal matrix as

$$\begin{pmatrix} a & -b \\ b & a \end{pmatrix} = a\begin{pmatrix} 1 & 0 \\ 0 & 1 \end{pmatrix} + b\begin{pmatrix} 0 & -1 \\ 1 & 0 \end{pmatrix}.$$

Notice that $\begin{pmatrix} 0 & -1 \\ 1 & 0 \end{pmatrix}$ is rotation through ninety degrees and thus

$$\begin{pmatrix} 0 & -1 \\ 1 & 0 \end{pmatrix}^2 = \begin{pmatrix} -1 & 0 \\ 0 & -1 \end{pmatrix} = -\begin{pmatrix} 1 & 0 \\ 0 & 1 \end{pmatrix}.$$

We write

$$\mathbb{1} \text{ for } \begin{pmatrix} 1 & 0 \\ 0 & 1 \end{pmatrix}$$

and

$$\mathrm{i} \text{ for } \begin{pmatrix} 0 & -1 \\ 1 & 0 \end{pmatrix}$$

so that

$$\begin{pmatrix} a & -b \\ b & a \end{pmatrix} = a\mathbb{1} + b\mathrm{i}$$

where

$$\mathrm{i}^2 = -\mathbb{1}.$$

In other words, we can identify the set of conformal matrices with the set of complex numbers.

The usual representation of a complex number as a point in the plane simply is the identification of the complex number with image of $\begin{pmatrix} 1 \\ 0 \end{pmatrix}$. For conformal matrices the point $\begin{pmatrix} a \\ b \end{pmatrix}$ determines the matrix $\begin{pmatrix} a & -b \\ b & a \end{pmatrix}$.

It is very easy to compute the $n$th power of a conformal matrix. Indeed, if we write

$$A = \begin{pmatrix} a & -b \\ b & a \end{pmatrix} = \begin{pmatrix} r & 0 \\ 0 & r \end{pmatrix} \begin{pmatrix} \cos\theta & -\sin\theta \\ \sin\theta & \cos\theta \end{pmatrix}$$

then, since $\begin{pmatrix} r & 0 \\ 0 & r \end{pmatrix}$ commutes with all $2 \times 2$ matrices,

$$A^n = \begin{pmatrix} r & 0 \\ 0 & r \end{pmatrix}^n \begin{pmatrix} \cos\theta & -\sin\theta \\ \sin\theta & \cos\theta \end{pmatrix}^n = \begin{pmatrix} r^n & 0 \\ 0 & r^n \end{pmatrix} \begin{pmatrix} \cos n\theta & -\sin n\theta \\ \sin n\theta & \cos n\theta \end{pmatrix}.$$

Thus

$$\begin{pmatrix} r\cos\theta & -r\sin\theta \\ r\sin\theta & r\cos\theta \end{pmatrix}^n = \begin{pmatrix} r^n\cos n\theta & -r^n\sin n\theta \\ r^n\sin n\theta & r^n\cos n\theta \end{pmatrix}.$$

In the language of complex numbers, this says that if

$$z = r(\cos\theta + \mathrm{i}\sin\theta)$$

then

$$z^n = r^n(\cos n\theta + \mathrm{i}\sin n\theta)$$

and is known as *DeMoivre's theorem.*

Another way of computing $A^n$ is to use the binomial formula: since

$$\begin{pmatrix} a & 0 \\ 0 & a \end{pmatrix} = a\begin{pmatrix} 1 & 0 \\ 0 & 1 \end{pmatrix}, \quad \text{and} \quad \begin{pmatrix} 0 & -b \\ b & 0 \end{pmatrix} = b\begin{pmatrix} 0 & -1 \\ 1 & 0 \end{pmatrix},$$

commute,

$$A^n = \left( a\begin{pmatrix} 1 & 0 \\ 0 & 1 \end{pmatrix} + b\begin{pmatrix} 0 & -1 \\ 1 & 0 \end{pmatrix} \right)^n$$

$$= a^n\begin{pmatrix} 1 & 0 \\ 0 & 1 \end{pmatrix} + na^{n-1}b\begin{pmatrix} 0 & -1 \\ 1 & 0 \end{pmatrix} + \tfrac{1}{2}n(n-1)a^{n-2}b^2\begin{pmatrix} 0 & -1 \\ 1 & 0 \end{pmatrix}^2 + \cdots.$$

But $\begin{pmatrix} 0 & -1 \\ 1 & 0 \end{pmatrix}^2 = \begin{pmatrix} -1 & 0 \\ 0 & -1 \end{pmatrix}$ so

$$A^n = \left( a^n - \binom{n}{2}a^{n-2}b^2 + \binom{n}{4}a^{n-4}b^4 + \cdots \right)\begin{pmatrix} 1 & 0 \\ 0 & 1 \end{pmatrix}$$

$$+ \left( \binom{n}{1}a^{n-1}b - \binom{n}{3}a^{n-3}b^3 + \cdots \right)\begin{pmatrix} 0 & -1 \\ 1 & 0 \end{pmatrix}.$$

In the next section we will provide an efficient algorithm for computing powers of any $2 \times 2$ matrix, not necessarily conformal. It will involve the notion of eigenvalue, a concept that plays a key role in quantum mechanics.

## 2.2. Eigenvectors and eigenvalues

Let $F$ be a linear transformation. We can ask whether $F$ carries some line through the origin into itself. (No non-trivial rotation has this property, for example, while any non-zero singular transformation carries its image into itself.)

If $\mathbf{v}$ is a non-zero vector lying on such a line, we must have

$$F(\mathbf{v}) = \lambda \mathbf{v}$$

for some real number $\lambda$. If this equation holds with $\mathbf{v} \neq \mathbf{0}$, $\lambda$ is called an *eigenvalue* of $f$ corresponding to the *eigenvector* $\mathbf{v}$. We can rewrite the above equation as

$$\left[F - \lambda\begin{pmatrix}1 & 0\\ 0 & 1\end{pmatrix}\right]\mathbf{v} = \mathbf{0}.$$

Since $\mathbf{v}$ is not zero, this implies that

$$\operatorname{Det}\left(F - \lambda\begin{pmatrix}1 & 0\\ 0 & 1\end{pmatrix}\right) = 0$$

which is an equation for $\lambda$. Explicitly, if $F = \begin{pmatrix}a & b\\ c & d\end{pmatrix}$, so that

$$F - \lambda\begin{pmatrix}1 & 0\\ 0 & 1\end{pmatrix} = \begin{pmatrix}a-\lambda & b\\ c & d-\lambda\end{pmatrix}$$

the preceding equation becomes

$$\operatorname{Det}\begin{pmatrix}a-\lambda & b\\ c & d-\lambda\end{pmatrix} = (a-\lambda)(d-\lambda) - bc = \lambda^2 - (a+d)\lambda + (ad-bc) = 0$$

or

$$\boxed{\lambda^2 - (\operatorname{tr} F)\lambda + \operatorname{Det} F = 0.}$$

The polynomial

$$P(X) = X^2 - (a+d)X + (ad-bc)$$

is called the *characteristic polynomial* of $\begin{pmatrix}a & b\\ c & d\end{pmatrix}$ and the equation

$$P(\lambda) = 0$$

is called the *characteristic equation.* It will have real roots

$$\lambda = \tfrac{1}{2}[(a+d) \pm \sqrt{\{(a+d)^2 - 4(ad-bc)\}}]$$

$$= \tfrac{1}{2}[(a+d) \pm \sqrt{\{(a-d)^2 + 4bc\}}]$$

if and only if

$$(a-d)^2 + 4bc \geqslant 0.$$

If this occurs, we know that

$$\begin{pmatrix}a-\lambda & b\\ c & d-\lambda\end{pmatrix}\begin{pmatrix}d-\lambda\\ -c\end{pmatrix} = 0$$

and

$$\begin{pmatrix}a-\lambda & b\\ c & d-\lambda\end{pmatrix}\begin{pmatrix}b\\ -(a-\lambda)\end{pmatrix} = 0$$

so that if $\begin{pmatrix} d-\lambda \\ -c \end{pmatrix}$ or $\begin{pmatrix} b \\ -(a-\lambda) \end{pmatrix}$ are non-zero they are eigenvectors and they both lie on the same line, since $\text{Det}\begin{pmatrix} d-\lambda & b \\ -c & -(a-\lambda) \end{pmatrix} = -\text{Det}\begin{pmatrix} a-\lambda & b \\ c & d-\lambda \end{pmatrix} = 0$. If they are both zero, then $a=\lambda$, $b=c=0$ and $d=\lambda$ so that

$$\begin{pmatrix} a & c \\ b & d \end{pmatrix} = \begin{pmatrix} \lambda & 0 \\ 0 & \lambda \end{pmatrix}, \quad F = \lambda I$$

and every (non-zero) vector in the plane is an eigenvector.

**Case 1. Real Distinct Roots**

If $(a-d)^2 + 4bc > 0$, so that there are two *distinct* real eigenvalues, $\lambda_1$ and $\lambda_2$, then $F \neq \lambda I$ and so there are only two lines through the origin left fixed, each spanned by an eigenvector $\mathbf{v}_1 = \begin{pmatrix} x_1 \\ y_1 \end{pmatrix}$ or $\mathbf{v}_2 = \begin{pmatrix} x_2 \\ y_2 \end{pmatrix}$. We have

$$\begin{pmatrix} a & b \\ c & d \end{pmatrix}\begin{pmatrix} x_1 \\ y_1 \end{pmatrix} = \lambda_1 \begin{pmatrix} x_1 \\ y_1 \end{pmatrix}$$

and

$$\begin{pmatrix} a & b \\ c & d \end{pmatrix}\begin{pmatrix} x_2 \\ y_2 \end{pmatrix} = \lambda_2 \begin{pmatrix} x_2 \\ y_2 \end{pmatrix}.$$

If we let $B$ be the matrix

$$B = \begin{pmatrix} x_1 & x_2 \\ y_1 & y_2 \end{pmatrix}$$

then $B$ is not singular since $\begin{pmatrix} x_1 \\ y_1 \end{pmatrix}$ and $\begin{pmatrix} x_2 \\ y_2 \end{pmatrix}$ do not lie on the same line, and we can combine the two equations for the eigenvectors to read

$$FB = B\Lambda$$

where $\Lambda$ is the matrix $\begin{pmatrix} \lambda_1 & 0 \\ 0 & \lambda_2 \end{pmatrix}$, or

$$\boxed{F = B\Lambda B^{-1}, \quad \Lambda = \begin{pmatrix} \lambda_1 & 0 \\ 0 & \lambda_2 \end{pmatrix},}$$

Conversely if $F = B\Lambda B^{-1}$ with $\Lambda = \begin{pmatrix} \lambda_1 & 0 \\ 0 & \lambda_2 \end{pmatrix}$, then

$$\begin{aligned} F\left[B\begin{pmatrix} 1 \\ 0 \end{pmatrix}\right] &= B\Lambda B^{-1}B\begin{pmatrix} 1 \\ 0 \end{pmatrix} \\ &= B\Lambda\begin{pmatrix} 1 \\ 0 \end{pmatrix} \\ &= B\lambda_1\begin{pmatrix} 1 \\ 0 \end{pmatrix} \\ &= \lambda_1 B\begin{pmatrix} 1 \\ 0 \end{pmatrix} \end{aligned}$$

so the first column of $B$ is an eigenvector of $F$ with eigenvalue $\lambda_1$ and, similarly, the second column is an eigenvector with eigenvalue $\lambda_2$.

**Case 2. Repeated Real Root**

If $(a-d)^2+4bc=0$, so that $P(X)=0$ has a double root, $\lambda$, the situation is a little more complicated. Consider the two matrices

$$\begin{pmatrix} 0 & 0 \\ 0 & 0 \end{pmatrix} \text{ and } \begin{pmatrix} 0 & 1 \\ 0 & 0 \end{pmatrix}.$$

For both of these matrices the characteristic polynomial is $P(X)=X^2$ so that $\lambda=0$ is a double root. Every non-zero vector in the plane is an eigenvector of the first matrix while only the vectors $\begin{pmatrix} x \\ 0 \end{pmatrix}$ are eigenvectors for the second. Notice, however, that both matrices satisfy the equation $F^2=0$, which we can write as $P(F)=0$, i.e., we substitute $F$ (as if it were a number) into its own characteristic polynomial and we get 0. We claim that this is a general fact, called the *Cayley–Hamilton theorem.*

*Given any matrix F whose characteristic polynomial is P(X) then*

$$P(F)=0$$

i.e.,

$$\begin{pmatrix} a & b \\ c & d \end{pmatrix}^2 - (a+d)\begin{pmatrix} a & b \\ c & d \end{pmatrix} + (ad-bc)\begin{pmatrix} 1 & 0 \\ 0 & 1 \end{pmatrix} = 0.$$

For our case of $2\times 2$ matrices this can be verified by direct calculation:

$$\begin{pmatrix} a & b \\ c & d \end{pmatrix}^2 = \begin{pmatrix} a^2+bc & ab+bd \\ ca+cd & cb+d^2 \end{pmatrix}$$

$$(a+d)\begin{pmatrix} a & b \\ c & d \end{pmatrix} = \begin{pmatrix} a^2+ad & ab+bd \\ ca+cd & ad+d^2 \end{pmatrix}$$

and

$$(ad-bc)\begin{pmatrix} 1 & 0 \\ 0 & 1 \end{pmatrix} = \begin{pmatrix} ad-bc & 0 \\ 0 & ad-bc \end{pmatrix}$$

so

$$\begin{pmatrix} a & b \\ c & d \end{pmatrix}^2 - (a+d)\begin{pmatrix} a & b \\ c & d \end{pmatrix} + (ad-bc)\begin{pmatrix} 1 & 0 \\ 0 & 1 \end{pmatrix} = 0.$$

If $P(X)$ has a double root,

$$P(X)=(X-\lambda)^2$$

so that

$$\left(F-\lambda\begin{pmatrix} 1 & 0 \\ 0 & 1 \end{pmatrix}\right)^2 = 0,$$

there are two possibilities:

$$F-\lambda\begin{pmatrix} 1 & 0 \\ 0 & 1 \end{pmatrix} = 0 \quad \text{so} \quad F = \begin{pmatrix} \lambda & 0 \\ 0 & \lambda \end{pmatrix}$$

or

$$F-\lambda\begin{pmatrix} 1 & 0 \\ 0 & 1 \end{pmatrix} \neq 0.$$

In this second case, let $\begin{pmatrix} x_1 \\ y_1 \end{pmatrix}$ be an eigenvector of $F$ and let $\begin{pmatrix} x_2 \\ y_2 \end{pmatrix}$ be some non-zero vector which is *not* an eigenvector of $F$. Then

$(F - \lambda I)\vec{y} = 0$
$\Rightarrow F(F - \lambda I)\vec{y} = \lambda I(F - \lambda I)\vec{y} = \lambda(F - \lambda I)\vec{y}$

$$\left[ F - \lambda \begin{pmatrix} 1 & 0 \\ 0 & 1 \end{pmatrix} \right] \begin{pmatrix} x_2 \\ y_2 \end{pmatrix}$$

is an eigenvector of $F$ since $\left( F - \begin{pmatrix} \lambda & 0 \\ 0 & \lambda \end{pmatrix} \right)^2 = 0$, and so is some non-zero multiple of $\begin{pmatrix} x_1 \\ y_1 \end{pmatrix}$. By multiplying $\begin{pmatrix} x_2 \\ y_2 \end{pmatrix}$ by a suitable non-zero constant, we can arrange that

$$\left[ F - \lambda \begin{pmatrix} 1 & 0 \\ 0 & 1 \end{pmatrix} \right] \begin{pmatrix} x_2 \\ y_2 \end{pmatrix} = \begin{pmatrix} x_1 \\ y_1 \end{pmatrix}.$$

Again the matrix

$$B = \begin{pmatrix} x_1 & x_2 \\ y_1 & y_2 \end{pmatrix}$$

is non-singular, and we can write the above equation as

$$FB = B \begin{pmatrix} \lambda & 1 \\ 0 & \lambda \end{pmatrix}$$

or

$$F = B \begin{pmatrix} \lambda & 1 \\ 0 & \lambda \end{pmatrix} B^{-1}.$$

Conversely any matrix of the form

$$F = B \begin{pmatrix} \lambda & 1 \\ 0 & \lambda \end{pmatrix} B^{-1}$$

has the property that $\left[ F - \lambda \begin{pmatrix} 1 & 0 \\ 0 & 1 \end{pmatrix} \right]^2 = 0$ but $F - \lambda \begin{pmatrix} 1 & 0 \\ 0 & 1 \end{pmatrix} \neq 0$ as can easily be checked.

## Case 3. Complex Roots

We still have to deal with the case of a transformation that has no real eigenvalues or eigenvectors. The most obvious example of such a transformation is a rotation through an angle that is not a multiple of $\pi$. Such a rotation clearly does not carry any non-zero vector into a multiple of itself. More generally, a conformal transformation, which may be viewed as a rotation of the plane followed by a uniform 'stretching', will have no real eigenvectors.

Consider, now, what happens if we try to find eigenvalues and eigenvectors for a conformal matrix

$$C = \begin{pmatrix} x & -y \\ y & x \end{pmatrix}.$$

The characteristic equation is

$$(x-\lambda)^2+y^2=0$$

so that

$$(\lambda-x)^2=-y^2$$
$$\lambda-x=\pm\mathrm{i}y$$

and

$$\lambda=x\pm\mathrm{i}y.$$

We previously observed, in section 2.1, that the conformal matrix $\begin{pmatrix} x & -y \\ y & x \end{pmatrix}$ can be used to represent the complex number $x+\mathrm{i}y$; now we see that this complex number is an eigenvalue of the matrix. Furthermore, given any pair of complex conjugate numbers, $x+\mathrm{i}y$ and $x-\mathrm{i}y$, there is a real conformal matrix that has these numbers as its eigenvalues. Of course, we cannot interpret these complex eigenvalues geometrically, since the associated eigenvectors have complex components and cannot be regarded as vectors in the real plane.

We will show that we can write a matrix $F$, whose eigenvalues are $x\pm\mathrm{i}y$, in the form

$$F=BCB^{-1},\quad C=\begin{pmatrix} x & -y \\ y & x \end{pmatrix}$$

*Proof.* Consider the matrix

$$G=F-xI.$$

The Cayley–Hamilton theorem says that

$$(F-xI)^2+y^2I=0.$$

Thus

$$G^2=-y^2I.$$

So pick any vector $\mathbf{v}_1\neq 0$ and define

$$\mathbf{v}_2=y^{-1}G\mathbf{v}_1$$

Then

$$G\mathbf{v}_2=y^{-1}G^2\mathbf{v}_1=-y\mathbf{v}_1$$

while, by definition

$$G\mathbf{v}_1=y\mathbf{v}_2.$$

Let $B$ be the matrix whose columns are $\mathbf{v}_1$ and $\mathbf{v}_2$. Then $\mathbf{v}_1=B\begin{pmatrix}1\\0\end{pmatrix}$ and $\mathbf{v}_2=B\begin{pmatrix}0\\1\end{pmatrix}$, so

$$GB\begin{pmatrix}1\\0\end{pmatrix}=yB\begin{pmatrix}0\\1\end{pmatrix}\quad\text{and}\quad GB\begin{pmatrix}0\\1\end{pmatrix}=-yB\begin{pmatrix}1\\0\end{pmatrix}\quad\text{so}\quad GB=B\begin{pmatrix}0 & -y\\ y & 0\end{pmatrix}.$$

Multiplying these equations by $B^{-1}$ shows that $G=B\begin{pmatrix}0 & -y\\ y & 0\end{pmatrix}B^{-1}$.

Thus

$$F = xI + B\begin{pmatrix} 0 & -y \\ y & 0 \end{pmatrix}$$

$$= B(xI)B^{-1} + B\begin{pmatrix} 0 & -y \\ y & 0 \end{pmatrix}B^{-1} \quad \text{since } B(xI) = (xI)B$$

$$= B\left[xI + \begin{pmatrix} 0 & -y \\ y & 0 \end{pmatrix}\right]B^{-1}.$$

So

$$F = B\begin{pmatrix} x & -y \\ y & x \end{pmatrix}B^{-1}$$

as was to be proved.

For example, we can make the convenient choice $\mathbf{v}_1 = \begin{pmatrix} 1 \\ 0 \end{pmatrix}$, though any other choice would have been equally suitable. Then, since $G\mathbf{v}_1 = y\mathbf{v}_2$, we have $\mathbf{v}_2 = y^{-1}G\mathbf{v}_1 = y^{-1}G\begin{pmatrix} 1 \\ 0 \end{pmatrix}$. Thus $\mathbf{v}_1$, the first column of $B$, is $\begin{pmatrix} 1 \\ 0 \end{pmatrix}$, while $\mathbf{v}_2$, the second column, is the first column of $G$ divided by $y$.

To summarize: if $F = \begin{pmatrix} a & b \\ c & d \end{pmatrix}$ has eigenvalues $x \pm iy$, with $y \neq 0$, then $F = BCB^{-1}$, where

$$B = \begin{pmatrix} 1 & (a-x)/y \\ 0 & c/y \end{pmatrix} \quad \text{and} \quad C = \begin{pmatrix} x & -y \\ y & x \end{pmatrix}.$$

Furthermore, $G = F - xI$ satisfies the equation $G^2 = -y^2 I$.

As in illustration of this decomposition, consider $F = \begin{pmatrix} 7 & -10 \\ 2 & -1 \end{pmatrix}$. Since $\operatorname{Tr} F = 6$ and $\operatorname{Det} F = 13$, the characteristic equation is $\lambda^2 - 6\lambda + 13 = 0$ and

$$\lambda = \frac{6 \pm \sqrt{(36-52)}}{2} = 3 \pm 2i.$$

Thus $x = 3$, $y = 2$. The matrix

$$G = F - 3I = \begin{pmatrix} 4 & -10 \\ 2 & -4 \end{pmatrix}$$

satisfies $G^2 = -4I$, as expected. To construct the second column of $B$, we just divide the first column of $G$ by $y$: $\frac{1}{2}\begin{pmatrix} 4 \\ 2 \end{pmatrix} = \begin{pmatrix} 2 \\ 1 \end{pmatrix}$. Hence $B = \begin{pmatrix} 1 & 2 \\ 0 & 1 \end{pmatrix}$ and $F = BCB^{-1} = \begin{pmatrix} 1 & 2 \\ 0 & 1 \end{pmatrix}\begin{pmatrix} 3 & -2 \\ 2 & 3 \end{pmatrix}\begin{pmatrix} 1 & -2 \\ 0 & 1 \end{pmatrix}$.

**Powers of a Matrix**

Suppose we are given a matrix $F$ and want to compute $F^n$ for many (or for large) values of $n$. (In the next section, we shall give an instance where this problem is of interest.)

**Case 1. Real Distinct Roots**

If

$$\Lambda = \begin{pmatrix} \lambda_1 & 0 \\ 0 & \lambda_2 \end{pmatrix}$$

then clearly

$$\Lambda^n = \begin{pmatrix} \lambda_1^n & 0 \\ 0 & \lambda_2^n \end{pmatrix}$$

So computing the powers of a diagonal matrix is reduced to computing the powers of real numbers. If

$$F = B\Lambda B^{-1}$$

then

$$F^2 = B\Lambda B^{-1}B\Lambda B^{-1} = B\Lambda^2 B^{-1}$$

and (by induction)

$$F^n = B\Lambda^n B^{-1}.$$

**Case 2. Repeated Real Root**

Next let us examine the matrix

$$\begin{pmatrix} \lambda & 1 \\ 0 & \lambda \end{pmatrix} = \begin{pmatrix} \lambda & 0 \\ 0 & \lambda \end{pmatrix} + \begin{pmatrix} 0 & 1 \\ 0 & 0 \end{pmatrix}.$$

Now $\begin{pmatrix} \lambda & 0 \\ 0 & \lambda \end{pmatrix}$ commutes with all $2 \times 2$ matrices and

$$\begin{pmatrix} 0 & 1 \\ 0 & 0 \end{pmatrix}^2 = 0$$

so, we may apply the binomial formula:

$$\begin{pmatrix} \lambda & 1 \\ 0 & \lambda \end{pmatrix}^n = \left(\begin{pmatrix} \lambda & 0 \\ 0 & \lambda \end{pmatrix} + \begin{pmatrix} 0 & 1 \\ 0 & 0 \end{pmatrix}\right)^n = \begin{pmatrix} \lambda & 0 \\ 0 & \lambda \end{pmatrix}^n + n\begin{pmatrix} \lambda & 0 \\ 0 & \lambda \end{pmatrix}^{n-1}\begin{pmatrix} 0 & 1 \\ 0 & 0 \end{pmatrix}$$

(as the remaining terms in the binomial formula vanish). Thus

$$\begin{pmatrix} \lambda & 1 \\ 0 & \lambda \end{pmatrix}^n = \begin{pmatrix} \lambda^n & n\lambda^{n-1} \\ 0 & \lambda^n \end{pmatrix}.$$

So if

$$F = B\begin{pmatrix} \lambda & 1 \\ 0 & \lambda \end{pmatrix}B^{-1}$$

then

$$F^n = B\begin{pmatrix} \lambda & 1 \\ 0 & \lambda \end{pmatrix}^n B^{-1} = B\begin{pmatrix} \lambda^n & n\lambda^{n-1} \\ 0 & \lambda^n \end{pmatrix}B^{-1}.$$

**Case 3. Complex Roots**

Finally, if

$$F = BCB^{-1}$$

where $C = \begin{pmatrix} x & -y \\ y & x \end{pmatrix}$ is a conformal matrix, then

$$F^n = BC^nB^{-1}$$

and where we can compute $C^n$ by either of the two methods given at the end of section 2.1.

Thus, for each of the three possibilities listed above (distinct real eigenvalues, repeated eigenvalues, complex eigenvalues), we have a simple method for computing the powers of a matrix $F$, once we have computed the eigenvalues and the change of basis matrix $B$.

Actually, for the last two cases, we do not have to compute $B$: for case 2,

$$(F - \lambda I) = N \text{ satisfies } N^2 = 0$$

so

$$\boxed{F^n = (\lambda I + N)^n = \lambda^n I + n\lambda^{n-1} N}$$

by the binomial formula.

For case 3, with eigenvalues $x \pm iy$,

$$F - xI = yH \quad \text{where} \quad H^2 = -I$$

so

$$F^n = (xI + yH)^n = x^n I + nx^{n-1}yH + \binom{n}{2}x^{n-2}y^2H^2 + \cdots$$

$$= \left(x^n - \binom{n}{2}x^{n-2}y^2 + \cdots\right)I + \left(\binom{n}{1}x^{n-1}y - \binom{n}{3}x^{n-3}y^3 + \cdots\right)H.$$

## 2.3. Markov processes

In this section we give an application of matrix multiplication to probability. We do not want to write a whole introductory treatise on the theory of probability. We just summarize the most basic facts: Probability assignments assign real numbers

$$0 \leqslant p(A) \leqslant 1, \quad 0 \leqslant p(B) \leqslant 1, \ldots$$

to 'events' $A, B$, etc., according to certain rules. These are

The probability of an event that is certain is 1;

The probability of an event that is impossible is 0;

If the event $A$ can occur in $k$ mutually exclusive ways (we write this as

$$A = A_1 \cup \cdots \cup A_k, \quad A_i \cap A_j = \varnothing, \quad i \neq j$$

then

$$p(A) = p(A_1) + \cdots + p(A_k).$$

In particular, if $A^c$ denotes the 'complementary event', the event that $A$ does not occur, then

$$A \cup A^c \text{ is certain (either } A \text{ will occur or not)}$$

and

$$A \cap A^c = \varnothing$$

so

$$p(A) + p(A^c) = 1.$$

One also has 'conditional probabilities':

$$p(B|A) = \text{the conditional probability of } B \text{ given } A.$$

Thus, if $A$ is the event 'it is raining today' and $B$ is the event 'it is clear tomorrow', then $p(B|A)$ is the probability that it will be clear tomorrow given that it is raining today. We then have the rule

$$p(A \cap B) = p(B|A)p(A)$$

i.e.,

> the probability of $A$ and $B$ equals the product of the conditional probability of $B$ given $A$ with the probability of $A$.

In particular, if $A_1, \ldots, A_k$ are mutually exclusive alternatives, $A_i \cap A_j = \varnothing$ and $B$ can occur only if one of the events $A_i$ occurs:

$$B = B \cap A_1 \cup \ldots \cup B \cap A_k$$

then

$$p(B) = p(B \cap A_1) + \cdots + p(B \cap A_k)$$

so

$$p(B) = p(B|A_1)p(A_1) + \cdots + p(B|A_k)p(A_k).$$

We shall now consider a system which can exist in one of two states; a switch might be on or off, or, in a game of badminton, 'state 1' might denote the situation where player number 1 is serving while 'state 2' is where player number 2 is serving. We envisage a situation in which in one 'step' there can be a 'transition' from one state to another. Thus, in our badminton example, at each 'step' in the process (at each point of the game), the system can stay in the same state (server makes the point and serves again) or make a transition from one state to the other (server loses the point and opponent gets to serve). For example, we can imagine that at some stage of the game if player 1 is serving, he has probability 0.8 of winning the point and probability 0.2 of losing, while if player 2 is to serve, then she has probability 0.7 of winning the point and probability 0.3 of losing. In a real game, the probability of a given player winning a point at some stage of the game will depend on a whole lot of factors (how encouraged or demoralized he is by the game up to that stage, how tired she is, etc.). We make the drastic assumption that none of these considerations matter, that all that matters is who are the opponents and who is serving. We can thus summarize the above probability assignments by the matrix

$$\begin{pmatrix} 0.8 & 0.3 \\ 0.2 & 0.7 \end{pmatrix}.$$

Thus 0.8 represents the conditional probability of the system being in state 1 *after* the step if it is in state 1 *before* the step, while 0.2 represents the conditional probability of being in state 2 after the step if the system was in state 1 before the step.

In general a (discrete time, two-state, stationary) Markov process is a process in which the states can change in discrete units of time, but where the probability of transition from one state to another depends only on the state the system is in, not on the past history of the system or on the time that the transition is taking place. Thus there are four 'transition probabilities' which can be arranged as a matrix

$$A = \begin{pmatrix} a & b \\ c & d \end{pmatrix}$$

where

$a =$ probability of transition from state 1 to state 1;
$b =$ probability of transition from state 2 to state 1;
$c =$ probability of transition from state 1 to state 2;
$d =$ probability of transition from state 2 to state 2.

Suppose that we do not know what state the system is in at a given time; all that we know is that there is probability $p$ that the system is in state 1 and probability $q = 1 - p$ that it is in state 2. This probability assignment can be represented by the vector

$$\mathbf{v} = \begin{pmatrix} p \\ q \end{pmatrix}.$$

After one step, the law for conditional probability says that

$$\begin{aligned} \left\{ \begin{matrix} \text{probability of} \\ \text{being in state 1} \\ \text{after the step} \end{matrix} \right\} &= \left\{ \begin{matrix} \text{trans. prob.} \\ \text{from state 1} \\ \text{to state 1} \end{matrix} \right\} \times \left\{ \begin{matrix} \text{prob. of} \\ \text{being in} \\ \text{state 1} \end{matrix} \right\} \\ &\quad + \left\{ \begin{matrix} \text{trans. prob.} \\ \text{from state 2} \\ \text{to state 1} \end{matrix} \right\} \times \left\{ \begin{matrix} \text{prob. of} \\ \text{being in} \\ \text{state 2} \end{matrix} \right\} \\ &= ap + bq \end{aligned}$$

and similarly the probability of being in state 2 after one step is

$$cp + dq.$$

In other words, the new 'probability vector' is

$$\begin{pmatrix} ap + bq \\ cp + dq \end{pmatrix} = \begin{pmatrix} a & b \\ c & d \end{pmatrix} \begin{pmatrix} p \\ q \end{pmatrix} = A\mathbf{v}.$$

Let us illustrate this in our badminton examples. Suppose we know that player 1 is to serve the first point. The vector

$$\mathbf{v}_0 = \begin{pmatrix} 1 \\ 0 \end{pmatrix}$$

then represents the initial probability vector at the beginning of the game. After the first point, the probability vector is

$$\mathbf{v}_1 = A\mathbf{v}_0 = \begin{pmatrix} 0.8 & 0.3 \\ 0.2 & 0.7 \end{pmatrix}\begin{pmatrix} 1 \\ 0 \end{pmatrix} = \begin{pmatrix} 0.8 \\ 0.2 \end{pmatrix}.$$

After the second point, it is

$$\mathbf{v}_2 = A\mathbf{v}_1 = \begin{pmatrix} 0.8 & 0.3 \\ 0.2 & 0.7 \end{pmatrix}\begin{pmatrix} 0.8 \\ 0.2 \end{pmatrix} = \begin{pmatrix} 0.7 \\ 0.3 \end{pmatrix} = A^2\mathbf{v}_0.$$

After the third point,

$$\mathbf{v}_3 = A\mathbf{v}_2 = \begin{pmatrix} 0.8 & 0.3 \\ 0.2 & 0.7 \end{pmatrix}\begin{pmatrix} 0.7 \\ 0.3 \end{pmatrix} = \begin{pmatrix} 0.65 \\ 0.35 \end{pmatrix} = A^3\mathbf{v}_0$$

and so on. In general, the effect of playing $n$ points is represented by the matrix $A^n$.

On thinking about this situation, you may realize that the probability vector after a large number of steps ought to be practically independent of the initial state: whether player 1 is serving for the fifteenth point is unlikely to depend strongly on which player served for the first point. This suspicion is confirmed by calculation: we find

$$A^2 = \begin{pmatrix} 0.8 & 0.3 \\ 0.2 & 0.7 \end{pmatrix}\begin{pmatrix} 0.8 & 0.3 \\ 0.2 & 0.7 \end{pmatrix} = \begin{pmatrix} 0.7 & 0.55 \\ 0.3 & 0.45 \end{pmatrix}$$

$$A^4 = \begin{pmatrix} 0.7 & 0.55 \\ 0.3 & 0.45 \end{pmatrix}^2 = \begin{pmatrix} 0.63 & 0.56 \\ 0.37 & 0.44 \end{pmatrix}$$

$$A^8 = \begin{pmatrix} 0.63 & 0.56 \\ 0.37 & 0.44 \end{pmatrix}^2 = \begin{pmatrix} 0.602 & 0.598 \\ 0.398 & 0.402 \end{pmatrix}$$

$$A^{16} = \begin{pmatrix} 0.602 & 0.598 \\ 0.398 & 0.402 \end{pmatrix}^2 = \begin{pmatrix} 0.600\,006 & 0.599\,994 \\ 0.399\,994 & 0.400\,006 \end{pmatrix}$$

and we might conjecture that

$$\lim_{n\to\infty} A^n = \begin{pmatrix} 0.6 & 0.6 \\ 0.4 & 0.4 \end{pmatrix} \text{ exactly.}$$

In fact it is easy to show in general that, as long as $b$ and $c$ do not both equal 0 or both equal 1, $\lim_{n\to\infty} A^n$ exists. We need only determine the eigenvalues and eigenvectors of $A$. Since $a + c = 1$, $b + d = 1$, we may write

$$A = \begin{pmatrix} 1-c & b \\ c & 1-b \end{pmatrix}.$$

Since $\operatorname{Tr} A = 2 - (b + c)$ and $\operatorname{Det} A = 1 - b - c + bc - bc = 1 - (b + c)$, the characteristic equation is

$$\lambda^2 - [2-(b+c)]\lambda + 1 - (b+c) = 0$$

or

$$(\lambda - 1)(\lambda - (1-b-c)) = 0.$$

The eigenvalues are $\lambda_1 = 1$, $\lambda_2 = 1-(b+c)$. Note that $|\lambda_2| \leqslant 1$, with equality only if $b=c=0$ or if $b=c=1$.

The eigenvectors are easily found by considering

$$A - \lambda_1 I = \begin{pmatrix} 1-c & b \\ c & 1-b \end{pmatrix} - \begin{pmatrix} 1 & 0 \\ 0 & 1 \end{pmatrix} = \begin{pmatrix} -c & b \\ c & -b \end{pmatrix}.$$

The kernel of this singular matrix consists of multiples of the eigenvector corresponding to $\lambda_1 = 1$: we normalize this vector so that its components sum to 1 and find

$$\mathbf{v}_1 = \frac{1}{b+c}\begin{pmatrix} b \\ c \end{pmatrix}.$$

The image of $A - \lambda_1 I$ consists of multiples of the eigenvector corresponding to $\lambda_2$: a convenient choice of this eigenvector is $\mathbf{v}_2 = \begin{pmatrix} -1 \\ 1 \end{pmatrix}$.

In terms of these eigenvectors and eigenvalues the operation of $A$ is easily visualized. The vector $\mathbf{v}_1$ lies on the line segment joining $\begin{pmatrix} 1 \\ 0 \end{pmatrix}$ to $\begin{pmatrix} 0 \\ 1 \end{pmatrix}$, and any vector on this segment is of the form $\mathbf{v} = \mathbf{v}_1 + \alpha\begin{pmatrix} -1 \\ 1 \end{pmatrix}$. Since $A\mathbf{v}_1 = \mathbf{v}_1$ and $A\begin{pmatrix} -1 \\ 1 \end{pmatrix} = \lambda_2\begin{pmatrix} -1 \\ 1 \end{pmatrix}$, we find

$$A\mathbf{v} = \mathbf{v}_1 + \alpha\lambda_2\begin{pmatrix} -1 \\ 1 \end{pmatrix}$$

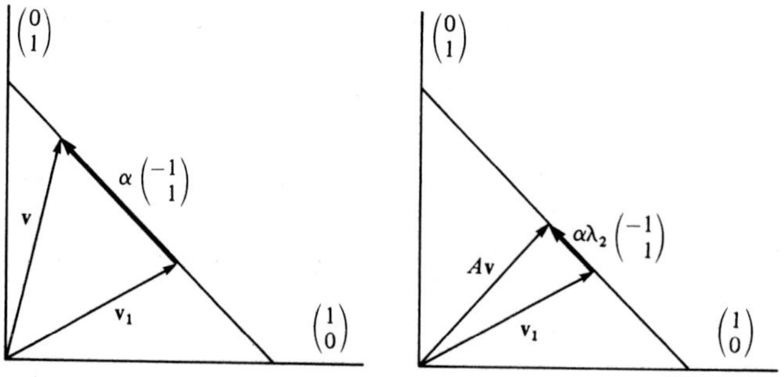


Figure 2.2

and more generally

$$A^n\mathbf{v} = \mathbf{v}_1 + \alpha(\lambda_2)^n\begin{pmatrix} -1 \\ 1 \end{pmatrix}.$$

Since $|\lambda_2| < 1$, it is clear that, *no matter what the vector* $\mathbf{v}$ *may be initially*, $\lim A^n\mathbf{v} = \mathbf{v}_1$.

To diagonalize $A$ explicitly, we write $A = B\Lambda B^{-1}$ where

$$B = \begin{pmatrix} b/(b+c) & -1 \\ c/(b+c) & 1 \end{pmatrix}, \quad B^{-1} = \begin{pmatrix} 1 & 1 \\ -c/(b+c) & b/(b+c) \end{pmatrix} \text{ and } \Lambda = \begin{pmatrix} 1 & 0 \\ 0 & \lambda_2 \end{pmatrix}.$$

Then

$$A^n = B\Lambda^n B^{-1}.$$

Since

$$\lim_{n\to\infty} \Lambda^n = \begin{pmatrix} 1 & 0 \\ 0 & 0 \end{pmatrix},$$

we find

$$\begin{aligned} \lim_{n\to\infty} A^n &= \begin{pmatrix} b/(b+c) & -1 \\ c/(b+c) & 1 \end{pmatrix}\begin{pmatrix} 1 & 0 \\ 0 & 0 \end{pmatrix}\begin{pmatrix} 1 & 1 \\ -c/(b+c) & b/(b+c) \end{pmatrix} \\ &= \begin{pmatrix} b/(b+c) & -1 \\ c/(b+c) & 1 \end{pmatrix}\begin{pmatrix} 1 & 1 \\ 0 & 0 \end{pmatrix} \\ &= \frac{1}{b+c}\begin{pmatrix} b & b \\ c & c \end{pmatrix}. \end{aligned}$$

Thus if $\mathbf{v} = \begin{pmatrix} p \\ q \end{pmatrix}$ with $p + q = 1$,

$$\lim_{n\to\infty} A^n\begin{pmatrix} p \\ q \end{pmatrix} = \begin{pmatrix} b/(b+c) \\ c/(b+c) \end{pmatrix}.$$

To summarize: if $A$ is a stochastic $2 \times 2$ matrix, that is $A = \begin{pmatrix} a & b \\ c & d \end{pmatrix}$ with $a \geqslant 0$, $b \geqslant 0$, $c \geqslant 0$, $d \geqslant 0$, $a + c = 1$, $b + d = 1$, then its eigenvalues are 1 and $1 - (b + c)$, with eigenvectors $\begin{pmatrix} b/(b+c) \\ c/(b+c) \end{pmatrix}$ and $\begin{pmatrix} -1 \\ 1 \end{pmatrix}$ respectively. If all its entries are strictly positive, then repeated action of $A$ causes the system to approach the limiting state $\begin{pmatrix} b/(b+c) \\ c/(b+c) \end{pmatrix}$. The 'discrepancy' between the current state and this limiting state is multiplied by $\lambda_2$ at each step. In the badminton example, with $A = \begin{pmatrix} 0.8 & 0.3 \\ 0.2 & 0.7 \end{pmatrix}$, $\lambda_2 = 0.5$, and the limiting state is $\begin{pmatrix} 0.6 \\ 0.4 \end{pmatrix}$. On wandering into a game after many points have been played, we expect to find player 1 serving 60% of the time.

The matrix $A = \begin{pmatrix} 0 & 1 \\ 1 & 0 \end{pmatrix}$ is a stochastic matrix which does not satisfy the strict positivity conditions. It is clear that

$$A^n = \begin{cases} I \text{ if } n \text{ is even,} \\ A \text{ if } n \text{ is odd.} \end{cases}$$

The meaning of the matrix $A$ is obvious. It represents a sure transition to the other state. There is no limit as $n \to \infty$. (Yet, in a certain average sense, we expect to find each state occupied about half the time.)

It is a straightforward matter to represent Markov processes for systems with more than two states by larger matrices – a three-state process by a $3 \times 3$ matrix, and so on. The entries in each column are non-negative and sum to unity. A typical $3 \times 3$ stochastic matrix is

$$A = \begin{pmatrix} 0.5 & 0 & 0.1 \\ 0.3 & 0.6 & 0 \\ 0.2 & 0.4 & 0.9 \end{pmatrix}.$$

The important features of the $2 \times 2$ case persist, with some differences. For instance

$$\begin{pmatrix} 0.5 & 0.3 & 0 & 0 \\ 0.5 & 0.7 & 0 & 0 \\ 0 & 0 & 0 & 1 \\ 0 & 0 & 1 & 0 \end{pmatrix}$$

represents a system in which it is impossible to get from the first two states to the last two and vice versa. A probability vector concentrated in the first two states will tend to a limit. A vector concentrated in the last two states (i.e., with first two components zero) will move around and its value will depend on whether $n$ is even or odd. It is not difficult to characterize when this kind of phenomenon can occur in terms of the matrix entries of $A$. With the exception of such cases, the $n$-dimensional case is the same as the two-dimensional one – the matrix has an eigenvalue of 1, with an associated eigenvector describing a limiting state, the other eigenvalues are all less than one, and $\lim_{n \to \infty} A^n$ is a singular matrix which transforms any probability vector into the eigenvector corresponding to $\lambda = 1$.

## Summary

A **Conformal matrices**

You should be able to identify a conformal matrix and describe in geometric terms the transformation that it represents.

You should be able to state and apply the isomorphism between conformal matrices and complex numbers.

B **Eigenvalues and eigenvectors**

You should be able to form the characteristic equation of a $2 \times 2$ matrix and use it to determine the eigenvalues of the matrix.

You should be able to determine eigenvectors corresponding to real eigenvalues of a $2 \times 2$ matrix and discribe the action of the matrix in terms of its eigenvectors and eigen-values.

C Similarity of matrices

Given a $2 \times 2$ matrix $A$, you should be able to construct a matrix $B$ so that $A = BCB^{-1}$, where $C$ is diagonal if $A$ has distinct real eigenvalues, $C$ is conformal if $A$ has complex eigenvalues, and $C$ is of the form $\begin{pmatrix} \lambda & 1 \\ 0 & \lambda \end{pmatrix}$ if $A$ has a repeated eigenvalue, but $A \neq \lambda I$. In each case you should be able to interpret the columns of $B$ geometrically.

D Markov processes

You should be able to write down the $n \times n$ matrix that represents a Markov process with $n$ states.

For a $2 \times 2$ matrix $A$ that represents a Markov process, you should be able to relate the eigenvalues and eigenvectors of $A$ to the behavior of the probabilities of the two states of the process.

---

## Exercises

2.1 Consider the conformal matrices

$$F_1 = \begin{pmatrix} 3 & -4 \\ 4 & 3 \end{pmatrix} \quad \text{and} \quad F_2 = \begin{pmatrix} 4 & 3 \\ -3 & 4 \end{pmatrix}.$$

(a) Write the complex numbers $z_1$ and $z_2$ which correspond to these two matrices.

(b) Express $F_1$ and $F_2$ each as the product of a multiple of the identity matrix and a rotation. Using the identity $e^{i\theta} = \cos\theta + i\sin\theta$, express $z_1$ and $z_2$ in 'polar form' $z = re^{i\theta}$.

(c) Calculate $F_1^{-1}$. Calculate $z_1^{-1}$, rationalizing the denominator. Compare.

(d) Calculate $F_1F_2$ and $F_2F_1$. Calculate $z_1z_2$ and compare.

2.2 Explicitly verify DeMoivre's theorem for the conformal matrices $F_1$ and $F_2$ of exercise 2.1; that is calculate $F_1^3$ and $F_2^2$.

2.3(a) Show that $R = \begin{pmatrix} 0.8 & -0.6 \\ 0.6 & 0.8 \end{pmatrix}$ represents a counterclockwise rotation through an angle of about 37°. Calculate $R^{-1}$.

(b) $S = \begin{pmatrix} 1 & 2 \\ 0 & 1 \end{pmatrix}$ represents a shear along the $+x$-axis. Calculate $S^{-1}$ and interpret it geometrically.

(c) Calculate $A = RSR^{-1}$ and interpret it geometrically. Do the same for $A^{-1}$, for $B = RS^{-1}R^{-1}$, and for $B^{-1}$.

2.4 Apply the diagonalization procedure to $F = \begin{pmatrix} -7 & 18 \\ -3 & 8 \end{pmatrix}$, as follows:

(a) Form the characteristic polynomial $P(\lambda)$ and set it equal to zero to find the eigenvalues of $F$. (Answer: $\lambda = 2, \lambda = -1$.)

(b) Check that $P(F)=0$, as promised by the Cayley–Hamilton theorem.
(c) Find an eigenvector for each eigenvalue. Let $y=1$ in each eigenvector.
(d) Form the matrices $B$ and $B^{-1}$, and confirm that $F=B\Lambda B^{-1}$.

2.5 Diagonalize the 'Lorentz transformation' matrix $L_2=\begin{pmatrix}\frac{5}{4} & \frac{3}{4}\\ \frac{3}{4} & \frac{5}{4}\end{pmatrix}$, expressing it as $L_2=B\Lambda B^{-1}$, where $B$ is a rotation and $\Lambda$ is diagonal. Interpret the result geometrically.

2.6 Find an invertible matrix $B$ and a diagonal matrix $D$ such that $B\begin{pmatrix}4 & -3\\ -1 & 2\end{pmatrix}B^{-1}=D$.

2.7 Diagonalize the matrix $F=\begin{pmatrix}-1 & 9\\ -1 & 5\end{pmatrix}$, which has a repeated eigenvalue, by the following procedure:

(a) Form the characteristic polynomial $P(\lambda)$ and find the eigenvalues.
(b) Find an eigenvector of $F$ of the form $\begin{pmatrix}x_1\\ 1\end{pmatrix}$.
(c) Form the matrix $G=F-\lambda I$. Show that the Cayley–Hamilton theorem implies that $G^2=0$, and confirm this explicitly. Find the image and kernel of $G$.
(d) Find a vector $\begin{pmatrix}x_2\\ 1\end{pmatrix}$ with the property that $G\begin{pmatrix}x_2\\ 1\end{pmatrix}=\begin{pmatrix}x_1\\ 1\end{pmatrix}$. Now form the matrices $B$ and $B^{-1}$ and check that you have succeeded in writing $F$ in the form $F=B\begin{pmatrix}\lambda & 1\\ 0 & \lambda\end{pmatrix}B^{-1}$.

2.8 Apply the 'diagonalization' procedure to the matrix

$$F=\begin{pmatrix}\frac{3}{5} & \frac{1}{5}\\ -\frac{4}{5} & \frac{7}{5}\end{pmatrix},$$

which has a repeated eigenvalue. Find the image and kernel of $G=F-\lambda I$, and describe geometrically the transformation represented by $F$.

2.9 Let $A$ be a $2\times 2$ matrix with eigenvalues $\lambda_1>\lambda_2>0$.

(a) Describe a procedure for calculating the matrix $G_n=\lambda_1^{-n}A^n$ easily by diagonalizing $A$. Show that the matrix $F=\lim_{n\to\infty}G_n$ is singular.
(b) Carry through this procedure for the matrix $A=\begin{pmatrix}3 & -2\\ 1 & 0\end{pmatrix}$, calculating $G_n$ and $F$ explicitly. Find the eigenvalues and eigenvectors of $A$, and find the image and kernel of the transformation $F$, and relate them to the eigenvectors of $A$.

2.10 For any matrix $A$, the trace of the matrix, $\operatorname{Tr} A$, is defined as the sum of the entries on the principal diagonal. Thus, if $A=\begin{pmatrix}a & b\\ c & d\end{pmatrix}$, $\operatorname{Tr} A=a+d$.

(a) Prove that if $A$ and $B$ are two $2\times 2$ matrices, $\operatorname{Tr}(AB)=\operatorname{Tr}(BA)$ even if $A$ and $B$ do not commute.
(b) Prove that $\operatorname{Tr} A$ equals the sum of the eigenvalues of $A$. Conclude that if $A=SBS^{-1}$, then $\operatorname{Tr} A=\operatorname{Tr} B$.
(c) Using the result of (a), prove that $\operatorname{Tr}(ABC)=\operatorname{Tr}(BCA)=\operatorname{Tr}(CAB)$.

2.11 Express the matrix $F = \begin{pmatrix} 3 & -1 \\ 5 & -1 \end{pmatrix}$, whose eigenvalues are complex, in the form $F = BCB^{-1}$ by the following procedure.

(a) Find the eigenvalues of $F$.

(b) Construct a conformal matrix $C$ with the same eigenvalues as $F$.

(c) Construct $B$ in the form $\begin{pmatrix} 1 & c \\ 0 & d \end{pmatrix}$.

2.12 Let $A = \begin{pmatrix} 4 & -5 \\ 1 & 0 \end{pmatrix}$. Find a conformal matrix $C$, and a matrix $S$ that represents a shear transformation, such that $A = SCS^{-1}$.

2.13 Let $F = \begin{pmatrix} a & b \\ c & d \end{pmatrix}$ be a matrix with real distinct eigenvalues $\lambda_1$ and $\lambda_2$. Let $x = \frac{1}{2}(\lambda_1 + \lambda_2)$, $y = \frac{1}{2}(\lambda_1 - \lambda_2)$.

(a) Show that $H = F - xI$ obeys the equation $H^2 = y^2 I$.

(b) Show that $S = \begin{pmatrix} x & y \\ y & x \end{pmatrix}$ has the same eigenvalues as $F$.

(c) Devise a procedure for constructing a matrix $B$, whose first column is $\begin{pmatrix} 1 \\ 0 \end{pmatrix}$, such that $H = B\begin{pmatrix} 0 & y \\ y & 0 \end{pmatrix}B^{-1}$ and $F = B\begin{pmatrix} x & y \\ y & x \end{pmatrix}B^{-1}$.

(d) Find a matrix $R$ such that

$$\begin{pmatrix} x & y \\ y & x \end{pmatrix} = R\begin{pmatrix} \lambda_1 & 0 \\ 0 & \lambda_2 \end{pmatrix}R^{-1}.$$

Prove that $BR\begin{pmatrix} 1 \\ 0 \end{pmatrix}$ and $BR\begin{pmatrix} 0 \\ 1 \end{pmatrix}$ are eigenvectors of $F$.

2.14 Let $F$ be a $2 \times 2$ matrix with distinct real eigenvalues $\lambda_1$ and $\lambda_2$. Define

$$P_1 = \frac{F - \lambda_2 I}{\lambda_1 - \lambda_2}, \quad P_2 = \frac{F - \lambda_1 I}{\lambda_2 - \lambda_1}.$$

Prove the following properties of $P_1$ and $P_2$:

(a) $P_1$ and $P_2$ are projections: $P_1^2 = P_1$, $P_2^2 = P_2$.

(b) $P_1 P_2 = P_2 P_1 = 0$.

(c) $F = \lambda_1 P_1 + \lambda_2 P_2$.

(d) $F^n = \lambda_1^n P_1 + \lambda_2^n P_2$.

(e) Calculate $P_1$ and $P_2$ explicitly for the case $F = \begin{pmatrix} 3 & 4 \\ -1 & -2 \end{pmatrix}$, and use the result to calculate $F^7$.

2.15 Let $F$ be a $2 \times 2$ matrix whose characteristic equation has roots $\lambda = x \pm iy$. We can alternatively write $x \pm iy = re^{\pm i\theta}$, where $r = \sqrt{(x^2 + y^2)}$ and $e^{\pm i\theta} = \cos\theta \pm i \sin\theta$. If $F$ is a conformal matrix, it rotates the plane through angle $\theta$ and stretches it uniformly by a factor of $r$. This problem explores the case where $F$ is not necessarily conformal.

(a) Show that $F^n$ is a multiple of the identity for integer $n$ if and only if $n\theta = m\pi$ for integer $m$, and that in this case

$$F^n = (-1)^m r^n I.$$

Hint: $F = BCB^{-1}$, where $C$ is conformal.

(b) Write $F = \begin{pmatrix} 3 & 7 \\ -1 & -1 \end{pmatrix}$ in the form $BCB^{-1}$, and thereby find the smallest integer $n$ for which $F^n$ is a multiple of the identity. Check your answer by direct multiplication.

(c) Show that your answer to (b) follows from the Cayley–Hamilton theorem.

(d) Find a 'square root' of $G = \begin{pmatrix} -2 & -15 \\ 3 & 10 \end{pmatrix}$, i.e., find a matrix $A$ such that $A^2 = G$. Reminder:

$$\cos^2 \tfrac{1}{2}\theta = \tfrac{1}{2}(1 + \cos\theta), \quad \sin^2 \tfrac{1}{2}\theta = \tfrac{1}{2}(1 - \cos\theta)$$

2.16 Modernistic composer Allie A. Tory constructs his two-tone works by the following Markov process:

1. If note $N-1$ was an F, then the probability $p_N$ that note $N$ is an F is $\frac{3}{4}$, while the probability $q_N$ that note $N$ is a G is $\frac{1}{4}$.
2. If note $N-1$ was a G, then the probability $p_N$ that note $N$ is an F is $\frac{1}{2}$, while the probability $q_N$ that note $N$ is a G is $\frac{1}{2}$.

(a) Construct a $2 \times 2$ matrix $A$ which transforms the probabilities

$$\begin{pmatrix} p_{N-1} \\ q_{N-1} \end{pmatrix}$$

into the probabilities

$$\begin{pmatrix} p_N \\ q_N \end{pmatrix}.$$

(b) Suppose that note 1 is an F. Use the matrix $A$ to find the probability that note 3 is an F.

(c) Determine the eigenvalues of $A$ and find an eigenvector of $A$ corresponding to each eigenvalue.

(d) Suppose that note 1 is an F, so that

$$\begin{pmatrix} p_1 \\ q_1 \end{pmatrix} = \begin{pmatrix} 1 \\ 0 \end{pmatrix}.$$

Show on a diagram the sequence of vectors

$$\begin{pmatrix} p_2 \\ q_2 \end{pmatrix}, \begin{pmatrix} p_3 \\ q_3 \end{pmatrix}, \ldots.$$

Determine the limit of this sequence, and interpret it in terms of the eigenvectors of $A$.

2.17 The quarterback of the Houston Eulers, who majored in probability theory in college, has devised a play-calling procedure with the following properties:

1. If play $N-1$ was a pass, then the probability $p_N$ that play $N$ is a pass is $\frac{1}{6}$, while the probability $q_N$ that play $N$ is a run is $\frac{5}{6}$.
2. If play $N-1$ was a run, then the probability $p_N$ that play $N$ is a pass is $p_N = \frac{2}{3}$, while the probability $q_N$ that play $N$ is a run is $q_N = \frac{1}{3}$.

(a) Construct the $2 \times 2$ matrix $A$ which transforms the probabilities

$$\begin{pmatrix} p_{N-1} \\ q_{N-1} \end{pmatrix}$$

into the probabilities

$$\begin{pmatrix} p_N \\ q_N \end{pmatrix}.$$

(b) Determine the eigenvalues of $A$ and find an eigenvector of $A$ corresponding to each eigenvalue. Illustrate on a diagram the action of $A$ on each eigenvector.

(c) No one knows how the quarterback decides what to do for play 1, but observation of game films shows that play 2 is a pass half the time, a run half the time. What are the probabilities

$$\begin{pmatrix} p_1 \\ q_1 \end{pmatrix}$$

for the first play?

2.18 Professor Constantine Bayes has been teaching his course 'Stochastic Methods in Classical Archaeology' for decades. It is widely known that Bayes selects examination questions by drawing colored balls from ancient Greek urns which he keeps in his office, but the contents of the urns are secret. However, by analyzing the pattern of Bayes' final examinations, which are on file in Lamont Library, students have learned the following:

1. If the final examination in year $N-1$ had a question on *statues*, the final examination in year $N$ will have a question on statues half the time, a question on pottery half the time.
2. If the final examination in year $N-1$ had a question on pottery, the final examination in year $N$ will have a question on statues $\frac{1}{4}$ of the time, a question on pottery $\frac{3}{4}$ of the time.

(a) Write the matrix $M$ which transforms the probabilities $\begin{pmatrix} p_{N-1} \\ q_{N-1} \end{pmatrix}$, for a statue question or pottery question respectively, into the probabilities $\begin{pmatrix} p_N \\ q_N \end{pmatrix}$ for the next year's final examination.

(b) Find the eigenvectors and eigenvalues of $M$, and write $M$ in the form $SDS^{-1}$, where $D$ is diagonal.

(c) By attending Bayes' office hours regularly, a student has finally learned details of his method. Bayes has two urns, but he uses them once for the hour examination, then once again for the final examination, so that the matrix $M$ represents *two* steps of a Markov process! By using your diagonalization of $M$, find the two possible matrices $N$ for *one* step of the process.

2.19 John and Eli are playing a game with a ball that can roll into one of two pockets labelled $H$ and $Y$. John wants to keep the ball in $H$ and Eli wants to keep it in $Y$. When it is John's turn to play, if he finds the ball in $H$ that is fine with him and he does nothing; but if he finds it in $Y$ he attempts to roll it into pocket $H$. This takes some skill; the probability that he succeeds is $\frac{2}{3}$, there being a $\frac{1}{3}$ chance that the ball will roll back into $Y$. When Eli's turn comes, he does nothing if the ball is in $Y$, but tries to get it there if he finds it in $H$. Eli is less skillful than John and his probability of succeeding in his effort is only $\frac{1}{2}$.

(a) Starting with the ball in $Y$ and John to play, what is the probability that the ball will be in $H$ after John's *second* play?

(b) Find a formula for the probability that the ball is in $H$ after John's $n$th play (i.e., after John has played $n$ times and Eli $(n-1)$ times).

(c) Suppose the game has been going on for a 'long time' and you look in just after Eli has played. What is the probability that the ball is now in $H$? How many turns constitutes a 'long time' if we want to be certain that this probability is correct within 0.001?

2.20 A bank has instituted a policy to prevent the tellers' lines from ever getting more than two persons long. If a third person arrives, all three customers are escorted into the manager's office to receive high-level personal service, and the teller starts again with no line. Furthermore, an armed guard at the entrance to the bank assures that no more than one customer per minute can enter (it takes that long for a really thorough search). As a result, the length of a teller's line is determined by the following Markov process, which describes what happens in a one minute interval of time.

1. If the line has zero customers, the probability is $\frac{1}{2}$ that one customer arrives, $\frac{1}{2}$ that no one arrives.
2. If the line has one customer, the probability is $\frac{1}{6}$ that the customer is served and leaves, $\frac{1}{3}$ that a second customer joins the line, and $\frac{1}{2}$ that nothing happens.
3. If the line has two customers, the probability is $\frac{1}{6}$ that one customer is served, $\frac{1}{6}$ that a third customer arrives and all three are taken to the manager, leaving no line, and $\frac{2}{3}$ that nothing happens.

(a) Construct the matrix $M$ which carries the probabilities $\begin{pmatrix} p_1 \\ p_2 \\ p_3 \end{pmatrix}$ for time $t$ into the probabilities for time $t+1$.

(b) At 9 am, when the bank opens, $p_1 = 1$. What are the probabilities $\begin{pmatrix} p_1 \\ p_2 \\ p_3 \end{pmatrix}$ at 9:03 am?

(c) Find the eigenvalues and eigenvectors of $M$.

(d) What is the limiting value of $\begin{pmatrix} p_1 \\ p_2 \\ p_3 \end{pmatrix}$ after the bank has been open for a long time? Estimate at what time the probabilities $p_1$, $p_2$, and $p_3$ will all be within 0.001 of these limiting values.

(e) On the average, how many customers per minute are served? How many are taken to the manager's office?

**The Ehrenfest model.** Suppose we have two boxes and $N$ balls. There can be $i$ balls in the first box and $N-i$ balls in the second. The state of the system is given by the integer $i$. So there are $n+1$ states: $i=0$, $i=1,\ldots,i=N$. At each instant of time, one of the $N$ balls is picked at random (i.e., with probability $1/N$) and moved from the box it is in to the other box. Thus $i$ can change to either $i-1$ or $i+1$ according as the ball picked was in the first or second box. The probabilities of these transitions are $iN^{-1}$ and $(N-i)N^{-1}$ respectively. Thus

$$\begin{aligned} p_{i-1,i} &= iN^{-1} \\ p_{i+1,1} &= (N-i)N^{-1} \\ p_{j,i} &= 0, \quad j \neq i-1 \text{ or } i+1. \end{aligned}$$

For example, if $N=4$, the $5\times 5$ transition matrix is

$$P = \begin{pmatrix} 0 & \frac{1}{4} & 0 & 0 & 0 \\ 1 & 0 & \frac{1}{2} & 0 & 0 \\ 0 & \frac{3}{4} & 0 & \frac{3}{4} & 0 \\ 0 & 0 & \frac{1}{2} & 0 & 1 \\ 0 & 0 & 0 & \frac{1}{4} & 0 \end{pmatrix}.$$

Notice that $P$ transforms any state with $i$ even into a state with $i$ odd and vice versa. Thus $P^2$ transforms even states into even states and odd states into odd states: $P^2$ has the form

$$\begin{aligned} p_{i-2,i} &= i(i-1)N^{-2} \\ p_{i,i} &= [i(N-i+1)+(N-i)(i+1)]N^{-2} \\ p_{i+2,i} &= (N-i)(N-i-1)N^{-2} \\ p_{ji} &= 0 \quad \text{if} \quad j \neq i-2, i, i+2. \end{aligned}$$

Thus, for $N=4$, squaring the preceding matrix gives

$$P^2 = \begin{pmatrix} \frac{1}{4} & 0 & \frac{1}{8} & 0 & 0 \\ 0 & \frac{5}{8} & 0 & \frac{3}{8} & 0 \\ \frac{3}{4} & 0 & \frac{3}{4} & 0 & \frac{3}{4} \\ 0 & \frac{3}{8} & 0 & \frac{5}{8} & 0 \\ 0 & 0 & \frac{1}{8} & 0 & \frac{1}{4} \end{pmatrix}.$$

Since transitions for $P^2$ are only between states of the same parity, we may as well consider the states $i=0,2,4$ and $i=1,3,5,\ldots$ separately. (In the above matrix, this means combining separately the matrices obtained by considering only the even–even positions and the odd–odd positions:

$$Q = \begin{pmatrix} \frac{1}{4} & \frac{1}{8} & 0 \\ \frac{3}{4} & \frac{3}{4} & \frac{1}{4} \\ 0 & \frac{1}{8} & \frac{3}{4} \end{pmatrix} \quad \text{and} \quad R = \begin{pmatrix} \frac{5}{8} & \frac{3}{8} \\ \frac{3}{8} & \frac{5}{8} \end{pmatrix}.$$

2.21.(a) For the matrix $Q$ show that $\begin{pmatrix} 1 \\ 6 \\ 1 \end{pmatrix}$ is an eigenvector with eigenvalue 1 and

for $R$ that $\begin{pmatrix} 1 \\ 1 \end{pmatrix}$ is an eigenvector with eigenvalue 1.

(b) Do the same computations for $N = 5$: Show that the 'even' eigenvector with eigenvalue 1 is proportional to $\begin{pmatrix}1\\10\\5\end{pmatrix}$ and the 'odd' one is proportional to $\begin{pmatrix}5\\10\\1\end{pmatrix}$.

(c) Prove in the general case that the 'even' and 'odd' eigenvectors with eigenvalue 1 are proportional to the vector whose entries are the even or odd binomial coefficients. In other words,

$$\underset{\text{even eigenvector}}{\begin{pmatrix}\binom{N}{0}\\ \binom{N}{2}\\ \binom{N}{4}\\ \vdots\end{pmatrix}} \quad \text{and} \quad \underset{\text{odd eigenvector.}}{\begin{pmatrix}\binom{N}{1}\\ \binom{N}{3}\\ \binom{N}{5}\\ \vdots\end{pmatrix}}$$

2.22. Fibonacci numbers.

The sequence $0, 1, 1, 2, 3, 5, 8, 13, \ldots$ is called the Fibonacci sequence. These numbers appear in the study of many interesting physical and mathematical problems ranging from plant growth to celestial mechanics (see, for example, D'Arcy Thompson's *On Growth and Form*). The recursion which generates the sequence is

$$x_{n+2} = x_{n+1} + x_n \quad x_0 = 0$$

(a) Compute the ratio $x_{n+1}/x_n$ for $n = 1$ up to 8 or so. Do you think this sequence has a limit?

(b) Find a matrix such that

$$\begin{pmatrix}x_{n+2}\\ x_{n+1}\end{pmatrix} = A\begin{pmatrix}x_{n+1}\\ x_n\end{pmatrix}.$$

Use $A$ to express $\begin{pmatrix}x_{n+2}\\ x_{n+1}\end{pmatrix}$ in terms of $\begin{pmatrix}x_1\\ x_0\end{pmatrix}$.

(c) Find an explicit expression for $x_n$ in terms of $x_1$ and $x_0$. (Hint: Diagonalize $A$.)

(d) Show that $\lim_{n\to\infty} x_{n+1}/x_n$ exists and compute its value.

(e) What does (d) tell you about the infinite continued fraction

$$1 + \cfrac{1}{1 + \cfrac{1}{1 + \cfrac{1}{1 + \cdots}}}$$

(f) Are there any values of $x_0$ and $x_1$ such that $\lim_{n\to\infty} x_{n+1}/x_n$ differs from the value obtained in (d)?

# 3

# Linear differential equations in the plane



The principal goal of Chapter 3 is to explain that a system of homogeneous linear differential equations with constant coefficients can be written as $d\mathbf{u}/dt = A\mathbf{u}$ where $A$ is a matrix and $\mathbf{u}$ is a vector, and that the solution can be written as $e^{At}\mathbf{u}_0$ where $\mathbf{u}_0$ gives the initial conditions. This of course requires us to explain what is meant by the exponential of a matrix. We also describe the qualitative behavior of solutions and the inhomogeneous case, including a discussion of resonance.

## 3.1. Functions of matrices

We have already encountered (in our discussion of the Cayley–Hamilton theorem) a 'polynomial in a matrix'. More generally, let $Q(X)$ be any polynomial. If

$$Q(X) = a_n X^n + a_{n-1} X^{n-1} + \cdots + a_1 X + a_0$$

then we define the matrix $Q(F)$ by

$$Q(F) = a_n F^n + a_{n-1} F^{n-1} + \cdots + a_1 F + a_0.$$

Now we can multiply two polynomials $(Q_1 Q_2)(X) = Q_1(X) Q_2(X)$ to obtain a third. Similarly

$$Q_1 Q_2(F) = Q_1(F) Q_2(F).$$

There is no problem with the fact that in general matrix multiplication is not commutative, since powers of a fixed matrix always commute with one another.

$$F^k F^l = F^{k+l} = F^l F^k$$

on account of the associative law. Similarly,

$$(Q_1 + Q_2)(F) = Q_1(F) + Q_2(F).$$

In short, there is no trouble in evaluating a polynomial function at a fixed matrix, and the usual algebraic laws are satisfied. We would like to consider some more general functions of matrices, and for this we need a slight digression about power series.

An expression of the form

$$R(X) = a_0 + a_1X + a_2X^2 + \cdots + a_nX^n + \cdots$$

where the $a_i$, $i = 0, 1, \ldots,$ are real numbers and $X$ is a symbol (as is $X^k$ for all $k$) is called a *formal power series*. We add two power series according to the rule

$$\begin{aligned}(a_0 + a_1X + a_2X^2 + \cdots) + (b_0 + b_1X + b_2X^2 + \cdots)\\ = (a_0 + b_0) + (a_1 + b_1)X + (a_2 + b_2)X^2 + \cdots,\end{aligned}$$

that is, we add the coefficients term by term. We multiply two power series by using the rule $X^k \cdot X^l = X^{k+l}$ and collecting coefficients:

$$\begin{aligned}(a_0 + a_1X + a_2X^2 + \cdots)(b_0 + b_1X + b_2X^2 + \cdots)\\ = a_0b_0 + (a_1b_0 + a_0b_1)X + (a_2b_0 + a_1b_1 + a_0b_2)X^2 + \cdots .\end{aligned}$$

Thus, for instance,

$$(1 + X + X^2 + \cdots)\cdot(1 + X + X^2 + \cdots) = 1 + 2X + 3X^2 + \cdots .$$

It is easy to check that all the usual rules for addition and multiplication of polynomials hold equally well for formal power series.

Let $t$ be any real number. We define the formal power series $\exp(tX)$ by

$$\exp(tX) = 1 + tX + \frac{1}{2!}t^2X^2 + \frac{1}{3!}t^3X^3 + \frac{1}{4!}t^4X^4 + \cdots . \tag{3.1}$$

Then

$$\begin{aligned}\exp(sX)\exp(tX) &= \left(1 + sX + \frac{1}{2!}s^2X^2 + \cdots\right)\left(1 + tX + \frac{1}{2!}t^2X^2 + \cdots\right)\\ &= 1 + (s + t)X + \frac{1}{2!}(s^2 + 2st + t^2)X^2 + \cdots\end{aligned}$$

where, on the right, the coefficient of $X^n$ is

$$\frac{1}{n!}\left(s^n + ns^{n-1}t + n\frac{(n-1)}{2}s^{n-2}t^2 + \cdots + t^n\right)$$

which, by the binomial theorem, is just $(1/n!)(s + t)^n$. Since

$$1 + (s + t)X + \frac{1}{2!}(s + t)^2X^2 + \frac{1}{3!}(s + t)^3X^3 + \cdots = \exp(s + t)X$$

by definition, we conclude that

$$\exp(sX)\exp(tX) = \exp(s + t)X \tag{3.2}$$

as an identity in formal power series.

In contrast to polynomials we cannot, in general, 'evaluate' a formal power series, $R(X)$ at a number, $r$, or at a matrix $F$. That is, if we try to substitute the real number $r$ for the symbol $X$ in

$$R(X) = a_0 + a_1X + a_2X^2 + \cdots$$

we get an 'infinite sum' of numbers

$$a_0 + a_1 r + a_2 r^2 + \cdots$$

which, as it stands, makes no sense. One way of trying to make sense of such an infinite sum is to chop off the end at some finite value, so as to get a finite sum and to hope that place where we chop it off makes little difference – provided that we go out far enough. We would then assign to $R(r)$ the value obtained as the 'limiting value' of the finite sum. Let us explain this procedure more precisely. For any integer $M$, define $R^M(r)$ to be the finite sum

$$R^M(r) = a_0 + a_1 r + a_2 r^2 + \cdots + a_M r^M.$$

We say that the power series $R(X)$ *converges* at the number $r$ if, for any positive number $\varepsilon$, no matter how small, we can find some large enough $M_0$ so that for any integers $M$ and $N > M_0$ we have

$$|R^M(r) - R^N(r)| < \varepsilon.$$

In other words, if we go far enough out, all the values $R^M(r)$ lie in some interval of length $\varepsilon$. Thus the further out we go, the closer the $R^M(r)$ cluster about some limiting value, and this limiting value is what we call $R(r)$.

We can now make essentially the same definition for matrices. For any matrix $F$ the expression

$$R^M(F) = a_0 + a_1 F + \cdots + a_M F^M$$

makes perfectly good sense. The difference

$$R^M(F) - R^N(F)$$

is again a matrix, and we shall take the condition

$$|R^M(F) - R^N(F)| < \varepsilon$$

to mean that each of the four entries of the matrix $R^M(F) - R^N(F)$ has absolute value less than $\varepsilon$. We say that $R(X)$ converges at $F$, if for any $\varepsilon > 0$ there is an $M_0$ such that $|R^M(F) - R^N(F)| < \varepsilon$ for $M$ and $N > M_0$. When this happens each of the entries of $R^M(F)$ clusters about some limiting value as we go out far enough. We thus get a matrix of limiting values and this limiting matrix is denoted by $R(F)$.

It is clear that if $R_1(X)$, $R_2(X)$ and $R_3(X)$ are formal power series such that

$$R_1(X)R_2(X) = R_3(X)$$

and *if* all three of these series converge at $F$ then

$$R_1(F)R_2(F) = R_3(F)$$

since we can replace each $R_1(X)$ by a finite approximation. Similarly for addition: if $R_1(X) + R_2(X) = R_3(X)$ and the series all converge at $F$, then $R_1(F) + R_2(F) = R_3(F)$.

## 3.2. The exponential of a matrix

We have a formal power series (3.1) for the exponential function and we know the identity (3.2). We will now prove that the power series for $\exp(tX)$ converges when we substitute a $2 \times 2$ matrix $A$ for the symbol $X$. As a first step, we review the proof that the series converges absolutely when we substitute any real number $k$ for $X$. We consider the power series

$$\exp y = 1 + y + \frac{1}{2!}y^2 + \frac{1}{3!}y^3 + \cdots \tag{3.3}$$

where $y = |tk|$. By making $y$ a positive number, we ensure that every term of the series is positive and thereby guarantee *absolute* convergence of $\exp tk$, once we prove that (3.3) converges.

To show that the series for $\exp y$ converges, we must demonstrate that the 'remainders' $r_{m,n}$ formed by summing terms from the $m$th to the $(m+n)$th term become and remain as small as we like when we choose a sufficiently large value of $m$. This is easily shown by comparing the remainder series

$$r_{m,n} = \frac{1}{m!}y^m + \frac{1}{m!(m+1)}y^{m+1} + \frac{1}{m!(m+1)(m+2)}y^{m+2} + \cdots + \frac{1}{m!(m+1)\ldots(m+n)}y^{m+n}$$

with the geometric series

$$s_{m,n} = \frac{1}{m!}y^m + \frac{1}{m!m}y^{m+1} + \frac{1}{m!m^2}y^{m+2} + \cdots + \frac{1}{m!}\frac{1}{m^n}y^{m+n}.$$

Clearly, $r_{m,n} \leqslant s_{m,n}$ for all $m$. But we can sum $s_{m,n}$ explicitly:

$$s_{m,n} = \frac{y^m}{m!}\left(1 + \frac{y}{m} + \left(\frac{y}{m}\right)^2 + \left(\frac{y}{m}\right)^3 + \cdots + \left(\frac{y}{m}\right)^n\right)$$

or

$$s_{m,n} = \left(1 - \left(\frac{y}{m}\right)^{n+1}\right)s_m$$

where

$$s_m = \frac{y^m}{m!}\frac{1}{1 - y/m}.$$

Thus

$$r_{m,n} \leqslant s_m.$$

Suppose we choose $m > 2y$, so that

$$\frac{1}{y - y/m} < \frac{1}{1 - \frac{1}{2}} = 2.$$

Then $s_m < 2y^m/m!$, and, whenever we increase $m$ by 1, we multiply $s_m$ by a factor which is less than $\frac{1}{2}$. Clearly, by choosing $m$ large enough, we can make $s_m$ as small as we like, and since $r_{m,n} \leqslant s_m$, we can thus make $r_{m,n}$ as small as we like. It follows that the series

$$\exp tk = 1 + tk + \frac{1}{2!}(tk)^2 + \frac{1}{3!}(tk)^3 + \cdots$$

converges absolutely. Incidentally, the well-known 'ratio-test', in which one proves absolute convergence of a power series

$$a_0 + a_1 y + a_2 y^2 + a_3 y^3 + \cdots$$

by showing that

$$\lim_{m \to \infty} \frac{a_{m+1} y^{m+1}}{a_m y^m} < 1,$$

relies on the argument just presented.

Suppose now that $A$ is a $2 \times 2$ matrix: $A = \begin{pmatrix} a & b \\ c & d \end{pmatrix}$, in which every entry is less in magnitude than $k/2$. Each entry in $A^2 = \begin{pmatrix} a & b \\ c & d \end{pmatrix}\begin{pmatrix} a & b \\ c & d \end{pmatrix}$ is the sum of two terms, each of which is smaller in magnitude than $(k/2)^2 = k^2/4$, and thus each entry in $A^2$ is smaller than $k^2/2$. By a similar argument, each entry in $A^3$ is less than $k^3/2$, and by induction we can prove that each entry in $A^m$ has absolute value less than $k^m/2$. Thus when we sum the series

$$\exp(tA) = I + tA + \frac{t^2}{2!}A^2 + \frac{t^3}{3!}A^3 + \cdots$$

each of the summands of the four entries in the resulting matrix is less than the corresponding summand of the series

$$1 + t\frac{k}{2} + \frac{t^2}{2!}\frac{k^2}{2} + \frac{t^3}{3!}\frac{k^3}{2} + \cdots.$$

It follows that, for any real number $t$ and any $2 \times 2$ matrix $A$, the series for $\exp(tA)$ converges. In fact, a similar argument, with $k/2$ replaced by $k/n$, shows that the series converges when $A$ is an $n \times n$ matrix.

It now follows that the fundamental identity for the exponential function

$$\exp(s + t)A = \exp(sA)\exp(tA)$$

holds for matrices.

You might ask, how about a more general identity of the form

$$\exp(A + B) \overset{?}{=} (\exp A)(\exp B)$$

where $A$ and $B$ are arbitrary matrices? To see what is involved, let us expand both sides

$$\begin{aligned}\exp(A + B) &= I + A + B + \tfrac{1}{2}(A + B)^2 + \cdots \\ &= I + A + B + \tfrac{1}{2}(A^2 + AB + BA + B^2) + \cdots\end{aligned}$$

while

$$\begin{aligned}(\exp A)(\exp B) &= (I + A + \tfrac{1}{2}A^2 + \cdots)(I + B + \tfrac{1}{2}B^2 + \cdots) \\ &= I + A + B + (\tfrac{1}{2}A^2 + AB + \tfrac{1}{2}B^2) + \cdots\end{aligned}$$

where $\cdots$ denotes a sum of terms of degree higher than 2 in $A$ and $B$. If we compare the quadratic terms, we see that they are *not* equal unless

$$\tfrac{1}{2}(AB + BA) = AB,$$

i.e., unless

$$AB = BA.$$

Thus, if the matrices $A$ and $B$ do not commute, there is no reason to expect that $\exp(A + B) = (\exp A)(\exp B)$ and, in fact, it will not, in general, be true. For a concrete example, take

$$A = \begin{pmatrix} 0 & 1 \\ 0 & 0 \end{pmatrix}.$$

In this case $A^2 = 0$, and hence all the higher order terms in $\exp A$ vanish and we have the simple expression

$$\exp A = I + A = \begin{pmatrix} 1 & 1 \\ 0 & 1 \end{pmatrix}.$$

Take

$$B = \begin{pmatrix} 0 & 0 \\ 1 & 0 \end{pmatrix} \quad \text{so} \quad B^2 = 0$$

and

$$\exp B = I + B = \begin{pmatrix} 1 & 0 \\ 1 & 1 \end{pmatrix}.$$

Then

$$(\exp A)(\exp B) = \begin{pmatrix} 1 & 1 \\ 0 & 1 \end{pmatrix}\begin{pmatrix} 1 & 0 \\ 1 & 1 \end{pmatrix} = \begin{pmatrix} 2 & 1 \\ 1 & 1 \end{pmatrix}.$$

On the other hand

$$A + B = \begin{pmatrix} 0 & 1 \\ 1 & 0 \end{pmatrix}$$

so

$$(A + B)^2 = \begin{pmatrix} 1 & 0 \\ 0 & 1 \end{pmatrix}.$$

Thus

$$\begin{aligned} \exp(A + B) &= I + (A + B) + \tfrac{1}{2}I + \frac{1}{3!}(A + B) + \frac{1}{4!}I + \cdots \\ &= \left(1 + \frac{1}{2!} + \frac{1}{4!} + \cdots\right)I + \left(1 + \frac{1}{3!} + \frac{1}{5!} + \cdots\right)(A + B). \end{aligned}$$

Now

$$1 + \frac{1}{2!} + \frac{1}{4!} + \cdots = \tfrac{1}{2}(\mathrm{e} + \mathrm{e}^{-1})$$

and

$$1 + \frac{1}{3!} + \frac{1}{5!} + \frac{1}{7!} \cdots = \tfrac{1}{2}(\mathrm{e} - \mathrm{e}^{-1})$$

so

$$\exp(A+B) = \frac{1}{2}\begin{pmatrix} e+e^{-1} & e-e^{-1} \\ e-e^{-1} & e+e^{-1} \end{pmatrix} \neq (\exp A)(\exp B).$$

The reason that we *do* have

$$\exp(s+t)A = \exp(sA)\exp(tA)$$

is that the matrices $sA$ and $tA$ commute.

Having shown that $\exp(tA)$ is well-defined by its power series, we next generalize the well-known formula

$$\frac{d}{dt}[\exp(tk)] = k\exp(tk). \tag{3.4}$$

We define the derivative of $\exp(tA)$ with respect to the real number $t$ by

$$\frac{d}{dt}[\exp(tA)] = \lim_{h\to 0}\frac{1}{h}[\exp((t+h)A) - \exp(tA)].$$

(The limit on the right-hand side of this equation means that each of the matrix entries tends to a limit.) Since $\exp((t+h)A) = \exp(hA)\exp(tA)$, we have

$$\frac{d}{dt}[\exp(tA)] = \lim_{h\to 0}\frac{1}{h}[\exp(hA) - I]\exp(tA).$$

But

$$\exp(hA) - I = hA + \frac{h^2A^2}{2!} + \frac{h^3A^3}{3!} + \cdots$$

so

$$\frac{1}{h}(\exp(hA) - I) = A + \frac{hA^2}{2!} + \frac{h^2A^3}{3!} + \cdots$$

and

$$\lim_{h\to 0}\frac{1}{h}[\exp(hA) - I] = A.$$

We have thus proved that

$$\boxed{\frac{d}{dt}[\exp(tA)] = A\exp(tA),}$$

which is just like (3.4) except that the multiplication on the right is now matrix multiplication.

Now let $\mathbf{v}_0 = \begin{pmatrix} x_0 \\ y_0 \end{pmatrix}$ be a fixed vector in the plane, and consider the time-dependent vector $\mathbf{v}(t)$ defined by

$$\mathbf{v}(t) = \exp(tA)\mathbf{v}_0 \quad \text{where} \quad A = \begin{pmatrix} a & b \\ c & d \end{pmatrix}.$$

We can define $\dot{\mathbf{v}}(t)$, the time derivative of $\mathbf{v}$, by

$$\dot{\mathbf{v}}(t) = \lim_{h\to 0}\frac{1}{h}[\mathbf{v}(t+h) - \mathbf{v}(t)].$$

If we write $\mathbf{v}(t)=\begin{pmatrix}x(t)\\ y(t)\end{pmatrix}$, this simply means $\dot{\mathbf{v}}(t)=\begin{pmatrix}\dot{x}(t)\\ \dot{y}(t)\end{pmatrix}$. Since $\mathbf{v}_0$ is constant, we have

$$\dot{\mathbf{v}}(t)=\lim_{h\to 0}\frac{1}{h}[\exp((t+h)A)-\exp(tA)]\mathbf{v}_0.$$

That is,

$$\dot{\mathbf{v}}(t)=\frac{\mathrm{d}}{\mathrm{d}t}\exp(tA)\mathbf{v}_0$$

so that

$$\dot{\mathbf{v}}(t)=A\exp(tA)\mathbf{v}_0=A\mathbf{v}(t).$$

We have shown that $\mathbf{v}(t)$ satisfies the differential equation

$$\boxed{\dot{\mathbf{v}}(t)=A\mathbf{v}(t),\quad \mathbf{v}(0)=\mathbf{v}_0}$$

Writing $\mathbf{v}(t)=\begin{pmatrix}x(t)\\ y(t)\end{pmatrix}$ we see that

$$\begin{pmatrix}\dot{x}(t)\\ \dot{y}(t)\end{pmatrix}=\begin{pmatrix}a & b\\ c & d\end{pmatrix}\begin{pmatrix}x(t)\\ y(t)\end{pmatrix};$$

that is,

$$\dot{x}(t)=ax(t)+by(t),$$
$$\dot{y}(t)=cx(t)+dy(t).$$

Thus

$$\begin{pmatrix}x(t)\\ y(t)\end{pmatrix}=\exp(tA)\begin{pmatrix}x_0\\ y_0\end{pmatrix}$$

is a solution to the *system of linear ordinary differential equations* written above.

In fact, it is easy to prove that *any* solution to the differential equation $\dot{\mathbf{v}}(t)=A\mathbf{v}(t)$ is of the form $\mathbf{v}(t)=\exp(tA)\mathbf{v}_0$. Simply consider the vector $\mathbf{w}(t)=\exp(-tA)\mathbf{v}(t)$. Then

$$\dot{\mathbf{w}}(t)=\frac{\mathrm{d}}{\mathrm{d}t}(\exp(-tA))\mathbf{v}(t)+\exp(-tA)\dot{\mathbf{v}}(t).$$

But

$$\frac{\mathrm{d}}{\mathrm{d}t}\exp(-tA)=-A\exp(-tA)$$

and by hypothesis

$$\dot{\mathbf{v}}(t)=A\mathbf{v}(t).$$

so

$$\dot{\mathbf{w}}(t)=-A\exp(-tA)\mathbf{v}(t)+\exp(-tA)A\mathbf{v}(t)=\mathbf{0}$$

since the matrices $A$ and $\exp(-tA)$ commute. It follows that $\mathbf{w}$ is a constant vector (call it $\mathbf{v}_0$) and we have

$$\mathbf{v}_0=\exp(-tA)\mathbf{v}(t)$$

or

$$\mathbf{v}(t) = \exp(tA)\mathbf{v}_0.$$

We thus see that the function $\exp(tA)$ determines the *general solution* to a system of differential equations with constant coefficients. It becomes important to discuss various methods for computing $\exp(tA)$.

## 3.3. Computing the exponential of a matrix

Suppose $F$ and $G$ are matrices which are related as

$$F = BGB^{-1}.$$

Then

$$F^k = BG^kB^{-1}$$

for any $k$, and hence it follows from the power series expansion of $\exp(tA)$ that

$$\exp(tF) = B\exp(tG)B^{-1}.$$

**Case 1. $F$ has distinct real eigenvalues.** We can now make use of our ability to diagonalize $2 \times 2$ matrices. Suppose that $F$ has distinct real eigenvalues $\lambda_1$ and $\lambda_2$. Then we can write

$$F = B\begin{pmatrix} \lambda_1 & 0 \\ 0 & \lambda_2 \end{pmatrix}B^{-1}.$$

Clearly

$$\begin{pmatrix} \lambda_1 & 0 \\ 0 & \lambda_2 \end{pmatrix}^n = \begin{pmatrix} \lambda_1^n & 0 \\ 0 & \lambda_2^n \end{pmatrix}$$

and it follows from the power series definition of the exponential function that

$$\exp\begin{pmatrix} \lambda_1 t & 0 \\ 0 & \lambda_2 t \end{pmatrix} = \begin{pmatrix} e^{\lambda_1 t} & 0 \\ 0 & e^{\lambda_2 t} \end{pmatrix}$$

so that

$$\exp(tF) = B\begin{pmatrix} e^{\lambda_1 t} & 0 \\ 0 & e^{\lambda_2 t} \end{pmatrix}B^{-1}.$$

As a concrete example of this technique, take $F = \begin{pmatrix} 7 & 4 \\ -8 & -5 \end{pmatrix}$. The characteristic polynomial of this matrix is $\lambda^2 - 2\lambda - 3 = 0$, so the eigenvalues are $\lambda_1 = 3$, $\lambda_2 = -1$. Considering $(F - 3I) = \begin{pmatrix} 4 & 4 \\ -8 & -8 \end{pmatrix}$, we find eigenvectors $\mathbf{v}_1 = \begin{pmatrix} 1 \\ -1 \end{pmatrix}$, the kernel of $F - 3I$, and $\mathbf{v}_2 = \begin{pmatrix} 1 \\ -2 \end{pmatrix}$, the image of $F - 3I$. Thus

$$B = \begin{pmatrix} 1 & -1 \\ -1 & 2 \end{pmatrix}, \quad B^{-1} = \begin{pmatrix} 2 & 1 \\ 1 & 1 \end{pmatrix},$$

and

$$F=\begin{pmatrix}1 & -1\\ -1 & 2\end{pmatrix}\begin{pmatrix}3 & 0\\ 0 & -1\end{pmatrix}\begin{pmatrix}2 & 1\\ 1 & 1\end{pmatrix}.$$

It follows immediately that

$$\exp(tF)=\begin{pmatrix}1 & -1\\ -1 & 2\end{pmatrix}\begin{pmatrix}e^{3t} & 0\\ 0 & e^{-t}\end{pmatrix}\begin{pmatrix}2 & 1\\ 1 & 1\end{pmatrix}$$

and we can multiply out the matrices to obtain

$$\exp(tF)=\begin{pmatrix}2e^{3t}-e^{-t} & e^{3t}-e^{-t}\\ -2e^{3t}+2e^{-t} & -e^{3t}+2e^{-t}\end{pmatrix}$$

Given any vector $\mathbf{v}_0$ which specifies *initial conditions*, that is the value of $\mathbf{v}$ at $t=0$, we can now write down the solution to $\dot{\mathbf{v}}(t)=F\mathbf{v}(t)$. Suppose, for example, that $\mathbf{v}_0=\begin{pmatrix}1\\ 1\end{pmatrix}$. Then

$$\mathbf{v}(t)=\exp(tF)\begin{pmatrix}1\\ 1\end{pmatrix}=\begin{pmatrix}3e^{3t}-2e^{-t}\\ -3e^{3t}+4e^{-t}\end{pmatrix}.$$

Differentiating each component, we find

$$\dot{\mathbf{v}}(t)=\begin{pmatrix}9e^{3t}+2e^{-t}\\ -9e^{3t}-4e^{-t}\end{pmatrix}$$

and we confirm that

$$F\mathbf{v}(t)=\begin{pmatrix}7 & 4\\ -8 & -5\end{pmatrix}\begin{pmatrix}3e^{3t}-2e^{-t}\\ -3e^{3t}+4e^{-t}\end{pmatrix}=\begin{pmatrix}9e^{3t}+2e^{-t}\\ -9e^{3t}-4e^{-t}\end{pmatrix}\text{ also.}$$

**Case 2. Repeated eigenvalues.** The method just described works when $F$ has *distinct* real eigenvalues. Suppose, instead, that $F$ has a repeated eigenvalue $\lambda$. Then either $F=\begin{pmatrix}\lambda & 0\\ 0 & \lambda\end{pmatrix}$, in which case $\exp(tF)=\begin{pmatrix}e^{\lambda t} & 0\\ 0 & e^{\lambda t}\end{pmatrix}$, or we can write

$$F=B\begin{pmatrix}\lambda & 1\\ 0 & \lambda\end{pmatrix}B^{-1}.$$

To exponentiate $\begin{pmatrix}\lambda t & t\\ 0 & \lambda t\end{pmatrix}$ we write

$$\begin{pmatrix}\lambda t & t\\ 0 & \lambda t\end{pmatrix}=\lambda tI+\begin{pmatrix}0 & t\\ 0 & 0\end{pmatrix}$$

and make use of the fact that, if matrices $C$ and $D$ commute, then

$$\exp(C+D)=(\exp C)(\exp D).$$

Since $\lambda I$ commutes with any matrix we have

$$\exp\begin{pmatrix}\lambda t & t\\ 0 & \lambda t\end{pmatrix}=\exp(\lambda tI)\exp\begin{pmatrix}0 & t\\ 0 & 0\end{pmatrix}.$$

But $\begin{pmatrix} 0 & t \\ 0 & 0 \end{pmatrix}^2 = \begin{pmatrix} 0 & 0 \\ 0 & 0 \end{pmatrix}$, so we have from the power series for the exponential

$$\exp\begin{pmatrix} 0 & 1 \\ 0 & 0 \end{pmatrix} t = \begin{pmatrix} 1 & 0 \\ 0 & 1 \end{pmatrix} + \begin{pmatrix} 0 & t \\ 0 & 0 \end{pmatrix} + \text{terms which are all zero,}$$

i.e., $\exp\begin{pmatrix} 0 & t \\ 0 & 0 \end{pmatrix} = \begin{pmatrix} 1 & t \\ 0 & 1 \end{pmatrix}$, and so

$$\exp\begin{pmatrix} \lambda t & t \\ 0 & \lambda t \end{pmatrix} = \begin{pmatrix} e^{\lambda t} & 0 \\ 0 & e^{\lambda t} \end{pmatrix}\begin{pmatrix} 1 & t \\ 0 & 1 \end{pmatrix} = \begin{pmatrix} e^{\lambda t} & te^{\lambda t} \\ 0 & e^{\lambda t} \end{pmatrix}.$$

It follows that

$$\text{if } F = B\begin{pmatrix} \lambda & 1 \\ 0 & \lambda \end{pmatrix}B^{-1}$$

$$\text{then } \exp(tF) = B\begin{pmatrix} e^{\lambda t} & te^{\lambda t} \\ 0 & e^{\lambda t} \end{pmatrix}B^{-1}.$$

As an example of this case, consider the system of equations

$$\dot{x} = x + y,$$
$$\dot{y} = -x + 3y.$$

The matrix $F = \begin{pmatrix} 1 & 1 \\ -1 & 3 \end{pmatrix}$ has characteristic equation $\lambda^2 - 4\lambda + 4 = 0$, with a double root $\lambda = 2$. Considering $(F - 2I) = \begin{pmatrix} -1 & 1 \\ -1 & 1 \end{pmatrix}$ we find the eigenvector $\mathbf{v}_1 = \begin{pmatrix} 1 \\ 1 \end{pmatrix}$ and the vector $\mathbf{v}_2 = \begin{pmatrix} 0 \\ 1 \end{pmatrix}$ for which $(F - 2I)\mathbf{v}_2 = \mathbf{v}_1$. Thus

$$F = \begin{pmatrix} 1 & 0 \\ 1 & 1 \end{pmatrix}\begin{pmatrix} 2 & 1 \\ 0 & 2 \end{pmatrix}\begin{pmatrix} 1 & 0 \\ -1 & 1 \end{pmatrix}.$$

Since $\exp\begin{pmatrix} 2t & t \\ 0 & 2t \end{pmatrix} = \begin{pmatrix} e^{2t} & te^{2t} \\ 0 & e^{2t} \end{pmatrix}$ we have

$$\exp(tF) = \begin{pmatrix} 1 & 0 \\ 1 & 1 \end{pmatrix}\begin{pmatrix} e^{2t} & te^{2t} \\ 0 & e^{2t} \end{pmatrix}\begin{pmatrix} 1 & 0 \\ -1 & 1 \end{pmatrix}$$

or

$$\exp(tF) = \begin{pmatrix} (1-t)e^{2t} & te^{2t} \\ -te^{2t} & (1+t)e^{2t} \end{pmatrix}.$$

If, for example, we wish to solve the above differential equations for initial conditions $x_0 = 2$, $y_0 = 1$, we just form

$$\exp(tF)\begin{pmatrix} 2 \\ 1 \end{pmatrix} = \begin{pmatrix} 2e^{2t} - te^{2t} \\ e^{2t} - te^{2t} \end{pmatrix}.$$

Then

$$\dot{x} = 4e^{2t} - e^{2t} - 2te^{2t} = x + y,$$
$$\dot{y} = 2e^{2t} - e^{2t} - 2te^{2t} = -x + 3y.$$

In fact, it is not really necessary to write $F = B\begin{pmatrix} \lambda & 1 \\ 0 & \lambda \end{pmatrix}B^{-1}$ in the case where $F$ has a repeated eigenvalue $\lambda$. By the Cayley–Hamilton theorem,

$$(F - \lambda I)^2 = 0$$

so the matrix $G = F - \lambda I$ is nilpotent. Now, since $\lambda I$ and $G$ commute,

$$\exp(tF) = \exp(\lambda t I)\exp(tG).$$

But $\exp(tG)$ is easily computed from the power series:

$$\exp(tG) = I + tG$$

since $(tG)^2$ and all subsequent terms are zero. So

$$\exp(tF) = \begin{pmatrix} e^{\lambda t} & 0 \\ 0 & e^{\lambda t} \end{pmatrix}(I + tG).$$

In the previous example,

$$F = \begin{pmatrix} 1 & 1 \\ -1 & 3 \end{pmatrix}; \quad G = F - 2I = \begin{pmatrix} -1 & 1 \\ -1 & 1 \end{pmatrix}$$

and

$$\exp(tF) = \begin{pmatrix} e^{2t} & 0 \\ 0 & e^{2t} \end{pmatrix}\left[\begin{pmatrix} 1 & 0 \\ 0 & 1 \end{pmatrix} + \begin{pmatrix} -t & t \\ -t & t \end{pmatrix}\right]$$
$$= \begin{pmatrix} (1-t)e^{2t} & te^{2t} \\ -te^{2t} & (1+t)e^{2t} \end{pmatrix}$$

as before.

**Case 3. Complex eigenvalues.** We have finally to deal with the case where $F$ has complex eigenvalues. We have proved that, if the eigenvalues of $F$ are $\alpha \pm i\beta$, we can write $F = BCB^{-1}$ where $C$ is the conformal matrix $C = \begin{pmatrix} \alpha & -\beta \\ \beta & \alpha \end{pmatrix}$. The problem is now to exponentiate $C$.

Notice that $C = \alpha I + \beta J$ where $J = \begin{pmatrix} 0 & -1 \\ 1 & 0 \end{pmatrix}$. The matrix $J$ satisfies $J^2 = -I$ and corresponds to the complex number $i = \sqrt{(-1)}$. Because $\alpha I$ and $\beta J$ commute, we have

$$\exp(tC) = \exp(t\alpha I)\exp(t\beta J).$$

Of course,

$$\exp(t\alpha I) = \begin{pmatrix} e^{\alpha t} & 0 \\ 0 & e^{\alpha t} \end{pmatrix}.$$

To compute $\exp(t\beta J)$, we use the power series

$$\exp(t\beta J) = I + t\beta J + \frac{1}{2!}(t\beta J)^2 + \frac{1}{3!}(t\beta J)^3 + \frac{1}{4!}(t\beta J)^4 + \cdots.$$

Since $J^2 = -I$, $J^3 = -J$, $J^4 = I$, $J^5 = J$, etc., we have

$$\begin{aligned}\exp(t\beta J) &= I + t\beta J - \frac{1}{2!}(t\beta)^2 I - \frac{1}{3!}(t\beta)^3 J + \frac{1}{4!}(t\beta)^4 I + \cdots \\ &= \left(1 - \frac{(t\beta)^2}{2!} + \frac{(t\beta)^4}{4!} + \cdots\right)I + \left(t\beta - \frac{(t\beta)^3}{3!} + \cdots\right)J.\end{aligned}$$

The coefficient of $I$ is the power series for $\cos\beta t$; the coefficient of $J$ is the power series for $\sin\beta t$, so we conclude that

$$\exp(t\beta J) = \cos\beta t\, I + \sin\beta t\, J = \begin{pmatrix}\cos\beta t & -\sin\beta t \\ \sin\beta t & \cos\beta t\end{pmatrix},$$

which is a time-dependent rotation matrix. Identifying $J$ with the complex number i, we see also that

$$e^{i\beta t} = \cos\beta t + i\sin\beta t.$$

Thus, if $C = \begin{pmatrix}\alpha & -\beta \\ \beta & \alpha\end{pmatrix}$,

$$\boxed{\exp(tC) = \begin{pmatrix}e^{\alpha t} & 0 \\ 0 & e^{\alpha t}\end{pmatrix}\begin{pmatrix}\cos\beta t & -\sin\beta t \\ \sin\beta t & \cos\beta t\end{pmatrix}}$$

and, if $F = BCB^{-1}$, we can calculate $\exp(tF)$ as

$$\exp(tF) = B\exp(tC)B^{-1}.$$

As an example, consider $F = \begin{pmatrix}3 & -10 \\ 2 & -5\end{pmatrix}$. The characteristic equation is $\lambda^2 + 2\lambda + 5 = 0$, with roots $\lambda = -1 \pm 2i$. So $F = BCB^{-1}$, where $C$ is the conformal matrix $\begin{pmatrix}-1 & -2 \\ 2 & -1\end{pmatrix}$. As described in section 2.2, we can choose $\begin{pmatrix}1 \\ 0\end{pmatrix}$ as the first column of $B$, and the second column of $B$ is then the first column of $F - \alpha I$ divided by $\beta$: that is $\frac{1}{2}\begin{pmatrix}3+1 \\ 2\end{pmatrix} = \begin{pmatrix}2 \\ 1\end{pmatrix}$. So $B = \begin{pmatrix}1 & 2 \\ 0 & 1\end{pmatrix}$ and $F = \begin{pmatrix}1 & 2 \\ 0 & 1\end{pmatrix} \times \begin{pmatrix}-1 & -2 \\ 2 & -1\end{pmatrix}\begin{pmatrix}1 & -2 \\ 0 & 1\end{pmatrix}$. We now calculate $\exp(tC)$ by the procedure just described:

$$\exp(tC) = \begin{pmatrix}e^{-t} & 0 \\ 0 & e^{-t}\end{pmatrix}\begin{pmatrix}\cos 2t & -\sin 2t \\ \sin 2t & \cos 2t\end{pmatrix}$$

and so

$$\exp(tF) = \begin{pmatrix}1 & 2 \\ 0 & 1\end{pmatrix}\exp(tC)\begin{pmatrix}1 & -2 \\ 0 & 1\end{pmatrix}.$$

Multiplying the matrices, we have

$$\exp(tF) = \begin{pmatrix} e^{-t}(\cos 2t + 2\sin 2t) & -5e^{-t}\sin 2t \\ e^{-t}\sin 2t & e^{-t}(\cos 2t - 2\sin 2t) \end{pmatrix}.$$

You can check for yourself that

$$\frac{d}{dt}\exp(tF) = F\exp(tF).$$

Again in this case, it is not really necessary to do the decomposition $F = BCB^{-1}$ explicitly. Suppose that $F$ has eigenvalues $\alpha \pm i\beta$ so that its characteristic equation is

$$(\lambda - \alpha)^2 + \beta^2 = 0.$$

Then, by the Cayley–Hamilton theorem,

$$(F - \alpha I)^2 + \beta^2 I = 0$$

so the traceless matrix $G = F - \alpha I$ satisfies $G^2 = -\beta^2 I$. Writing $F = \alpha I + G$, we see that

$$\exp(tF) = \exp(\alpha t I)\exp(tG).$$

Again we exponentiate $tG$ by using the power series:

$$\exp(tG) = I + tG + \frac{t^2G^2}{2!} + \frac{t^3G^3}{3!} + \frac{t^4G^4}{4!} + \cdots.$$

Taking advantage of the fact that $G^2 = -\beta^2 I$, we obtain

$$\exp(tG) = I + tG - \frac{\beta^2t^2}{2!}I - \frac{\beta^2t^3G}{3!} + \frac{\beta^4t^4}{4!}I + \cdots.$$

The coefficient of $I$ is again the power series for $\cos\beta t$. The coefficient of $G$ is

$$t - \frac{\beta^2t^3}{3!} + \frac{\beta^4t^5}{5!} = \frac{1}{\beta}\left(\beta t - \frac{\beta^3t^3}{3!} + \frac{\beta^5t^5}{5!} + \cdots\right) = (\sin\beta t)/\beta.$$

We conclude that

$$\exp(tG) = \cos\beta t\, I + [(\sin\beta t)/\beta]G$$

and

$$\exp(tF) = \begin{pmatrix} e^{\alpha t} & 0 \\ 0 & e^{\alpha t} \end{pmatrix}\exp(tG).$$

Returning to the example $F = \begin{pmatrix} 3 & -10 \\ 2 & -5 \end{pmatrix}$, for which $\alpha = -1$, $\beta = 2$, we form $G = F + I = \begin{pmatrix} 4 & -10 \\ 2 & -4 \end{pmatrix}$, a traceless matrix satisfying $G^2 = -4I$. Then

$$\exp(tG) = \begin{pmatrix} \cos 2t & 0 \\ 0 & \cos 2t \end{pmatrix} + \tfrac{1}{2}\sin 2t\begin{pmatrix} 4 & -10 \\ 2 & -4 \end{pmatrix}$$

$$\exp(tG) = \begin{pmatrix} \cos 2t + 2\sin 2t & -5\sin 2t \\ \sin 2t & \cos 2t - 2\sin 2t \end{pmatrix}.$$

On multiplying by $\exp(\alpha tI) = \begin{pmatrix} e^{-t} & 0 \\ 0 & e^{-t} \end{pmatrix}$ we obtain the earlier result for $\exp(tF)$, but with much less effort.

## 3.4. Differential equations and phase portraits

We have seen that the differential equation $\dot{\mathbf{v}} = A\mathbf{v}$, with the initial condition $\mathbf{v}(0) = \mathbf{v}_0$, has the unique solution

$$\mathbf{v}(t) = \exp(tA)\mathbf{v}_0.$$

This solution $\mathbf{v}(t)$ defines a function from the *time axis* to a two-dimensional vector space. Because $\exp(tA)$ is defined for negative $t$ as well as positive $t$, the domain of $\mathbf{v}(t)$ is $-\infty < t < \infty$. By plotting the point whose position vector is $\mathbf{v}$ for all values of $t$, we obtain a *solution curve* for the differential equation. This curve is like the path of a particle which moves in a plane, and the vector $\dot{\mathbf{v}}(t) = A\mathbf{v}(t)$, which is like the velocity vector for that particle, is tangent to the path. Through each point in the plane there passes a unique solution curve, and the effect of the transformation $\exp(tA)$ is represented by moving $t$ units along the solution curve.

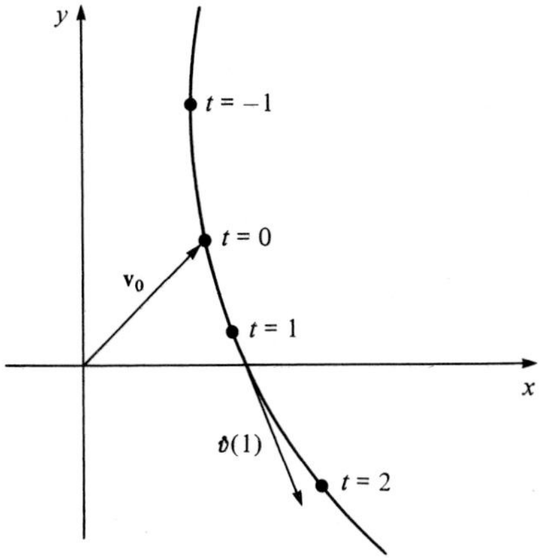


**Figure 3.1**

By plotting a whole family of solution curves, we can create a *phase portrait* which conveys the important features of the solutions of the differential equation. Although there are many different matrices $A$ which could appear in the differential equation $\dot{\mathbf{v}} = A\mathbf{v}$, there are only a limited number of different *types* of phase portraits. To be specific, if matrices $A$ and $F$ are conjugate, so that $A = BFB^{-1}$, then the solution curves for $\dot{\mathbf{v}} = A\mathbf{v}$ are obtained from those for $\dot{\mathbf{w}} = F\mathbf{w}$ by the linear transformation $\mathbf{v}(t) = B\mathbf{w}(t)$. The proof is simple: since $B$ is constant, $\dot{\mathbf{v}}(t) = B\dot{\mathbf{w}}(t)$, and it follows that, if $\dot{\mathbf{w}} = F\mathbf{w}$, then

$$\dot{\mathbf{v}} = B\dot{\mathbf{w}} = BF\mathbf{w} = BFB^{-1}\mathbf{v} = A\mathbf{v}.$$

Thus the phase portraits for $\dot{\mathbf{v}} = A\mathbf{v}$ are essentially the same as those for $\dot{\mathbf{w}} = F\mathbf{w}$ if $A$ and $F$ are conjugate. We can therefore determine possible phase portraits (up to a linear transformation) by considering the different possibilities for the eigenvalues of $A$.

We note first that if $\mathbf{v}_0$ is an eigenvector of $A$, with eigenvalue $\lambda$, then

$$\begin{aligned}\mathbf{v}(t) = \exp(tA)\mathbf{v}_0 &= \left[1 + tA + \frac{1}{2!}(tA)^2 + \cdots\right]\mathbf{v}_0 \\ &= \left[1 + t\lambda + \frac{1}{2!}(t\lambda)^2 + \cdots\right]\mathbf{v}_0 = e^{\lambda t}\mathbf{v}_0.\end{aligned}$$

So in this case the solution curve is the straight line through the origin on which $\mathbf{v}_0$ lies. If $\lambda$ is positive, $\mathbf{v}(t)$ moves away from the origin as $t$ becomes large and positive; if $\lambda$ is negative, $\mathbf{v}(t)$ moves in toward the origin as $t \to \infty$. If $\lambda = 0$ then $\mathbf{v}(t) = \mathbf{v}_0$ for all $t$, so each point on the line through $\mathbf{v}_0$ stays fixed.

We can now enumerate all possible cases.

**Case 1. A is a multiple of the identity matrix.**

*Case 1a.* $A = \begin{pmatrix} 0 & 0 \\ 0 & 0 \end{pmatrix}$, $\exp(tA) = \begin{pmatrix} 1 & 0 \\ 0 & 1 \end{pmatrix}$. So all points stay fixed.

*Case 1b.* $A = \begin{pmatrix} \lambda & 0 \\ 0 & \lambda \end{pmatrix}$, $\exp(tA) = \begin{pmatrix} e^{\mu} & 0 \\ 0 & e^{\mu} \end{pmatrix}$

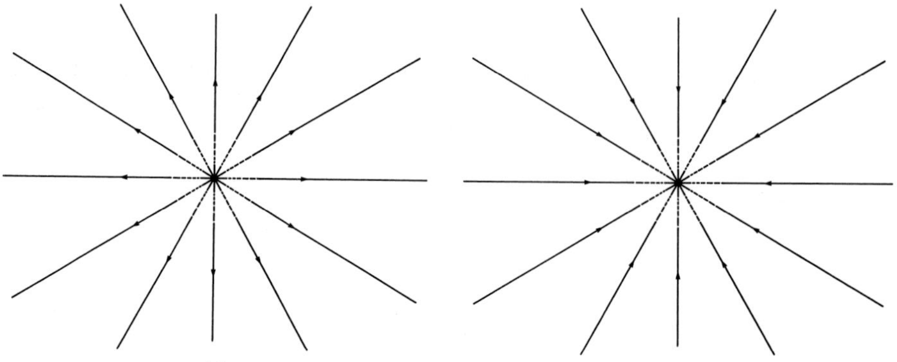

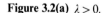

(a) (b)

**Figure 3.2(a)** $\lambda > 0$. **Figure 3.2(b)** $\lambda < 0$

Every vector is an eigenvector, so all solution curves are straight lines through the origin. If $\lambda > 0$, each point moves exponentially away from the origin as $t \to \infty$; if $\lambda < 0$ each point moves towards the origin.

**Case 2. A has real distinct eigenvalues $\lambda_1$ and $\lambda_2$.**

Then

$$A = B\begin{pmatrix} \lambda_1 & 0 \\ 0 & \lambda_2 \end{pmatrix}B^{-1}$$

*Case 2a.* $\lambda_1$ and $\lambda_2$ both positive, $\lambda_1 > \lambda_2$.
In the special case where $B = I$, the solution curves are as illustrated in figure 3.3(a). The $x$- and $y$-axes are solution curves.

Since

$$\begin{pmatrix} x \\ y \end{pmatrix} = \begin{pmatrix} e^{\lambda_1 t} x_0 \\ e^{\lambda_2 t} y_0 \end{pmatrix}$$

and $e^{\lambda_1 t} > e^{\lambda_2 t}$, $x/y$ becomes larger and larger as $t \to \infty$. As $t \to -\infty$, curves become tangent to the $y$-axis.

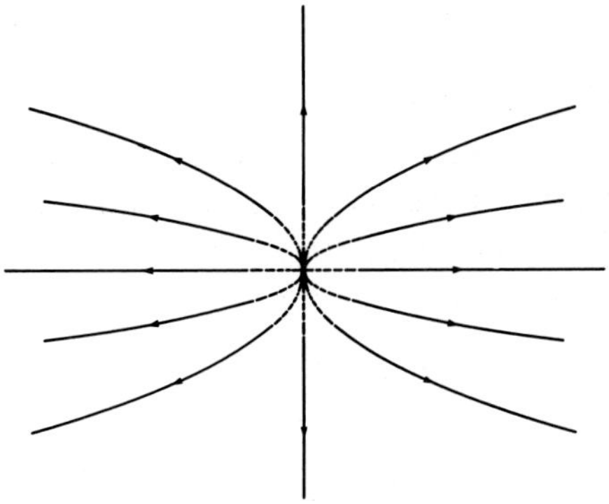

(a)

**Figure 3.3(a)** $A = \begin{pmatrix} \lambda_1 & 0 \\ 0 & \lambda_2 \end{pmatrix}$

More generally, let $\mathbf{v}_1$ and $\mathbf{v}_2$ be the eigenvectors of $A$. Lines along $\mathbf{v}_1$ and $\mathbf{v}_2$ are solution curves. Since $\lambda_1 > \lambda_2$, other curves become parallel to $\mathbf{v}_1$ as $t \to +\infty$, tangent to $\mathbf{v}_2$ as $t \to -\infty$, as illustrated in figure 3.3(b)

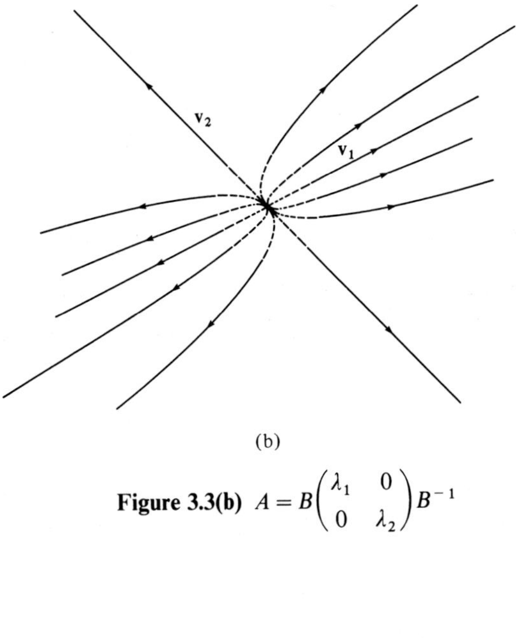


(b)

**Figure 3.3(b)** $A = B\begin{pmatrix} \lambda_1 & 0 \\ 0 & \lambda_2 \end{pmatrix}B^{-1}$

*Case 2b.* $\lambda_1$ and $\lambda_2$ both negative, $|\lambda_1| < |\lambda_2|$. This is similar to Case 2a, but all arrows are reversed. As $t \to \infty$, all solution curves approach the origin.

*Case 2c.* $\lambda_1$ positive, $\lambda_2$ negative. In the special case where $B = I$, the $x$-axis and $y$-axis are again both solution curves. As $t$ increases, $x$ becomes larger, $y$ smaller. Other solution curves approach the $x$-axis as $t \to \infty$, the $y$-axis as $t \to -\infty$, as illustrated in figure 3.4(a).

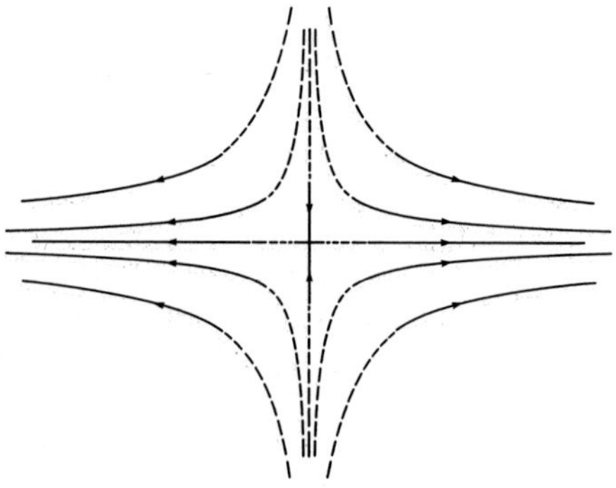

(a)

**Figure 3.4(a)** $A = \begin{pmatrix} \lambda_1 & 0 \\ 0 & \lambda_2 \end{pmatrix}, \begin{pmatrix} x \\ y \end{pmatrix} = \begin{pmatrix} e^{\lambda_1 t} x_0 \\ e^{\lambda_2 t} y_0 \end{pmatrix}.$

More generally, suppose eigenvector $\mathbf{v}_1$ corresponds to eigenvalue $\lambda_1 > 0$, while eigenvector $\mathbf{v}_2$ corresponds to eigenvalue $\lambda_2 < 0$. Points along line through $\mathbf{v}_1$ move *out*, those along line through $\mathbf{v}_2$ move *in*. Solution curves approach the line through $\mathbf{v}_1$ as $t \to +\infty$, the line through $\mathbf{v}_2$ as $t \to -\infty$, as illustrated in figure 3.4(b).

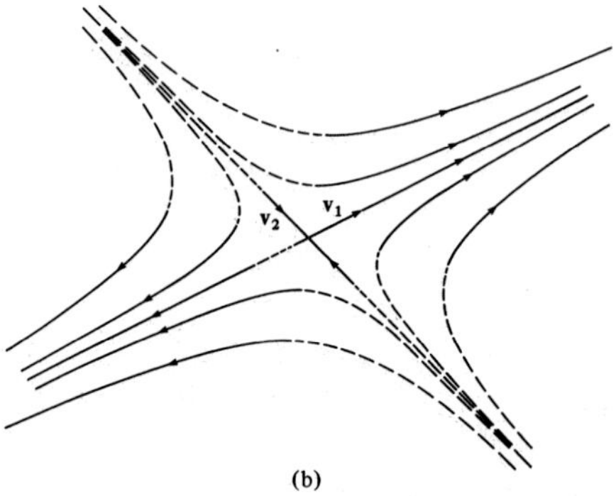


(b)

**Figure 3.4(b)** $A = B \begin{pmatrix} \lambda_1 & 0 \\ 0 & \lambda_2 \end{pmatrix} B^{-1}$

*Case 2d.* $\lambda_1$ positive, $\lambda_2$ zero.

In the special case where $B = I$ the $y$-axis is held fixed. Lines parallel to the $x$-axis are solution curves. Points move away from the $y$-axis as $t \to +\infty$, toward it as

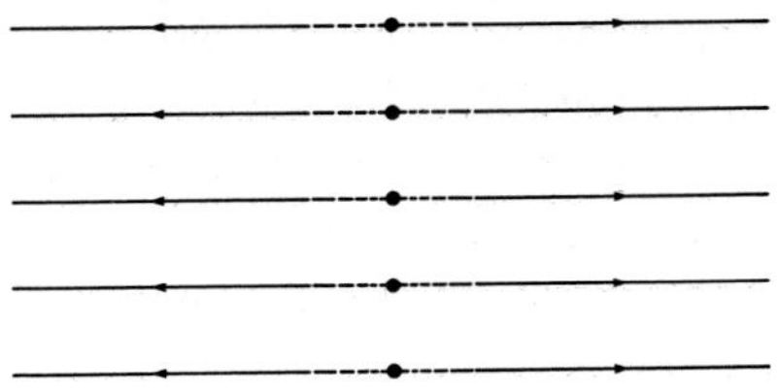

(a)

**Figure 3.5(a)** $A = \begin{pmatrix} \lambda_1 & 0 \\ 0 & 0 \end{pmatrix}, \begin{pmatrix} x \\ y \end{pmatrix} = \begin{pmatrix} e^{\lambda_1 t} x_0 \\ y_0 \end{pmatrix}$

$t \to -\infty$. More generally, suppose eigenvector $\mathbf{v}_1$ corresponds to $\lambda_1 > 0$, while $\mathbf{v}_2$ corresponds to $\lambda_2 = 0$. The line through the origin along $\mathbf{v}_2$ is held fixed. Lines parallel to $\mathbf{v}_1$ are solution curves. Points move away from the line through $\mathbf{v}_2$ as $t \to \infty$, toward this line as $t \to -\infty$.

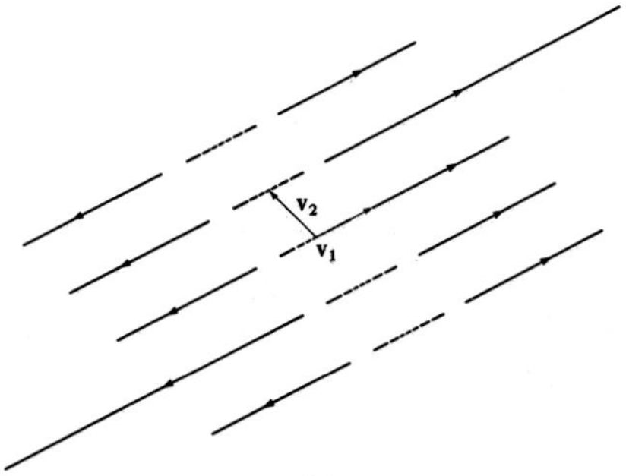


(b)

**Figure 3.5(b)** $A = B\begin{pmatrix} \lambda_1 & 0 \\ 0 & 0 \end{pmatrix}B^{-1}$

*Case 2e.* $\lambda_1$ is negative, $\lambda_2$ zero. Just reverse all arrows in the preceding case. As $t \to +\infty$, entire plane is projected onto the line of eigenvector $\mathbf{v}_2$.

**Case 3. Repeated eigenvalue $\lambda$, but $A \neq \lambda I$.**

$$A = B\begin{pmatrix} \lambda & 1 \\ 0 & \lambda \end{pmatrix}B^{-1}; \exp(tA) = Be^{\lambda t}\begin{pmatrix} 1 & t \\ 0 & 1 \end{pmatrix}B^{-1}.$$

In this case there is only one eigenvector $\mathbf{v}_1$.

*Case 3a.* $\lambda > 0$.

In the special case where $B = I$, the eigenvector $\mathbf{v}_1$ lies along the $x$-axis. $x$-axis is a solution curve; points on it move out. As $t \to \infty$, $x$ becomes much greater than $y$; curve becomes parallel to $x$-axis. As $t \to -\infty$, $x$ becomes opposite in sign to $y$. All curves cross the $y$-axis for $t = -x_0/y_0$.

$$\begin{pmatrix} x \\ y \end{pmatrix} = e^{\lambda t}\begin{pmatrix} x_0 + ty_0 \\ y_0 \end{pmatrix}$$

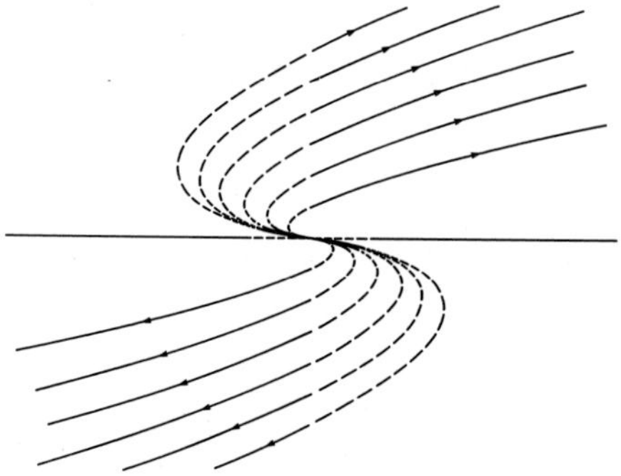

(a)

**Figure 3.6(a)** $A = \begin{pmatrix} \lambda & 1 \\ 0 & \lambda \end{pmatrix}$

More generally, the line through the origin along $\mathbf{v}_1$ is a solution curve; all other curves become parallel to this line as $t = \infty$. All curves cross *every* other line through the origin.

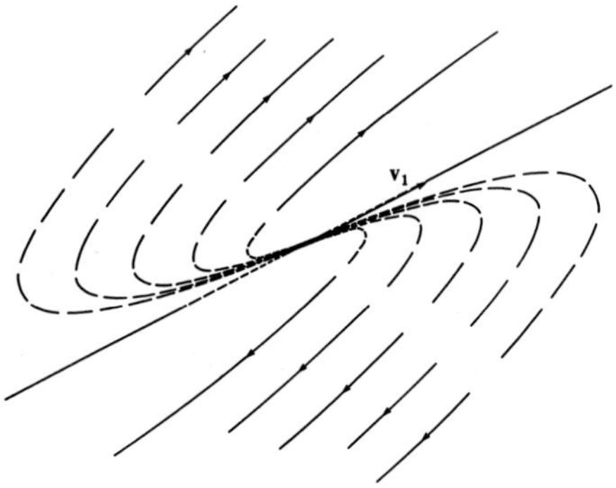


(b)

**Figure 3.6(b)** $A = B\begin{pmatrix} \lambda & 1 \\ 0 & \lambda \end{pmatrix}B^{-1}$

*Case 3b.* $\lambda < 0$. Just reverse all arrows in the preceding case.
*Case 3c.* $\lambda = 0$.

In the special case where $B = I$, the $x$-axis stays fixed. Lines parallel to it are solution curves. For any $t$, $\exp(tA)$ is a shear transformation.

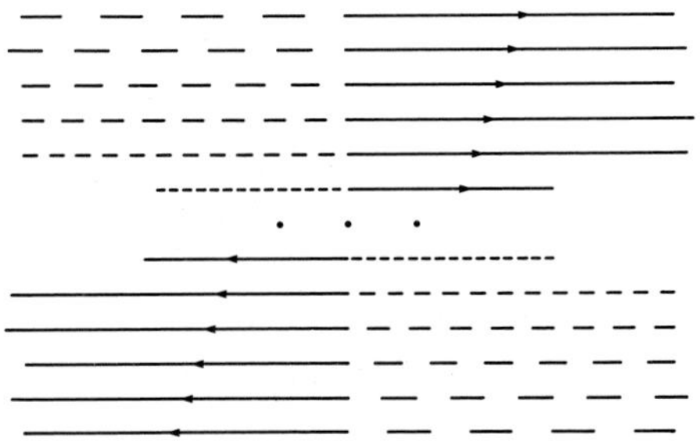

(a)

**Figure 3.7(a)** $A = \begin{pmatrix} 0 & 1 \\ 0 & 0 \end{pmatrix}, \exp(tA) = \begin{pmatrix} 1 & t \\ 0 & 1 \end{pmatrix}.$

More generally, the line through the origin along eigenvector $\mathbf{v}_1$ stays fixed; the plane is sheared parallel to this line.

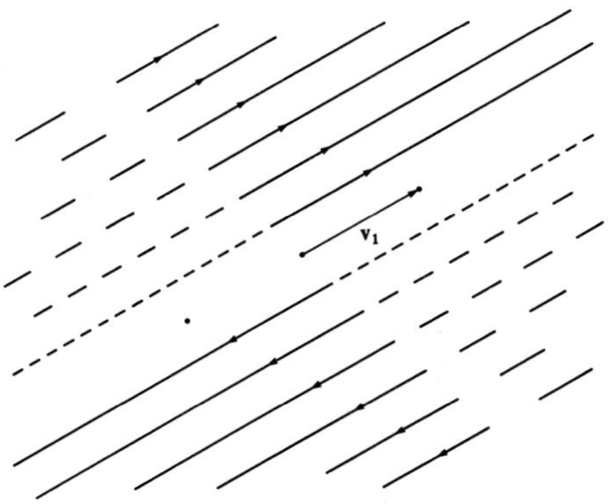


(b)

**Figure 3.7(b)** $A = B\begin{pmatrix} \lambda & 1 \\ 0 & \lambda \end{pmatrix}B^{-1}$

**Case 4. *A* has complete eigenvalues $\alpha \pm 1\beta$. In this case there are no eigenvectors.**

*Case 4a.* $\alpha = 0$.

In the special case where $B = I$, the solution curves are circles centered at the origin. The transformation $\exp(tA)$ is a rotation. The solution is periodic with period $2\pi/\beta$.

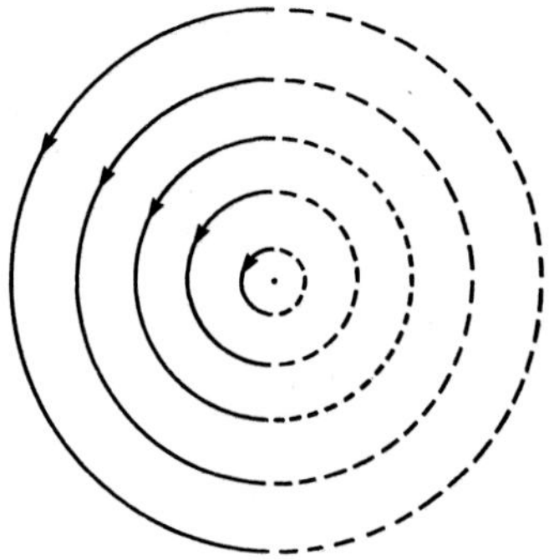

(a)

**Figure 3.8(a)** $A = \begin{pmatrix} 0 & -\beta \\ \beta & 0 \end{pmatrix}, \exp(tA) = \begin{pmatrix} \cos \beta t & -\sin \beta t \\ \sin \beta t & \cos \beta t \end{pmatrix}$.

More generally, the solution curves are ellipses. The solution is periodic with period $2\pi/\beta$.

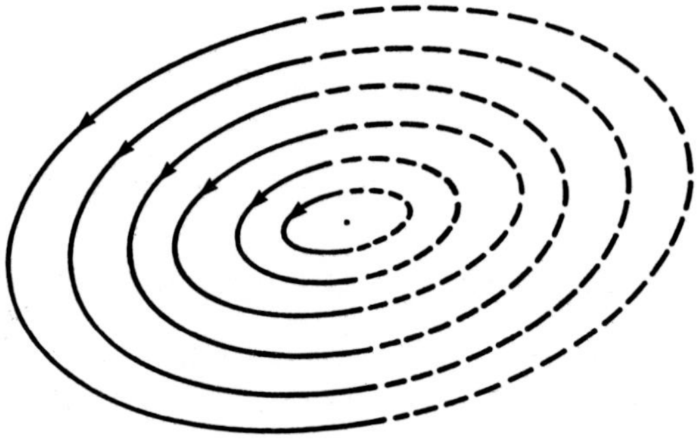

(b)

**Figure 3.8(b)** $A = B\begin{pmatrix} 0 & -\beta \\ \beta & 0 \end{pmatrix}B^{-1}$

*Case 4b.* $\alpha > 0$. Solution curves are counter-clockwise spirals. Points move out as $t' \to \infty$, move into origin as $t \to -\infty$.

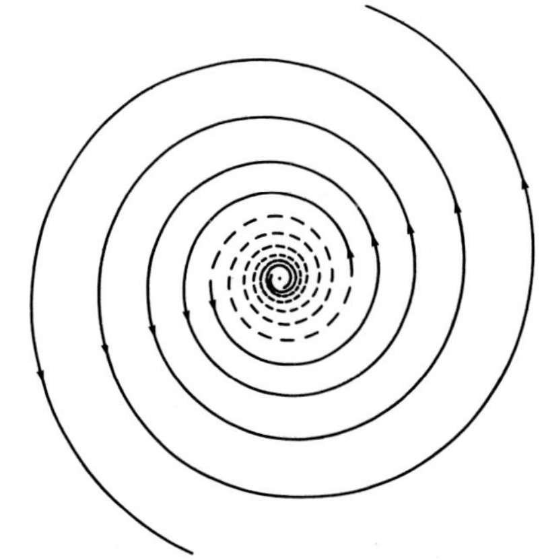

(a)

**Figure 3.9(a)** $A = \begin{pmatrix} \alpha & -\beta \\ \beta & \alpha \end{pmatrix}$, $\exp(tA) = e^{\alpha t}\begin{pmatrix} \cos\beta t & -\sin\beta t \\ \sin\beta t & \cos\beta t \end{pmatrix}$.

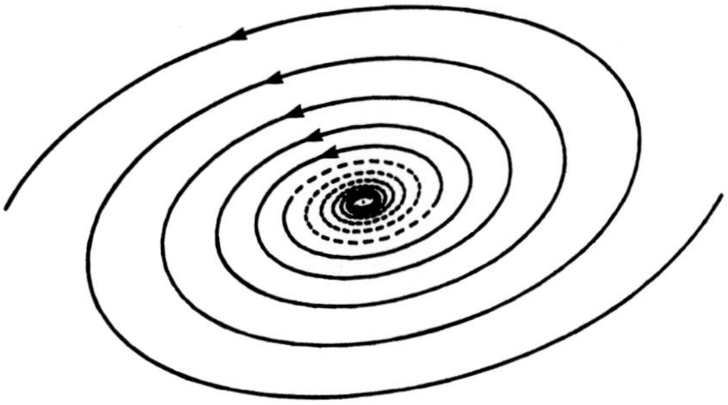

(b)

**Figure 3.9(b)** $A = B\begin{pmatrix} \alpha & -\beta \\ \beta & \alpha \end{pmatrix}B^{-1}$

*Case 4c.* $\alpha < 0$. Same as above, but spiraling into the origin as $t \to \infty$.

## 3.5. Applications of differential equations

The best-known physical system which gives rise to a differential equation of the sort we have just been considering consists of a mass $M$ which moves under the

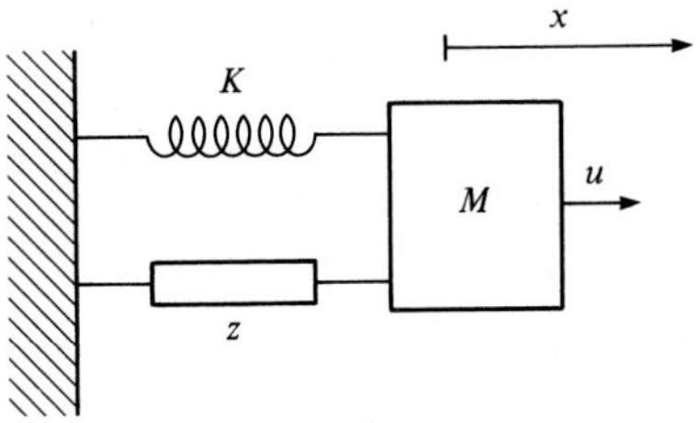


**Figure 3.10**

influence of a spring which exerts a force, $-kx$ and proportional to the displacement, $x$, a 'linear dashpot' which exerts a force $-zu$, proportional to the velocity $u$ of the mass. Newton's second law says that $M\dot{u} = -kx - zu$ while the definition of velocity gives $\dot{x} = u$. The state of the system is completely described by specifying the vector $\mathbf{w} = \begin{pmatrix} x \\ u \end{pmatrix}$ whose components are the position and velocity of the mass. Since

$$\dot{x} = u$$
$$\dot{u} = -\frac{k}{M}x - \frac{z}{M}u$$

we can write this differential equation as

$$\dot{\mathbf{w}} = A\mathbf{w}$$

where

$$A = \begin{pmatrix} 0 & 1 \\ -\omega_0^2 & -\Gamma \end{pmatrix}, \quad \omega_0^2 = \frac{k}{M}, \quad \Gamma = \frac{z}{M}.$$

The character of the motion is determined by the eigenvalues of $A$. Since the characteristic equation is

$$\lambda^2 + \Gamma\lambda + \omega_0^2 = 0,$$

we have

$$\lambda = \frac{-\Gamma \pm \sqrt{(\Gamma^2 - 4\omega_0^2)}}{2}.$$

There are four distinct possibilities:

**1.** If $\Gamma = 0$ (no friction), $\lambda = \pm i\omega_0$, and the oscillator is *undamped*. The phase portrait corresponds to case 4a; the solution curves are ellipses in the $xu$-plane. The equation of one of these ellipses is $\frac{1}{2}Kx^2 + \frac{1}{2}Mu^2 = \text{constant}$, which implies conservation of energy. The period of the motion is $2\pi/\omega_0$.

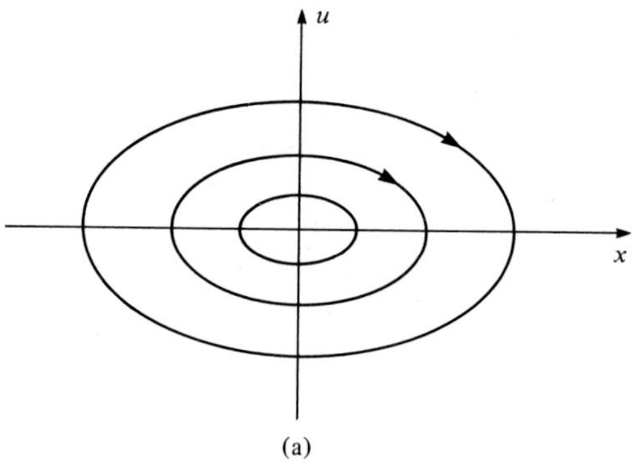


(a)

**Figure 3.11(a)**

**2.** If $\Gamma > 0$ but $\Gamma^2 - 4\omega_0^2 < 0$, the eigenvalues are

$$\lambda = -\frac{\Gamma}{2} \pm i\sqrt{\left(\omega_0^2 - \frac{1}{4}\Gamma^2\right)}.$$

The oscillator is called *underdamped*. The phase portrait corresponds to case 4c; the solution curves spiral into the origin.

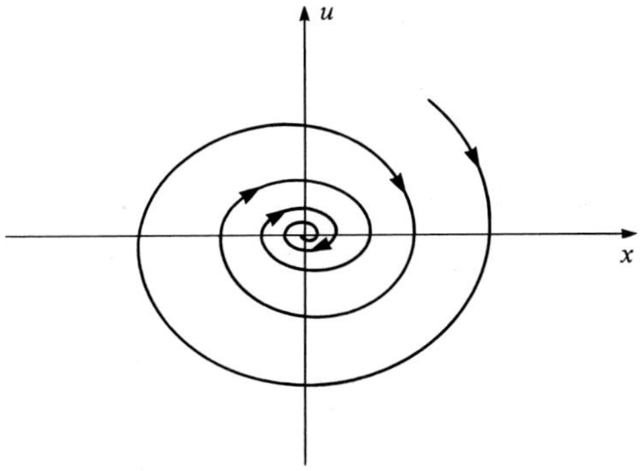


**Figure 3.11(b)**

**3.** If $\Gamma^2 = 4\omega_0^2$ the characteristic equation has a double root $\lambda = -\Gamma/2$. The oscillator is *critically damped* and the phase portrait corresponds to case 3b. On looking at

$$A - \lambda I = \begin{pmatrix} \Gamma/2 & 1 \\ -\Gamma^2/4 & -\Gamma/2 \end{pmatrix}$$

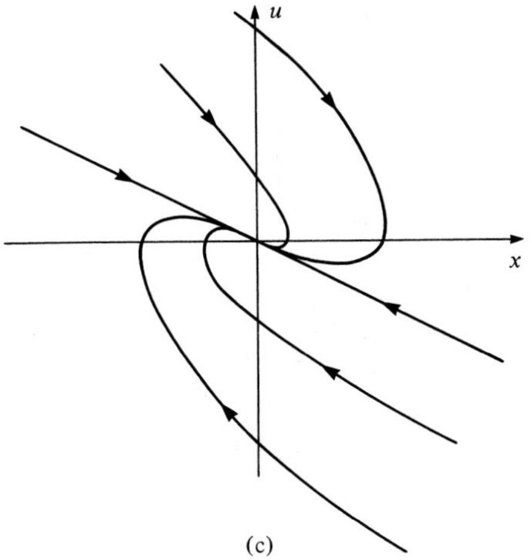


(c)

**Figure 3.11(c)**

we see that the one eigenvector of $A$ is $\mathbf{w}_1 = \begin{pmatrix} 1 \\ -\Gamma/2 \end{pmatrix}$. This implies that if

$$\begin{pmatrix} x_0 \\ u_0 \end{pmatrix} = \begin{pmatrix} x_0 \\ -(\Gamma/2)x_0 \end{pmatrix}$$

then $x$ and $u$ remain proportional throughout the subsequent motion.

4. If $\Gamma^2 > 4\omega_0^2$ the characteristic equation has real negative roots

$$\lambda = -\Gamma/2 \pm \sqrt{(\Gamma^2/4 - \omega_0^2)}.$$

The oscillator is *overdamped*, and the phase portrait corresponds to case 2b.

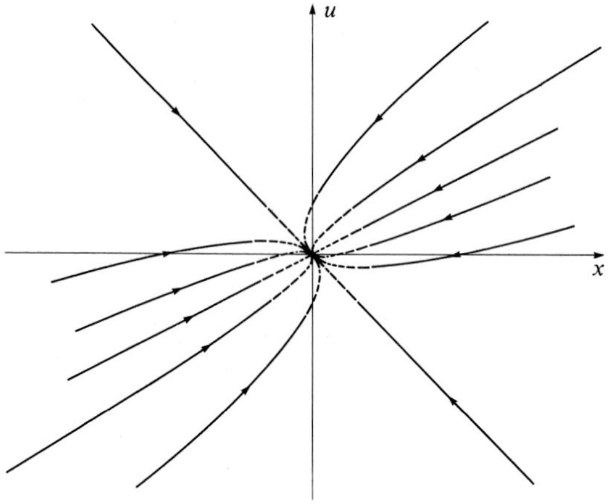


**Figure 3.11(d)**

In order to understand better these four types of motion, let us pick a typical trajectory and plot the $x$-coordinate as a function of $t$:

**Case 1.**

We have

$$\begin{pmatrix} 0 & 1 \\ -\omega_0^2 & 0 \end{pmatrix} = \begin{pmatrix} \omega_0^{-1/2} & 0 \\ 0 & \omega_0^{1/2} \end{pmatrix} \begin{pmatrix} 0 & \omega_0 \\ -\omega_0 & 0 \end{pmatrix} \begin{pmatrix} \omega_0^{1/2} & 0 \\ 0 & \omega_0^{-1/2} \end{pmatrix}$$

so

$$\exp\begin{pmatrix} 0 & 1 \\ -\omega_0^2 & 0 \end{pmatrix} = \begin{pmatrix} \omega_0^{-1/2} & 0 \\ 0 & \omega_0^{1/2} \end{pmatrix} \begin{pmatrix} \cos\omega_0 t & \sin\omega_0 t \\ -\sin\omega_0 t & \cos\omega_0 t \end{pmatrix} \begin{pmatrix} \omega_0^{1/2} & 0 \\ 0 & \omega_0^{-1/2} \end{pmatrix}$$

and $x$ as a function of $t$ will be of the form

$$x = m\cos\omega_0 t + n\sin\omega_0 t$$

where $m$ and $n$ depend on the initial condition. Writing

$$m^2 + n^2 = \rho^2$$

and defining $\phi$ by

$$m = \rho\sin\phi, \quad n = \rho\cos\phi$$

we can write

$$x(t) = \rho \cos(\omega_0 t + \phi).$$

The dependence is sinusoidal with frequency $\omega_0$. The phase, $\phi$, and the amplitude, $\rho$, depend on the initial conditions.

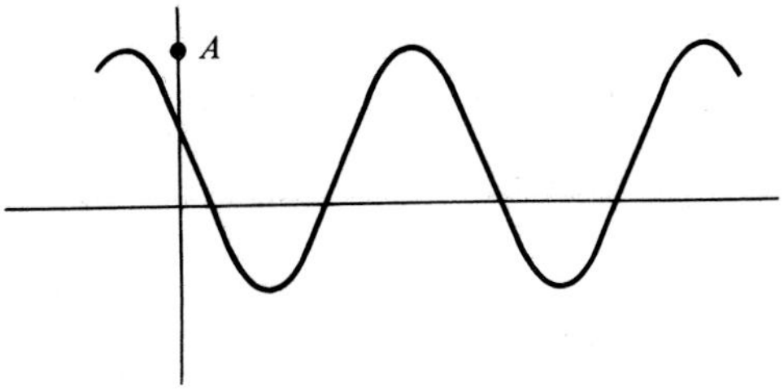


**Figure 3.12**

**Case 2.** Here we know that

$$\exp tA = B\begin{pmatrix} e^{-\Gamma t}\cos\omega_1 t & -e^{-\Gamma t}\sin\omega_1 t \\ e^{-\Gamma t}\sin\omega_1 t & e^{-\Gamma t}\cos\omega_1 t \end{pmatrix} B^{-1}$$

where

$$\omega_1 = \tfrac{1}{2}\sqrt{(4\omega_0^2 - \Gamma^2)}.$$

Thus the $x$ dependence on $t$ will be of the form

$$x(t) = \rho e^{-\Gamma t}\cos(\omega_1 t + \phi)$$

where $\rho$ and $\phi$ depend on the initial conditions. This is an *exponentially damped sinusoidal curve.*

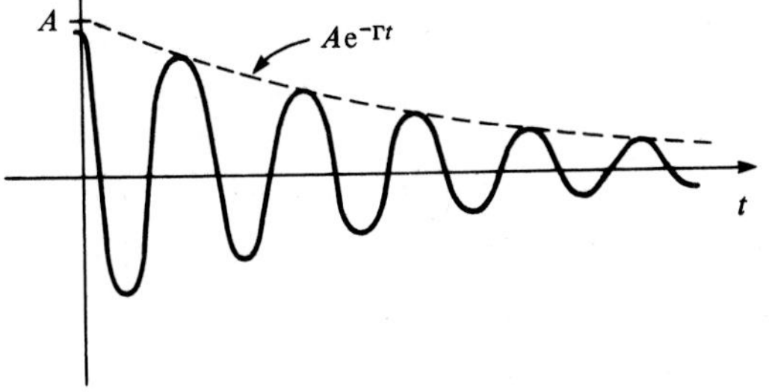


**Figure 3.13**

**Cases 3 and 4**

There is no sinusoidal component, $x$ decays exponentially with $t$ and crosses the $x = 0$ line at most one time.

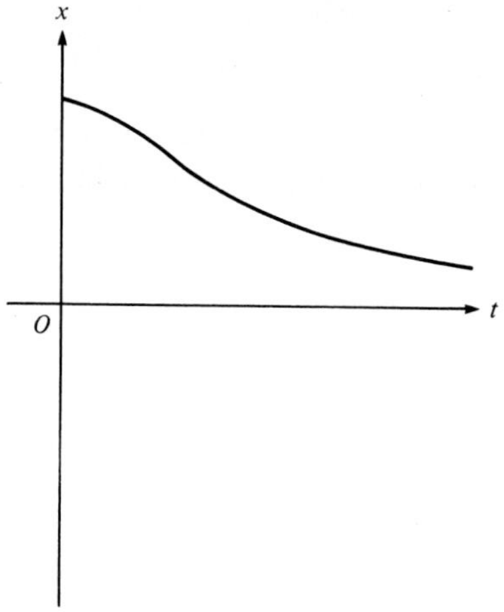


Figure 3.14

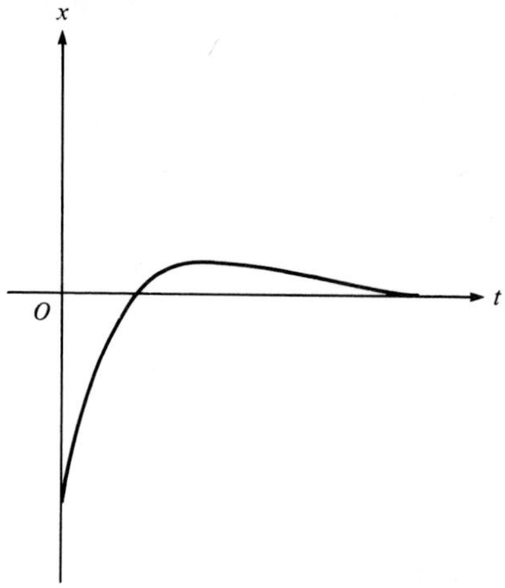


Figure 3.15

**Forced oscillation**

Frequently the differential equations which arise in solving physics problems are not *homogeneous* equations of the form $\dot{\mathbf{v}} - A\mathbf{v} = 0$ but *inhomogeneous* equations of the form

$$\dot{\mathbf{v}} - A\mathbf{v} = \mathbf{b}(t),$$

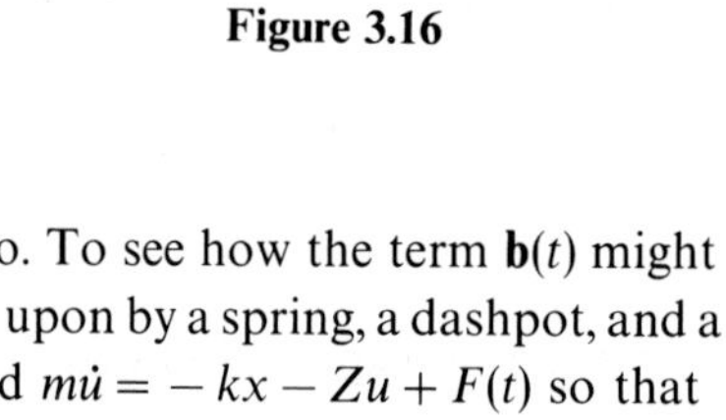

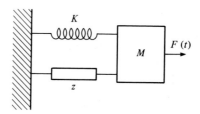

Figure 3.16

with $\mathbf{b}$ not identically zero. To see how the term $\mathbf{b}(t)$ might arise, consider a *driven oscillator*, a mass $M$ acted upon by a spring, a dashpot, and a motor which supplies a force $F(t)$. Then $\dot{x} = u$ and $m\dot{u} = -kx - Zu + F(t)$ so that

$$\begin{pmatrix} \dot{x} \\ \dot{u} \end{pmatrix} - \begin{pmatrix} 0 & 1 \\ -k/M & -Z/M \end{pmatrix} \begin{pmatrix} x \\ u \end{pmatrix} = \begin{pmatrix} 0 \\ F(t)/M \end{pmatrix}.$$

To solve the equation $\dot{\mathbf{v}} - A\mathbf{v} = \mathbf{b}(t)$ we generalize the method called *variation of parameters*, a well-known technique for solving linear differential equations in a single variable. Recall that for one dependent variable, to solve

$$\dot{x} - kx = b(t)$$

we take advantage of the fact that $e^{kt}$ is a solution of the homogeneous equation $\dot{x} - kx = 0$ in order to write $x(t) = e^{kt}u(t)$. Substituting this trial solution into $\dot{x} - kx = b(t)$, we obtain

$$ke^{kt}u(t) + e^{kt}\dot{u}(t) - ke^{kt}u(t) = b(t)$$

so that

$$\dot{u}(t) = e^{-kt}b(t).$$

Then, integrating once with respect to time, we find

$$u(t) = \int_0^t e^{-ks}b(s)\mathrm{d}s$$

and finally

$$x(t) = e^{kt}u(t) = e^{kt}\int_0^t e^{-ks}b(s)\mathrm{d}s$$

or

$$x(t) = \int_0^t e^{k(t-s)}b(s)\mathrm{d}s,$$

a solution which satisfies the initial condition $x(0) = 0$. To obtain the general solution, satisfying the initial condition $x(0) = x_0$, we simply add on the appropriate solution to the homogeneous equation, $e^{kt}x_0$, so that the solution to $\dot{x} - kx = b(t)$ with $x(0) = x_0$ is

$$x(t) = \int_0^t e^{k(t-s)}b(s)\mathrm{d}s + e^{kt}x_0.$$

Exactly the same approach works when we set out to solve $\dot{\mathbf{v}} - A\mathbf{v} = \mathbf{b}(t)$. Since $\exp(tA)\mathbf{v}_0$ is a solution of the homogeneous equation $\dot{\mathbf{v}} - A\mathbf{v} = \mathbf{0}$, we replace the constant vector $\mathbf{v}_0$ by a function $\mathbf{w}(t)$, and try the solution $\mathbf{v}(t) = \exp(tA)\mathbf{w}(t)$. Substituting into $\dot{\mathbf{v}} - A\mathbf{v} = \mathbf{b}(t)$, we obtain

$$A\exp(tA)\mathbf{w}(t) + \exp(tA)\dot{\mathbf{w}}(t) - A\exp(tA)\mathbf{w}(t) = \mathbf{b}(t)$$

so that

$$\dot{\mathbf{w}}(t) = \exp(-tA)\mathbf{b}(t).$$

Integrating, we have

$$\mathbf{w}(t) = \int_0^t \exp(-sA)\mathbf{b}(s)\mathrm{d}s$$

and finally

$$\mathbf{v}(t) = \exp(tA)\mathbf{w}(t),$$

$$\mathbf{v}(t) = \int_0^t \exp[(t-s)A]\mathbf{b}(s)\mathrm{d}s.$$

This solution clearly satisfies $\mathbf{v}(0) = \begin{pmatrix} 0 \\ 0 \end{pmatrix}$; to find a solution satisfying any other initial conditions, we add on the general solution to the homogeneous equation, obtaining

$$\mathbf{v}(t) = \int_0^t \exp[(t-s)A]\mathbf{b}(s)\mathrm{d}s + \exp(tA)\mathbf{v}_0$$

which satisfies $\mathbf{v}(0) = \mathbf{v}_0$.

This general solution is not the one which is usually found in discussions of the forced oscillator in physics textbooks. There it is usually assumed that the *driving term* $\mathbf{b}(t)$ is sinusoidal with fixed frequency $\omega$; for example,

$$\mathbf{b}(t) = \begin{pmatrix} \sin \omega t + \cos \omega t \\ 3 \cos \omega t \end{pmatrix}.$$

Then $\mathbf{b}(t)$ satisfies $\ddot{\mathbf{b}} = -\omega^2 \mathbf{b}$, and we can use integration by parts to evaluate the integral in the more general solution which we obtained above. The trick is the same one used to evaluate antiderivatives like $\int e^{-ks} \sin s \, \mathrm{d}s$: integrate by parts twice to get an equation for the unknown integral.

Let

$$\mathbf{v} = \int_0^t \exp[(t-s)A]\mathbf{b}(s)\,\mathrm{d}s.$$

Integrating once by parts, and assuming that $A$ is non-singular, we have

$$\mathbf{v} = [-\exp[(t-s)A]A^{-1}\mathbf{b}(s)]_0^t + \int_0^t \exp[(t-s)A]A^{-1}\dot{\mathbf{b}}(s)\,\mathrm{d}s.$$

Integrating again by parts, we have

$$\mathbf{v} = [-\exp[(t-s)A]A^{-1}\mathbf{b}(s)]_0^t - [\exp(t-s)A^{-2}\dot{\mathbf{b}}(s)]_0^t$$
$$+ \int_0^t \exp[(t-s)A]A^{-2}\ddot{\mathbf{b}}(s)\mathrm{d}s.$$

Replacing $\ddot{\mathbf{b}}(s)$ by $-\omega^2\mathbf{b}(s)$, we see that the last term is just $-A^{-2}\omega^2\mathbf{v}$. Thus

$$\mathbf{v} + A^{-2}\omega^2\mathbf{v} = -[\exp[(t-s)A][A^{-1}\mathbf{b}(s) + A^{-2}\dot{\mathbf{b}}(s)]]_0^t$$

or

$$(A^2 + \omega^2 I)\mathbf{v} = -[\exp[(t-s)A][\dot{\mathbf{b}}(s) + A\mathbf{b}(s)]]_0^t.$$

Unless the eigenvalues of $A$ are $\pm i\omega$, in which case $(A^2 + \omega^2 I)$ would be singular, we can multiply both sides by $(A^2 + \omega^2 I)^{-1}$ to obtain the explicit solution

$$\mathbf{v} = -(A^2 + \omega^2 I)^{-1}[\dot{\mathbf{b}}(t) + A\mathbf{b}(t) - \exp(tA)(\dot{\mathbf{b}}(0) + A\mathbf{b}(0))].$$

The term involving $\exp(tA)$ serves only to guarantee that $\mathbf{v}(0) = 0$; if $\Gamma > 0$, then $\exp tA$ times any vector tends to 0 as $t \to +\infty$. Thus, for large $t$, we can drop this term, obtaining the *steady-state* solution

$$\mathbf{v} = -(A^2 + \omega^2 I)^{-1}[\dot{\mathbf{b}}(t) + A\mathbf{b}(t)].$$

To check this result, notice that

$$\dot{\mathbf{v}} = -(A^2 + \omega^2 I)^{-1}[\ddot{\mathbf{b}}(t) + A\dot{\mathbf{b}}(t)]$$

and

$$A\mathbf{v} = -(A^2 + \omega^2 I)^{-1}[A\dot{\mathbf{b}}(t) + A^2\mathbf{b}(t)]$$

so that

$$\dot{\mathbf{v}} - A\mathbf{v} = (A^2 + \omega^2 I)^{-1}[\omega^2\mathbf{b}(t) + A^2\mathbf{b}(t)] = \mathbf{b}(t).$$

This check shows that the result is correct even if $A$ is a singular matrix!

Suppose, for example, we wish to solve

$$\begin{pmatrix} \dot{x} \\ \dot{y} \end{pmatrix} - \begin{pmatrix} -1 & -2 \\ 2 & -1 \end{pmatrix}\begin{pmatrix} x \\ y \end{pmatrix} = \begin{pmatrix} \sin 3t \\ \cos 3t \end{pmatrix}.$$

Here

$$A = \begin{pmatrix} -1 & -2 \\ 2 & -1 \end{pmatrix}, \quad \mathbf{b}(t) = \begin{pmatrix} \sin 3t \\ \cos 3t \end{pmatrix}$$

and $\ddot{\mathbf{b}}(t) = -3^2\mathbf{b}(t)$, so $\omega = 3$. Then

$$A^2 + \omega^2 I = \begin{pmatrix} -1 & -2 \\ 2 & -1 \end{pmatrix}\begin{pmatrix} -1 & -2 \\ 2 & -1 \end{pmatrix} + \begin{pmatrix} 9 & 0 \\ 0 & 9 \end{pmatrix} = \begin{pmatrix} 6 & 4 \\ -4 & 6 \end{pmatrix},$$

$$(A^2 + \omega^2 I)^{-1} = \frac{1}{52}\begin{pmatrix} 6 & -4 \\ 4 & 6 \end{pmatrix} = \frac{1}{26}\begin{pmatrix} 3 & -2 \\ 2 & 3 \end{pmatrix}$$

and

$$\mathbf{v} = -\frac{1}{26}\begin{pmatrix} 3 & -2 \\ 2 & 3 \end{pmatrix}\left[\begin{pmatrix} 3\cos 3t \\ -3\sin 3t \end{pmatrix} + \begin{pmatrix} -\sin 3t - 2\cos 3t \\ 2\sin 3t - \cos 3t \end{pmatrix}\right]$$

$$= -\frac{1}{26}\begin{pmatrix} 3 & -2 \\ 2 & 3 \end{pmatrix}\begin{pmatrix} \cos 3t - \sin 3t \\ -\cos 3t - \sin 3t \end{pmatrix} = -\frac{1}{26}\begin{pmatrix} \sin 3t - 5\cos 3t \\ 5\sin 3t + \cos 3t \end{pmatrix}.$$

This is the steady-state solution to the original differential equation. Notice that the components of $\mathbf{v}$ are again sinusoidal functions of $t$ with the same frequency as the forcing term. However, both the amplitude of the wave form of the components of $\mathbf{v}$ and its phase (the location of the crests and troughs) have been changed.

Let us examine what the steady-state behavior is for the case of the physical system described at the beginning of the section, with sinusoidal forcing term, so

$$A = \begin{pmatrix} 0 & 1 \\ -\omega_0^2 & -\Gamma \end{pmatrix}, \quad \mathbf{b}(t) = \begin{pmatrix} 0 \\ \sin \omega t \end{pmatrix}.$$

Then

$$A^2 = \begin{pmatrix} -\omega_0^2 & -\Gamma \\ \Gamma\omega_0^2 & \Gamma^2 - \omega_0^2 \end{pmatrix}$$

so

$$A^2 + \omega^2 I = \begin{pmatrix} +\omega^2 - \omega_0^2 & -\Gamma \\ \Gamma\omega_0^2 & \Gamma^2 + \omega^2 - \omega_0^2 \end{pmatrix}$$

and

$$\operatorname{Det}(A^2 + \omega^2 I) = (\omega^2 - \omega_0^2)^2 + \Gamma^2(\omega^2 - \omega_0^2) + \Gamma^2\omega_0^2 = (\omega^2 - \omega_0^2)^2 + \Gamma^2\omega^2.$$

Therefore

$$(A^2+\omega^2 I)^{-1}=\frac{1}{(\omega^2-\omega_0^2)^2+\Gamma^2\omega^2}\begin{pmatrix}\Gamma^2+\omega^2-\omega_0^2 & \Gamma\\ -\Gamma\omega_0^2 & \omega^2-\omega_0^2\end{pmatrix}$$

and our formula for the steady-state behavior of the system is given by

$$-(A^2+\omega^2 I)^{-1}(\dot{\mathbf{b}}(t)+A\mathbf{b}(t))$$
$$=\frac{-1}{(\omega^2-\omega_0^2)^2+\Gamma^2\omega^2}\begin{pmatrix}\Gamma^2+\omega^2-\omega_0^2 & \Gamma\\ -\Gamma\omega_0^2 & \omega^2-\omega_0^2\end{pmatrix}\left(\begin{pmatrix}0\\ \omega\cos\omega t\end{pmatrix}+\begin{pmatrix}\sin\omega t\\ -\Gamma\sin\omega t\end{pmatrix}\right)$$

Thus the $x$-component of the motion is

$$x(t)=\frac{-1}{(\omega^2-\omega_0^2)^2+\Gamma^2\omega^2}[\omega\Gamma\cos\omega t+(\omega^2-\omega_0^2)\sin\omega t]$$
$$=\rho\sin(\omega t+\phi)$$

where

$$\rho=\frac{1}{\sqrt{((\omega^2-\omega_0^2)^2+\Gamma^2\omega^2)}},\quad \phi=\arcsin(-\Gamma\omega/\rho)=\arccos(\omega_0^2-\omega^2)/\rho.$$

Notice that, if $\Gamma$ is small, the amplification factor is large for $\omega$ near $\omega_0$. This phenomenon is known as *resonance.* Notice also how the *phase shift*, $\phi$, changes from 0, for small values of $\omega^2$, to $-\pi/2$ for $\omega^2=\omega_0^2$, to $\arccos(-1)=-\pi$ for large values of $\omega^2$.

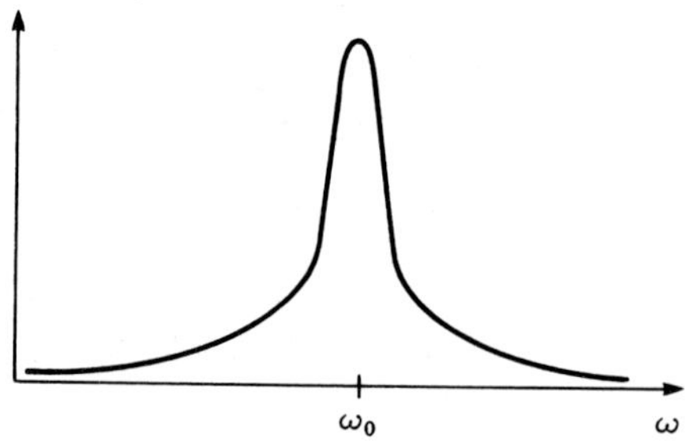


**Figure 3.17.** Response curve

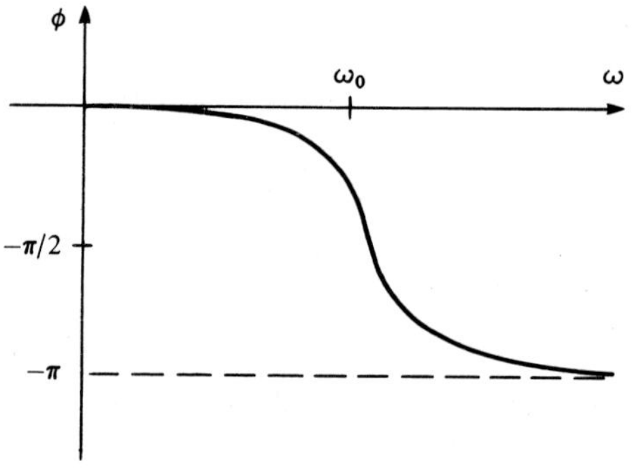


**Figure 3.18.** Phase shift graph

## Summary

A. The exponential of a matrix

You should be able to express the exponential function $e^{tX}$ as a formal power series in the matrix $X$ and to evaluate this function in cases where the powers of $X$ have a particularly simple form.

You should be able to explain the meaning of the derivative $(d/dt)e^{tX}$ and to show that it equals $Xe^{tX}$.

B. Linear differential equations

You should be able to show that every solution to $\dot{\mathbf{v}} = A\mathbf{v}$ is of the form $\mathbf{v} = e^{At}\mathbf{v}_0$.

You should be able to calculate $e^{At}$ for any $2 \times 2$ matrix $A$ and thereby to solve $\dot{\mathbf{v}} = A\mathbf{v}$ for given initial conditions.

By determining the eigenvalues of a $2 \times 2$ matrix $A$, you should be able to identify or sketch a phase portrait that represents solution curves for $\dot{\mathbf{v}} = A\mathbf{v}$.

C. Inhomogeneous equations and the harmonic oscillator

You should be able to convert the second-order differential equation that describes a harmonic oscillator to the form $\dot{\mathbf{v}} - A\mathbf{v} = \mathbf{b}$, where $A$ is a $2 \times 2$ matrix and $\mathbf{b}$ a time-dependent vector.

You should be able to solve the above equation and relate the solution to properties of the behavior of an oscillator such as damping and resonance.

## Exercises

3.1.(a) Write the power series expansions for $(1 - X)^{-1}$ and for $(1 - X)^{-2}$.

(b) Multiply these two series and compare the general term with the series for $(1 - X)^{-3}$.

3.2.(a) Let $F = \begin{pmatrix} \frac{1}{4} & \frac{1}{4} \\ \frac{1}{4} & \frac{1}{4} \end{pmatrix}$. Prove that $F^2 = \frac{1}{2}F$ and that $F^n = F/2^{n-1}$. Using this result, evaluate the series expansion of $(I - F)^{-1}$. Compute the inverse directly, and compare.

(b) Try to evaluate $\begin{pmatrix} \frac{3}{2} & \frac{1}{2} \\ \frac{1}{2} & \frac{3}{2} \end{pmatrix}^{-1}$ by writing it as $(I + P)^{-1}$ where $P$ is the projection $\begin{pmatrix} \frac{1}{2} & \frac{1}{2} \\ \frac{1}{2} & \frac{1}{2} \end{pmatrix}$ and using the series expansion of $(1 + X)^{-1}$. Notice that although the inverse exists, the series fails to converge.

3.3.(a) The matrix $N_{\pi/4} = \begin{pmatrix} -\frac{1}{2} & \frac{1}{2} \\ -\frac{1}{2} & \frac{1}{2} \end{pmatrix}$ has the property that $N_{\pi/4}^2 = 0$. Taking advantage of this property, evaluate the matrix $F(t) = \exp(tN_{\pi/4})$ and check explicitly that $F'(t) = N_{\pi/4}F(t)$.

(b) The matrix $P_{\pi/4} = \begin{pmatrix} \frac{1}{2} & \frac{1}{2} \\ \frac{1}{2} & \frac{1}{2} \end{pmatrix}$ has the property that $P_{\pi/4}^2 = P_{\pi/4}$. Taking advantage of this property, evaluate $G(t) = \exp(tP_{\pi/4})$ and check that $G'(t) = P_{\pi/4}G(t)$.

3.4. Suppose that a matrix $P$ satisfies the equation $P^2 = 3P$.

(a) What are the eigenvalues of $P$? Explain your reasoning.

(b) Using the power series for the exponential, show that $\exp(tP)$ can be expressed in the form

$$\exp(tP) = I + g(t)P.$$

Find an expression for the function $g(t)$.

3.5. Suppose that $B$ is a $2 \times 2$ matrix which has a repeated eigenvalue $\lambda$.

(a) Show that the matrix $N = B - \lambda I$ is nilpotent (i.e., $N^2 = 0$).

(b) By writing $B = N + \lambda I$ and using the series for the exponential function, show that

$$\exp(tB) = (I + tN)\exp(t\lambda I).$$

3.6. Use exercise 3.5 to solve the system of equations

$$\dot{x}(t) = x(t) - y(t)$$

$$\dot{y}(t) = x(t) + 3y(t)$$

for arbitrary initial conditions $\begin{pmatrix} x_0 \\ y_0 \end{pmatrix}$.

3.7. Calculate $\exp(tA)$ for the following matrices, and verify that $(\mathrm{d}/\mathrm{d}t)\exp(tA) = A\exp(tA)$:

(a) $A = \begin{pmatrix} -4 & 5 \\ -2 & 3 \end{pmatrix}$.

(b) $A = \begin{pmatrix} -1 & 9 \\ -1 & 5 \end{pmatrix}$. (Hint: $A = 2I + N$ where $N$ is nilpotent.)

(c) $A = \begin{pmatrix} 3 & -1 \\ 5 & -1 \end{pmatrix}$.

3.8. Let $A$ be a $2 \times 2$ matrix which has two distinct real eigenvalues $\lambda_1$ and $\lambda_2$, with associated eigenvectors $v_1$ and $v_2$.

(a) Show that the matrix $P_1 = (A - \lambda_2 I)/(\lambda_1 - \lambda_2)$ is a projection onto the line determined by the eigenvector $\mathbf{v}_1$: $P_1^2 = P_1$, the image of $P_1$ is the set of $\lambda\mathbf{v}_1$ and the kernel of $P_1$ is the set of $\lambda\mathbf{v}_2$.

(b) Similarly $P_2 = (A - \lambda_1 I)/(\lambda_2 - \lambda_1)$ is a projection onto the line determined by $\mathbf{v}_2$. Show that $P_1P_2 = P_2P_1 = 0$, that $P_1 + P_2 = I$, and that $\lambda_1 P_1 + \lambda_2 P_2 = A$.

(c) By using the power series for the exponential, show that

$$\exp(t\lambda_1 P_1 + t\lambda_2 P_2) = \mathrm{e}^{\lambda_1 t}P_1 + \mathrm{e}^{\lambda_2 t}P_2.$$

(d) Use this result to solve the equations

$$\dot{x}(t) = -4x(t) + 5y(t),$$

$$\dot{y}(t) = -2x(t) + 3y(t)$$

for arbitrary initial conditions $\begin{pmatrix} x_0 \\ y_0 \end{pmatrix}$.

3.9. Let $A$ be a $2 \times 2$ matrix whose trace is 0 and whose determinant is 1.

(a) Write down the characteristic equation of $A$, and state what this implies about $A^2$.

(b) Using the power series expansion of the exponential function, develop an expression for $\exp(tA)$ of the form

$$\exp(tA) = F(t)I + G(t)A$$

where $F(t)$ and $G(t)$ involve trigonometric functions of $t$.

(c) The solution curve for the equation $\dot{\mathbf{v}} = A\mathbf{v}$, with initial condition $\mathbf{v} = \mathbf{v}_0$, is an ellipse as shown in figure 3.19. Prove that all chords joining $\exp(tA)\mathbf{v}_0$ to $\exp(-tA)\mathbf{v}_0$ are parallel to $A\mathbf{v}_0$ and that the midpoint of each such chord lies on the diameter of the ellipse on which $\mathbf{v}_0$ lies.

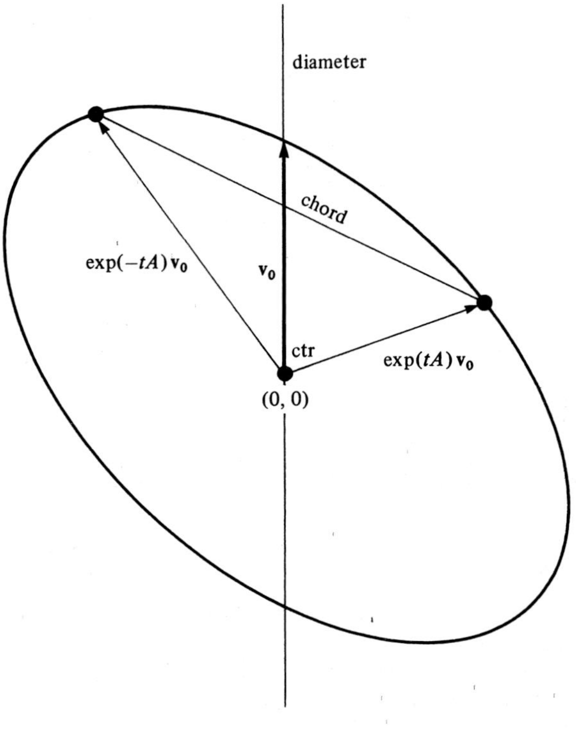


**Figure 3.19**

3.10. Suppose that $G$ is a matrix whose trace is zero and whose determinant is $-\beta^2$.

(a) According to the Cayley–Hamilton theorem, what does $G^2$ equal?

(b) Using the power series for the exponential function, show that $\exp G + \exp(-G)$ is a multiple of the identity matrix. Find a function $f$ such that

$$\exp(G) + \exp(-G) = f(\beta)I.$$

(c) By multiplying the above identity by $\exp G$ and applying the Cayley–Hamilton theorem, show that $\operatorname{Det}(\exp(G)) = 1$, and find an expression for the trace of $\exp G$.

(d) Let $F = \lambda I + G$. Using the above results, show that $\operatorname{Det}(\exp F) = e^{(\operatorname{tr} F)}$.

3.11. For each of the following differential equations, determine which of the phase portraits given in cases 1 through 4c best represents the nature of the general solution, then solve the equation completely for initial conditions $\begin{pmatrix} x_0 \\ y_0 \end{pmatrix} = \begin{pmatrix} 3 \\ 1 \end{pmatrix}$ at $t = 0$.

(a) $\dot{x} = -4y,$
$\dot{y} = x - 4y.$

(b) $\dot{x} = x - 2y,$
$\dot{y} = -2x + 4y.$

(c) $\dot{x} = 4x - 5y,$
$\dot{y} = 4x - 4y.$

(d) $\dot{x} = 2x + y,$
$\dot{y} = -x + 4y.$

(e) $\dot{x} = x - 5y,$
$\dot{y} = 2x - 5y.$

(f) $\dot{x} = -2x + 4y,$
$\dot{y} = -x + 2y.$

3.12. For each of the following differential equations, determine which of the phase portraits given in cases 1 through 4c best represents the general solution, then solve the equation completely for initial conditions $\begin{pmatrix} x_0 \\ y_0 \end{pmatrix} = \begin{pmatrix} -2 \\ 1 \end{pmatrix}$ at $t = 0$.

(a) $\dot{x} = 3y,$
$\dot{y} = x - 2y.$

(b) $\dot{x} = -x + y,$
$\dot{y} = -5x + 3y.$

(c) $\dot{x} = 3x + y,$
$\dot{y} = -x + y.$

(d) $\dot{x} = -5x + 4y,$
$\dot{y} = -8x + 7y.$

(e) $\dot{x} = -4x - 2y,$
$\dot{y} = 5x + 2y$

(f) $\dot{x} = x + 2y,$
$\dot{y} = 2x - 4y.$

3.13. By generalizing what you know about calculating and using the exponential of a $2 \times 2$ matrix to the $3 \times 3$ case, solve the differential equations

$$\dot{x} = y,$$
$$\dot{y} = z,$$
$$\dot{z} = -6x - 11y - 6z$$

for initial conditions $\begin{pmatrix} x \\ y \\ z \end{pmatrix} = \begin{pmatrix} 1 \\ 2 \\ -1 \end{pmatrix}$ at $t = 0$.

(Note: The one tricky new step is inverting a $3 \times 3$ matrix. If you regard this as the problem of solving three sets of simultaneous linear equations, you can do it by brute force.)

3.14. By generalizing the techniques which you already know. Solve the equations

$$\dot{x} = x + y - z,$$

$$\dot{y} = -x + 5y + z,$$
$$\dot{z} = -2x + 2y + 4z,$$

for initial conditions

$$\begin{pmatrix} x \\ y \\ z \end{pmatrix} = \begin{pmatrix} 1 \\ 1 \\ 1 \end{pmatrix}.$$

3.15. By introducing the variable $v = \dot{x}$, convert the second-order differential equation

$$\ddot{x} + 4\dot{x} + 5x = 0$$

to a pair of first-order equations, then solve these equations for arbitrary initial conditions $\begin{pmatrix} x_0 \\ v_0 \end{pmatrix}$.

3.16.(a) The differential equation for a critically damped harmonic oscillator, expressed in units chosen so that $\omega_0^2 = 1$, is

$$\ddot{x} + 2\dot{x} + x = 0.$$

Solve this equation by matrix methods, introducing $v = \dot{x}$ as a new variable. Write down the solution for initial conditions $\begin{pmatrix} x \\ v \end{pmatrix} = \begin{pmatrix} x_0 \\ 0 \end{pmatrix}$ and for $\begin{pmatrix} x \\ v \end{pmatrix} = \begin{pmatrix} 0 \\ v_0 \end{pmatrix}$, and sketch phase portraits of these and other solutions. Show that $x = 0$ or $v = 0$ can occur at most once.

(b) One way of solving the above equation without having to contend with a repeated eigenvalue is first to solve $\ddot{x} + 2\dot{x} + (1 - \varepsilon^2)x = 0$, which leads to a matrix with distinct real eigenvalues, then let $\varepsilon \to 0$. (Physically, this corresponds to using a slightly weaker spring.) Carry through the procedure, first finding solutions for initial conditions $\begin{pmatrix} x_0 \\ 0 \end{pmatrix}$ and $\begin{pmatrix} 0 \\ v_0 \end{pmatrix}$, then letting $\varepsilon \to 0$. Show what happens to the phase portraits as $\varepsilon > 0$.

(c) Another alternative is first to solve $\ddot{x} + 2\dot{x} + (1 + \varepsilon^2)x = 0$, which leads to a matrix with complex eigenvalues, then let $\varepsilon \to 0$. Do this, again showing what happens to the solutions for initial conditions $\begin{pmatrix} x_0 \\ 0 \end{pmatrix}$ and $\begin{pmatrix} 0 \\ v_0 \end{pmatrix}$, and to the phase portraits, as $\varepsilon \to 0$.

3.17. Consider the function $\cos tx$.

(a) Show, by use of formal power series, that

$$\frac{d^2}{dt^2}(\cos tx) = -x^2 \cos tx$$

and that

$$\frac{d}{dt}(\cos tx) = 0 \quad \text{for} \quad t = 0.$$

(b) Suppose that $\begin{pmatrix} \ddot{x} \\ \ddot{y} \end{pmatrix} = -B\begin{pmatrix} x \\ y \end{pmatrix}$, where $B$ is a matrix which has a square root $A$. Show that $\cos tA\begin{pmatrix} x(0) \\ y(0) \end{pmatrix}$ is a solution to the second-order system of equations

$$\frac{d^2}{dt^2}v(t) = -A^2 v(t)$$

with initial conditions $v(0) = v_0$ and $dv/dt(0) = 0$.

(c) Let $B = \begin{pmatrix} \frac{5}{2} & -\frac{3}{2} \\ -\frac{3}{2} & \frac{5}{2} \end{pmatrix}$. Find a matrix $A$, with positive eigenvalues, such that $A^2 = B$. (Hint: diagonalize $B$.)

(d) For the matrix $A$ which you have just constructed, compute the matrix $\cos(tA)$. (Hint: You have already diagonalized $A$. Use procedure similar to that for computing $\exp(tA)$.)

(e) Use the above results to solve the equations

$$\ddot{x} = -\tfrac{5}{2}x + \tfrac{3}{2}y,$$
$$\ddot{y} = \tfrac{3}{2}x - \tfrac{5}{2}y$$

for initial conditions $\dot{x}(0) = \dot{y}(0) = 0$, $x(0) = x_0$, $y(0) = y_0$.

3.18. Consider the system of differential equations

$$\dot{x} = 4\beta x - y$$
$$\dot{y} = 9x + \beta y$$

where $\beta$ is a real-valued parameter.

(a) Solve the system for arbitrary initial conditions and $\beta = 0$.

(b) Find two critical values of the parameter, $\beta_1 < 0$ and $\beta_2 > 0$, at which the nature of the solution changes. Discuss the solutions for $\beta = \beta_1$ and $\beta = \beta_2$.

(c) Draw phase portraits which describe qualitatively the nature of the solutions for $\beta < \beta_1$, $\beta_1 < \beta < \beta_2$, and $\beta > \beta_2$.

3.19. Let $A = \begin{pmatrix} -2 & 1 \\ 2 & -1 \end{pmatrix}$.

(a) Find matrices $D$ and $B$ so that $A = BDB^{-1}$.

(b) Construct the solution to the differential equation $\dot{\mathbf{v}} = A\mathbf{v}$ for arbitrary initial conditions $\mathbf{v}_0 = \begin{pmatrix} x_0 \\ y_0 \end{pmatrix}$ when $t = 0$. Please remember that $e^0 = 1$.

(c) Sketch a phase portrait for the equation $\dot{\mathbf{v}} = A\mathbf{v}$. Determine the image and kernel of the matrix

$$F = \lim_{t\to\infty} \exp(At),$$

and explain their significance in relation to the phase portrait.

(d) By using the trial solution $\mathbf{v} = \exp(At)\mathbf{w}$, construct a solution to the differential equation $\dot{\mathbf{v}} - A\mathbf{v} = \begin{pmatrix} 1 \\ 2 \end{pmatrix}$.

3.20.(a) By introducing $u = \dot{x}$ as a new variable, convert

$$\ddot{x} + 2\dot{x} - 3x = 3\sin 2t + 2\cos 2t$$

to an equation of the form

$$\begin{pmatrix} \dot{x} \\ \dot{u} \end{pmatrix} - A\begin{pmatrix} x \\ u \end{pmatrix} = \mathbf{b}(t).$$

(b) Solve this equation for initial conditions $x(0)=0$, $u(0)=0$ by using the results developed in section 3.4.

3.21. When an undamped oscillator is acted upon by a force at the natural frequency of the oscillator, conventional methods of solution fail because no steady state is ever achieved. The formula developed in the notes,

$$\mathbf{v}(t)=\int_0^t \exp[(t-s)A]\mathbf{b}(s)\,\mathrm{d}s,$$

works fine, however. Use it to solve

$$\begin{pmatrix}\dot{x}\\ \dot{u}\end{pmatrix}-\begin{pmatrix}0 & 1\\ \omega^2 & 0\end{pmatrix}\begin{pmatrix}x\\ u\end{pmatrix}=\begin{pmatrix}0\\ \sin\omega t\end{pmatrix}$$

for initial conditions $x(0)=0$, $u(0)=0$.

# 4

# Scalar products



Chapter 4 is devoted the study of scalar products and quadratic forms. It is rich in physical applications, including a discussion of normal modes and a detailed treatment of special relativity.

## 4.1. The Euclidean scalar product

In an affine plane, as you will recall, we have only a very restricted notion of length: we can compare lengths of segments of parallel lines, but not lengths of segments along lines which are not parallel. For example, it is meaningful to say that the length of $QR$ (or $Q'R'$) is twice the length of $PQ$ in figure 4.1, but we cannot compare the length of $PQ'$ with that of $PQ$.

**Figure 4.1**

A Euclidean plane is an affine plane endowed with a *distance function* which assigns to every pair of points a non-negative real number, $D(P,Q)$, called the distance between them. This distance function is compatible with the limited notion of length in affine geometry; e.g., $D(Q',R') = 2D(P,Q)$ in figure 4.1, but it also permits us to compare lengths of nonparallel segments such as $PQ$ and $PQ'$. In the Euclidean plane $\mathbb{R}^2$, the distance function is defined by the well-known formula

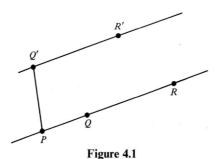

$$D(P, Q) = \sqrt{[(x_Q - x_P)^2 + (y_Q - y_P)^2]}.$$

A *Euclidean* transformation $f: \mathbb{R}^2 \to \mathbb{R}^2$ is an affine transformation which preserves this distance function: i.e., $D(f(P), f(Q)) = D(P, Q)$.

Turning our attention to the Euclidean vector space of displacements in the Euclidean plane, we see that the distance function provides a way of assigning a length to each vector: the length is simply the distance from 'head' to 'tail'. We denote the length of a vector $\mathbf{v}$ by $\|\mathbf{v}\|$. Clearly, if $\mathbf{v} = \begin{pmatrix} x \\ y \end{pmatrix}$, then $\|\mathbf{v}\| = \sqrt{(x^2 + y^2)}$.

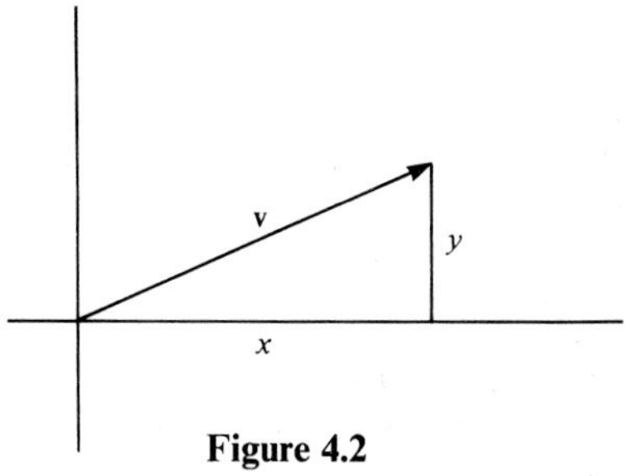


**Figure 4.2**

In general, the linear transformations of the vector space $\mathbb{R}^2$ do not preserve the lengths of vectors. Those linear transformations which do preserve length are called *orthogonal* transformations: they are all either rotations about the origin or reflections in lines through the origin.

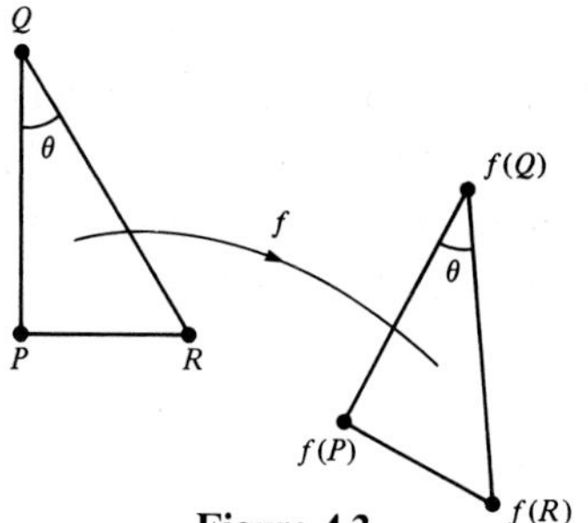


**Figure 4.3**

Since a Euclidean transformation of the plane preserves length, it carries every triangle into a congruent triangle and hence preserves *angles* as well as lengths. In particular, the notion of 'perpendicular' makes good sense in Euclidean geometry (though not in affine geometry). We say that two vectors $\mathbf{v}$ and $\mathbf{w}$ are *perpendicular* or *orthogonal* if the triangle which they define satisfies the Pythagorean theorem: i.e., if

$$\|\mathbf{v}\|^2 + \|\mathbf{w}\|^2 = \|\mathbf{v} - \mathbf{w}\|^2.$$

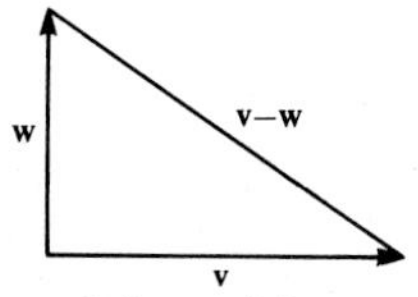


**Figure 4.4**

In terms of length and angle, we can now define the Euclidean *scalar product* of two vectors. If $\mathbf{v} = \begin{pmatrix} x \\ y \end{pmatrix}$ and $\mathbf{v}' = \begin{pmatrix} x' \\ y' \end{pmatrix}$ are two vectors, their scalar product, $(\mathbf{v}, \mathbf{v}')$ is defined as

$$(\mathbf{v}, \mathbf{v}') = \| \mathbf{v} \| \, \| \mathbf{v}' \| \cos \theta$$

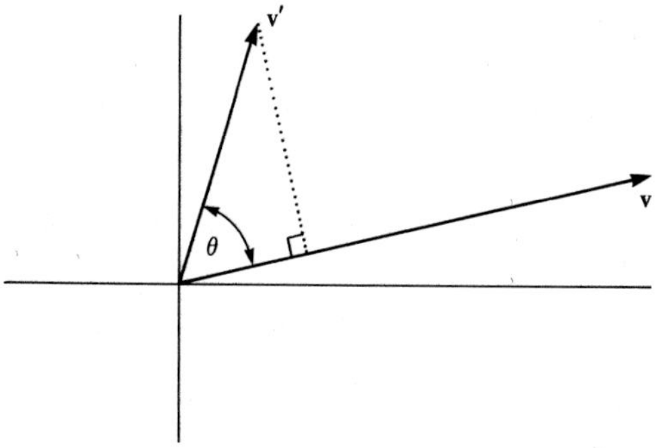


**Figure 4.5**

where $\theta$ is the angle between the two vectors. Geometrically this means the following: we take the projection of $\mathbf{v}'$ onto the line through $\mathbf{v}$; we then multiply the length of this projected vector with the length of $\mathbf{v}$ with a plus or minus sign according as to whether the projected vector points in the same or in the opposite direction as $\mathbf{v}$. Since the scalar product is defined entirely in terms of the Euclidean geometry of the plane any linear transformation which preserves length must also preserve the scalar product: any such linear transformation preserves lengths of vectors and the angle between them, hence preserves their scalar product. In other words, if $M$ is any orthogonal transformation then

$$(M\mathbf{v}, M\mathbf{v}') = (\mathbf{v}, \mathbf{v}')$$

for any pair of vectors $\mathbf{v}$ and $\mathbf{v}'$. We shall give a more algebraic proof of this fact (cf. eq. (4.1)) below. Conversely, since $(\mathbf{v}, \mathbf{v}) = \| \mathbf{v} \|^2$, any $M$ which satisfies the above equation for all $\mathbf{v}$ and $\mathbf{v}'$ is certainly orthogonal. Suppose we hold $\mathbf{v}$ fixed and consider $(\mathbf{v}, \mathbf{v}')$ as a function of $\mathbf{v}'$. We claim that $(\mathbf{v}, \mathbf{v}')$ is a linear function of $\mathbf{v}'$; i.e. that

$(\mathbf{v}, a\mathbf{v}' + b\mathbf{w}') = a(\mathbf{v}, \mathbf{v}') + b(\mathbf{v}, \mathbf{w}')$ *for any numbers $a$ and $b$ and any vectors $\mathbf{v}'$ and $\mathbf{w}'$.*

We can see this most simply as follows. Suppose that we first consider the special case where $\mathbf{v} = \begin{pmatrix} c \\ 0 \end{pmatrix}$ lies on the $x$-axis. Then $(\mathbf{v}, \mathbf{v}') = cx'$ for $\mathbf{v}' = \begin{pmatrix} x' \\ y' \end{pmatrix}$. This expression clearly depends linearly on $\mathbf{v}'$, so we have verified the above assertion for this special case. But now let $\mathbf{v}$ be any vector. We can find a rotation $M$ which moves $\mathbf{v}$ to the $x$-axis. But $(\mathbf{v}, \mathbf{v}') = (M\mathbf{v}, M\mathbf{v}')$ depends linearly on $M\mathbf{v}'$; and $M\mathbf{v}'$ depends linearly on $\mathbf{v}'$ so we are done. To repeat the argument in more detail:

$$
\begin{aligned}
(\mathbf{v}, a\mathbf{v}' + b\mathbf{w}') &= (M\mathbf{v}, M(a\mathbf{v}' + b\mathbf{w}')) && \text{since } M \text{ is orthogonal} \\
&= (M\mathbf{v}, aM\mathbf{v}' + bM\mathbf{w}') && \text{since } M \text{ is linear} \\
&= a(M\mathbf{v}, M\mathbf{v}') + b(M\mathbf{v}, M\mathbf{w}') && \text{because we have verified this in the special case that } M\mathbf{v} \text{ lies on the } x\text{-axis} \\
&= a(\mathbf{v}, \mathbf{v}') + b(\mathbf{v}, \mathbf{w}') && \text{since } M \text{ is orthogonal.}
\end{aligned}
$$

Since the scalar product $(\mathbf{v}, \mathbf{v}')$ is symmetrical in $\mathbf{v}$ and $\mathbf{v}'$, we see that $(\mathbf{v}, \mathbf{v}')$ is also linear as a function of $\mathbf{v}$ when we hold $\mathbf{v}'$ fixed. These two facts allow us to write down the formula for the scalar product: write $\mathbf{v} = \begin{pmatrix} x \\ y \end{pmatrix} = x\begin{pmatrix} 1 \\ 0 \end{pmatrix} + y\begin{pmatrix} 0 \\ 1 \end{pmatrix}$ and $\mathbf{v}' = \begin{pmatrix} x' \\ y' \end{pmatrix} = x'\begin{pmatrix} 1 \\ 0 \end{pmatrix} + y'\begin{pmatrix} 0 \\ 1 \end{pmatrix}$. Now the scalar product of $\begin{pmatrix} 1 \\ 0 \end{pmatrix}$ with $\begin{pmatrix} 0 \\ 1 \end{pmatrix}$ vanishes since the vectors are orthogonal, and each of these basis vectors has length one. So,

$$
\begin{aligned}
(\mathbf{v}, \mathbf{v}') &= x\left(\begin{pmatrix} 1 \\ 0 \end{pmatrix}, \mathbf{v}'\right) + y\left(\begin{pmatrix} 0 \\ 1 \end{pmatrix}, \mathbf{v}'\right) \quad \text{using the linearity in } \mathbf{v} \\
&= xx'\left(\begin{pmatrix} 1 \\ 0 \end{pmatrix}, \begin{pmatrix} 1 \\ 0 \end{pmatrix}\right) + xy'\left(\begin{pmatrix} 1 \\ 0 \end{pmatrix}, \begin{pmatrix} 0 \\ 1 \end{pmatrix}\right) + yx'\left(\begin{pmatrix} 1 \\ 0 \end{pmatrix}, \begin{pmatrix} 0 \\ 1 \end{pmatrix}\right) + yy'\left(\begin{pmatrix} 0 \\ 1 \end{pmatrix}, \begin{pmatrix} 0 \\ 1 \end{pmatrix}\right) \\
&\qquad \text{using linearity in } \mathbf{v}' \\
&= xx' + yy' \quad \text{since} \quad \left(\begin{pmatrix} 1 \\ 0 \end{pmatrix}, \begin{pmatrix} 1 \\ 0 \end{pmatrix}\right) = 1 = \left(\begin{pmatrix} 0 \\ 1 \end{pmatrix}, \begin{pmatrix} 0 \\ 1 \end{pmatrix}\right) \quad \text{and} \\
&\qquad \left(\begin{pmatrix} 1 \\ 0 \end{pmatrix}, \begin{pmatrix} 0 \\ 1 \end{pmatrix}\right) = 0 = \left(\begin{pmatrix} 0 \\ 1 \end{pmatrix}, \begin{pmatrix} 1 \\ 0 \end{pmatrix}\right).
\end{aligned}
$$

We have thus found a convenient formula for the scalar product of two vectors in the plane:

$$(\mathbf{v}, \mathbf{v}') = xx' + yy'.$$

We can summarize the important properties of the Euclidean scalar product as follows:

(1) Symmetry: $(\mathbf{v}, \mathbf{v}') = (\mathbf{v}', \mathbf{v})$.
(2) Bilinearity: $(\mathbf{v}, a\mathbf{v}' + b\mathbf{w}') = a(\mathbf{v}, \mathbf{v}') + b(\mathbf{v}, \mathbf{w}')$.
(3) Positive definiteness: $(\mathbf{v}, \mathbf{v}) \geqslant 0$, and $(\mathbf{v}, \mathbf{v}) = 0$ only if $\mathbf{v} = 0$.

Using these properties, it is easy to express the scalar product in terms of length. Just consider

$$\|\mathbf{v} - \mathbf{w}\|^2 = (\mathbf{v} - \mathbf{w}, \mathbf{v} - \mathbf{w}).$$

Because the scalar product is linear in each factor,

$$\|\mathbf{v}-\mathbf{w}\|^2 = (\mathbf{v},\mathbf{v}) - (\mathbf{v},\mathbf{w}) - (\mathbf{w},\mathbf{v}) + (\mathbf{w},\mathbf{w}).$$

But since $(\mathbf{v},\mathbf{v}) = \|\mathbf{v}\|^2$, $(\mathbf{w},\mathbf{w}) = \|\mathbf{w}\|^2$ and $(\mathbf{w},\mathbf{v}) = (\mathbf{v},\mathbf{w})$, we have

$$2(\mathbf{v},\mathbf{w}) = \|\mathbf{v}\|^2 + \|\mathbf{w}\|^2 - \|\mathbf{v}-\mathbf{w}\|^2$$

and so

$$(\mathbf{v},\mathbf{w}) = \tfrac{1}{2}\{\|\mathbf{v}\|^2 + \|\mathbf{w}\|^2 - \|\mathbf{v}-\mathbf{w}\|^2\}. \tag{4.1}$$

This formula makes it clear that the Euclidean scalar product follows immediately from the Euclidean notion of length. If you write $(\mathbf{v},\mathbf{w}) = \|\mathbf{v}\|\,\|\mathbf{w}\|\cos\theta$ and look at figure 4.6 you will see that (4.1) is nothing more than the 'law of cosines' in disguise.

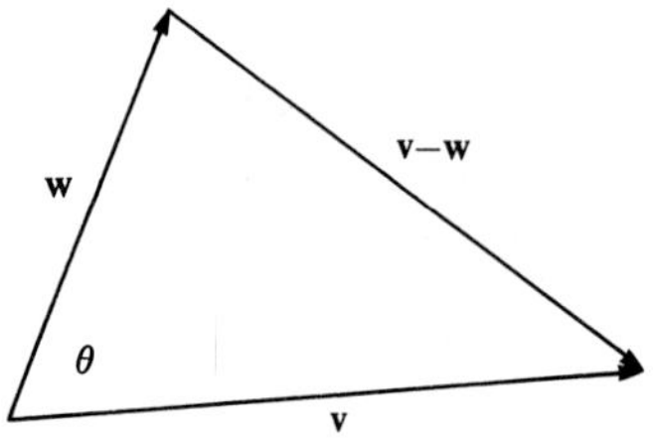


**Figure 4.6**

## 4.2. The Gram–Schmidt process

Let $V$ be an abstract two-dimensional vector space and suppose that we are given a positive-definite scalar product, $(\ ,\ )_V$ on $V$. That is, suppose we are given a function which assigns to each pair of vectors $\mathbf{v}_1, \mathbf{v}_2$ in $V$ a real number $(\mathbf{v}_1, \mathbf{v}_2)_V$ and which satisfies the conditions of symmetry, bilinearity and positive definiteness. We claim that there exists a linear isomorphism* $L: V \to \mathbb{R}^2$ such that

$$(\mathbf{v}_1, \mathbf{v}_2)_V = (L\mathbf{v}_1, L\mathbf{v}_2).$$

In other words, by the correct choice of a basis on $V$, we can arrange that the scalar product $(\ ,\ )_V$ on $V$ looks just like the Euclidean scalar product $(\ ,\ )$ on $\mathbb{R}^2$. To prove this, choose some non-zero vector $\mathbf{w}$ in $V$. Since $\|\mathbf{w}\|_V^2 = (\mathbf{w},\mathbf{w})_V > 0$, the vector

$$\mathbf{e}_1 = \frac{\mathbf{w}}{\|\mathbf{w}\|_V}$$

has unit length, i.e.,

$$\|\mathbf{e}_1\|^2 = (\mathbf{e}_1, \mathbf{e}_1)_V = \frac{1}{\|\mathbf{w}\|_V^2}(\mathbf{w},\mathbf{w})_V = 1.$$

* We recall that the word isomorphism means that L is linear, is one to one, surjective (and therefore has a linear inverse), see §1.12.

Now let $\mathbf{u}$ be any vector in $V$ which is linearly independent of $\mathbf{e}_1$. We know that such a $\mathbf{u}$ exists since $V$ is two-dimensional. Let

$$\mathbf{u}_2 = \mathbf{u} - (\mathbf{u}, \mathbf{e}_1)_V \mathbf{e}_1.$$

We observe that $\mathbf{u}_2$ is perpendicular to $\mathbf{e}_1$, that is,

$$(\mathbf{u}_2, \mathbf{e}_1)_V = 0.$$

Indeed,

$$\begin{aligned}(\mathbf{u}_2, \mathbf{e}_1)_V &= (\mathbf{u} - (\mathbf{u}, \mathbf{e}_1)_V \mathbf{e}_1, \mathbf{e}_1)_V \\ &= (\mathbf{u}, \mathbf{e}_1)_V - (\mathbf{u}, \mathbf{e}_1)_V (\mathbf{e}_1, \mathbf{e}_1)_V \\ &= (\mathbf{u}, \mathbf{e}_1)_V - (\mathbf{u}, \mathbf{e}_1)_V = 0\end{aligned}$$

since $(\mathbf{e}_1, \mathbf{e}_1)_V = 1$. Also $\mathbf{u}_2 \neq 0$, for otherwise $\mathbf{e}_1$ and $\mathbf{u}$ would be linearly dependent. Now set

$$\mathbf{e}_2 = \frac{1}{\|\mathbf{u}_2\|_V} \mathbf{u}_2.$$

Then $(\mathbf{e}_2, \mathbf{e}_1)_V = (1/\|\mathbf{u}_2\|_V)(\mathbf{u}_2, \mathbf{e}_1)_V = 0$, and $\|\mathbf{e}_2\|_V = 1$. We will use $\mathbf{e}_1, \mathbf{e}_2$ as our basis of $V$. The most general vector in $V$ can be written as

$$\mathbf{v} = x\mathbf{e}_1 + y\mathbf{e}_2.$$

Notice that

$$x = (\mathbf{v}, \mathbf{e}_1)$$

since $(\mathbf{e}_2, \mathbf{e}_1) = 0$ and $(\mathbf{e}_1, \mathbf{e}_1) = 1$. Similarly

$$y = (\mathbf{v}, \mathbf{e}_2).$$

Suppose that

$$\mathbf{v}_1 = x_1\mathbf{e}_1 + y_1\mathbf{e}_2$$

and

$$\mathbf{v}_2 = x_2\mathbf{e}_1 + y_2\mathbf{e}_2$$

so that the map $L: V \to \mathbb{R}^2$ by our basis satisfies

$$L\mathbf{v}_1 = \begin{pmatrix} x_1 \\ y_1 \end{pmatrix} \quad \text{and} \quad L\mathbf{v}_2 = \begin{pmatrix} x_2 \\ y_2 \end{pmatrix}.$$

Then

$$\begin{aligned}(\mathbf{v}_1, \mathbf{v}_2)_V &= (x_1\mathbf{e}_1 + y_1\mathbf{e}_2, x_2\mathbf{e}_1 + y_2\mathbf{e}_2)_V \\ &= x_1x_2 + y_1y_2 \quad \text{as} \quad (\mathbf{e}_1, \mathbf{e}_2)_V = (\mathbf{e}_2, \mathbf{e}_1)_V = 0 \\ &\qquad\qquad\qquad \text{and} \quad (\mathbf{e}_1, \mathbf{e}_1)_V = (\mathbf{e}_2, \mathbf{e}_2)_V = 1 \\ &= (L\mathbf{v}_1, L\mathbf{v}_2).\end{aligned}$$

This is what we wanted to prove.

On $\mathbb{R}^3$ we can define the Euclidean scalar product by

$$\left( \begin{pmatrix} x_1 \\ y_1 \\ z_1 \end{pmatrix}, \begin{pmatrix} x_2 \\ y_2 \\ z_2 \end{pmatrix} \right) = x_1x_2 + y_1y_2 + z_1z_2.$$

Again, it is clear that if $\mathbf{v} = \begin{pmatrix} x \\ y \\ z \end{pmatrix}$ then

$$\| \mathbf{v} \|^2 = (\mathbf{v}, \mathbf{v})$$

represents the square of the Euclidean length of the vector $\mathbf{v}$. The argument given above shows that we can recover the scalar product from the length by the same formula, (4.1):

$$(\mathbf{v}, \mathbf{w}) = \tfrac{1}{2}(\| \mathbf{v} \|^2 + \| \mathbf{w} \|^2 - \| \mathbf{v} - \mathbf{w} \|^2).$$

So any rotation of three-dimensional space preserves the scalar product. In particular, if we are given two vectors, $\mathbf{v}$ and $\mathbf{w}$, we can rotate the plane that they span into the $z = 0$ plane. For vectors in that plane, the scalar product reduces to the scalar product for $\mathbb{R}^2$. For such vectors we know that

$$(\mathbf{v}, \mathbf{w}) = \| \mathbf{v} \| \, \| \mathbf{w} \| \cos \theta$$

and hence (since both sides are invariant under rotation) it is true for all pairs of vectors.

A vector space $V$ is called three-dimensional if every four vectors are linearly dependent but there *are* three vectors which are linearly independent. Thus given *any* four vectors $\mathbf{v}_1, \mathbf{v}_2, \mathbf{v}_3, \mathbf{v}_4$ we can find four numbers $a_1, a_2, a_3, a_4$ not all zero such that

$$a_1\mathbf{v}_1 + a_2\mathbf{v}_2 + a_3\mathbf{v}_3 + a_4\mathbf{v}_4 = \mathbf{0}$$

but there exist three vectors $\mathbf{u}, \mathbf{v}, \mathbf{w}$ such that

$$a\mathbf{u} + b\mathbf{v} + c\mathbf{w} = \mathbf{0}$$

is *not* true unless $a = b = c = 0$. Suppose that $V$ has a positive-definite scalar product $( \ , \ )_V$. We can now repeat the argument given above for the two-dimensional case. Pick some non-zero vector. By multiplying by a scalar, we can arrange that it has unit length. Call it $\mathbf{e}_1$. Choose some vector $\mathbf{u}$ so that $\mathbf{e}_1$ and $\mathbf{u}$ are linearly independent. Set

$$\mathbf{u}_2 = \mathbf{u} - (\mathbf{u}, \mathbf{e}_1)_V \mathbf{e}_1$$

and

$$\mathbf{e}_2 = \frac{1}{\| \mathbf{u}_2 \|_V} \mathbf{u}_2.$$

Then $\mathbf{e}_1$ and $\mathbf{e}_2$ satisfy

$$\| \mathbf{e}_1 \|_V = \| \mathbf{e}_2 \|_V = 1 \quad \text{and} \quad (\mathbf{e}_1, \mathbf{e}_2)_V = 0.$$

The set of all vectors of the form $x\mathbf{e}_1 + y\mathbf{e}_2$ is isomorphic to $\mathbb{R}^2$ and hence is a two-dimensional vector space. Thus it can not be all of $V$. (We can not find three linearly independent vectors in this set.) Thus there must be some vector $\mathbf{w}$ in $V$ which is not of the form $x\mathbf{e}_1 + y\mathbf{e}_2$. Thus

$$\mathbf{w}_3 = \mathbf{w} - (\mathbf{w}, \mathbf{e}_1)_V \mathbf{e}_1 - (\mathbf{w}, \mathbf{e}_2)_V \mathbf{e}_2$$

is not zero. Set

$$\mathbf{e}_3 = \frac{1}{\|\mathbf{w}_3\|_V}\mathbf{w}_3.$$

Then

$$\|\mathbf{e}_1\|_V = \|\mathbf{e}_2\|_V = \|\mathbf{e}_3\|_V = 1$$

and

$$(\mathbf{e}_1, \mathbf{e}_2)_V = (\mathbf{e}_1, \mathbf{e}_3)_V = (\mathbf{e}_2, \mathbf{e}_3)_V = 0.$$

If $\mathbf{v}$ is any vector in $V$, we claim that

$$\mathbf{v} - (\mathbf{v}, \mathbf{e}_1)_V\mathbf{e}_1 - (\mathbf{v}, \mathbf{e}_2)_V\mathbf{e}_2 - (\mathbf{v}, \mathbf{e}_3)_V\mathbf{e}_3 = \mathbf{0}.$$

Indeed, by the same argument as before, we set

$$\mathbf{v}_4 = \mathbf{v} - (\mathbf{v}, \mathbf{e}_1)\mathbf{e}_1 - (\mathbf{v}, \mathbf{e}_2)\mathbf{e}_2 - (\mathbf{v}, \mathbf{e}_3)\mathbf{e}_3$$

then

$$(\mathbf{v}_4, \mathbf{e}_1)_V = (\mathbf{v}_4, \mathbf{e}_2)_V = (\mathbf{v}_4, \mathbf{e}_3)_V = 0.$$

But this means that if $\mathbf{v}_4 \neq \mathbf{0}$ the vectors $\mathbf{e}_1, \mathbf{e}_2, \mathbf{e}_3, \mathbf{v}_4$ would be linearly independent: indeed, taking the scalar product of

$$a_1\mathbf{e}_1 + a_2\mathbf{e}_2 + a_3\mathbf{e}_3 + a_4\mathbf{v}_4 = \mathbf{0}$$

with $\mathbf{e}_1, \mathbf{e}_2$ and $\mathbf{e}_3$ shows that $a_1 = 0$, $a_2 = 0$, $a_3 = 0$. Thus, if $\mathbf{v}_4 \neq \mathbf{0}$, $a_4 = 0$. This contradicts the assumption that $V$ is three-dimensional.

Thus every vector in $V$ can be written as

$$\mathbf{v} = x\mathbf{e}_1 + y\mathbf{e}_2 + z\mathbf{e}_3 \quad \text{where} \quad x = (\mathbf{v}, \mathbf{e}_1), y = (\mathbf{v}, \mathbf{e}_2), z = (\mathbf{v}, \mathbf{e}_3).$$

Just as in the two-dimensional case, we can define the map

$$L: V \to \mathbb{R}^3$$

$$L(\mathbf{v}) = \begin{pmatrix} x \\ y \\ z \end{pmatrix} \quad \text{if} \quad \mathbf{v} = x\mathbf{e}_1 + y\mathbf{e}_2 + z\mathbf{e}_3.$$

This map is a linear, one-to-one, map of $V$ onto $\mathbb{R}^3$ and

$$(\mathbf{u}, \mathbf{v})_V = (L\mathbf{u}, L\mathbf{v})_{\mathbb{R}^3}.$$

It is clear that we can prove the same sort of result in four, five, $\ldots, n$ dimensions. On $\mathbb{R}^n$ define the *Euclidean scalar product*

$$\left(\begin{pmatrix} x_1 \\ \vdots \\ x_n \end{pmatrix}, \begin{pmatrix} w_1 \\ \vdots \\ w_n \end{pmatrix}\right) = x_1w_1 + \cdots + x_nw_n.$$

A vector space $V$ is called $n$-dimensional if there exist $n$ linearly independent vectors but every collection of $n+1$ vectors is linearly dependent. We shall study the general theory of $n$-dimensional vector spaces in Chapter 10. If $V$ is $n$-dimensional and has a positive-definite scalar product, then we can find an *orthonormal basis*

$\mathbf{e}_1, \ldots, \mathbf{e}_n$. That is, we can find $n$ vectors $\mathbf{e}_1, \ldots, \mathbf{e}_n$ such that

$$\|\mathbf{e}_1\|_V = \|\mathbf{e}_2\|_V = \cdots = \|\mathbf{e}_n\|_V = 1$$

and

$$(\mathbf{e}_i, \mathbf{e}_j)_V = 0 \quad i \neq j.$$

Every vector $\mathbf{v}$ in $V$ can be written as

$$\mathbf{v} = x_1\mathbf{e}_1 + \cdots + x_n\mathbf{e}_n, \quad x_i = (\mathbf{v}, \mathbf{e}_i)$$

and thus define the map $L: V \to \mathbb{R}^n$

$$L\mathbf{v} = \begin{pmatrix} x_1 \\ \vdots \\ x_n \end{pmatrix}.$$

Then

$$(\mathbf{v}, \mathbf{w})_V = (L\mathbf{v}, L\mathbf{w}).$$

In fact, if we start with $n$ independent vectors $\mathbf{v}_1, \ldots, \mathbf{v}_n$, we can get $\mathbf{e}$s by the algorithm

$$\begin{aligned}
\mathbf{e}_1 &= \frac{1}{\|\mathbf{v}_1\|}\mathbf{v}_1,\\
\mathbf{u}_2 &= \mathbf{v}_2 - (\mathbf{v}_2, \mathbf{e}_1)\mathbf{e}_1,\\
\mathbf{e}_2 &= \frac{1}{\|\mathbf{u}_2\|}\mathbf{u}_2,\\
\mathbf{u}_3 &= \mathbf{v}_3 - (\mathbf{v}_3, \mathbf{e}_1)\mathbf{e}_1 - (\mathbf{v}_3, \mathbf{e}_2)\mathbf{e}_2,\\
\mathbf{e}_3 &= \frac{1}{\|\mathbf{u}_3\|}\mathbf{u}_3,
\end{aligned}$$

etc. This algorithm is known as the *Gram–Schmidt orthonormalization* procedure.

As a first example of the Gram–Schmidt process, let us apply it (or begin to apply it) to vectors $\mathbf{v}_1, \mathbf{v}_2, \mathbf{v}_3, \mathbf{v}_4$ in $\mathbb{R}^4$ where

$$\mathbf{v}_1 = \begin{pmatrix} 1 \\ 1 \\ 1 \\ -1 \end{pmatrix}, \quad \mathbf{v}_2 = \begin{pmatrix} 3 \\ 3 \\ 5 \\ 3 \end{pmatrix}.$$

(We will only carry it to the first two steps so $\mathbf{v}_3$ and $\mathbf{v}_4$ are irrelevant.) The scalar product in $\mathbb{R}^4$ is the 'usual' one. That is, we are assuming that

$$\begin{pmatrix} 1 \\ 0 \\ 0 \\ 0 \end{pmatrix}, \begin{pmatrix} 0 \\ 1 \\ 0 \\ 0 \end{pmatrix}, \begin{pmatrix} 0 \\ 0 \\ 1 \\ 0 \end{pmatrix}, \begin{pmatrix} 0 \\ 0 \\ 0 \\ 1 \end{pmatrix}$$

form an orthonormal basis.

The first step is to convert $\mathbf{v}_1$ to a unit vector:

$$(\mathbf{v}_1, \mathbf{v}_1) = 1^2 + 1^2 + 1^2 + (-1)^2 = 4$$

so

$$\sqrt{(\mathbf{v}_1, \mathbf{v}_1)} = 2$$

and

$$\mathbf{e}_1 = \tfrac{1}{2}\mathbf{v}_1 = \begin{pmatrix} \frac{1}{2} \\ \frac{1}{2} \\ \frac{1}{2} \\ -\frac{1}{2} \end{pmatrix}.$$

Next we subtract the component of $\mathbf{v}_2$ along $\mathbf{e}_1$:

$$\mathbf{w}_2 = \begin{pmatrix} 3 \\ 3 \\ 5 \\ 3 \end{pmatrix} - \left( \begin{pmatrix} 3 \\ 3 \\ 5 \\ 3 \end{pmatrix}, \begin{pmatrix} \frac{1}{2} \\ \frac{1}{2} \\ \frac{1}{2} \\ -\frac{1}{2} \end{pmatrix} \right) \begin{pmatrix} \frac{1}{2} \\ \frac{1}{2} \\ \frac{1}{2} \\ -\frac{1}{2} \end{pmatrix}$$

$$\mathbf{w}_2 = \begin{pmatrix} 3 \\ 3 \\ 5 \\ 3 \end{pmatrix} - (\tfrac{3}{2} + \tfrac{3}{2} + \tfrac{5}{2} - \tfrac{3}{2}) \begin{pmatrix} \frac{1}{2} \\ \frac{1}{2} \\ \frac{1}{2} \\ -\frac{1}{2} \end{pmatrix} = \begin{pmatrix} 1 \\ 1 \\ 3 \\ 5 \end{pmatrix}.$$

Finally we convert $\mathbf{w}_2$ to a unit vector:

$$(\mathbf{w}_2, \mathbf{w}_2) = \left( \begin{pmatrix} 1 \\ 1 \\ 3 \\ 5 \end{pmatrix}, \begin{pmatrix} 1 \\ 1 \\ 3 \\ 5 \end{pmatrix} \right) = 1^2 + 1^2 + 3^2 + 5^2 = 36$$

so

$$\mathbf{e}_2 = \frac{\mathbf{w}_2}{6} = \frac{1}{6}\begin{pmatrix} 1 \\ 1 \\ 3 \\ 5 \end{pmatrix}.$$

As a check, note that $\left( \begin{pmatrix} 1 \\ 1 \\ 1 \\ -1 \end{pmatrix}, \begin{pmatrix} 1 \\ 1 \\ 3 \\ 5 \end{pmatrix} \right) = 0.$

Now we can easily write any vector $\mathbf{v}$ in $\mathbb{R}^4$ as the sum of a vector $\pi\mathbf{v}$ which is a linear combination of $\mathbf{e}_1$ and $\mathbf{e}_2$ (and hence of $\mathbf{v}_1$ and $\mathbf{v}_2$) and a vector which is perpendicular to both $\mathbf{e}_1$ and $\mathbf{e}_2$. Consider, for example,

$$\mathbf{v} = \begin{pmatrix} 4 \\ 0 \\ -1 \\ 7 \end{pmatrix}.$$

Define $\pi\mathbf{v}$ by

$$\pi\mathbf{v} = (\mathbf{v}, \mathbf{e}_1)\mathbf{e}_1 + (\mathbf{v}, \mathbf{e}_2)\mathbf{e}_2$$

$$\pi\mathbf{v} = \frac{1}{4}\left( \begin{pmatrix} 4 \\ 0 \\ -1 \\ 7 \end{pmatrix}, \begin{pmatrix} 1 \\ 1 \\ 1 \\ -1 \end{pmatrix} \right) \begin{pmatrix} 1 \\ 1 \\ 1 \\ -1 \end{pmatrix} + \frac{1}{36}\left( \begin{pmatrix} 4 \\ 0 \\ -1 \\ 7 \end{pmatrix}, \begin{pmatrix} 1 \\ 1 \\ 3 \\ 5 \end{pmatrix} \right) \begin{pmatrix} 1 \\ 1 \\ 3 \\ 5 \end{pmatrix}$$

$$\pi\mathbf{v} = \tfrac{1}{4}(-4)\begin{pmatrix}1\\1\\1\\-1\end{pmatrix} + \tfrac{1}{36}\cdot 36\begin{pmatrix}1\\1\\3\\5\end{pmatrix} = \begin{pmatrix}-1\\-1\\-1\\1\end{pmatrix} + \begin{pmatrix}1\\1\\3\\5\end{pmatrix} = \begin{pmatrix}0\\0\\2\\6\end{pmatrix}.$$

Then you can check that

$$\mathbf{v} - \pi\mathbf{v} = \begin{pmatrix}4\\0\\-1\\7\end{pmatrix} - \begin{pmatrix}0\\0\\2\\6\end{pmatrix} = \begin{pmatrix}4\\0\\-3\\1\end{pmatrix}$$

is orthogonal to $\mathbf{e}_1$ and $\mathbf{e}_2$, either by verifying that it is orthogonal to the original basis vectors $\mathbf{v}_1$ and $\mathbf{v}_2$ or to the orthonormal vectors $\mathbf{e}_1$ and $\mathbf{e}_2$. We say that the transformation $\pi$ sending $\mathbf{v}$ into $\pi\mathbf{v}$ is *orthogonal projection* onto the subspace $W$ spanned by $\mathbf{v}_1$ and $\mathbf{v}_2$.

As a second example of the Gram–Schmidt process, consider the (four-dimensional) space of polynomials of degree $\leqslant 3$, with scalar product

$$(\mathbf{f}, \mathbf{g}) = \int_{-1}^{1} f(t)g(t)\mathrm{d}t.$$

(Check that this defines a scalar product!) We start with the ordered basis

$$\mathbf{v}_1 = 1, \quad \mathbf{v}_2 = t, \quad \mathbf{v}_3 = t^2, \quad \mathbf{v}_4 = t^3.$$

If we started with different basis elements, or even the same elements in a different order, we would end up with a different orthonormal basis. We first calculate

$$(\mathbf{v}_1, \mathbf{v}_1) = \int_{-1}^{1} \mathrm{d}t = 2$$

and convert $\mathbf{v}_1$ to a unit vector:

$$\mathbf{e}_1 = \mathbf{v}_1/\sqrt{(\mathbf{v}_1, \mathbf{v}_1)} = 1/\sqrt{2}.$$

We next calculate

$$(\mathbf{e}_1, \mathbf{v}_2) = \int_{-1}^{1} (t/\sqrt{2})\mathrm{d}t = 0$$

and conclude that $\mathbf{v}_2$ is already orthogonal to $\mathbf{e}_1$. Since

$$(\mathbf{v}_2, \mathbf{v}_2) = \int_{-1}^{1} t^2\mathrm{d}t = 2/3$$

we have $\mathbf{e}_2 = t/\sqrt{(2/3)} = \sqrt{(3/2)}\,t$.

Next we calculate $\mathbf{w}_3$:

$$\begin{aligned}\mathbf{w}_3 &= \mathbf{v}_3 - \mathbf{e}_1(\mathbf{e}_1, \mathbf{v}_3) - \mathbf{e}_2(\mathbf{e}_2, \mathbf{v}_3)\\ &= t^2 - \frac{1}{2}\int_{-1}^{1} t^2\mathrm{d}t - \frac{3}{2}t\int_{-1}^{1} t^3\mathrm{d}t\\ &= t^2 - \tfrac{1}{3}.\end{aligned}$$

Since

$$(\mathbf{w}_3, \mathbf{w}_3) = \int_{-1}^{1} (t^2 - \tfrac{1}{3})^2 dt = \tfrac{8}{45}$$

the third normalized basis vector is

$$\mathbf{e}_3 = \frac{t^2 - \frac{1}{3}}{\sqrt{(\frac{8}{45})}} = \sqrt{(\tfrac{45}{8})}(t^2 - \tfrac{1}{3}) = \sqrt{(\tfrac{5}{8})}(3t^2 - 1).$$

Finally, we calculate $\mathbf{w}_4$:

$$\begin{aligned}\mathbf{w}_4 &= \mathbf{v}_4 - \mathbf{e}_1(\mathbf{e}_1, \mathbf{v}_4) - \mathbf{e}_2(\mathbf{e}_2, \mathbf{v}_4) - \mathbf{e}_3(\mathbf{e}_3, \mathbf{v}_4)\\ &= t^3 - \tfrac{1}{2}\int_{-1}^{1} t^3 dt - \tfrac{3}{2}t\int_{-1}^{1} t^4 dt - \tfrac{5}{8}(3t^2 - 1)\int_{-1}^{1}(3t^5 - t^3)dt\\ &= t^2 - 0 - \tfrac{3}{2}t\cdot\tfrac{2}{5} - 0 = t^3 - \tfrac{3}{5}t.\end{aligned}$$

Dividing by $\sqrt{(\mathbf{w}_4, \mathbf{w}_4)}$ we obtain finally

$$\mathbf{e}_4 = \frac{\mathbf{w}_4}{\sqrt{(\mathbf{w}_4, \mathbf{w}_4)}} = \sqrt{\tfrac{7}{8}}(5t^3 - 3t).$$

Clearly, proceeding in this manner, we could construct a sequence of orthogonal polynomials of higher and higher degree. These polynomials, known as the Legendre polynomials, will appear naturally in the solution of problems in electrostatics using spherical polar coordinates. Indeed, it is usually true in physical applications that vector spaces of functions, which frequently arise as solutions to differential equations, have orthogonal bases which arise naturally from physical considerations. For this reason it is rarely necessary in practice to carry out the tedious Gram–Schmidt process.

## 4.3. Quadratic forms and symmetric matrices

In sections 4.1 and 4.2 we have studied the *Euclidean* scalar product which satisfied three conditions: it was bilinear, symmetric, and positive-definite. We now want to investigate more general 'scalar products', which are not necessarily positive-definite. They play a central role in the theory of relativity.

We return to $\mathbb{R}^2$. Suppose that we are given a scalar product, $\langle\ ,\ \rangle$ on $\mathbb{R}^2$, which is not necessarily positive-definite. Thus we assume that $\langle\ ,\ \rangle$ is

$$\text{bilinear: } \langle \mathbf{v}, a\mathbf{u} + b\mathbf{w}\rangle = a\langle \mathbf{v}, \mathbf{u}\rangle + b\langle \mathbf{v}, \mathbf{w}\rangle$$

and

$$\text{symmetric: } \langle \mathbf{u}, \mathbf{v}\rangle = \langle \mathbf{v}, \mathbf{u}\rangle$$

for all vectors $\mathbf{u}, \mathbf{v}, \mathbf{w}$ and all real numbers $a$ and $b$. We wish to compare $\langle\ ,\ \rangle$ with the Euclidean scalar product $(\ ,\ )$. We begin with the following elementary lemma.

Let $l: \mathbb{R}^2 \to \mathbb{R}$ be a linear map. Then there is a unique vector $\mathbf{w}$ such that

$$l(\mathbf{v}) = (\mathbf{v}, \mathbf{w}) \quad \text{for all } \mathbf{v} \text{ in } \mathbb{R}^2.$$

Indeed, $l$ is given by a $1 \times 2$ matrix $(a \ \ b)$, i.e.,

$$l\begin{pmatrix} x \\ y \end{pmatrix} = ax + by \quad \text{for any } \mathbf{v} = \begin{pmatrix} x \\ y \end{pmatrix} \in \mathbb{R}^2.$$

Then take

$$\mathbf{w} = \begin{pmatrix} a \\ b \end{pmatrix}$$

so

$$(\mathbf{v}, \mathbf{w}) = \left( \begin{pmatrix} x \\ y \end{pmatrix}, \begin{pmatrix} a \\ b \end{pmatrix} \right) = ax + by$$

as desired, and it is clear that $\mathbf{w}$ is the unique vector in $\mathbb{R}^2$ with this property. Now consider $\langle \mathbf{u}, \mathbf{v} \rangle$ as a function of $\mathbf{v}$ for fixed $\mathbf{u}$. This is a linear function of $\mathbf{v}$, hence there is a vector $\mathbf{w}$ such that

$$\langle \mathbf{u}, \mathbf{v} \rangle = (\mathbf{v}, \mathbf{w}) \quad \text{for all } \mathbf{v} \in V.$$

The vector $\mathbf{w}$ depends on $\mathbf{u}$, so we should write $\mathbf{w}(\mathbf{u})$ in the above equation. To repeat, $\mathbf{w}(\mathbf{u})$ is that vector whose *Euclidean* scalar product with any $\mathbf{v}$ equals $\langle \mathbf{u}, \mathbf{v} \rangle$. Let $\mathbf{u}_1$ and $\mathbf{u}_2$ be two vectors, and $\mathbf{w}(\mathbf{u}_1)$ and $\mathbf{w}(\mathbf{u}_2)$ their corresponding $\mathbf{w}$s. Now

$$\begin{aligned} \langle a\mathbf{u}_1 + b\mathbf{u}_2, \mathbf{v} \rangle &= \langle \mathbf{v}, a\mathbf{u}_1 + b\mathbf{u}_2 \rangle \quad \text{by symmetry} \\ &= a\langle \mathbf{v}, \mathbf{u}_1 \rangle + b\langle \mathbf{v}, \mathbf{u}_2 \rangle \quad \text{by bilinearity} \\ &= a\langle \mathbf{u}_1, \mathbf{v} \rangle + b\langle \mathbf{u}_2, \mathbf{v} \rangle \quad \text{by symmetry} \\ &= a(\mathbf{v}, \mathbf{w}(\mathbf{u}_1)) + b(\mathbf{v}, \mathbf{w}(\mathbf{u}_2)) \\ &= (\mathbf{v}, a\mathbf{w}(\mathbf{u}_1) + b(\mathbf{w}(\mathbf{u}_2))). \end{aligned}$$

Thus $\mathbf{w}(a\mathbf{u}_1 + b\mathbf{u}_2) = a\mathbf{w}(\mathbf{u}_1) + b\mathbf{w}(\mathbf{u}_2)$. In other words, $\mathbf{w}$ depends linearly on $\mathbf{u}$. Thus we can write $\mathbf{w}(\mathbf{u}) = A\mathbf{u}$, where $A$ is a linear transformation. Going back to the definition of $\mathbf{w} = A\mathbf{u}$, we see that

$$\boxed{\langle \mathbf{u}, \mathbf{v} \rangle = (\mathbf{v}, A\mathbf{u})}$$

for all $\mathbf{u}$ and $\mathbf{v}$ in $\mathbb{R}^2$. So far we have only used the fact that $\langle \mathbf{u}, \mathbf{v} \rangle$ is bilinear, i.e., linear in $\mathbf{u}$ when $\mathbf{v}$ is fixed and linear in $\mathbf{v}$ when $\mathbf{u}$ is fixed. (This is how we used the symmetry of $\langle \ , \ \rangle$.) Now let us use the fact that $\langle \ , \ \rangle$ is symmetric. Since

$$\langle \mathbf{u}, \mathbf{v} \rangle = \langle \mathbf{v}, \mathbf{u} \rangle$$

this implies that

$$(\mathbf{v}, A\mathbf{u}) = (\mathbf{u}, A\mathbf{v})$$

and, since $(\mathbf{u}, \mathbf{v}) = (\mathbf{v}, \mathbf{u})$, that

$$(\mathbf{v}, A\mathbf{u}) = (A\mathbf{v}, \mathbf{u})$$

for all $\mathbf{u}$ and $\mathbf{v}$ in $V$. Let us see what this says for the matrix $A$.

For any matrix $B$, the expression $(B\mathbf{v}, \mathbf{u})$ is linear in $\mathbf{v}$ and $\mathbf{u}$ separately. Thus, by our preceding argument, there is a unique linear transformation, call it $B^{\mathrm{T}}$, the *transpose* of $B$, such that

$$(B\mathbf{v}, \mathbf{u}) = (\mathbf{v}, B^{\mathrm{T}}\mathbf{u})$$

for $\mathbf{v}, \mathbf{u}$ in $V$. To see what $B^{\mathrm{T}}$ is, suppose

$$\mathbf{v} = \begin{pmatrix} x \\ y \end{pmatrix}, \quad \mathbf{u} = \begin{pmatrix} x' \\ y' \end{pmatrix} \quad \text{and} \quad B = \begin{pmatrix} e & f \\ g & h \end{pmatrix}.$$

Then

$$\begin{aligned}(B\mathbf{v}, \mathbf{u}) &= (ex + fy)x' + (gx + hy)y' \\ &= exx' + fyx' + gxy' + hyy' \\ &= x(ex' + gy') + y(fx' + hy')\end{aligned}$$

so

$$B^{\mathrm{T}}\begin{pmatrix} x' \\ y' \end{pmatrix} = \begin{pmatrix} ex' + gy' \\ fx' + hy' \end{pmatrix}$$

or

$$B^{\mathrm{T}} = \begin{pmatrix} e & g \\ f & h \end{pmatrix}.$$

In other words, the transpose of a matrix is obtained by flipping the matrix along the diagonal.

Then our symmetry condition says that

$$\boxed{A = A^{\mathrm{T}},}$$

in other words, $A$ is a *symmetric* matrix. Thus $A$ has the form

$$A = \begin{pmatrix} a & b \\ b & c \end{pmatrix}.$$

If we set

$$Q(\mathbf{v}) = \langle \mathbf{v}, \mathbf{v} \rangle$$

then, as in section 4.1,

$$\langle \mathbf{u}, \mathbf{v} \rangle = \tfrac{1}{2}(Q(\mathbf{u}) + Q(\mathbf{v}) - Q(\mathbf{u} - \mathbf{v}))$$

and

$$Q(\mathbf{v}) = (A\mathbf{v}, \mathbf{v}) = ax^2 + 2bxy + cy^2$$

if

$$\mathbf{v} = \begin{pmatrix} x \\ y \end{pmatrix}.$$

A function $Q$ of this type is called a *quadratic form*. Thus by the preceding formulas, each quadratic form $Q$ determines a scalar product $\langle\ ,\ \rangle$, and every scalar product determines a quadratic form.

The coefficients $a, 2b, c$ of the quadratic polynomial $Q(\mathbf{v})$ give us the matrix $A$, which is just another way of saying that $Q$ determines $A$ and hence also $\langle\ ,\ \rangle$.

The characteristic polynomial of $A$ is

$$x^2-(a+c)x+ac-b^2$$

and

$$(a+c)^2-4(ac-b^2)=(a-c)^2+4b^2\geqslant 0.$$

This expression, $(a-c)^2+4b^2$, is called the *discriminant* of the quadratic form $Q$. The discriminant can equal zero if and only if

$$a=c \quad\text{and}\quad b=0$$

so

$$A=\begin{pmatrix} a & 0\\ 0 & a\end{pmatrix}=aI$$

and

$$\langle\mathbf{u},\mathbf{v}\rangle=a(\mathbf{u},\mathbf{v}).$$

In this case, $\langle\ ,\ \rangle$ is just a scalar multiple of $(\ ,\ )$.

Suppose that $A$ has two distinct eigenvalues, $\lambda_1\neq\lambda_2$ corresponding to eigenvectors $\mathbf{v}_1$ and $\mathbf{v}_2$. We claim that $\mathbf{v}_1$ and $\mathbf{v}_2$ are orthogonal, i.e., that $(\mathbf{v}_1,\mathbf{v}_2)=0$. The proof is easy:

$$\begin{aligned}
(A\mathbf{v}_1,\mathbf{v}_2)&=(\mathbf{v}_1,A\mathbf{v}_2) &&\text{because } A \text{ is symmetric;}\\
(\lambda_1\mathbf{v}_1,\mathbf{v}_2)&=(\mathbf{v}_1,\lambda_2\mathbf{v}_2) &&\text{because } \mathbf{v}_1 \text{ and } \mathbf{v}_2 \text{ are eigenvectors;}\\
\lambda_1(\mathbf{v}_1,\mathbf{v}_2)&=\lambda_2(\mathbf{v}_1,\mathbf{v}_2) &&\text{because the scalar product is linear;}\\
(\mathbf{v}_1,\mathbf{v}_2)&=0 &&\text{because } \lambda_1\neq\lambda_2.
\end{aligned}$$

Conversely, suppose that we start with an eigenvector $\mathbf{v}_1$ of $A$ corresponding to the eigenvalue $\lambda_1$. Let $\mathbf{v}_2$ be a non-zero vector orthogonal to $\mathbf{v}_1$, so

$$(\mathbf{v}_1,\mathbf{v}_2)=0.$$

Then

$$(\mathbf{v}_1,A\mathbf{v}_2)=(A\mathbf{v}_1,\mathbf{v}_2)=\lambda_1(\mathbf{v}_1,\mathbf{v}_2)=0$$

so $A\mathbf{v}_2$ is again orthogonal to $\mathbf{v}_1$. But there is only one line perpendicular to $\mathbf{v}_1$, and $\mathbf{0}\neq\mathbf{v}_2$ lies on it. Hence $A\mathbf{v}_2$ must be some multiple of $\mathbf{v}_2$, i.e., $A\mathbf{v}_2=\lambda_2\mathbf{v}_2$ for some eigenvalue $\lambda_2$.

We have thus shown that any symmetric matrix $A$ has two *orthogonal* eigenvectors, $\mathbf{v}_1$ and $\mathbf{v}_2$. By multiplying $\mathbf{v}_1$ and $\mathbf{v}_2$ by suitable scalars, we can arrange that $\mathbf{v}_1$ and $\mathbf{v}_2$ both have length 1, and that the matrix $\begin{pmatrix} x_1 & x_2\\ y_1 & y_2\end{pmatrix}$, where $\mathbf{v}_1=\begin{pmatrix} x_1\\ y_1\end{pmatrix}$ and $\mathbf{v}_2=\begin{pmatrix} x_2\\ y_2\end{pmatrix}$, is a rotation.

Thus $A=R\begin{pmatrix} \lambda_1 & 0\\ 0 & \lambda_2\end{pmatrix}R^{-1}$ for some suitable rotation $R$. Since an orthogonal matrix $M$ satisfies $(M\mathbf{v},M\mathbf{v})=(\mathbf{v},M^{\mathrm{T}}M\mathbf{v})=(\mathbf{v},\mathbf{v})$ for all $\mathbf{v}$. We see that $M^{\mathrm{T}}=M^{-1}$, and we can equivalently write

$$\boxed{A=R\begin{pmatrix} \lambda_1 & 0\\ 0 & \lambda_2\end{pmatrix}R^{T}.}$$

Suppose that we have chosen our eigenvectors so that $\lambda_1 > \lambda_2$. Then the eigenvector $\mathbf{v}_1$, which has been chosen to have unit length, can be characterized, among all vectors $\mathbf{v}$ of unit length, as one for *which* $Q(v)$ *assumes its maximum value*, while $\mathbf{v}_2$ is the vector of unit length for which $Q(\mathbf{v}, \mathbf{v})$ assumes its minimum value: i.e.,

$$\boxed{Q(\mathbf{v}_1) \geqslant Q(\mathbf{v}) \geqslant Q(\mathbf{v}_2)}$$

for any $\mathbf{v}$ with $(\mathbf{v}, \mathbf{v}) = 1$. To prove this statement, we write $\mathbf{v} = \mathbf{v}_1 \cos\theta + \mathbf{v}_2 \sin\theta$. Clearly, since the eigenvectors $\mathbf{v}_1$ and $\mathbf{v}_2$ are orthogonal and have unit length,

$$(\mathbf{v}, \mathbf{v}) = (\mathbf{v}_1, \mathbf{v}_1)\cos^2\theta + (\mathbf{v}_2, \mathbf{v}_2)\sin^2\theta = 1.$$

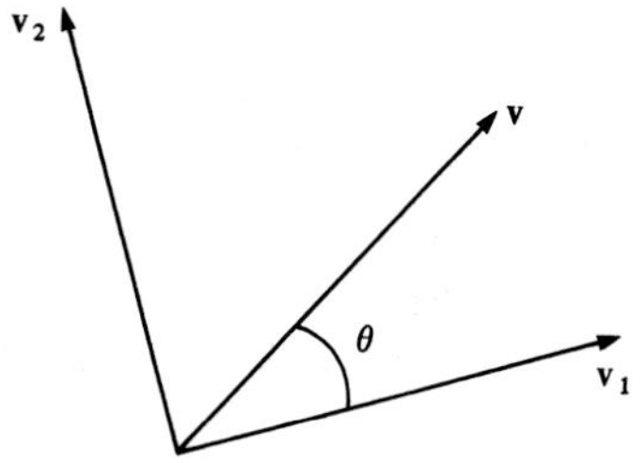


**Figure 4.7**

Then

$$\begin{aligned} Q(\mathbf{v}) &= (A\mathbf{v}, \mathbf{v}) \\ &= (A\mathbf{v}_1 \cos\theta + A\mathbf{v}_2 \sin\theta, \mathbf{v}_1 \cos\theta + \mathbf{v}_2 \sin\theta) \\ &= (\lambda_1 \mathbf{v}_1 \cos\theta + \lambda_2 \mathbf{v}_2 \sin\theta, \mathbf{v}_1 \cos\theta + \mathbf{v}_2 \sin\theta) \\ &= \lambda_1(\mathbf{v}_1, \mathbf{v}_1)\cos^2\theta + \lambda_2(\mathbf{v}_2, \mathbf{v}_2)\sin^2\theta, \text{ since } (\mathbf{v}_1, \mathbf{v}_2) = 0 \\ &= \lambda_1 \cos^2\theta + \lambda_2 \sin^2\theta \\ &= \lambda_1 - (\lambda_1 - \lambda_2)\sin^2\theta. \end{aligned}$$

Clearly $Q(\mathbf{v})$ achieves its maximum value when $\sin^2\theta = 0$, (when $\mathbf{v} = \pm\mathbf{v}_1$) and its minimum value when $\sin^2\theta = 1$ (when $\mathbf{v} = \pm\mathbf{v}_2$).

It is now apparent how to draw the graph of $Q(\mathbf{v}) = (A\mathbf{v}, \mathbf{v}) = \text{constant}$. We can diagonalize $A$ by a rotation $R$:

$$A = R\begin{pmatrix} \lambda_1 & 0 \\ 0 & \lambda_2 \end{pmatrix} R^{-1}$$

so that

$$Q(\mathbf{v}) = \left( R\begin{pmatrix} \lambda_1 & 0 \\ 0 & \lambda_2 \end{pmatrix} R^{-1}\mathbf{v}, \mathbf{v} \right).$$

Since $R$ is orthogonal, $R^{\mathrm{T}} = R^{-1}$, and we have

$$Q(\mathbf{v}) = \left( \begin{pmatrix} \lambda_1 & 0 \\ 0 & \lambda_2 \end{pmatrix} R^{-1}\mathbf{v}, R^{-1}\mathbf{v} \right).$$

If we write $\begin{pmatrix} x' \\ y' \end{pmatrix} = R^{-1}\begin{pmatrix} x \\ y \end{pmatrix} = R^{-1}\mathbf{v}$, then

$$Q(\mathbf{v}) = \left( \begin{pmatrix} \lambda_1 & x' \\ \lambda_2 & y' \end{pmatrix}, \begin{pmatrix} x' \\ y' \end{pmatrix} \right) = \lambda_1 x'^2 + \lambda_2 y'^2.$$

If $\lambda_1$ and $\lambda_2$ are both positive, the graph of $Q(\mathbf{v}, \mathbf{v}) = k$ is an ellipse if $k > 0$, the origin only if $k = 0$, empty if $k < 0$. If $\lambda_1$ and $\lambda_2$ are both negative, the graph is an ellipse if $k < 0$. If $\lambda_1$ and $\lambda_2$ have opposite signs, the graph of $Q(\mathbf{v})$ is a hyperbola, which degenerates to two straight lines if $k = 0$. The vertices of the ellipse or hyperbola, where the distance from the origin is a local extremum, lie along the lines determined by the eigenvectors of $A$.

Suppose, for example, that $A = \begin{pmatrix} 9 & 2 \\ 2 & 6 \end{pmatrix}$, so that

$$Q(\mathbf{v}) = 9x^2 + 4xy + 6y^2.$$

The eigenvalues of $A$ are $\lambda_1 = 10$, $\lambda_2 = 5$, with associated eigenvectors $\begin{pmatrix} 2 \\ 1 \end{pmatrix}$ and $\begin{pmatrix} -1 \\ 2 \end{pmatrix}$. We can write $A = R\begin{pmatrix} 10 & 0 \\ 0 & 5 \end{pmatrix}R^{-1}$, where $R$ is the rotation

$$R = \frac{1}{\sqrt{5}}\begin{pmatrix} 2 & -1 \\ 1 & 2 \end{pmatrix}.$$

By introducing new coordinates

$$\begin{pmatrix} x' \\ y' \end{pmatrix} = R^{-1}\begin{pmatrix} x \\ y \end{pmatrix} = \frac{1}{\sqrt{5}}\begin{pmatrix} 2 & 1 \\ -1 & 2 \end{pmatrix}\begin{pmatrix} x \\ y \end{pmatrix}$$

i.e.,

$$x' = \frac{1}{\sqrt{5}}(2x + y),$$

$$y' = \frac{1}{\sqrt{5}}(-x + 2y),$$

we can write

$$Q(\mathbf{v}) = 10x'^2 + 5y'^2.$$

The graph of $Q(\mathbf{v}) = 1$, i.e., of

$$10x'^2 + 5y'^2 = 1$$

is an ellipse of minor axis $\sqrt{\frac{1}{10}}$, major axis $\sqrt{\frac{1}{5}}$. The axes coincide with the eigenvectors of $A$: $\begin{pmatrix} 2 \\ 1 \end{pmatrix}$ and $\begin{pmatrix} -1 \\ 2 \end{pmatrix}$.

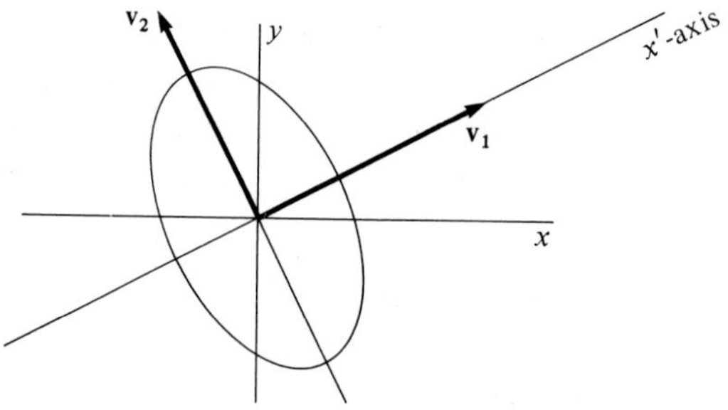


**Figure 4.8**

Suppose we allow not only rotations as changes of coordinates but also non-orthogonal transformations such as $x'' = \alpha x'$ and $y'' = \beta y'$. Then, in terms of $x''$ and $y''$, we have

$$Q(\mathbf{v}) = \frac{\lambda_1}{\alpha^2} x''^2 + \frac{\lambda_2}{\beta^2} y''^2.$$

If $\lambda_1 \neq 0$, we can choose $\alpha^2 = |\lambda_1|$ so that $\lambda_1/\alpha^2 = \pm 1$ and similarly for $\lambda_2$. We have thus proved:

Let $Q$ be any quadratic form in $\mathbb{R}^2$. We can then find coordinates $x''$ and $y''$ such that $Q$ has one of the following expressions:

$$Q(\mathbf{v}) = \begin{cases} x''^2 + y''^2 \\ x''^2 \\ 0 \\ -x''^2 \\ x''^2 - y''^2 \\ -x''^2 - y''^2. \end{cases}$$

If there are two plus signs, $Q(\mathbf{v})$ has a minimum at $\mathbf{v} = \mathbf{0}$; if two minus signs, a maximum. If there is one plus sign, one minus sign, then $Q(\mathbf{v})$ has neither a maximum nor a minimum, but rather a *saddle point*, as suggested in figure 4.10.

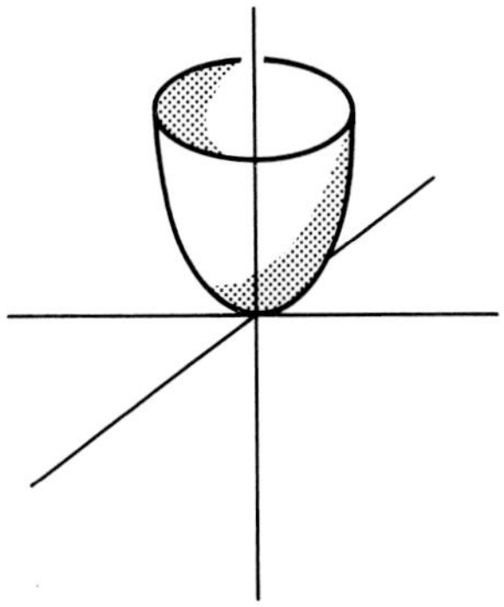

**Figure 4.9**

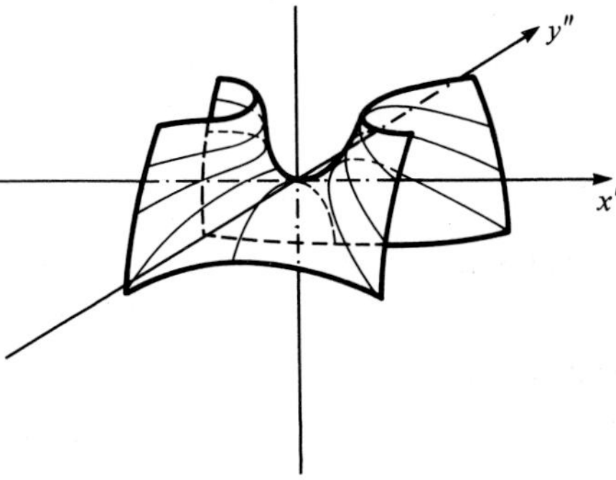


**Figure 4.10**

## 4.4. Normal modes

One of the most important applications of the results of the preceding section is to the theory of coupled oscillators. To explain what is involved, consider the following mechanical system. We have two undamped oscillators which we connect by a spring with spring constant $k$. The equations of motion, from Newton's laws, are

$$m_1\ddot{x}_1 = -k_1x_1 - k(x_1 - x_2),$$
$$m_2\ddot{x}_2 = -k_2x_2 - k(x_2 - x_1)$$

or

$$T\begin{pmatrix}\ddot{x}_1\\ \ddot{x}_2\end{pmatrix} = -H\begin{pmatrix}x_1\\ x_2\end{pmatrix}$$

where the symmetric matrices $T$ and $H$ are

$$T = \begin{pmatrix} m_1 & 0 \\ 0 & m_2 \end{pmatrix}, \quad H = \begin{pmatrix} k_1 + k & -k \\ -k & k_2 + k \end{pmatrix}.$$

**Figure 4.11** Uncoupled oscillators

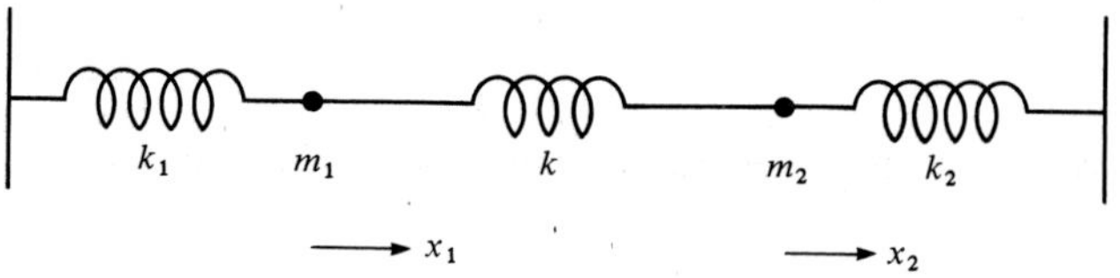


**Figure 4.12** Coupled oscillators

Our strategy will be to try to *simultaneously* diagonalize $T$ and $H$, so as to 'uncouple' the equations. Let us discuss the general case. We want to consider *two* symmetric matrices $T$ and $H$ where $T$ is positive-definite. Our first claim is that we can find a positive-definite matrix $B$ such that

$$T = B^2.$$

Indeed, if $T$ is diagonal, as in our example, set

$$B = \begin{pmatrix} m_1^{1/2} & 0 \\ 0 & m_2^{1/2} \end{pmatrix}.$$

Otherwise, we can find a rotation $R_\theta$ such that

$$T = R_\theta \Delta R_\theta^{\mathrm{T}} \quad \text{where } \Delta \text{ is a diagonal matrix.}$$

Write $\Delta = C^2$ with $C$ positive-definite. Then

$$B = R_\theta C R_\theta^{\mathrm{T}}$$

is symmetric, positive-definite, and satisfies $B^2 = T$. Now define

$$\mathbf{w} = B\mathbf{v}$$

so

$$\mathbf{v} = B^{-1}\mathbf{w}.$$

Then

$$\ddot{\mathbf{v}} = B^{-1}\ddot{\mathbf{w}}$$

and the equation $T\ddot{\mathbf{v}} = -H\mathbf{v}$ becomes

$$TB^{-1}\ddot{\mathbf{w}} = -HB^{-1}\mathbf{w}$$

or, since $T = B^2$

$$B\ddot{\mathbf{w}} = -HB^{-1}\mathbf{w}$$

or

$$\ddot{\mathbf{w}} = -A\mathbf{w} \quad \text{where } A = B^{-1}HB^{-1}.$$

Note that $A$ is again symmetric, so we have reduced the problem to the case where $T = I$. (The astute reader may have noticed that, from a geometric point of view, we have simply passed to a coordinate system in which the *quadratic function* associated to $T$ takes on the normal form $x^2 + y^2$.)

To solve the equation $\ddot{\mathbf{w}} = -A\mathbf{w}$, all we have to do is to find the eigenvalues and eigenvectors of $A$. Suppose that $\mathbf{v}_1$ is an eigenvector of $A$ with eigenvalue $\omega_1^2 > 0$. Then, for any choice of amplitude $\rho$ and phase $\alpha$, the function

$$\mathbf{w}(t) = \rho\cos(\omega_1 t + \alpha)\mathbf{v}_1$$

is clearly a solution. Similarly for the second eigenvector and eigenvalue giving $\rho\cos(\omega_2 t + \alpha)$. These are called the *normal modes* of oscillation of the vibrating system.

Suppose that

$$A = RDR^{-1}$$

where $D$ is a diagonal matrix. Then writing

$$\mathbf{w} = R\mathbf{u}$$

we have

$$\ddot{\mathbf{w}} = R\ddot{\mathbf{u}} = -RDR^{-1}R\mathbf{u}$$

or

$$\ddot{\mathbf{u}} = -D\mathbf{u}.$$

Since $D$ is diagonal, this is just two separate differential equations for each of the components. Assume that the eigenvalues of $A$ are both positive – say $\omega_1^2$ and $\omega_2^2$. Then the general solution of

$$\ddot{\mathbf{u}} = -D\mathbf{u}$$

is

$$\begin{pmatrix} u_1 \\ u_2 \end{pmatrix} = \begin{pmatrix} \rho_1\cos(\omega_1 t + \alpha_1) \\ \rho_2\cos(\omega_2 t + \alpha_2) \end{pmatrix}.$$

If

$$\mathbf{v}_1 = R\begin{pmatrix} 1 \\ 0 \end{pmatrix}, \quad \mathbf{v}_2 = R\begin{pmatrix} 0 \\ 1 \end{pmatrix}$$

are the two eigenvectors of $A$, we see that the most general solution of $\ddot{\mathbf{w}} = -A\mathbf{w}$ is

$$\boxed{\mathbf{w} = \rho_1\cos(\omega_1 t + \alpha_1)\mathbf{v}_1 + \rho_2\cos(\omega_2 t + \alpha_2)\mathbf{v}_2.}$$

Thus the general solution is a 'superposition' of normal modes.

Let us illustrate this result in the case of two *identical* coupled springs. We thus assume that $m_1 = m_2$ and $k_1 = k_2$. In the absence of the coupling, the equation of each spring would be

$$\ddot{x} = -\omega_0^2 x, \quad \omega_0^2 = k_1/m_1 = k_2/m_2.$$

In the presence of the coupling, it is

$$\ddot{x}_1 = -(\omega_0^2 + s)x_1 + sx_2, \quad \ddot{x}_2 = sx_1 - (\omega_0^2 + s)x_2, \quad s = k/m$$

or

$$\ddot{\mathbf{v}} = -A\mathbf{v}, \quad A = \begin{pmatrix} \omega_0^2 + s & -s \\ -s & \omega_0^2 + s \end{pmatrix}.$$

By symmetry we see that the eigenvectors of $A$ are

$$\begin{pmatrix} 1 \\ 1 \end{pmatrix} \text{ with eigenvalue } \omega_0^2$$

and

$$\begin{pmatrix} 1 \\ -1 \end{pmatrix} \text{ with eigenvalue } \omega_0^2 + 2s.$$

These are the two *normal modes of oscillation* in this case. The first corresponds to the bobs moving in tandem, the second to their moving in opposite directions.

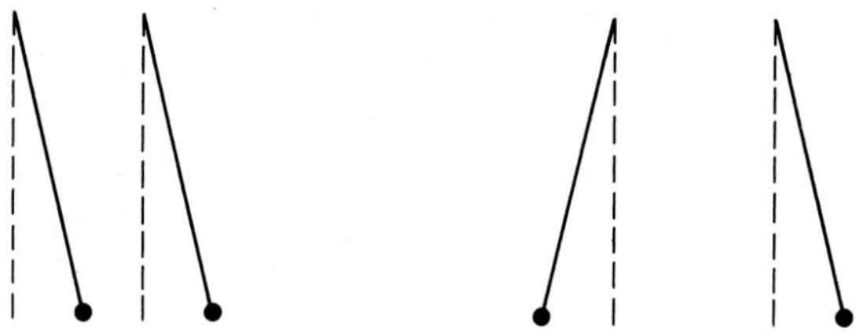

**Figure 4.13**

$$\omega' = (\omega_0^2 + 2s)^{1/2} \sim \omega_0 + s/\omega_0, \text{if } s/\omega_0 \text{ is small.}$$

Then the general solution for our differential equation is

$$x_1 = \rho_1 \cos(\omega_0 t + \alpha_1) + \rho_2 \cos(\omega' t + \alpha_2),$$
$$x_2 = \rho_1 \cos(\omega_0 t + \alpha_1) - \rho_2 \cos(\omega' t + \alpha_2).$$

Let us examine the particular solution where we excite one spring and let it go at time $t = 0$. Thus we wish to consider the initial conditions.

$$x_1(0) = C, \quad \dot{x}_1(0) = 0,$$
$$x_2(0) = 0, \quad \dot{x}_2(0) = 0.$$

Substituting into the above equations, we see that $\rho_1 = \rho_2 = \frac{1}{2}C$ and $\alpha_1 = \alpha_2 = 0$. The particular solution is this:

$$x_1 = \tfrac{1}{2}C(\cos\omega_0 t + \cos\omega' t),$$
$$x_2 = \tfrac{1}{2}C(\cos\omega_0 t - \cos\omega' t).$$

Recall that $\cos\alpha = \frac{1}{2}(e^{i\alpha} + e^{-i\alpha})$ and therefore

$$\cos\alpha + \cos\beta = 2\cos\left(\frac{\alpha - \beta}{2}\right)\cos\left(\frac{\alpha + \beta}{2}\right)$$

and similarly

$$\cos\alpha - \cos\beta = -2\sin\left(\frac{\alpha-\beta}{2}\right)\sin\left(\frac{\alpha+\beta}{2}\right).$$

Substituting $\alpha = \omega_0 t$ and $\beta = \omega' t$, we see that our particular solution is given by

$$x_1 = C\cos(\omega' - \omega_0)t\cos\omega_0 t,$$
$$x_2 = -C\sin(\omega' - \omega_0)t\sin\omega_0 t.$$

In the case of small coupling $\omega' - \omega_0$ is a small quantity. If we graph the motion of both springs, we get figure 4.14. The oscillators of each spring (with natural frequency $\omega_0$) are modulated. The *beats* are determined by the modulating factors $\cos(\omega' - \omega_0)t$, $\sin(\omega' - \omega_0)t$. The energy alternates between the two springs; when one oscillates with maximum amplitude, the other is at rest. This phenomenon is known as *resonance.*

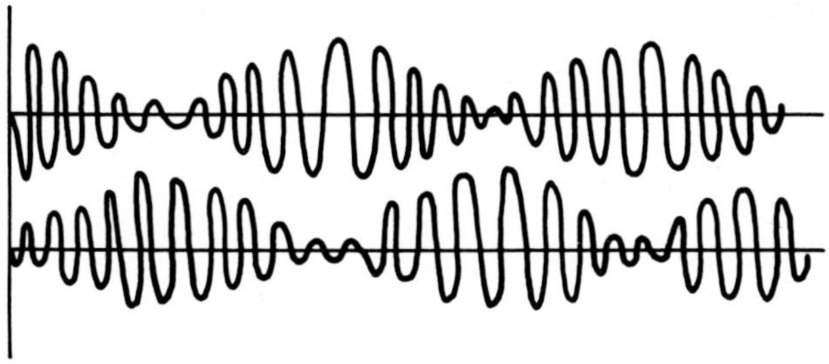

**Figure 4.14**

In case the two springs are not identical, but are only slightly 'out of tune', the behavior is similar. There will still be modulated harmonic motion at both springs. The second spring will come to rest at periodic intervals, but the first will continue to oscillate even when the second is oscillating at maximum amplitude. Imperfect 'tuning' results in an incomplete transfer of energy from the first spring to the second. We will leave the details, which are a straightforward, if somewhat messy, calculation of eigenvectors and eigenvalues of $A$, as an exercise to the reader.

## 4.5. Normal modes in higher dimensions

Let $V$ be an $n$-dimensional vector space equipped with a positive-definite scalar product. Let $\langle\ ,\ \rangle$ be some other, not necessarily positive-definite, scalar product. An examination of the argument given in section 4.3 will show that there exists a linear transformation $A: V \to V$ such that

$$\langle \mathbf{u}, \mathbf{v}\rangle = (A\mathbf{u}, \mathbf{v}) \quad \text{for all } \mathbf{u}, \mathbf{v} \text{ in } \mathbf{V}$$

and $A$ is symmetric in that

$$(A\mathbf{u}, \mathbf{v}) = (\mathbf{u}, A\mathbf{v}).$$

In fact, we know from section 4.2 that we can find an isomorphism of $V$ with $\mathbb{R}^n$ so that ( , ) is carried over into the Euclidean scalar product. Then the arguments of section 4.3 work without any change to show that $A$ is a symmetric matrix.

We claim that we can turn the argument of section 4.3 around to show that $A$ has $n$ mutually perpendicular eigenvectors. Indeed, consider the quadratic form

$$Q(\mathbf{v}) = \langle \mathbf{v}, \mathbf{v} \rangle = (A\mathbf{v}, \mathbf{v})$$

restricted to the *unit sphere*

$$\{\mathbf{v} \mid \|\mathbf{v}\| = 1\}.$$

This function is continuous and is bounded. Indeed, if all the entries $A_{ij}$ of $A$ satisfy

$$|A_{ij}| \leqslant M$$

for some number $M$, then if

$$\mathbf{v} = \begin{pmatrix} x_1 \\ \vdots \\ x_n \end{pmatrix}$$

we have $\|\mathbf{v}\|^2 = \Sigma x_i^2 = 1$ so $|x_i| \leqslant 1$ for all $i$ and

$$(A\mathbf{v}, \mathbf{v}) = \sum A_{ij} x_i x_j$$

so

$$|(A\mathbf{v}, \mathbf{v})| \leqslant \sum |A_{ij}| \leqslant nM.$$

Let $\mathbf{v}$ be a point on the unit sphere where $Q(\mathbf{v})$ takes on its maximum value. (At this juncture, we are really using some deep properties of the real number system which guarantee that there will indeed exist a point on the sphere where $Q$ takes on its maximum value). We claim that $\mathbf{v}$ is an eigenvector of $A$. Indeed, define the vector $\mathbf{w}$ by

$$\mathbf{w} = A\mathbf{v} - (A\mathbf{v}, \mathbf{v})\mathbf{v}.$$

We will show that $\mathbf{w} = \mathbf{0}$ if $Q$ takes its maximum at $\mathbf{v}$. Since $(\mathbf{v}, \mathbf{v}) = 1$, the vector $\mathbf{w}$ is perpendicular to $\mathbf{v}$,

$$(\mathbf{w}, \mathbf{v}) = 0,$$

and hence

$$(A\mathbf{v}, \mathbf{w}) = \|\mathbf{w}\|^2.$$

Then, for any real number $s$

$$\|\mathbf{v} + s\mathbf{w}\|^2 = (\mathbf{v} + s\mathbf{w}, \mathbf{v} + s\mathbf{w}) = \|\mathbf{v}\|^2 + s^2\|\mathbf{w}\|^2 = 1 + s^2\|\mathbf{w}\|^2$$

and

$$(A(\mathbf{v} + s\mathbf{w}), \mathbf{v} + s\mathbf{w}) = (A\mathbf{v}, \mathbf{v}) + s(A\mathbf{w}, \mathbf{v}) + s(A\mathbf{v}, \mathbf{w}) + s^2(A\mathbf{w}, \mathbf{w})$$

or, since $(A\mathbf{v}, \mathbf{w}) = (\mathbf{w}, A\mathbf{v})$,

$$\begin{aligned} (A(\mathbf{v} + s\mathbf{w}), \mathbf{v} + s\mathbf{w}) &= (A\mathbf{v}, \mathbf{v}) + 2s(A\mathbf{v}, \mathbf{w}) + s^2(A\mathbf{w}, \mathbf{w}) \\ &= (A\mathbf{v}, \mathbf{v}) + 2s\|\mathbf{w}\|^2 + s^2(A\mathbf{w}, \mathbf{w}). \end{aligned}$$

Let us rescale the vector $\mathbf{v}+s\mathbf{w}$ so as to make it of unit length: replace it by

$$\mathbf{u}=\frac{1}{\|\mathbf{v}+s\mathbf{w}\|}(\mathbf{v}+s\mathbf{w}).$$

Then

$$f(s)\overset{\text{def}}{=}(A\mathbf{u},\mathbf{u})=\frac{1}{\|\mathbf{v}+s\mathbf{w}\|^2}(A(\mathbf{v}+s\mathbf{w}),\mathbf{v}+s\mathbf{w})$$

$$=\frac{1}{1+s^2\|\mathbf{w}\|^2}((A\mathbf{v},\mathbf{v})+2s\|\mathbf{w}\|^2+s^2(A\mathbf{w},\mathbf{w})).$$

This expression is a differentiable function of $s$. By hypothesis, it has a maximum at $s=0$. We conclude that its derivative, $f'(0)$, at $s=0$ must vanish. But $f'(0)=2\|\mathbf{w}\|^2$. So $\|\mathbf{w}\|^2=0$ and hence $\mathbf{w}=\mathbf{0}$. Thus

$$\boxed{A\mathbf{v}=(A\mathbf{v},\mathbf{v})\mathbf{v}.}$$

In other words, $\mathbf{v}$ is an eigenvector of $A$ with eigenvalue $(A\mathbf{v},\mathbf{v})$. Call this eigenvector $\mathbf{v}_1$ and the eigenvalue $(A\mathbf{v}_1,\mathbf{v}_1)=\lambda_1$.

Now consider the space of all vectors $\mathbf{z}$ in $V$ which are perpendicular to $\mathbf{v}_1$. Thus we look at all $\mathbf{z}$ such that

$$(\mathbf{z},\mathbf{v}_1)=0.$$

For such $\mathbf{z}$,

$$(A\mathbf{z},\mathbf{v}_1)=(\mathbf{z},A\mathbf{v}_1)=\lambda_1(\mathbf{z},\mathbf{v}_1)=0,\quad \lambda_1=(A\mathbf{v}_1,\mathbf{v}_1).$$

Consider the set of all $\mathbf{z}$ of unit length, that is the set of all $\mathbf{z}$ such that

$$\|\mathbf{z}\|=1,\quad (\mathbf{z},\mathbf{v}_1)=0$$

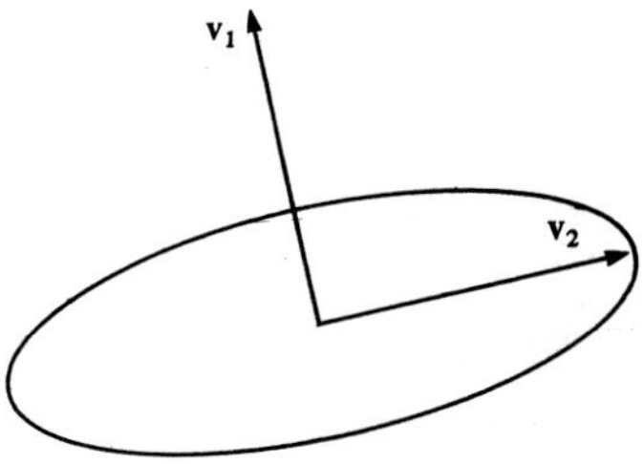


**Figure 4.15**

Let $\mathbf{v}_2$ be a point where $Q$ takes a maximum among these vectors. Write

$$A\mathbf{v}_2=(A\mathbf{v}_2,\mathbf{v}_2)\mathbf{v}_2+\mathbf{w}_2.$$

Since $(\mathbf{v}_2,\mathbf{v}_1)=0$ and $(A\mathbf{v}_2,\mathbf{v}_1)=0$, we see that $(\mathbf{w}_2,\mathbf{v}_1)=0$. As before, we conclude that $(\mathbf{w}_2, \mathbf{v}_2) = 0$, then that

$$\frac{1}{\|\mathbf{v}_2 + s\mathbf{w}_2\|^2}(A(\mathbf{v}_2 + s\mathbf{w}_2), (\mathbf{v}_2 + s\mathbf{w}_2))$$

$$= \frac{1}{1 + s^2\|\mathbf{w}_2\|^2}((A\mathbf{v}_2, \mathbf{v}_2) + 2s\|\mathbf{w}_2\|^2 + s^2(A\mathbf{w}_2, \mathbf{w}_2))$$

has a maximum at $s = 0$ and hence that $\mathbf{w}_2 = \mathbf{0}$, and $\mathbf{v}_2$ is an eigenvector of $A$.

We keep proceeding in this manner: Look at all $\mathbf{z}$ satisfying $(\mathbf{z}, \mathbf{v}_1) = (\mathbf{z}, \mathbf{v}_2) = 0$ and $\|\mathbf{z}\| = 1$, etc. At each stage, we produce a new eigenvector of $A$, perpendicular to all the previous ones. When does it all come to an end? When we run out of non-zero vectors perpendicular to $\mathbf{v}_1, \ldots, \mathbf{v}_k$. This can happen only if $k = n$. Indeed, $k$ can not be $> n$ since then $\mathbf{v}_1, \ldots, \mathbf{v}_{n+1}$ would be mutually perpendicular and hence linearly independent. This contradicts the assumption that $V$ has no $n+1$ linearly independent vectors (one of the hypotheses is the assumption that $V$ is $n$-dimensional). On the other hand, if $k < n$, the equations

$$\begin{gathered}(\mathbf{v}_1, \mathbf{w}) = 0\\ \vdots\\ (\mathbf{v}_k, \mathbf{w}) = 0\end{gathered}$$

in $\mathbb{R}^n$ are a system of $k$ homogeneous linear equations in $n$ unknowns. This always has a solution. We will prove this general fact among others in Chapter 10. Here is a proof for the existence of $\mathbf{w} \neq 0$. If the $n$th component of $\mathbf{v}_k$ is $\neq 0$, i.e.,

$$\mathbf{v}_k = \begin{pmatrix} x_1 \\ \vdots \\ x_n \end{pmatrix} \quad \text{with } x_n \neq 0,$$

the last equation is

$$x_1\mathbf{w}_1 + \cdots + x_n\mathbf{w}_n = \mathbf{0}$$

which we can solve for $\mathbf{w}_n$ in terms of $\mathbf{w}_1, \ldots, \mathbf{w}_{n-1}$:

$$\mathbf{w}_n = \frac{-1}{x_n}(x_1\mathbf{w}_1 + \cdots x_{n-1}\mathbf{w}_{n-1}).$$

Substituting this into the preceding equation gives $k-1$ equations in $n-1$ unknowns and we can proceed by induction. If the $n$th component of *any* of the vectors $\mathbf{v}_1, \ldots, \mathbf{v}_k$ does not vanish, we can still do the same – just use the $\mathbf{v}_j$ with non-vanishing $n$th component to solve for $\mathbf{w}_n$. If the $n$th components of *all* the $\mathbf{v}_1, \ldots, \mathbf{v}_n$ vanish, then the vector

$$\mathbf{w} = \begin{pmatrix} 0 \\ \vdots \\ 1 \end{pmatrix}$$

is a solution (all the first $n-1$ components vanish).

So we must keep on going until $k = n$.

**Normal Modes as Waves**

Let us now work out an interesting $n$-dimensional example. We shall imagine a sequence of identical mass points, each one connected to its nearest neighbor by a spring, with all the springs identical as well. Thus the force acting on the $i$th mass point is

$$-(k(x_i - x_{i+1}) + k(x_i - x_{i-1})).$$

Newton's equations then say

$$m\ddot{x}_i = -k(2x_i - x_{i-1} - x_{i+1})$$

We will also assume that the first and last point are also connected by the same spring: so we can imagine the points arranged in a circle.

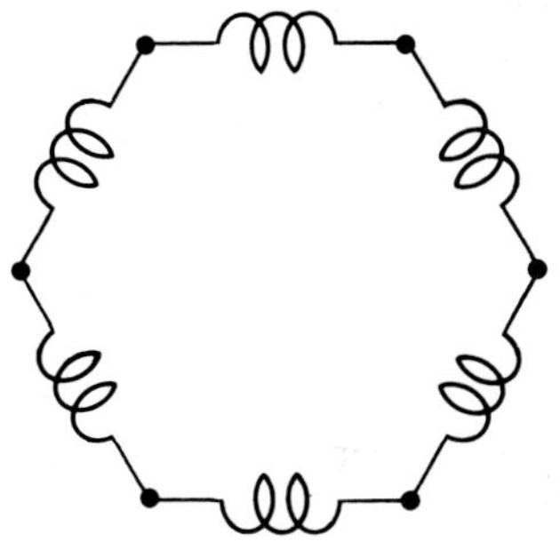

**Figure 4.16**

Thus with $\omega^2 = k/m$, the equations are

$$\ddot{x} = -\omega^2 A x$$

where $A$ is the matrix

$$\begin{pmatrix} 2 & -1 & 0 & \cdots & 0 & -1 \\ -1 & 2 & -1 & \cdots & & \\ 0 & -1 & 2 & \cdots & & \\ \vdots & \vdots & \vdots & & & -1 \\ -1 & 0 & & & -1 & 2 \end{pmatrix}.$$

Our problem is to find the eigenvalues and eigenvectors of $A$. Before describing the general solution, let us work out a few low-dimensional cases, beginning with the case $n = 3$.

$$\begin{pmatrix} 2 & -1 & -1 \\ -1 & 2 & -1 \\ -1 & -1 & 2 \end{pmatrix} \begin{pmatrix} 1 \\ 1 \\ 1 \end{pmatrix} = \begin{pmatrix} 0 \\ 0 \\ 0 \end{pmatrix}$$

so $\begin{pmatrix} 1 \\ 1 \\ 1 \end{pmatrix}$ is an eigenvector with eigenvalue 0. We know the other eigenvectors must be orthogonal to $\begin{pmatrix}1\\1\\1\end{pmatrix}$. So let us try $\begin{pmatrix}1\\-1\\0\end{pmatrix}$. Then

$$\begin{pmatrix}2&-1&-1\\-1&2&-1\\-1&-1&2\end{pmatrix}\begin{pmatrix}1\\-1\\0\end{pmatrix}=\begin{pmatrix}3\\-3\\0\end{pmatrix}=3\begin{pmatrix}1\\-1\\0\end{pmatrix}.$$

So $\begin{pmatrix}1\\-1\\0\end{pmatrix}$ and similarly $\begin{pmatrix}0\\1\\-1\end{pmatrix}$ are eigenvectors with eigenvalue 3. Thus 0 and 3 are the eigenvalues, with 3 occurring with multiplicity 2.

Now to $n=4$:

$$A=\begin{pmatrix}2&-1&0&-1\\-1&2&-1&0\\0&-1&2&-1\\-1&0&-1&2\end{pmatrix}.$$

Then

$$A\begin{pmatrix}1\\1\\1\\1\end{pmatrix}=\begin{pmatrix}0\\0\\0\\0\end{pmatrix}$$

as before. Also

$$A\begin{pmatrix}1\\-1\\1\\-1\end{pmatrix}=4\begin{pmatrix}1\\-1\\1\\-1\end{pmatrix}.$$

The remaining eigenvectors must be orthogonal to these. Let us try

$$\begin{pmatrix}1\\0\\-1\\0\end{pmatrix}.$$

Then

$$\begin{pmatrix}2&-1&0&-1\\-1&2&-1&0\\0&-1&2&-1\\-1&0&-1&2\end{pmatrix}\begin{pmatrix}1\\0\\-1\\0\end{pmatrix}=2\begin{pmatrix}1\\0\\-1\\0\end{pmatrix}$$

and similarly

$$\begin{pmatrix}2&-1&0&-1\\-1&2&-1&0\\0&-1&2&-1\\-1&0&-1&2\end{pmatrix}\begin{pmatrix}0\\1\\0\\-1\end{pmatrix}=2\begin{pmatrix}0\\1\\0\\-1\end{pmatrix}.$$

Thus 0 and 4 are eigenvalues occurring once and 2 occurs twice.

In order to deal with the $n$-dimensional case, we shall introduce some methodology of far reaching significance. Notice that the problem is invariant under the 'rotation' sending the first point into the $n$th, the second into the first, etc., with the $n$th into the $(n-1)$st. This is the matrix:

$$S = \begin{pmatrix} 0 & 1 & 0 & \cdots & \cdots & 0 \\ 0 & 0 & 1 & 0 & \cdots & \\ \vdots & & & & & \\ 0 & \cdots & \cdots & \cdots & \cdots & 1 \\ 1 & 0 & \cdots & 0 & \cdots & 0 \end{pmatrix}.$$

It is easy to check that

$$SA = AS.$$

We shall find eigenvectors of $S$. If $S\mathbf{w} = \lambda\mathbf{w}$, then $SA\mathbf{w} = AS\mathbf{w} = A(\lambda\mathbf{w}) = \lambda A\mathbf{w}$. So if $\mathbf{w}$ is an eigenvector of $S$ with eigenvalue $\lambda$; so is $A\mathbf{w}$. We will find $n$ distinct eigenvalues of $S$. Then if $S\mathbf{w} = \lambda\mathbf{w}$, $A\mathbf{w}$ will have to be a multiple of $\mathbf{w}$ – hence an eigenvector of $A$.

The 'eigenvalues' of $S$ that we will find will be complex numbers and the 'eigenvectors' will have complex entries. Both the real and imaginary parts of these eigenvectors will be eigenvectors of $A$. Here are the details:

Let

$$\tau = e^{2\pi i/n}$$

so

$$\tau^n = 1.$$

Then

$$S\begin{pmatrix} 1 \\ \vdots \\ 1 \end{pmatrix} = \begin{pmatrix} 1 \\ \vdots \\ 1 \end{pmatrix},$$

$$S\begin{pmatrix} 1 \\ \tau \\ \tau^2 \\ \vdots \\ \tau^{n-1} \end{pmatrix} = \begin{pmatrix} \tau \\ \tau^2 \\ \\ \vdots \\ 1 \end{pmatrix} = \begin{pmatrix} \tau \\ \tau^2 \\ \\ \vdots \\ \tau^n \end{pmatrix} = \tau\begin{pmatrix} 1 \\ \\ \\ \vdots \\ \tau^{n-1} \end{pmatrix}$$

$$S\begin{pmatrix} 1 \\ \tau^2 \\ \tau^4 \\ \tau^6 \\ \vdots \\ \tau^{2(n-1)} \end{pmatrix} = \tau^2\begin{pmatrix} 1 \\ \tau^2 \\ \tau^4 \\ \tau^6 \\ \vdots \\ \tau^{2(n-1)} \end{pmatrix}$$

etc. The eigenvalues $1, \tau, \tau^2, \ldots, \tau^{n-1}$ are all distinct. Thus each of the eigenvectors of $S$ must be an eigenvector of $A$. Let us call these 'eigenvectors' $\mathbf{e}_1 \ldots \mathbf{e}_n$. We know that

$$A\mathbf{e}_k = \lambda_k \mathbf{e}_k$$

for some $\lambda_k$, which we must now compute.

Now the entry of the second row of $A\mathbf{e}_k$ is

$$(-1 \quad 2 \quad -1 \quad 0\ldots 0)\begin{pmatrix}1\\ \tau^k\\ \tau^{2k}\\ \vdots\end{pmatrix}$$

$$= -1 + 2\tau^k - \tau^{2k} = (-\tau^{-k} + 2 - \tau^k)\tau^k$$
$$= 2(1 - \cos(2\pi k/n))\tau^k.$$

We conclude that the $k$th eigenvalue is

$$\lambda_k = 2(1 - \cos(2\pi k/n)).$$

This is the same eigenvalue for $k$ and for $n-k$. We may thus get real eigenvectors by adding and subtracting the eigenvectors for $k$ and for $n-k$. Thus

$$\begin{pmatrix}1\\ \cos(2\pi k/n)\\ \cos(4\pi k/n)\\ \cos(6\pi k/n)\\ \vdots\end{pmatrix} \text{ and } \begin{pmatrix}0\\ \sin(2\pi k/n)\\ \sin(4\pi k/n)\\ \sin(6\pi k/n)\\ \vdots\end{pmatrix}$$

are orthogonal eigenvectors with eigenvalue

$$2(1 + \cos(2\pi k/n)).$$

If $n = 2m$ is even, then the second column vanishes for $k = m$. Otherwise all the vectors do not vanish. We can thus consider each normal mode of the system as a sine or cosine 'wave' of compression of the system.

## 4.6. Special relativity

In this section we wish to study in some detail the geometry of a two-dimensional vector space with a quadratic form $Q(\mathbf{v}) = \langle \mathbf{v}, \mathbf{v}\rangle$ which takes on both positive and negative values. As we know, we can identify this space with $\mathbb{R}^2$ and the quadratic form $Q$ with

$$Q\begin{pmatrix}t\\ x\end{pmatrix} = t^2 - x^2.$$

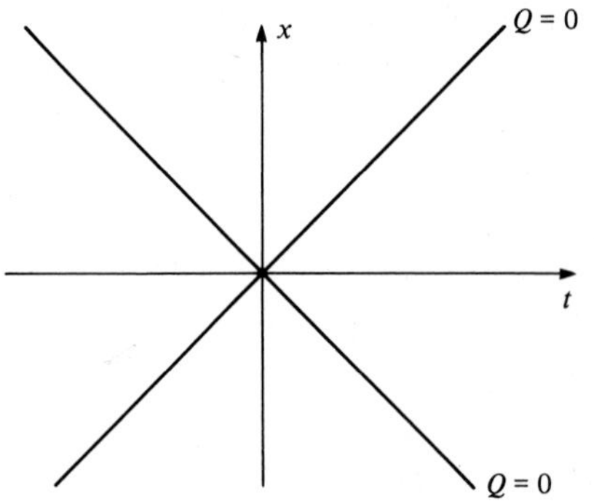


**Figure 4.17**

We shall see that the geometry of this space gives a good model for understanding special relativity. We use the word *model* in the following sense. Our ordinary space is three-dimensional. Therefore, if we add *time* as an *additional dimension*, we get a four-dimensional *spacetime*. In our model, we shall imagine that space is one-dimensional, so that our spacetime becomes two-dimensional instead of four, and we will be able to draw all the geometric constructs. Actually, most of what we have to say works in the honest four-dimensional world, with little modification from our two-dimensional model.

The first postulate of special relativity is to keep Newton's law which asserts that particles not subject to any forces will move along straight lines. Thus the geometry of our spacetime singles out the straight lines among all possible curves. Our spacetime is the affine plane with, perhaps, some additional geometrical structure.

The second postulate is that the speed of light is a finite absolute constant. Thus, at each point of spacetime there are two well-defined lines representing light moving to the right or to the left. The spatial and temporal invariance of the speed of light says that translating $P$ into $Q$ will carry the two light rays through $P$ into the two light rays through $Q$.

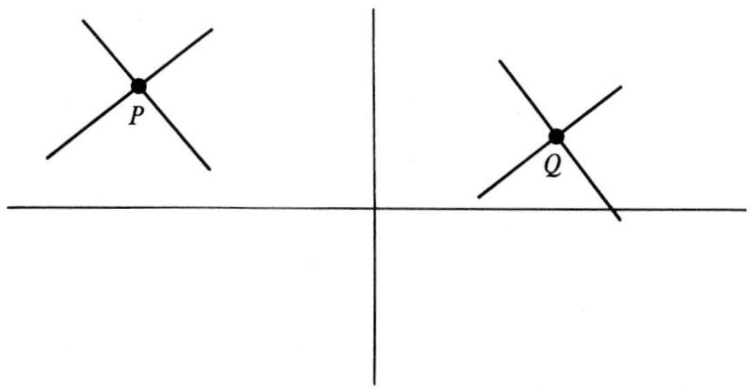


**Figure 4.18**

We want to investigate those affine transformations that carry light rays into light rays. Since translations do, we are reduced to investigating which linear transformations preserve the light rays through the origin. We are thus given two lines $x = \pm ct$, and ask for the linear transformations which preserve these lines. In doing our computations, it will be convenient to introduce *natural units* of length and time so that the speed of light is unity. For example, we could measure $t$ in years and $x$ in light-years. Or, if we choose a nanosecond ($10^{-9}$ seconds) as the unit of time, then the corresponding unit of length is one foot to remarkable accuracy. So we could introduce natural units by measuring $t$ in nanoseconds and $x$ in feet.

We thus are interested in studying those linear transformations which preserve the figure given by the pair of lines $x = t$ and $x = -t$:

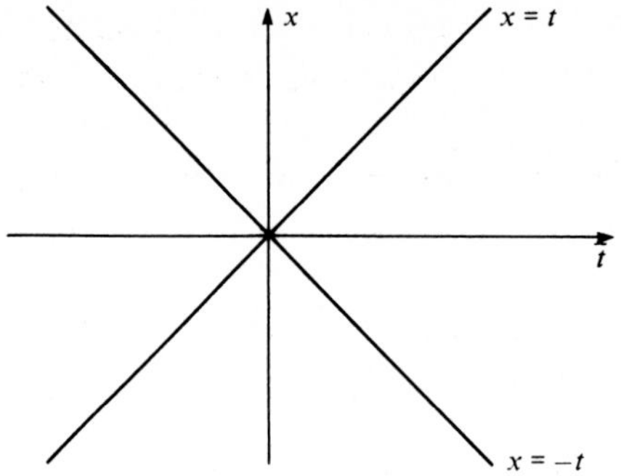


**Figure 4.19**

To repeat: having determined these linear transformations, *we will have determined all transformations of spacetime which preserve straight lines and preserve the speed of light.*

In fact, we wish, at least temporarily, to exclude certain kinds of transformations. For example, the reflections

$$\begin{pmatrix} t \\ x \end{pmatrix} \to \begin{pmatrix} -t \\ x \end{pmatrix} \quad \text{and} \quad \begin{pmatrix} t \\ x \end{pmatrix} \to \begin{pmatrix} t \\ -x \end{pmatrix}$$

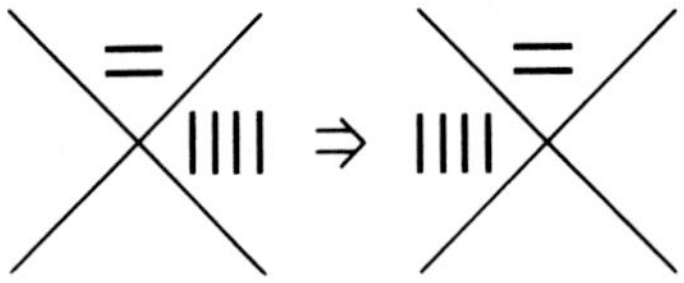

**Figure 4.20**

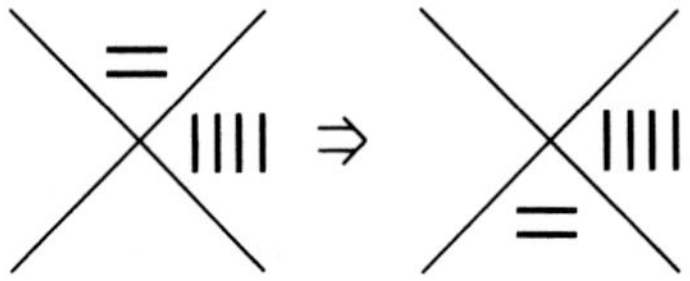

**Figure 4.21**

and the inversion

$$\begin{pmatrix} t \\ x \end{pmatrix} \to \begin{pmatrix} -t \\ -x \end{pmatrix}$$

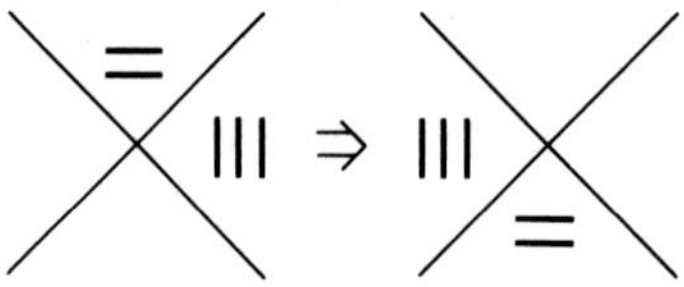

**Figure 4.22**

all send the pair of lines $x = \pm t$ into $x = \pm t$, possibly interchanging the lines. They interchange the various four regions of the plane as shown. So, by multiplying by one of them, we can arrange that the transformations we wish to study preserve each of the four regions. Thus we are looking at linear transformations, $F$, of the plane that preserve each of the lines $x = t$ and $x = -t$ and the *forward region* $t^2 > x^2$, $t > 0$.

To study such transformations, we might as well pass to coordinates in which these lines become the coordinate axes:

$$\begin{pmatrix} p \\ q \end{pmatrix} = R^{-1}\begin{pmatrix} t \\ x \end{pmatrix}, \quad R = \begin{pmatrix} 1 & -1 \\ 1 & 1 \end{pmatrix}, \quad R^{-1} = \frac{1}{2}\begin{pmatrix} 1 & 1 \\ -1 & 1 \end{pmatrix}.$$

this seems rather confused. shouldn't it be $\begin{pmatrix} p \\ q \end{pmatrix} = \frac{1}{2}\begin{pmatrix} 1 & -1 \\ 1 & 1 \end{pmatrix}\begin{pmatrix} t \\ x \end{pmatrix}$ the matrices have evidently been switched

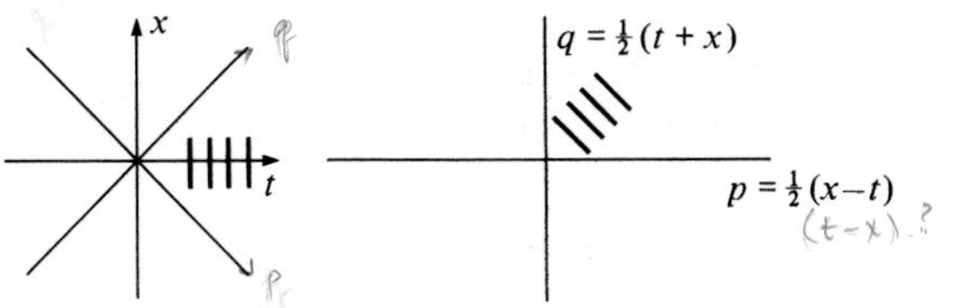


**Figure 4.23**

Thus $R^{-1}FR$ preserves the coordinate axes and the positive quadrant. Thus $R^{-1}FR$ is a transformation which preserves the coordinate axes, hence a diagonal matrix, *and* preserves the first quadrant. Thus

$$R^{-1}FR = \begin{pmatrix} a & 0 \\ 0 & d \end{pmatrix}, \quad a > 0, \quad d > 0.$$

Let us write $ad = s^2$ and $a/d = r^2$ so

$$R^{-1}FR = \begin{pmatrix} s & 0 \\ 0 & s \end{pmatrix}\begin{pmatrix} r & 0 \\ 0 & r^{-1} \end{pmatrix}.$$

Therefore we have proved that

$$F = SL_r$$

where

$$S = \begin{pmatrix} s & 0 \\ 0 & s \end{pmatrix}$$

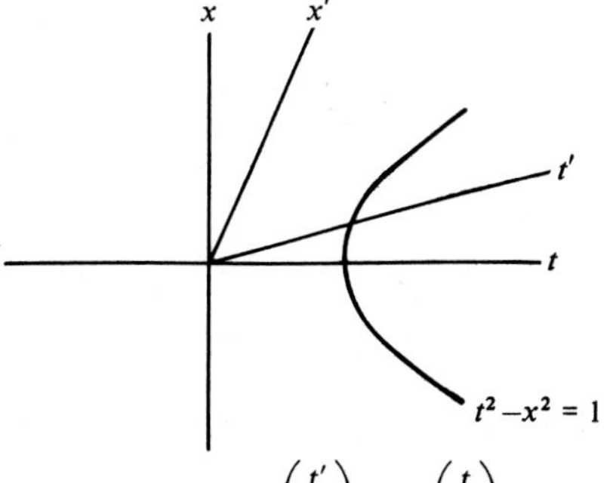


**Figure 4.24.** $\begin{pmatrix} t' \\ x' \end{pmatrix} = L_r\begin{pmatrix} t \\ x \end{pmatrix}$

is a scale transformation and

$$L_r = R\begin{pmatrix} r & 0 \\ 0 & r^{-1} \end{pmatrix}R^{-1} = \frac{1}{2}\begin{pmatrix} r+r^{-1} & r-r^{-1} \\ r-r^{-1} & r+r^{-1} \end{pmatrix}. \tag{4.2}$$

The transformation $L_r$ is called a *proper Lorentz transformation* with parameter $r$. We claim that

$$\langle L_r\mathbf{v}_1, L_r\mathbf{v}_2 \rangle = \langle \mathbf{v}_1, \mathbf{v}_2 \rangle \tag{4.3}$$

for any pair of vectors $\mathbf{v}_1, \mathbf{v}_2$. Indeed, by the analogue of (4.1) for the scalar product $\langle\ ,\ \rangle$, it is sufficient to prove that

$$Q(\mathbf{v}) = Q(L_r\mathbf{v})$$

Now

$$Q(\mathbf{v}) = t^2 - x^2 = -4pq$$

and if

$$\mathbf{v}' = L_r\mathbf{v} = \begin{pmatrix} t' \\ x' \end{pmatrix}, \quad \text{then } \begin{pmatrix} p' \\ q' \end{pmatrix} = \begin{pmatrix} r & p \\ r^{-1} & q \end{pmatrix}$$

and

$$Q(\mathbf{v}') = -4p'q' = -4pq = Q(\mathbf{v})$$

which is what we wanted to prove.

The effect of the scale transformation $S$ is to multiply all lengths and time measurements by a factor of $s$. The existence of atomic clocks, along with definite spectral lines, shows that the transformation $S$, for $s \neq 1$, is not a symmetry of nature.

A linear transformation $A$ which preserves the quadratic form $Q$ in the sense that

$$Q(A\mathbf{v}) = Q(\mathbf{v}) \quad \text{for all } \mathbf{v} \text{ in } V$$

is called a *Lorentz transformation*. Such an $A$ must carry the *light cone* (also called the *null cone*)

$$\{\mathbf{v} \mid Q(\mathbf{v}) = 0\}$$

into itself, i.e., preserve the set $\{x = \pm t\}$. If, in addition, $A$ carries the forward region into itself, it must be a proper Lorentz transformation

$$A = L_r,$$

for some $r$.

The proper Lorentz transformations can be characterized among all Lorentz transformations by the property that they can be continuously deformed to the identity through a family of Lorentz transformations. Indeed, let $A(t)$ be a family of Lorentz transformations with $A(0) = I$, $A(1) = A$. Let $\mathbf{v}$ be some point in the forward region. Then $A(t)\mathbf{v}$ can not cross the null cone since $Q(A(t)\mathbf{v}) = A(\mathbf{v}) > 0$. Similarly $\operatorname{Det} A = \pm 1$ for any Lorentz transformation since $A$ times a matrix of the form $\begin{pmatrix} 0 & 1 \\ 1 & 0 \end{pmatrix}$ or $\begin{pmatrix} \pm 1 & 0 \\ 0 & \pm 1 \end{pmatrix}$ *is* a proper Lorentz transformation – and $\operatorname{Det} L = 1$ for a proper Lorentz transformation $L$. Thus since $\operatorname{Det} A(t)$ varies continuously with $t$ and $\operatorname{Det} A(0) = 1$, we must have $\operatorname{Det} A(t) \equiv 1$ so $\operatorname{Det} A = 1$. Thus, if $A$ can be continuously deformed to the identity, $A$ must be proper. On the other hand, if $A = L_r$, just set $A(t) = L_{tr}$ so $A(0) = I$ and $A(1) = A$.

The product of two proper Lorentz transformations is again a proper Lorentz transformation. Indeed, if

$$L_r = R\begin{pmatrix} r & 0 \\ 0 & r^{-1} \end{pmatrix} R^{-1} \quad \text{and} \quad L_{r'} = R\begin{pmatrix} r' & 0 \\ 0 & r'^{-1} \end{pmatrix} R^{-1},$$

then

$$L_r L_{r'} = R\begin{pmatrix} r & 0 \\ 0 & r^{-1} \end{pmatrix} R^{-1} R \begin{pmatrix} r' & 0 \\ 0 & r'^{-1} \end{pmatrix} R^{-1}$$

so

$$L_r L_{r'} = L_{rr'}. \tag{4.4}$$

It is convenient to write $r = e^{\alpha}$ and set

$$L^{\alpha} = L_{e^{\alpha}} = \frac{1}{2}\begin{pmatrix} e^{\alpha} + e^{-\alpha} & e^{\alpha} - e^{-\alpha} \\ e^{\alpha} - e^{-\alpha} & e^{\alpha} + e^{-\alpha} \end{pmatrix}.$$

Then

$$L^{\alpha} \cdot L^{\alpha'} = L^{\alpha + \alpha'}. \tag{4.5}$$

Sometimes, the hyperbolic functions

$$\cosh \alpha = \tfrac{1}{2}(e^{\alpha} + e^{-\alpha})$$

and

$$\sinh \alpha = \tfrac{1}{2}(e^{\alpha} - e^{-\alpha})$$

are used so

$$L^{\alpha} = \begin{pmatrix} \cosh \alpha & \sinh \alpha \\ \sinh \alpha & \cosh \alpha \end{pmatrix}.$$

Then the Lorentz transformations $L^{\alpha}$ look very much like the rotations $R_{\theta}$:

$$L^{\alpha} = \begin{pmatrix} \cosh \alpha & \sinh \alpha \\ \sinh \alpha & \cosh \alpha \end{pmatrix} \quad \text{while} \quad R_{\theta} = \begin{pmatrix} \cos \theta & -\sin \theta \\ \sin \theta & \cos \theta \end{pmatrix}.$$

We have the multiplication formulas

$$L^{\alpha_1} \cdot L^{\alpha_2} = L^{\alpha_1 + \alpha_2} \quad \text{while} \quad R_{\theta_1} \cdot R_{\theta_2} = R_{\theta_1 + \theta_2}.$$

as we let $\alpha$ vary, the point $L_{\alpha}\mathbf{v}$ moves along a hyperbola, except in the limiting case where $\mathbf{v}$ lies on the light cone, in which case $L^{\alpha}\mathbf{v}$ moves in or out along the light cone (unless $\mathbf{v} = \mathbf{0}$ when $L^{\alpha}\mathbf{v} \equiv \mathbf{0}$ for all $\alpha$). It is for this reason that the functions cosh and sinh are called hyperbolic functions, with cosh called the *hyperbolic cosine* and sinh the *hyperbolic sine*. As we let $\theta$ vary, the point $R_{\theta}\mathbf{v}$ moves along a circle, except for $\mathbf{v} = \mathbf{0}$ which stays fixed. This is why cos and sin are called *circular* functions.

A *Euclidean motion* of the plane is a transformation of the form $\begin{pmatrix} x \\ y \end{pmatrix} \mapsto R\begin{pmatrix} x \\ y \end{pmatrix} + \begin{pmatrix} a \\ b \end{pmatrix}$ where $R$ is an orthogonal transformation; in other words, a Euclidean transformation is the composite of a translation and an orthogonal transformation. Euclidean geometry is the study of properties of subsets of the plane which are invariant under all Euclidean transformations. A *Poincaré transformation* of the plane is a transformation of the form $\begin{pmatrix} x \\ y \end{pmatrix} \mapsto L\begin{pmatrix} x \\ y \end{pmatrix} + \begin{pmatrix} a \\ b \end{pmatrix}$ where $L$ is a Lorentz transformation. The geometry of special relativity is concerned with those properties which are invariant under all Poincaré transformations. To be parallel to the $y$-axis is *not* a Euclidean property of a line, $l$: if $l$ is parallel to the $y$-axis, then $Rl$ will not be parallel to the $y$-axis if $R$ is a rotation other than through 0° or 180°. Similarly, to be parallel to the $x$-axis is *not* an admissible property of a line in special relativity; if $l$ is parallel to the $x$-axis, then $Ll$ will not be, for any proper Lorentz transformation $L$ other than the identity. This last assertion is usually formulated by saying that 'the notion of simultaneity does not make sense for spatially separated points in the theory of special relativity'.

Similarly, the notion of a particle 'being at rest' makes no sense. We might want to say that the line $x = 0$, the $t$-axis, represents a stationary particle at the origin. But the Lorentz transformation $L_r$ carries this line into the line through the origin and

$$L\begin{pmatrix} 1 \\ 0 \end{pmatrix} = \frac{1}{2}\begin{pmatrix} r + r^{-1} \\ r - r^{-1} \end{pmatrix}.$$

Thus $L$ applied to the $t$-axis is the line

$$x = vt \quad \text{where} \quad v = \frac{r - r^{-1}}{r + r^{-1}} = \frac{r^2 - 1}{r^2 + 1}.$$

This now looks like the line of a particle moving with constant velocity $v$. We can solve the equation

$$v = (r^2 - 1)/(r^2 + 1)$$

for $r$ in terms of $v$

$$r = \sqrt{((1 + v)/(1 - v))}$$

as can easily be checked. We can, if we like, use $v$ as a parameter to describe $L$: define

$$L(v) = L_r = L_{e^\alpha}$$

where

$$r = \sqrt{((1 + v)/(1 - v))} = e^\alpha$$

$$v = \frac{r - r^{-1}}{r + r^{-1}} = \frac{e^\alpha - e^{-\alpha}}{e^\alpha + e^{-\alpha}} = \frac{\sinh \alpha}{\cosh \alpha} = \tanh \alpha$$

Notice that

$$L(v)L(v') = L_{rr'}, \quad r = \sqrt{((1 + v)/(1 - v))}, \quad r' = \sqrt{((1 + v')/(1 - v'))}.$$

But

$$rr' = \left[\frac{1 + \dfrac{v + v'}{1 + vv'}}{1 - \dfrac{v + v'}{1 + vv'}}\right]^{1/2}$$

so

$$L(v)L(v') = L\left(\frac{v + v'}{1 + vv'}\right). \tag{4.7}$$

This is the *addition of velocity law* in special relativity.

We are thus using three different parametrizations of the same proper Lorentz transformation:

$$L_r = L^{\alpha} = L(v)$$

where

$$r = e^{\alpha} = \sqrt{((1 + v)(1 - v))}.$$

The formula for multiplying two of them is given by equations (4.4), (4.5), or (4.7), depending on the parametrization.

We have shown that the linear transformations of special relativity preserve the quadratic form $Q(\mathbf{v})$. But we have not given a direct physical interpretation of $Q(\mathbf{v})$. Here is one involving only light rays and clocks: Consider the points $t_1$ and $t_2$ on the $t$-axis which are joined to $\begin{pmatrix} t \\ x \end{pmatrix}$ by light rays (lines parallel to $t = x$ and $t = -x$).

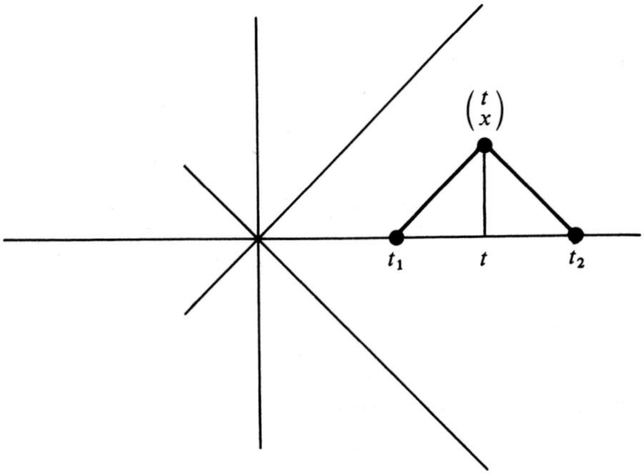


**Figure 4.25**

Then

$$t - t_1 = x \quad \text{or} \quad t_1 = t - x$$

and

$$t_2 - t = x \quad \text{or} \quad t_2 = t + x$$

so

$$t_1 t_2 = Q(\mathbf{v}) \quad \text{for} \quad \mathbf{v} = \begin{pmatrix} t \\ x \end{pmatrix}.$$

Point $\begin{pmatrix} 0 \\ 0 \end{pmatrix}$, at rest or in uniform motion, wishes to communicate with $\mathbf{v}$. It records the time $t_1$ when a light signal emitted at $t$ will reach $\mathbf{v}$ and records the time $t_2$ when the return signal, issued immediately is received. The product, $t_1 t_2$, is the *Minkowski distance* $Q(\mathbf{v})$ between the two events. Notice that if $\mathbf{v}$ lies on the line $x = 0$ then $t_1 = t_2 = t$ since the transmission will take no time at all. If $\mathbf{v}$ lies on a light ray through $\begin{pmatrix} 0 \\ 0 \end{pmatrix}$, then $t_1 = 0$. If $Q(\mathbf{v}) < 0$, then $t_1 < 0$ and $t_2 > 0$.

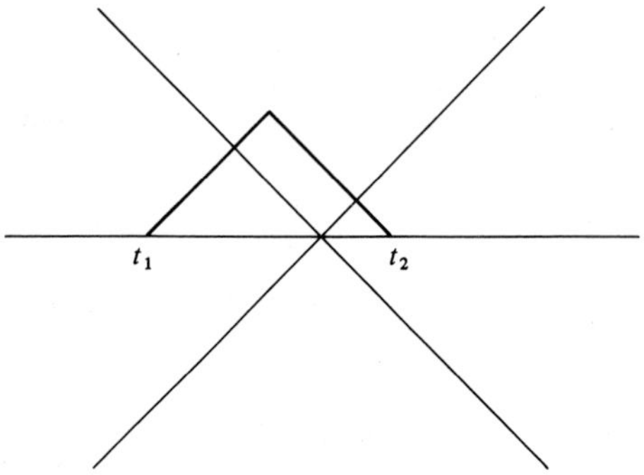


**Figure 4.26**

Here is another important property of the geometry of Minkowski space: Recall that in Euclidean geometry, we have the *triangle inequality*

$$\|\mathbf{u} + \mathbf{v}\| \leqslant \|\mathbf{u}\| + \|\mathbf{v}\|,$$

with equality only if $\mathbf{u}$ and $\mathbf{v}$ lie on the same line and point in the same direction. This is illustrated in figure 4.27.

$$c = a + b, \quad a = \|\mathbf{u}\|, \quad b = \|\mathbf{v}\|.$$

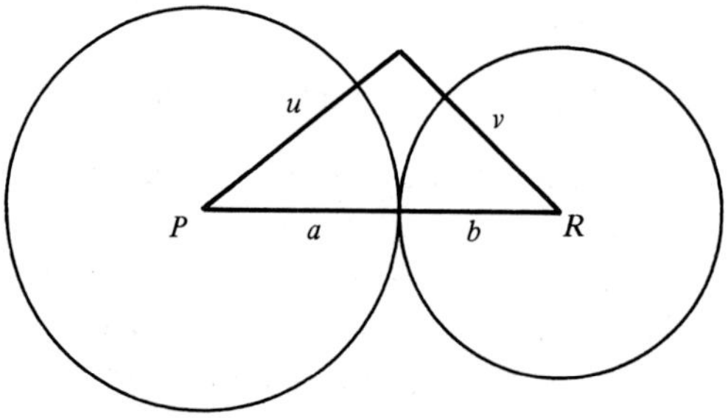


**Figure 4.27**

The broken path is clearly longer than the straight line. This shows that 'the straight line is the shortest distance between two points' in Euclidean geometry.

Now let us consider a similar diagram in our spacetime geometry, where the circles

$$\|A-P\|^2=a^2 \quad \text{and} \quad \|B-R\|^2=b^2$$

are replaced by hyperbolas $Q(A-P)=a^2$ and $Q(B-R)=b^2$. But now for any segments $l$ and $m$ which give a broken path from $P$ to $R$ we have

$$Q(l)<a^2 \quad \text{and} \quad Q(m)<b^2.$$

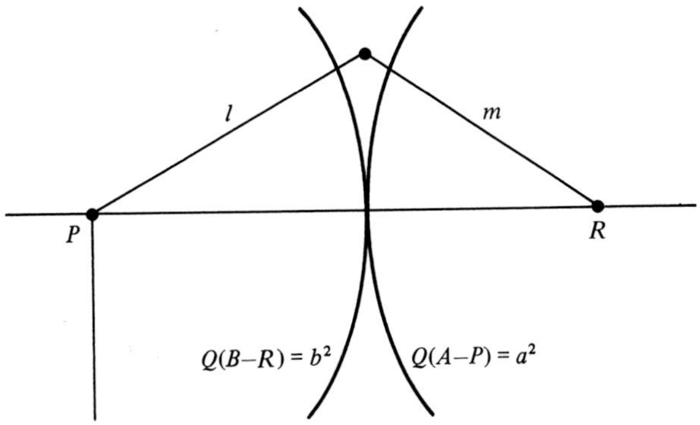


**Figure 4.28**

Now $Q(l)$ is just the square of the length of time elapsed on a clock moving uniformly along the line of $l$. Thus, we have the *reverse triangle inequality*.

> The time measured by a clock moving uniformly from $P$ to $R$ will be *longer* than the time measured by a clock moving along any broken path joining $P$ to $R$.

This is called the twin effect. The twin moving along the broken paths (if he survives the bumps) will be younger than the twin moving uniformly from $P$ to $R$. This is sometimes known as the *twin paradox*. It is, of course, no paradox, just an immediate corollary of the reverse triangle inequality.

## 4.7. The Poincaré group and the Galilean group

So far we have been describing the transformations of Euclidean geometry and of special relativity in terms of natural units. The points of spacetime are sometimes called *events*. They record when and where something happens. If we record the total events of a single human consciousness (say roughly 70 years measured in seconds) and several hundred or thousand meters measured in seconds we get a set of events which is enormously stretched out in one particular time direction compared to the space directions, by a factor of something like $10^{18}$. Being very

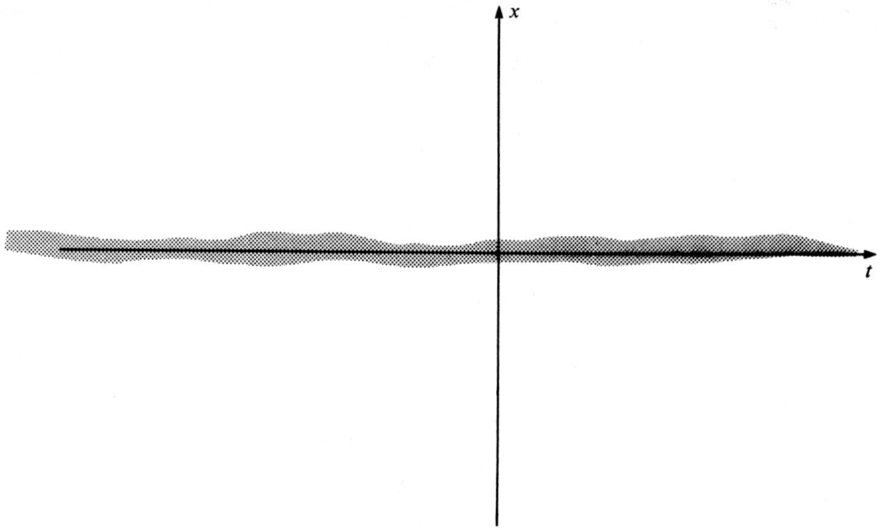


**Figure 4.29**

skinny in the space direction as compared with the time direction we tend to have a preferred splitting of spacetime with space and time picked out; and to measure distances in space with much smaller units (such as meters) than the units we use (such as seconds) to measure time. Of course, if we use a small unit the corresponding numerical value of the measurement will be large; that is in terms of human or 'ordinary' units, the space distances will be greatly magnified in comparison to the time differences. This suggests that we consider variables $T$ and $X$ related to the natural units $t$ and $x$ by $T=t$ and $X=cx$, or

$$\begin{pmatrix} T \\ X \end{pmatrix} = \begin{pmatrix} 1 & 0 \\ 0 & c \end{pmatrix} \begin{pmatrix} t \\ x \end{pmatrix}$$

where $c$ is a large number. The light cone $|x|=|t|$ goes over into $c^{-1}|X|=|T|$ or $|X|=c|T|$.

We say that 'the speed of light is $c$ in ordinary units'. Similarly, the hyperbola $t^2-x^2=k$ goes over into the curve $T^2-c^2X^2=k$; the 'timelike hyperbolas' corresponding to $k>0$ look very flattened out, almost like vertical straight lines for small values of $X$.

Let us see how to express a Lorentz transformation in terms of ordinary units. We do this as follows: we pick a point $\begin{pmatrix} T \\ X \end{pmatrix}$, find the point $\begin{pmatrix} t \\ x \end{pmatrix} = \begin{pmatrix} 1 & 0 \\ 0 & c^{-1} \end{pmatrix} \begin{pmatrix} T \\ X \end{pmatrix}$ that it corresponds to then apply the Lorentz transformation $L$ to $\begin{pmatrix} t \\ x \end{pmatrix}$ to obtain $L\begin{pmatrix} 1 & 0 \\ 0 & c^{-1} \end{pmatrix} \begin{pmatrix} T \\ X \end{pmatrix}$ and then express this new vector in ordinary units by multi-

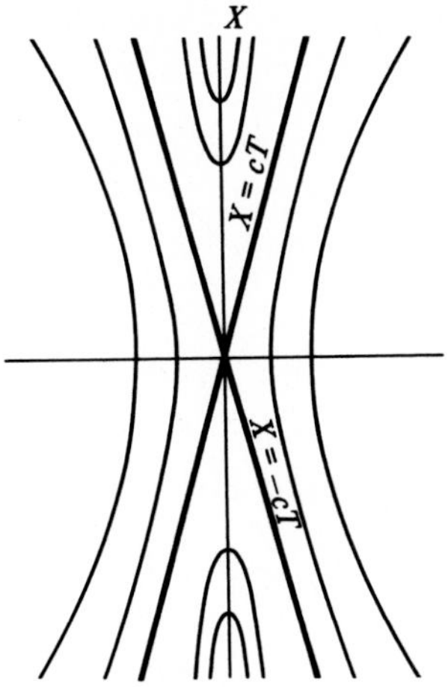


**Figure 4.30**

plying by the matrix $\begin{pmatrix} 1 & 0 \\ 0 & c \end{pmatrix}$ to get

$$\begin{pmatrix} 1 & 0 \\ 0 & c \end{pmatrix} L \begin{pmatrix} 1 & 0 \\ 0 & c^{-1} \end{pmatrix} \begin{pmatrix} T \\ X \end{pmatrix}.$$

Thus, in ordinary units a Lorentz transformation sends the vector $\begin{pmatrix} T \\ X \end{pmatrix}$ into the vector $M\begin{pmatrix} T \\ X \end{pmatrix}$ where $M$ is the matrix

$$M = \begin{pmatrix} 1 & 0 \\ 0 & c \end{pmatrix} L \begin{pmatrix} 1 & 0 \\ 0 & c^{-1} \end{pmatrix}.$$

Let us take $L = L^{\alpha}$ and carry out the multiplication so as to obtain

$$M = \begin{pmatrix} \cosh\alpha & c^{-1}\sinh\alpha \\ c\sinh\alpha & \cosh\alpha \end{pmatrix}.$$

This is the expression for any $\alpha$. Let us look at Lorentz transformations, $L$, for which $\alpha$ is 'small': Let $\tanh\alpha = \sinh\alpha/\cosh\alpha = v/c$, where we think of $c$ as being a very large velocity and $v$ an ordinary sized velocity, so that $\alpha$ is very small. Now

$$M = \begin{pmatrix} \cosh\alpha & c^{-1}\tanh\alpha\cosh\alpha \\ c\tanh\alpha\cosh\alpha & \cosh\alpha \end{pmatrix}$$

or

$$M = \cosh\alpha \begin{pmatrix} 1 & v/c^2 \\ v & 1 \end{pmatrix}.$$

Now

$$\cosh\alpha = \frac{1}{(1-\tanh^2\alpha)^{1/2}} = (1 - v^2/c^2)^{-1/2}$$

so

$$\cosh\alpha = 1 + \tfrac{1}{2}v^2/c^2 + \cdots.$$

Substituting into the above expression for $M$ we get

$$M = \left(1 + \frac{1}{2}\frac{v^2}{c^2} + \cdots\right)\begin{pmatrix} 1 & v/c^2 \\ v & 1 \end{pmatrix}$$

or

$$M = \begin{pmatrix} 1 & 0 \\ v & 1 \end{pmatrix} + E$$

where the entries of $E$ are all of order $c^{-2}$. The matrix

$$G_v = \begin{pmatrix} 1 & 0 \\ v & 1 \end{pmatrix}$$

is called a *velocity transformation* corresponding to velocity $v$. It preserves the lines $T = \text{constant}$; in fact $G_v\begin{pmatrix} T \\ X \end{pmatrix} = \begin{pmatrix} T \\ X + vT \end{pmatrix}$. We thus see that the velocity transformations can be regarded as 'limiting cases' of Lorentz transformations. When considering the velocity of light to be very large, the timelike hyperbolas go over into vertical straight lines, and Lorentz transformations with small values of $\alpha$ become velocity transformations.

The collection of velocity transformations also forms a group, $G_{v_1}G_{v_2} = G_{v_1+v_2}$ as can easily be checked; and this group preserves the notion of simultaneity.

A transformation of the form $\begin{pmatrix} T \\ X \end{pmatrix} v \to G_v\begin{pmatrix} T \\ X \end{pmatrix} + \begin{pmatrix} a \\ b \end{pmatrix}$ is known as a *Galilean transformation*. Thus a Galilean transformation is a translation composed with a velocity transformation. Newtonian mechanics was based on the geometry of Galilean relativity – those concepts invariant under all Galilean transformations. It was the genius of Lorentz, Poincaré and Einstein to recognize that our notion of simultaneity is only approximately valid, over small distances and velocities, and that the velocity transformation $G_v = \begin{pmatrix} 1 & 0 \\ v & 1 \end{pmatrix}$ must be regarded as an approximation to the Lorentz transformation:

$$\begin{pmatrix} 1/\sqrt{(1 - v^2/c^2)} & v/c^2\sqrt{(1 - v^2/c^2)} \\ v/\sqrt{(1 - v^2/c^2)} & 1/\sqrt{(1 - v^2/c^2)} \end{pmatrix}$$

(expressed in ordinary units).

## 4.8. Momentum, energy and mass

The passage from the Galilean group to the Poincaré group required a reformulation of the basic concepts of mechanics. The outline for such a theory was pointed out by Poincaré in his address to the World's Fair in St Louis in 1904 and was carried out by him, and, independently, by Einstein, in their fundamental papers in 1905. We will describe some of the ideas here.

In classical mechanics there are two principles which are useful in describing the motion of particles – the conservation of momentum and the conservation of energy. For example, suppose we are studying the collision of two particles, $A$ and $B$. Let $p_A$ denote the momentum of particle $A$ before the collision, and $p'_A$ denote its momentum after the collision. Similarly for particle $B$. The law of conservation of momentum says that

$$p_A + p_B = p'_A + p'_B \qquad \text{conservation of momentum}.$$

The collision is called elastic if the total kinetic energy is conserved. An example of an *in*elastic collision is one where the particles get stuck together upon impact. Conversely, if two particles are initially in contact, and at rest, say, with an explosive charge between them, when the charge is exploded the particles will move apart. This can be regarded as a reverse 'collision': if we ran a film of it backwards, it would look like two particles colliding and sticking together. Total kinetic energy is not conserved – the total kinetic energy was zero before the explosion and positive after the particles were set in motion. The energy released by the explosion was converted into kinetic energy. Similarly, we believe that when two particles collide and stick together, kinetic energy is converted into energy of some other form; heat or potential energy. For an inelastic collision one still has the law of conservation of momentum. In an *elastic* collision, there is no exchange between kinetic and other forms of energy so the total kinetic energy is conserved:

$$E_A + E_B = E'_A + E'_B \qquad \text{conservation of energy}$$

where $E_A$ denotes the kinetic energy of particle $A$ before the collision, $E'_A$ its kinetic energy after the collision, etc.

It turns out that the laws of conservation of momentum and of energy hold in special relativity just as they do in Newtonian mechanics. What must be changed is the definition of momentum and of energy:

In Newtonian mechanics, the momentum of a particle is defined as

$$\mathbf{p} = m\mathbf{v}$$

where $\mathbf{v}$ is the velocity of a moving particle and $m$ is its *mass*. The velocity (and hence the momentum) is a vector in three-dimensional space. In our model universe it will be considered as one-dimensional. (Alternatively, we can consider particles *constrained* to move on a line.) The mass can, in principle, be defined by the following series of experiments. Suppose we have a collection of objects – say little balls made of different materials. We consider two held together at rest and then pulled apart by an explosion set off between them or by a spring released between them. One object will then move to the right and the other to the left. If the two objects are identical –

**Figure 4.31**

the same size balls made of the same material, say – we would expect that the motion will be completely symmetrical. For example, if there are reflecting barriers placed at equal distances from the point of explosion, we expect that the two objects will bounce back and collide with one another at precisely the initial point of explosion. We can perform the experiment and observe that this is indeed the case. Next let us take two balls made of the same material but of different sizes. Say the larger ball is on the right. We will then observe that the point of collision will be to the right of center – the smaller ball will have travelled further. We can then perform the same experiment with balls of differing materials. For example, we will find that if we use two balls of the same diameter, one of lead on the right and one of aluminum on the left, the point of collision will be to the right. On the other hand, if we take a very small ball of lead on the right with our fixed size ball of aluminum on the left, we will find that the point of collision will be to the left. Assuming that we have enough sizes of balls of lead, we will find a lead ball which exactly matches the aluminum ball.

We can now compare lead balls with copper balls, say. Suppose we found an aluminum ball that matches a lead ball (in the sense that the point of recollision is at the center) and a copper ball that matches the lead ball. We can than compare the aluminum ball with the copper ball. It is an *experimental fact* that the aluminum ball will match the copper ball. This is a *law of nature*, not an assertion in logic.* But we can now *define* the notion of mass by declaring that two objects have the *same mass* if they match in our explosion–collision experiment. The law of nature referred to above is then the assertion that this notion of mass is well defined – if $A$ has the same mass as $B$ and $B$ has the same mass as $C$, then $A$ has the same mass as $C$. We can observe the following law of nature: If $A_1$ matches $B_1$ and $A_2$ matches $B_2$, then performing the experiment with the two balls; $A_1$ and $A_2$ against $B_1$ and $B_2$, will show that $A_1$ and $A_2$ match $B_1$ and $B_2$. (Alternatively, we could also observe that the

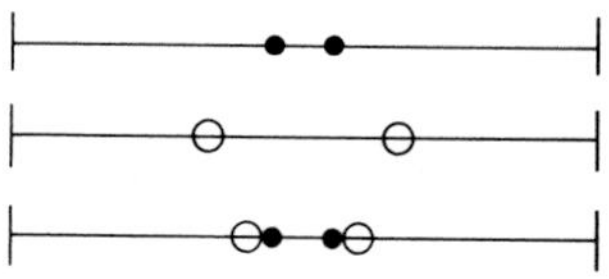

**Figure 4.32**

mass of a ball of the same material is proportional to its volume: if a ball of radius $r_A$ of lead matches a ball of radius $r_B$ of copper, then a ball of radius $3r_A$ of lead will match 27 balls of radius $r_B$ of copper.) This allows us to introduce units of mass: having fixed one object say a lead ball of volume 1 cm$^3$, we can then compare any other object with a multiple of our given object (a lead ball of volume $m$) and this assigns, in a well-defined way, a numerical value to any mass. Originally, in the metric system the gram was taken to be the mass of the 1 cubic centimeter of water of

* We will also find as a law of nature that turning the apparatus around – that is, interchanging right and left – will not affect the matching or non-matching properties of objects.

4 °C. Since water at 4 °C is difficult to work with in our collision experiments, we might want to define the gram as a mass of a ball of copper whose volume is 0.11 ... $cm^3$. It is interesting to observe that in the above series of experiments we did not need any clocks.

We now return to the conservation of momentum. In Newtonian mechanics this laws says that if we define momentum by

$$\mathbf{p} = m\mathbf{v}$$

then the total momentum is conserved. (In fact, it is not hard to show that this version of the law of conservation of momentum is a consequence of our definition of mass and of the assumption that the laws of nature are invariant under the *Galilean* group. See Feynman's *Lectures on Physics, I*, Chapter 10 for a very lucid presentation of this argument.) In special relativity this definition of momentum makes no sense because *velocity makes no sense*! After all, velocity is defined as

$$v = \frac{dx}{dt}$$

and this presupposes that we have chosen $x$ and $t$ axes and have decided to parameterize the curve describing the motion of the particle by $t$ – that is why we are writing the curve as $x(t)$. If we apply a Lorentz transformation, we will get different $t'$- and $x'$-axes and hence a different velocity, $v'$. Let us put the problem another way. Suppose we decide to parametrize the curve describing the motion of the particle in spacetime by some neutral third parameter, $s$. For example, $s$ might be the reading on some internal clock that the particle might be carrying along with it on its motion. Thus the curve in our space time plane is given by

$$\mathbf{u}(s) = \begin{pmatrix} t(s) \\ x(s) \end{pmatrix}.$$

At some instant $s_0$, we can compute the tangent vector

$$\frac{d\mathbf{u}}{ds} = \begin{pmatrix} dt/ds \\ dx/ds \end{pmatrix} = \mathbf{w} = \begin{pmatrix} a \\ b \end{pmatrix}.$$

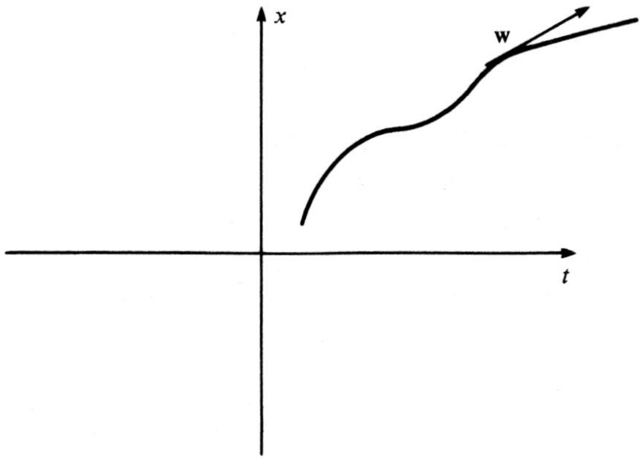


**Figure 4.33**

The velocity $v = dx/dt$ is then given by

$$v = b/a,$$

in the $t, x$ coordinate system. It is clear that for a vector

$$\mathbf{w} = \begin{pmatrix} a \\ b \end{pmatrix}$$

the ratio $v = b/a$ makes no sense in that if we replace $\mathbf{w}$ by

$$\mathbf{w}' = L\mathbf{w}$$

and write

$$\mathbf{w}' = \begin{pmatrix} a' \\ b' \end{pmatrix}, \quad v' = b'/a'$$

then (unless $v = \pm 1$ or $L = I$) $v'$ will not be equal to $v$. The one property of $\mathbf{w}$ that *is* conserved is

$$Q(\mathbf{w}) = a^2 - b^2.$$

The condition $Q(\mathbf{w}) > 0$ is the same as the condition $|v| < 1$. Since 1 is the speed of light in our units, it *does* make sense to say that the velocity $v$ is less than the speed of light.

It is an experimental fact that all particles with positive rest mass (defined below) move at speeds less than the speed of light – that, for them, $Q(\mathbf{w}) > 0$.

So let us call

$$Q(\mathbf{w}) = \mu^2 > 0$$

so

$$a^2 - b^2 = \mu^2$$

and let

$$b/a = v$$

in a particular spacetime splitting. Notice that the equation $b/a = v$ only determines the ratio of $b$ to $a$. (This is a reflection of the fact that we have not really specified the mysterious parameter $s$ in the curve $\mathbf{u}(s)$.) But we can solve the two equations $a^2 - b^2 = \mu^2$ and $b/a = v$ to get

$$a = \frac{\mu}{\sqrt{(1 - v^2)}},$$

$$b = \frac{\mu v}{\sqrt{(1 - v^2)}},$$

in a given spacetime splitting. For small values of $v$ we have the Taylor expansion

$$\frac{1}{\sqrt{(1 - v^2)}} = 1 + \tfrac{1}{2}v^2 - \cdots$$

so

$$a \doteq \mu + \tfrac{1}{2}\mu v^2 + \cdots,$$
$$b \doteq \mu v + \tfrac{1}{2}\mu v^3 + \cdots.$$

Notice that the expression for $b$ looks very much like $p = mv$ *if we identify* $\mu$ *with* $m$, and ignore the higher order terms in $v$. In the same way the second term in the expression for $a$ looks just like the expression for kinetic energy in Newtonian mechanics. We are thus led to the following modification of the definitions of energy and momentum. Associated to any object there is a definite value of $\mu$. To avoid confusion this value is denoted by $m_0$ and called the *rest mass* of the object. If the object happens to be at rest in some space time splitting, this rest mass coincides (up to a choice of units, of course) with the rest mass defined experimentally above. Suppose that $m_0 > 0$ (as we have been implicitly assuming). Then, when the object is in motion, its *energy-momentum vector* is defined to be the unique vector

$$\mathbf{w} = \begin{pmatrix} E \\ p \end{pmatrix}$$

such that

$$Q(\mathbf{w}) = E^2 - p^2 = m_0^2$$

and

$\mathbf{w}$ is a scalar multiple of $\dot{\mathbf{u}} = \dfrac{\mathrm{d}\mathbf{u}}{\mathrm{d}s}$ where $\mathbf{u}(s)$ is the curve describing the motion of the object in spacetime.

In terms of a given spacetime splitting where

$$\mathbf{u}(s) = \begin{pmatrix} t(s) \\ x(s) \end{pmatrix}$$

so

$$\dot{\mathbf{u}}(s) = \begin{pmatrix} \dot{t}(s) \\ \dot{x}(s) \end{pmatrix}$$

and

$$v = \frac{\mathrm{d}x}{\mathrm{d}t} = \frac{\dot{x}(s)}{\dot{t}(s)}$$

we have

$$p = \frac{m_0 v}{\sqrt{(1 - v^2)}}$$

and

$$E = \frac{m_0}{\sqrt{(1 - v^2)}}.$$

In particular, if the object is at rest in a spacetime splitting so that $v = 0$, then

$$p = 0 \quad \text{and} \quad E = m_0$$

in that system of coordinates.

The law of conservation of energy-momentum now says that

$$\begin{pmatrix} E_A \\ p_A \end{pmatrix} + \begin{pmatrix} E_B \\ p_B \end{pmatrix} = \begin{pmatrix} E'_A \\ p'_A \end{pmatrix} + \begin{pmatrix} E'_B \\ p'_B \end{pmatrix}$$

at any collision – a conservation law for *vectors* in spacetime.

We have written all of the above equations in terms of natural units where the speed of light is one and $v$ is a number, so an expression such as $\surd(1 - v^2)$ makes sense. If we use 'psychological units', then $v$ is not a number but a velocity expressed in cm/s, for example. So an expression such as $\surd(1 - v^2)$ makes no sense as it stands. We must replace it by $\surd(1 - (v^2/c^2))$. To make $p$ look as it should in the small $v$ approximation, we must write

$$p = \frac{m_0 \mathbf{v}/\mathbf{c}}{\surd(1 - v^2/c^2)}.$$

Similarly, to make the units of $E$ and the kinetic energy term come out right, we must write

$$E = \frac{m_0 c^2}{\surd(1 - v^2/c^2)}.$$

This is the appropriate rescaling. For the particle at rest, we get the famous Einstein mass–energy relation

$$E = m_0 c^2.$$

## 4.9. Antisymmetric forms

We have considered two kinds of scalar product between vectors in the plane, the Euclidean scalar product defined by

$$(\mathbf{w}, \mathbf{w}') = xx' + yy' \quad \text{where} \quad \mathbf{w} = \begin{pmatrix} x \\ y \end{pmatrix} \quad \text{and} \quad \mathbf{w}' = \begin{pmatrix} x' \\ y' \end{pmatrix}$$

and the Lorentz scalar product

$$(\mathbf{v}, \mathbf{v}') = tt' - xx' \quad \text{where} \quad \mathbf{v} = \begin{pmatrix} t \\ x \end{pmatrix} \quad \text{and} \quad \mathbf{v}' = \begin{pmatrix} t' \\ x' \end{pmatrix}.$$

Both of these scalar products are bilinear, that is, when one variable is held fixed, we get a linear function of the other:

$$(a\mathbf{w}_1 + b\mathbf{w}_2, \mathbf{w}') = a(\mathbf{w}_1, \mathbf{w}') + b(\mathbf{w}_2, \mathbf{w}')$$

and so on. Also, both of these scalar products are symmetric:

$$(\mathbf{w}, \mathbf{w}') = (\mathbf{w}', \mathbf{w})$$

and

$$(\mathbf{v}, \mathbf{v}') = (\mathbf{v}', \mathbf{v}).$$

We now introduce a third kind of product between two vectors in the plane which is bilinear, but anti-symmetric: we define

$$\omega(\mathbf{v}, \mathbf{v}') = qp' - q'p = \operatorname{Det}\begin{pmatrix} q & q' \\ p & p' \end{pmatrix} \quad \text{where} \quad \mathbf{v} = \begin{pmatrix} q \\ p \end{pmatrix} \quad \text{and} \quad \mathbf{v}' = \begin{pmatrix} q' \\ p' \end{pmatrix}.$$

Here

$$\omega(\mathbf{v}, \mathbf{v}') = -\omega(\mathbf{v}', \mathbf{v})$$

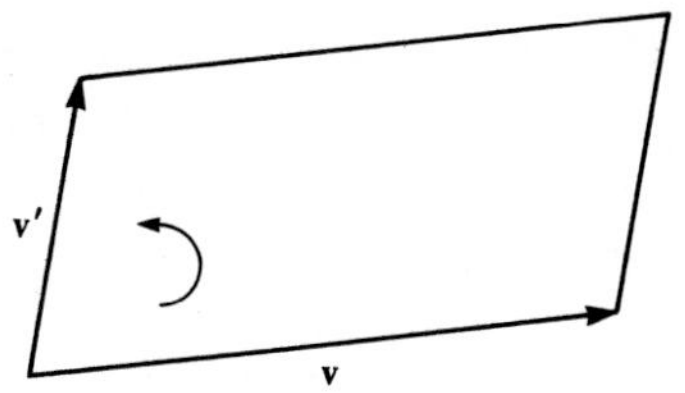


**Figure 4.34.**

which is what we mean by *anti-symmetric*. The geometric meaning of $\omega(\mathbf{v}, \mathbf{v}')$ is clear; it is the oriented area of the parallelogram spanned by $\mathbf{v}$ and $\mathbf{v}'$. It is also clear that $\omega(\mathbf{v}, \mathbf{v}')$ is bilinear. Such an $\omega$ is called a symplectic scalar product.

A linear transformation, $A$, is called *symplectic* if it preserves the scalar product $\omega$. Thus $A$ is symplectic if and only if

$$\omega(A\mathbf{v}, A\mathbf{v}') = \omega(\mathbf{v}, \mathbf{v}')$$

for all $\mathbf{v}$ and $\mathbf{v}'$. The matrix whose columns are $A\mathbf{v}$ and $A\mathbf{v}'$ is just the product of the matrix $A$ with the matrix $\begin{pmatrix} q & q' \\ p & p' \end{pmatrix}$. Therefore

$$\omega(A\mathbf{v}, A\mathbf{v}') = \operatorname{Det} A\begin{pmatrix} q & q' \\ p & p' \end{pmatrix} = (\operatorname{Det} A)\left(\operatorname{Det}\begin{pmatrix} q & q' \\ p & p' \end{pmatrix}\right) = \operatorname{Det} A\,\omega(\mathbf{v}, \mathbf{v}').$$

Thus $A$ is symplectic if and only if $\operatorname{Det} A = 1$. Any symplectic matrix clearly has an inverse which is again symplectic and the product of two symplectic matrices is again symplectic. Thus the collection of all $2 \times 2$ symplectic matrices forms a *group*, called the (two-dimensional) symplectic group. The symplectic group plays a very important role in the study of optics, as we shall see in Chapter 9.

---

## Summary

### A Euclidean scalar product

You should be able to list and apply the properties of a Euclidean scalar product.

You should be able to write down the transpose of a matrix and to apply the transpose operation in connection with scalar products and Euclidean transformations.

Given a vector space of 2 or more dimensions, with a Euclidean scalar product, you should know how to use the Gram–Schmidt process to construct an orthonormal basis and to find the orthogonal projection onto a subspace.

### B Quadratic forms

You should be able to express a quadratic form $Q(\mathbf{v}, \mathbf{v})$ in terms of a symmetric matrix $A$ and relate maximum and minimum values of $Q$ to the eigenvectors and eigenvalues of $A$.

Given a quadratic form $Q$ on the plane, you should be able to introduce coordinates $x'$ and $y'$ so that

$$Q = \lambda_1 x'^2 + \lambda_2 y'^2$$

and to use these coordinates as an aid in graphing $Q = \text{constant}$.

C Coupled oscillations

You should be able to reduce the problem of two coupled oscillators to the form $\ddot{\mathbf{w}} = -A\mathbf{w}$ and to solve for the normal modes in terms of the eigenvectors and eigenvalues of $A$.

D Lorentz scalar product

You should be able to calculate the Lorentz scalar product of two vectors, identifying Lorentz transformations that preserve this scalar product, and apply these concepts to the special theory of relativity.

---

## Exercises

4.1.(a) Using the three properties of the scalar product (symmetry, linearity, positive-definiteness), prove the *Cauchy–Schwartz inequality*

$$(\mathbf{v}, \mathbf{w}) \leqslant \sqrt{((\mathbf{v}, \mathbf{v})(\mathbf{w}, \mathbf{w}))}$$

for any pair of vectors $\mathbf{v}$ and $\mathbf{w}$.
(Hint: Consider $(\mathbf{v} - \alpha\mathbf{w}, \mathbf{v} - \alpha\mathbf{w})$. This is a quadratic polynomial in $\alpha$, but it can not have any real roots unless $\mathbf{v} = \alpha\mathbf{w}$.)

(b) Prove the *triangle inequality*

$$\|\mathbf{v} + \mathbf{w}\| \leqslant \|\mathbf{v}\| + \|\mathbf{w}\|$$

(where $\|\mathbf{v}\|^2 = (\mathbf{v}, \mathbf{v})$, etc.) (Hint: square both sides and use (*a*))

4.2.(a) Let $\mathbf{v}$ and $\mathbf{v}'$ be two vectors in the plane. Show that a rotation $R_\theta$ through an angle $\theta$ for which

$$\cos\theta = \frac{(\mathbf{v}, \mathbf{v}')}{\sqrt{((\mathbf{v}, \mathbf{v})(\mathbf{v}', \mathbf{v}'))}}$$

will carry $\mathbf{v}$ into a multiple of $\mathbf{v}'$. Determine the angle between $\begin{pmatrix} 2 \\ 1 \end{pmatrix}$ and $\begin{pmatrix} 4 \\ -2 \end{pmatrix}$.

(b) Let $\mathbf{v}$ and $\mathbf{v}'$ be two vectors in two-dimensional spacetime which are either both spacelike, both forward timelike, or both backward timelike. Show that a proper Lorentz transformation $L_\alpha$ for which

$$\cosh\alpha = \frac{\{\mathbf{v}, \mathbf{v}'\}}{\sqrt{(\{\mathbf{v}, \mathbf{v}\}\{\mathbf{v}', \mathbf{v}'\})}}$$

will carry $\mathbf{v}$ into a multiple of $\mathbf{v}'$.

Use this result to find a Lorentz transformation which carries $\begin{pmatrix} 5 \\ 4 \end{pmatrix}$ into a multiple of $\begin{pmatrix} 5 \\ 3 \end{pmatrix}$.

What goes wrong if $\mathbf{v}$ is spacelike but $\mathbf{v}'$ is timelike? If $\mathbf{v}$ is forward timelike but $\mathbf{v}'$ is backward timelike? If $\mathbf{v}$ or $\mathbf{v}'$ is lightlike?

4.3. For practice with the Lorentz scalar product, consider the following vectors in two-dimensional spacetime. (The first coordinate is $t$; the second is $x$.)

$$\mathbf{v}_1 = \begin{pmatrix} -2 \\ 2 \end{pmatrix}, \quad \mathbf{v}_2 = \begin{pmatrix} -1 \\ 3 \end{pmatrix}, \quad \mathbf{v}_3 = \begin{pmatrix} 1 \\ 3 \end{pmatrix}, \quad \mathbf{v}_4 = \begin{pmatrix} 1 \\ 1 \end{pmatrix}$$

$$\mathbf{v}_5 = \begin{pmatrix} 3 \\ -1 \end{pmatrix}, \quad \mathbf{v}_6 = \begin{pmatrix} 3 \\ 1 \end{pmatrix}.$$

(a) Calculate the Lorentz scalar product $\{\mathbf{v}, \mathbf{v}\}$ of each vector with itself. Plot each vector on a spacetime diagram and identify each as spacelike, forward or backward lightlike, or forward or backward timelike.

(b) Calculate the Lorentz scalar products $\{\mathbf{v}_2, \mathbf{v}_3\}$, $\{\mathbf{v}_5, \mathbf{v}_6\}$, and $\{\mathbf{v}_3, \mathbf{v}_6\}$.

(c) Calculate the vectors $\mathbf{w}_1, \ldots, \mathbf{w}_6$ which result from applying the Lorentz transformation:

$$L_2 = \begin{pmatrix} \frac{5}{4} & \frac{3}{4} \\ \frac{3}{4} & \frac{5}{4} \end{pmatrix}$$

to each of the vectors $\mathbf{v}_1 \ldots \mathbf{v}_6$. Plot the transformed vectors on the spacetime diagram.

(d) Calculate $\{\mathbf{w}_2, \mathbf{w}_2\}$, $\{\mathbf{w}_6, \mathbf{w}_6\}$, $\{\mathbf{w}_2, \mathbf{w}_3\}$, $\{\mathbf{w}_5, \mathbf{w}_6\}$, and $\{\mathbf{w}_3, \mathbf{w}_6\}$. All these scalar products should be the same as for the corresponding $\mathbf{v}$ vectors.

4.4. Let $S$ be a symmetric matrix with positive eigenvalues. Define a new scalar product $[\mathbf{v}, \mathbf{w}]_S$ by the equation $[\mathbf{v}, \mathbf{w}]_S = (S\mathbf{v}, \mathbf{w})$.

(a) Show that this scalar product is symmetric, bilinear, and positive definite.

(b) Show that a matrix $C$ preserves this scalar product (i.e., $[C\mathbf{v}, C\mathbf{w}]_S = [\mathbf{v}, \mathbf{w}]_S$ if and only if $C^{\mathrm{T}}SC = S$).

(c) Describe a procedure for constructing a matrix $B$ with the property that if $\mathbf{v}' = B^{-1}\mathbf{v}$, $\mathbf{w}' = B^{-1}\mathbf{w}$, then $[\mathbf{v}, \mathbf{w}]_S = (\mathbf{v}', \mathbf{w}')$. Explain how, given one matrix $B$ with this property, you could construct many others.

4.5 In the preceding problem, let $S = \begin{pmatrix} 3.7 & 0.9 \\ 0.9 & 1.3 \end{pmatrix}$.

(a) Find a vector $\mathbf{v}$ which is orthogonal to $\mathbf{w} = \begin{pmatrix} 1 \\ -1 \end{pmatrix}$ under the scalar product defined by $S$, so that $[\mathbf{v}, \mathbf{w}]_S = 0$.

(b) Construct a matrix $B$ with the properties described in 4.4 (c), and verify that with $\mathbf{v}$ and $\mathbf{w}$ as in part (a), $(B^{-1}\mathbf{v}, B^{-1}\mathbf{w}) = 0$.

(c) Construct an *orthogonal projection* matrix $P$, satisfying $P^2 = P$, whose image consists of multiples of $\mathbf{w} = \begin{pmatrix} 1 \\ -1 \end{pmatrix}$ and which satisfies $[\mathbf{w}, P\mathbf{v}]_S = [\mathbf{w}, \mathbf{v}]_S$ for all vectors $\mathbf{v}$.

(d) Construct a matrix $C$, satisfying $C^2 = -I$, which preserves the scalar product defined by $S$. $\left(\text{Hint: } R = \begin{pmatrix} 0 & -1 \\ 1 & 0 \end{pmatrix} \text{ satisfies } R^2 = -I \text{ and}\right.$

preserves the ordinary scalar product.$\Big)$

4.6. Apply the following procedure to the quadratic form

$$Q(\mathbf{v}) = 8x^2 + 12xy + 17y^2:$$

(a) Write $Q$ in the form $(A\mathbf{v}, \mathbf{v})$ where $A$ is a symmetric matrix.
(b) Find the eigenvalues of $A$.
(c) Express $A$ in the form

$$A = R_\theta \begin{pmatrix} \lambda_1 & 0 \\ 0 & \lambda_2 \end{pmatrix} R_\theta^{-1}.$$

(d) Find coordinates $x'$ and $y'$ such that $Q$ can be expressed in the form

$$Q(\mathbf{v}) = 20x'^2 + 5y'^2.$$

(e) Sketch a graph of the equation $Q(\mathbf{v}) = 20$. Indicate both the $xy$-axes and $x'y'$-axes on the sketch.

4.7.(a) Determine the eigenvalues $\lambda_1$ and $\lambda_2$ of the matrix $S = \begin{pmatrix} 9 & 2 \\ 2 & 6 \end{pmatrix}$, and find eigenvectors $\mathbf{v}_1$ and $\mathbf{v}_2$ associated with these two eigenvalues.
(b) Construct a rotation matrix $R$ such that $S = R\Lambda R^{-1}$, where $\Lambda$ is diagonal. Be sure that $R$ represents a rotation!
(c) Find new coordinates $x'$ and $y'$, linear functions of $x$ and $y$, such that

$$9x^2 + 4xy + 6y^2 = \lambda_1 x'^2 + \lambda_2 y'^2.$$

4.8.(a) Determine the eigenvalues and eigenvectors of the matrix

$$A = \begin{pmatrix} 10 & 6 \\ 6 & 10 \end{pmatrix}.$$

(b) Construct a rotation matrix $R$ and a diagonal matrix $\Lambda$ such that $A = R\Lambda R^{-1}$.
(c) Sketch the graph of the equation $10x^2 + 12xy + 10y^2 = 24$.

4.9.(a) Find coordinates $x'$ and $y'$ such that the quadratic form

$$Q(\mathbf{v}) = -x^2 + 6xy + 7y^2$$

can be expressed in the form

$$Q(\mathbf{v}) = \lambda_1 x'^2 + \lambda_2 y'^2.$$

Identify and sketch the graph of $Q(\mathbf{v}) = 40$.
(b) Let $x$ and $y$ lie on the unit circle, so that $x = \cos\theta$, $y = \sin\theta$. Find the values of $\theta$ for which $Q$ achieves its maximum and minimum values, and calculate those maximum and minimum values. What is the relationship of these answers to the answers to part (a)?

4.10. Suppose that $M$ and $K$ are both *symmetric* $2 \times 2$ matrices.

(a) Construct an example to show that $M^{-1}K$ is not necessarily symmetric.
(b) Describe how to construct a *symmetric* matrix $B$ such that $B^2 = M^{-1}$. Show that the matrix $S = BKB$ is symmetric, and hence can be written as $S = R\Lambda R^{-1}$, where $R$ is a rotation and $\Lambda = \begin{pmatrix} \lambda_1 & 0 \\ 0 & \lambda_2 \end{pmatrix}$.

(c) Show that if $A = BR$, then $M^{-1}K = A\Lambda A^{-1}$. This proves that $M^{-1}K$ has real eigenvalues.

(d) Define new coordinates $x'$ and $y'$ by $\begin{pmatrix} x' \\ y' \end{pmatrix} = A^{-1}\begin{pmatrix} x \\ y \end{pmatrix}$. Show that, if $\mathbf{v} = \begin{pmatrix} x \\ y \end{pmatrix}$, then $(M\mathbf{v}, \mathbf{v}) = x'^2 + y'^2$, while $(K\mathbf{v}, \mathbf{v}) = \lambda_1 x'^2 + \lambda_2 y'^2$.

(Hints: $B$ is symmetric, so $(B\mathbf{v}, \mathbf{w}) = (\mathbf{v}, B\mathbf{w})$. $R$ is orthogonal, so $(R\mathbf{v}, \mathbf{w}) = (\mathbf{v}, R^{-1}\mathbf{w})$.)

4.11.(a) Show that, if $A = \begin{pmatrix} 0 & \alpha \\ \alpha & 0 \end{pmatrix}$, $\exp(tA)$ is a Lorentz transformation.

(b) In relativistic mechanics, the total energy $E$ and the linear momentum $p$ of a particle of mass $m$ moving along a line form a vector $\mathbf{v} = \begin{pmatrix} E \\ p \end{pmatrix}$ with $\langle \mathbf{v}, \mathbf{v} \rangle = E^2 - p^2 = m^2$. If the particle moves so that its acceleration is always $\alpha$ according to an observer who sees the particle as instantaneously at rest, then $E$ and $p$ are related by

$$\frac{dE}{d\tau} = \alpha p, \quad \frac{dp}{d\tau} = \alpha E,$$

where $\tau$ is time as measured by a clock carried along with the particle. Solve these equations to determine

$$\begin{pmatrix} E(\tau) \\ p(\tau) \end{pmatrix}$$

for initial conditions $\begin{pmatrix} E_0 \\ p_0 \end{pmatrix}$ when $\tau = 0$.

4.12. Suppose that distances along two perpendicular axes in the plane are measured in units which differ by a large factor $c$. For example, in considering straight lines which might be drawn along a straight superhighway which is 1000 kilometers long (along $x$) but only 1000 centimeters wide (along $y$), we might wish to define new 'ordinary' coordinates by $X = x$ and $Y = cy$, where $c = 10^5$, so that $X$ is measured in kilometers while $Y$ is measured in centimeters. Construct the matrix that represents a rotation through angle $\theta$ in terms of coordinates $X$ and $Y$, and show that for lines whose slope $Y/X$ in ordinary coordinates is a number of the order of unity, the rotation matrix becomes a shear matrix in the limit $c \to \infty$. Explain this phenomenon geometrically by considering what happens to the circles $x^2 + y^2 = k$.
(Note: After working this problem, reread the discussion of the limit $c \to \infty$ for Lorentz transformations, in section 4.3.)

4.13. Calculate the symplectic scalar product $\omega(\mathbf{v}_1, \mathbf{v}_2)$ for the vectors

$$\mathbf{v}_1 = \begin{pmatrix} 2 \\ 1 \end{pmatrix}, \quad \mathbf{v}_2 = \begin{pmatrix} 1 \\ 3 \end{pmatrix}.$$

Confirm explicitly that this scalar product is preserved under the action of the symplectic matrix

$$A = \begin{pmatrix} 5 & 3 \\ -2 & -1 \end{pmatrix}.$$

4.14. Consider the system of springs and masses shown in figure 4.35.

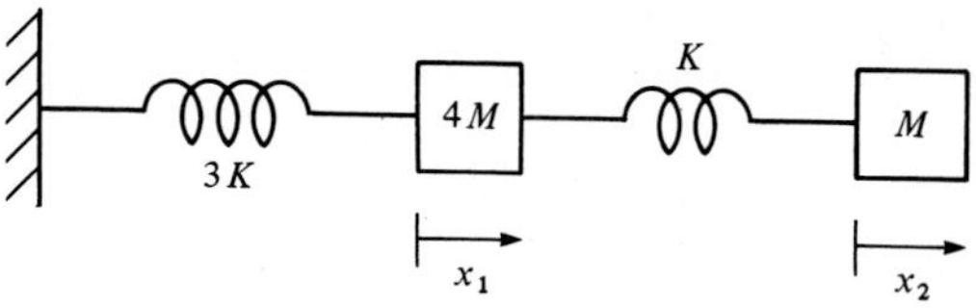


**Figure 4.35**

(a) Show that, if $x_1$ and $x_2$ represent displacements to the right of equilibrium, then the motion of this system is governed by

$$T\begin{pmatrix}\ddot{x}_1\\ \ddot{x}_2\end{pmatrix} = -H\begin{pmatrix}x_1\\ x_2\end{pmatrix}$$

where

$$T=\begin{pmatrix}4 & 0\\ 0 & 1\end{pmatrix} \quad \text{and} \quad H=\begin{pmatrix}4 & -1\\ -1 & 1\end{pmatrix}.$$

(b) Let $B$ be the diagonal matrix with positive entries satisfying $B^2 = T$. Construct the matrix $A = B^{-1}HB^{-1}$, find its eigenvalues and eigenvectors, and use them to determine the general solution to $\ddot{\mathbf{w}} = -A\mathbf{w}$.

(c) Describe the normal modes of the system by specifying the frequency of each in terms of $\omega_0 = \sqrt{(K/M)}$ and by specifying the ratio $x_2/x_1$.

4.15. Consider the system of masses and springs shown in figure 4.36. Let $x_1$ and $x_2$ denote displacements to the right of equilibrium.

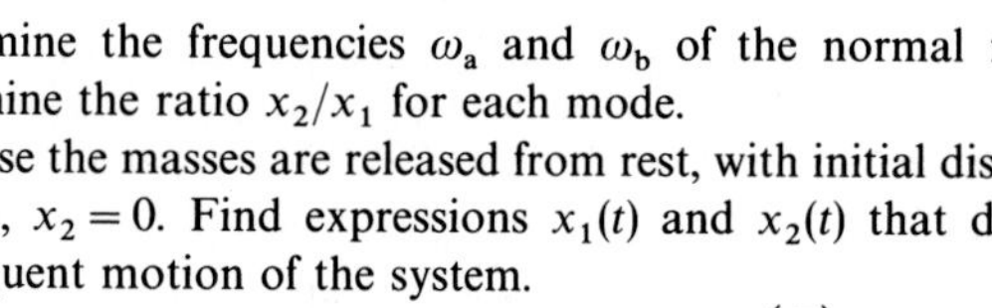

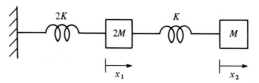

**Figure 4.36**

(a) Determine the frequencies $\omega_a$ and $\omega_b$ of the normal modes and determine the ratio $x_2/x_1$ for each mode.

(b) Suppose the masses are released from rest, with initial displacements $x_1 = A$, $x_2 = 0$. Find expressions $x_1(t)$ and $x_2(t)$ that describe the subsequent motion of the system.

4.16. A particle whose energy-momentum vector is $\begin{pmatrix}E\\ p\end{pmatrix}$ is subjected to a Lorentz transform represented by the matrix

$$\frac{1}{2}\begin{pmatrix}r+r^{-1} & r-r^{-1}\\ r-r^{-1} & r+r^{-1}\end{pmatrix}.$$

Show that the sum of its energy and momentum is multiplied by $r$, while their difference is divided by $r$. Interpret this result in terms of eigenvectors and eigenvalues of $L$.

4.17. A particle of mass 15 (arbitrary units) moving at velocity $u = \frac{12}{13}$ (in units where $c = 1$) collides with a stationary particle whose mass is 6 units, and the two combine to form a single particle.

(a) Determine the energy-momentum vector $\begin{pmatrix} E \\ p \end{pmatrix}$ for each of the colliding particles and for the single particle formed in the collision. Thereby determine the mass and velocity of the particle that is formed.

(b) Using the Lorentz transformation matrix $L = \begin{pmatrix} \frac{5}{3} & -\frac{4}{3} \\ -\frac{4}{3} & \frac{5}{3} \end{pmatrix}$, which corresponds to a velocity of $\frac{4}{5}c$, determine the energy-momentum vector $\begin{pmatrix} E' \\ p' \end{pmatrix}$ for each particle as viewed from a frame of reference moving to the right at speed $\frac{4}{5}c$.

4.18. Suppose that two particles have energy-momentum vectors $\mathbf{w}_1 = \begin{pmatrix} E_1 \\ p_1 \end{pmatrix}$ and $\mathbf{w}_2 = \begin{pmatrix} E_2 \\ p_2 \end{pmatrix}$ respectively, where $m_1 = E_1^2 - p_1^2$, $m_2 = E_2^2 - p_2^2$.

(a) Write the Lorentz scalar product of these two vectors as $\{\mathbf{w}_1, \mathbf{w}_2\} = m_1 m_2 \cosh\alpha$. Show that $v = \tanh\alpha = \sqrt{(\cosh^2\alpha - 1)}/\cosh\alpha$ represents the speed of one of these particles in a frame of reference where the other is at rest.

(b) Determine $v$ for the case where

$$\mathbf{w}_1 = \begin{pmatrix} 13 \\ -5 \end{pmatrix}, \quad \mathbf{w}_2 = \begin{pmatrix} 5 \\ 3 \end{pmatrix}$$

and for the case where

$$\mathbf{w}_1 = \begin{pmatrix} 13 \\ 5 \end{pmatrix}, \quad \mathbf{w}_2 \begin{pmatrix} 5 \\ 3 \end{pmatrix}.$$

4.19 In units where $c$ is not numerically equal to 1, the matrix that represents a Lorentz transformation acting on $\begin{pmatrix} t \\ x \end{pmatrix}$ is

$$L = \begin{pmatrix} \cosh\alpha & (1/c)\sinh\alpha \\ c\sinh\alpha & \cosh\alpha \end{pmatrix}.$$

(a) Show that the matrix that transforms $\begin{pmatrix} E \\ p \end{pmatrix}$ is the *transpose* of this matrix.

(b) Show that the same matrix $L$ will serve to transform energy-momentum if we represent it as a *row* vector, i.e.,

$$(E', p') = (E, p)L.$$

4.20. A photon has energy and momentum that are equal in magnitude (in units where $c = 1$). That is, its energy-momentum vector is of the form $E\begin{pmatrix} 1 \\ 1 \end{pmatrix}$ or $E\begin{pmatrix} 1 \\ -1 \end{pmatrix}$, depending on its direction of motion.

(a) Suppose that a stationary particle of mass $2m$ decays into a particle of mass $m$ plus a photon. Use conservation of energy-momentum to determine the speed of the particle of mass $m$ and the energy of the photon.

(b) Use the Lorentz transformation to describe this decay process in a frame of reference where the particle of $2m$ is initially moving at speed $\frac{3}{5}$.

4.21. A photon of energy $E_\gamma$, whose energy-momentum vector is $\begin{pmatrix} E_\gamma \\ E_\gamma \end{pmatrix}$ in units where $c = 1$, collides with a stationary particle of mass $m_1$ to form a single particle of mass $m_2$. Show that

$$E_\gamma = \frac{m_2^2 - m_1^2}{2m_1}.$$

4.22. Using the scalar product $(f,g) = \int_0^\infty f(t)g(t)\,dt$, construct an orthonormal basis for the space of functions which satisfy the differential equation $\ddot{x} + 3\dot{x} + 2x = 0$.

4.23. Construct an orthonormal basis for the subspace of $\mathbb{R}^4$ spanned by the three vectors

$$\mathbf{v}_1 = \begin{pmatrix} 2 \\ 2 \\ 1 \\ 0 \end{pmatrix}, \quad \mathbf{v}_2 = \begin{pmatrix} 4 \\ 0 \\ 1 \\ 0 \end{pmatrix}, \quad \mathbf{v}_3 = \begin{pmatrix} 0 \\ 4 \\ 1 \\ 2 \end{pmatrix}.$$

4.24. Define a scalar product on $\mathbb{R}^2$ by $(\mathbf{v}_1, \mathbf{v}_2) = 4x_1x_2 + y_1y_2$. Construct a $2 \times 2$ matrix $P$ which projects any vector $\mathbf{v}$ orthogonally (with respect to the above scalar product) onto the line $\begin{pmatrix} 1 \\ 1 \end{pmatrix}$. Show that $(I - P)\mathbf{v}$ is orthogonal to $P\mathbf{v}$.

# 5

# Calculus in the plane



Chapters 5 and 6 present the basic facts of the differential calculus. In Chapter 5 we define the differential of a map from one vector space to another, and discuss its basic properties, in particular the chain rule. We give some physical applications such as Kepler motion and the Born approximation. We define the concepts of directional and partial derivatives, and linear differential forms.

## Introduction

Our first goal is to develop the theory of the differential calculus for four types of functions:

(i) functions from $\mathbb{R}^1 \to \mathbb{R}^2$,
(ii) functions from $\mathbb{R}^2 \to \mathbb{R}^1$,
(iii) functions from $\mathbb{R}^1 \to \mathbb{R}^1$, and
(iv) functions from $\mathbb{R}^2 \to \mathbb{R}^2$.

Functions from $\mathbb{R}^1 \to \mathbb{R}^2$ can be visualized as curves in the plane: The *graph* of a function from $\mathbb{R}^2 \to \mathbb{R}^1$ can be visualized as a surface in three-space. Functions from $\mathbb{R}^1 \to \mathbb{R}^1$ are familiar from first-year calculus. We studied *linear* functions from one plane to another in Chapter 1.

**Figure 5.1** A function from $\mathbb{R}^1$ to $\mathbb{R}^2$

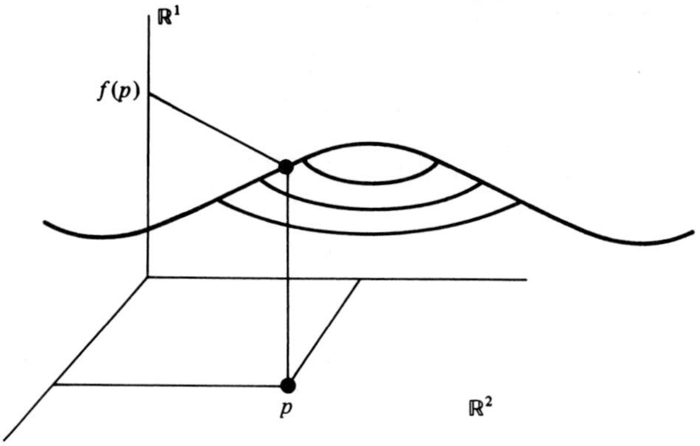


**Figure 5.2** A function from $\mathbb{R}^2$ to $\mathbb{R}^1$

We now want to extend that study to include nonlinear functions from one plane to another: In order not to have to consider the various cases separately, we will introduce some uniform notation when we develop the theory. In what follows we will let $V, W, Z$, etc. stand for either $\mathbb{R}^1$ or $\mathbb{R}^2$. So when we write

$$f: V \to W$$

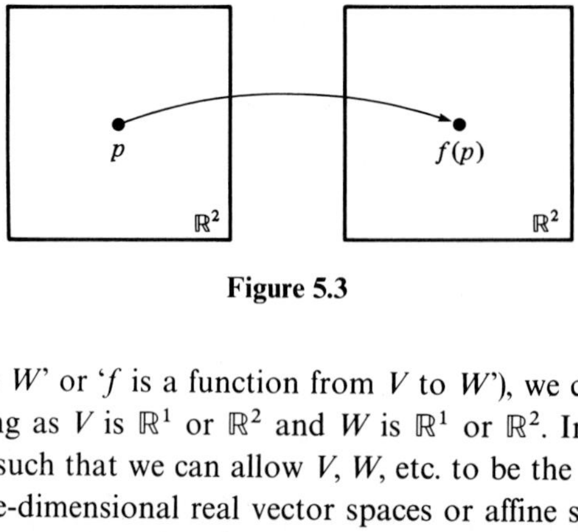


**Figure 5.3**

(read: '$f$ maps $V$ to $W$' or '$f$ is a function from $V$ to $W$'), we can be in any of the four cases according as $V$ is $\mathbb{R}^1$ or $\mathbb{R}^2$ and $W$ is $\mathbb{R}^1$ or $\mathbb{R}^2$. In fact, our notation and proofs will be such that we can allow $V$, $W$, etc. to be the spaces $\mathbb{R}^n$, or, more generally, any finite-dimensional real vector spaces or affine spaces (when we get to learn what these spaces are in Chapter 10). In fact, we shall illustrate some of these more general computations in this chapter, even though we will not have made all of the formal definitions.

We begin by pointing out a fact that the reader is probably aware of by now, as an easy generalization of the discussion in Chapter 1: a linear map from $\mathbb{R}^p$ to $\mathbb{R}^q$ is given by a matrix with $q$ rows and $p$ columns. Thus

$$A = \begin{pmatrix} 5 & 2 & 1 & 0 & 1 \\ 4 & 1 & -1 & 0 & 3 \\ 3 & 0 & 1 & 1 & 0 \end{pmatrix}$$

gives the linear map from $\mathbb{R}^5$ to $\mathbb{R}^3$ with

$$A\begin{pmatrix}1\\0\\0\\0\\0\end{pmatrix}=\begin{pmatrix}5\\4\\3\end{pmatrix},\quad A\begin{pmatrix}0\\1\\0\\0\\0\end{pmatrix}=\begin{pmatrix}2\\1\\0\end{pmatrix},\ \text{etc.}$$

If

$$B=\begin{pmatrix}1&5&9\\2&6&10\\3&7&11\\4&8&12\end{pmatrix}$$

so that $B$ maps $\mathbb{R}^3\to\mathbb{R}^4$, then $BA$ maps $\mathbb{R}^5\to\mathbb{R}^4$ and so is a matrix with four rows and five columns whose entries are computed according to the usual rules of matrix multiplication.

$$BA=\begin{pmatrix}\cdot&\cdot&\cdot&\cdot&\cdot\\\cdot&\cdot&x&\cdot&\cdot\\\cdot&\cdot&\cdot&\cdot&\cdot\\\cdot&\cdot&\cdot&\cdot&\cdot\end{pmatrix}$$

$x=(2)(1)+(6)(-1)+(10)(1)$ in our example.

In particular, a linear map from $\mathbb{R}^p\to\mathbb{R}^1=\mathbb{R}$ (usually just called a linear function) is given by a matrix with one row and $p$ columns. This is usually called a row vector. Thus

$$\mathbf{l}=(1,2,3,4)$$

is the linear map from $\mathbb{R}^4\to\mathbb{R}$ such that

$$\mathbf{l}\begin{pmatrix}1\\0\\0\\0\end{pmatrix}=1,\quad \mathbf{l}\begin{pmatrix}0\\1\\0\\0\end{pmatrix}=2,\ \text{etc.}$$

Evaluated on any vector, we have

$$\mathbf{l}\begin{pmatrix}x\\y\\z\\w\end{pmatrix}=x+2y+3z+4w.$$

So again, the value of the row vector

$$\mathbf{l}=(a,b,c,d)$$

on the column vector

$$\mathbf{v}=\begin{pmatrix}x\\y\\z\\w\end{pmatrix}$$

is given by the usual rule of matrix multiplication – this time with just one entry:

$$\mathbf{l}(\mathbf{v}) = ax + by + cz + dw.$$

If $A: \mathbb{R}^p \to \mathbb{R}^q$ and $\mathbf{l}: \mathbb{R}^q \to \mathbb{R}$, then $\mathbf{l} \circ A: \mathbb{R}^p \to \mathbb{R}$ is again given by matrix multiplication: a $1 \times q$ matrix times a $q \times p$ matrix. For example, if $q = 3$ and $p = 5$ and

$$\mathbf{l} = (1, 2, 3)$$

$$A = \begin{pmatrix} 5 & 2 & 1 & 0 & 1 \\ 4 & 1 & -1 & 0 & 3 \\ 3 & 0 & 1 & 1 & 0 \end{pmatrix}$$

then

$$A = (1, 2, 3)\begin{pmatrix} 5 & 2 & 1 & 0 & 1 \\ 4 & 1 & -1 & 0 & 3 \\ 3 & 0 & 1 & 1 & 0 \end{pmatrix} = (22, 4, 2, 3, 7).$$

One final bit of notational reminder from section 4.1. On the space $\mathbb{R}^k$ we have the Euclidean scalar product ( , ) and associated norm $\| \ \|$ given by

$$\| \mathbf{v} \|^2 = (\mathbf{v}, \mathbf{v}) = x_1^2 + \cdots + x_k^2$$

when

$$\begin{pmatrix} x_1 \\ \vdots \\ x_k \end{pmatrix}.$$

The triangle inequality says that

$$\| \mathbf{u} + \mathbf{v} \| \leqslant \| \mathbf{u} \| + \| \mathbf{v} \|.$$

## 5.1. Big 'oh' and little 'oh'

In the theory of the differential calculus of one variable, a function $f$ is said to have a *derivative* $A$ at a point $x$ if $f$ is defined in some neighborhood of $x$ and the difference quotient,

$$\frac{f(x+v) - f(x)}{v},$$

defined for all sufficiently small $v \neq 0$, tends to the limit $A$ as $v \to 0$. We would like to generalize this definition to maps $f: V \to W$. Our first obstacle is that division by a vector makes no sense, so we cannot use the notion of a difference quotient. So we consider rather

$$f(\mathbf{x} + \mathbf{v}) - f(\mathbf{x}) = A\mathbf{v} + \phi(\mathbf{v}). \tag{5.1}$$

The condition that $A$ be the derivative of $f$ at $x$ is that the error term $\phi(\mathbf{v})$ go to zero 'faster than $\mathbf{v}$'. We can give a precise meaning to the assertion in quotation marks by requiring that

$$\lim \frac{\| \phi(\mathbf{v}) \|}{\| \mathbf{v} \|} = 0 \quad \text{as} \quad \| \mathbf{v} \| \to 0, \tag{5.2}$$

or, to be even more precise, this means that

Given any $\varepsilon > 0$ there exists a $\delta > 0$ such that

$$\|\phi(\mathbf{v})\| \leqslant \varepsilon \|\mathbf{v}\| \tag{5.3}$$

for all $\mathbf{v}$ such that $\|\mathbf{v}\| \leqslant \delta$.

In (5.2) and (5.3), the expression $\|\mathbf{v}\|$ denotes the length of the vector $\mathbf{v}$ in the space $V$ and perhaps we should make this explicit by writing $\|\mathbf{v}\|_V$. Similarly, $\|\phi(\mathbf{v})\|$ denotes the length of the vector $\phi(\mathbf{v})$ in the space $W$, so to emphasize this point, we might want to write $\|\phi(\mathbf{v})\|_W$. We would then write the first inequality in (5.3) as

$$\|\phi(\mathbf{v})\|_W \leqslant \varepsilon \|\mathbf{v}\|_V.$$

Since these subscripts would tend to clutter up the notation, we will not use them, but stick to the notation (5.2) and (5.3).

For example, suppose that $f$, and hence $\phi$, is a map from $\mathbb{R}^2$ to $\mathbb{R}^1$. Suppose that we write the most general vector $\mathbf{v}$ in $\mathbb{R}^2$ as $\mathbf{v} = \begin{pmatrix} x \\ y \end{pmatrix}$ and, for typographical simplicity, write $\phi(\mathbf{v})$ as $\phi(x, y)$. Then $\|\mathbf{v}\| = (x^2 + y^2)^{1/2}$ and $\|\phi(\mathbf{v})\| = |\phi(x, y)|$. In this case, condition (5.3) reads:

Given any $\varepsilon > 0$ there exists a $\delta > 0$ such that

$$|\phi(x, y)| < \varepsilon(x^2 + y^2)^{1/2}$$

for all $x$ and $y$ such that

$$(x^2 + y^2)^{1/2} < \delta.$$

A function $\phi: V \to W$ which is defined in some ball about the origin and which satisfies (5.3) is said to be 'little oh of $\mathbf{v}$'. In symbols, we write '$\phi$ is $o(\mathbf{v})$' or, with some abuse of notation, $\phi = o(\mathbf{v})$. Thus we would write the condition that $A$ is the derivative of $f$ at $\mathbf{x}$ as

$$f(\mathbf{x} + \mathbf{v}) - f(\mathbf{x}) = A\mathbf{v} + \phi(\mathbf{v}) \quad \text{where} \quad \phi(\mathbf{v}) \text{ is } o(\mathbf{v}) \quad \text{or 'where } \phi(\mathbf{v}) = o(\mathbf{v})\text{'}$$

or, even more succinctly, as

$$f(\mathbf{x} + \mathbf{v}) - f(\mathbf{x}) = A\mathbf{v} + o(\mathbf{v}). \tag{5.4}$$

This last version is logically a bit sloppy but is the one that we will frequently use for convenience. The expression $o(\mathbf{v})$ in (5.4) really stands for 'some function $\phi(\mathbf{v})$ which is $o(\mathbf{v})$'. In many cases we are not interested in the error functions $\phi$, we just want to know that they satisfy (5.3). So it is convenient not to have to introduce a separate symbol for each function $\phi$ that arises.

To get some feeling for the concept of $o(\mathbf{v})$, let us prove the following lemma:

*Suppose that $\phi: V \to W$ is a linear transformation and that $\phi(\mathbf{v}) = o(\mathbf{v})$. Then $\phi \equiv 0$.* (5.5)

*Proof.* Suppose that $\phi(\mathbf{v}) = B\mathbf{v}$. Then $\phi(r\mathbf{v}) = r\phi(\mathbf{v})$ for any real number $r$. For any $\varepsilon > 0$, choose the $\delta$ so that $\|\phi(\mathbf{v})\| \leqslant \varepsilon \|\mathbf{v}\|$ when $\|\mathbf{v}\| \leqslant \delta$. Now for any vector $\mathbf{w}$, choose $r = \|\mathbf{w}\|/\delta$, and write

$$\mathbf{w} = r\mathbf{w}' \quad \text{if } \mathbf{w} \neq \mathbf{0}.$$

Then $\|\mathbf{w}'\| = \delta$ so

$$\|\phi(\mathbf{w})\| = r\|\phi(\mathbf{w}')\| \leqslant r\varepsilon\|\mathbf{w}'\| = \varepsilon r\delta = \varepsilon\|\mathbf{w}\|.$$

So

$$\|\phi(\mathbf{w})\| \leqslant \varepsilon\|\mathbf{w}\| \quad \text{for } \mathit{all}\ \mathbf{w} \neq \mathbf{0}$$

(and this is clearly true for $\mathbf{w} = \mathbf{0}$ as well, since $\phi(\mathbf{0}) = \mathbf{0}$ if $\phi$ is a linear map). But this inequality is to hold for all $\varepsilon$. So $\phi \equiv 0$.

From (5.5) it follows that if (5.4) holds, then the $A$ occurring in (5.4) is uniquely determined. Indeed, suppose that

$$f(\mathbf{x} + \mathbf{v}) - f(\mathbf{x}) = A\mathbf{v} + \phi(\mathbf{v})$$

and

$$f(\mathbf{x} + \mathbf{v}) - f(\mathbf{x}) = A'\mathbf{v} + \phi'(\mathbf{v})$$

where both $\phi$ and $\phi'$ are $o(\mathbf{v})$. Then

$$(A' - A)\mathbf{v} = \phi(\mathbf{v}) - \phi'(\mathbf{v}).$$

But, we claim, the sum or difference of two functions that are both $o(\mathbf{v})$ is again $o(\mathbf{v})$. Indeed, for any $\varepsilon > 0$, we can find $\delta_1 > 0$ and $\delta_2 > 0$ such that

$$\|\phi(\mathbf{v})\| \leqslant \tfrac{1}{2}\varepsilon\|\mathbf{v}\| \quad \text{for} \quad \|\mathbf{v}\| \leqslant \delta_1$$

and

$$\|\phi'(\mathbf{v})\| \leqslant \tfrac{1}{2}\varepsilon\|\mathbf{v}\| \quad \text{for} \quad \|\mathbf{v}\| \leqslant \delta_2.$$

Then choosing $\delta$ to be the smaller of the two numbers $\delta_1$ and $\delta_2$, we obtain, by the triangle inequality,

$$\|\phi(\mathbf{v}) \pm \phi'(\mathbf{v})\| \leqslant \|\phi(\mathbf{v})\| + \|\phi'(\mathbf{v})\| \leqslant \varepsilon\|\mathbf{v}\|.$$

The linear transformation $A' - A$ is $o(\mathbf{v})$ and hence must vanish; in other words, $A = A'$.

A function $f$ which satisfies (5.4) for some (and hence a unique) $A$ is said to be *differentiable* at $\mathbf{x}$. The unique linear transformation $A$ is then called the *differential of* $f$ at $\mathbf{x}$ and will be denoted by $\mathrm{d}f_{\mathbf{x}}$. To repeat, the differential of $f$ at $\mathbf{x}$ is the unique linear map from $V$ to $W$ which approximates the actual change in $f$ at $\mathbf{x}$ for small $v$ in the sense that

$$f(\mathbf{x} + \mathbf{v}) - f(\mathbf{x}) = \mathrm{d}f_{\mathbf{x}}[\mathbf{v}] + o(\mathbf{v}).$$

In order to prove the basic theorems about the differential calculus, we will need to assemble some facts about functions that are $o(\mathbf{v})$, and for this it is convenient to introduce some more notation.

A subset $S$ of $V$ is called a *neighborhood* of $\mathbf{0}$ if it contains some ball about the origin, i.e., if, for some $\delta > 0$, it contains the set of all $\mathbf{v}$ with $\|\mathbf{v}\| \leqslant \delta$. Clearly, the intersection of two neighborhoods is again a neighborhood (just take the smaller of the two balls, it is contained in the intersection). Similarly, we can talk of the neighborhood of any point $\mathbf{x}$. It will be a set which contains some ball about $\mathbf{x}$, i.e. which contains a set of the form $\{\mathbf{y} | \|\mathbf{y}-\mathbf{x}\| \leqslant \delta\}$.

If $A$ is an invertible linear transformation from $V$ to $W$ (so, in particular, $V$ and $W$ have the same dimension), then we can find constants $k_1$ and $k_2 > 0$ so that

$$\|A\mathbf{v}\| \leqslant k_1 \|\mathbf{v}\|$$

and

$$\|A^{-1}\mathbf{w}\| \leqslant k_2 \|\mathbf{w}\|$$

or, setting $\mathbf{w} = A\mathbf{v}$,

$$k_2^{-1} \|\mathbf{v}\| \leqslant \|A\mathbf{v}\|.$$

Thus the image of any ball of radius $r$ is contained in a ball of radius $k_1 r$ and contains a ball of radius $k_2^{-1} r$. In particular, $A$ carries neighborhoods into neighborhoods as does $A^{-1}$.

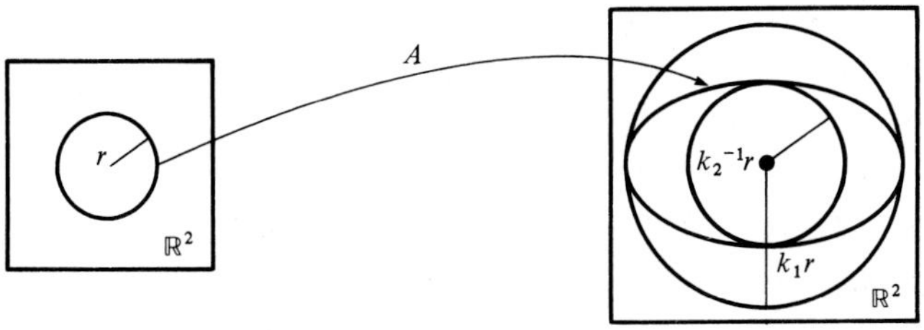


**Figure 5.4**

Let us now return to the general case where $V$ and $W$ do not necessarily have the same dimension. We will let $o(V, W)$ denote the space of all functions which are $o(\mathbf{v})$. Thus, a function $\phi$ belongs to $o(V, W)$ if $\phi = o(\mathbf{v})$. In detail:

> $\phi \in o(V, W)$ if $\phi$ is defined in some neighborhood of the origin and satisfies (5.3).

We say that a function $\psi$ is $O(\mathbf{v})$ (read as '$\psi$ is big oh of $\mathbf{v}$') if $\psi$ is defined in some neighborhood of $\mathbf{0}$ and there is some constant $k > 0$ such that

$$\|\psi(\mathbf{v})\| \leqslant k \|\mathbf{v}\|$$

for all $\mathbf{v}$ in this neighborhood. For example, any linear map is automatically $O(\mathbf{v})$. Also, clearly any function which is $o(\mathbf{v})$ is certainly $O(\mathbf{v})$. We let $O(V, W)$ denote the space of all functions which are $O(\mathbf{v})$. Finally, we let $I(V, W)$ denote the space of functions defined near $\mathbf{0}$ which tend to 0 as $\mathbf{v} \to \mathbf{0}$. Thus

> $\chi \in I(V, W)$ if $\chi$ is defined in some neighborhood of the origin and, for every $\varepsilon > 0$, there is a $\delta > 0$ such that
>
> $$\|\chi(\mathbf{v})\| < \varepsilon \quad \text{when} \quad \|\mathbf{v}\| < \delta.$$

Clearly

$$o(V, W) \subset O(V, W) \subset I(V, W).$$

If for example we take $V = W = \mathbb{R}^1$ and define

$$\phi(x) = x^2$$
$$\psi(x) = x$$

and

$$\chi(x) = |x|^{1/2},$$

then

$$\phi \in o(V, W),$$
$$\psi \in O(V, W) \quad \text{but} \quad \psi \notin o(V, W)$$

and

$$\chi \in I(V, W) \quad \text{but} \quad \chi \notin O(V, W)$$

so the above inclusions are strict.

We have proved that the sum of two functions in $o(V, W)$ is again in $o(V, W)$. The same proof shows that the sum of two functions in $O(V, W)$ is in $O(V, W)$ and similarly for $I(V, W)$.

We now study the behavior of these spaces under composition. Let $X$ be a third space. We will prove the following three useful facts:

If $\psi_1 \in O(V, W)$ and $\psi_2 \in O(W, X)$, then $\psi_2 \circ \psi_1 \in O(V, X)$, (5.6)

If $\psi_1 \in O(V, W)$ and $\psi_2 \in o(W, X)$, then $\psi_2 \circ \psi_1 \in o(V, X)$, (5.7)

If $\psi_1 \in o(V, W)$ and $\psi_2 \in O(W, X)$, then $\psi_2 \circ \psi_1 \in o(V, X)$. (5.8)

*Proof*. If $\|\psi_1(\mathbf{v})\| \leqslant k_1 \|v\|$ for $\|\mathbf{v}\| \leqslant \delta_1$ and $\|\psi_2(\mathbf{w})\| \leqslant k_2 \|\mathbf{w}\|$ when $\|\mathbf{w}\| \leqslant \delta_2$, then $\psi_2 \circ \psi_1$ will be defined for $\|\mathbf{v}\| \leqslant \delta$ where $\delta$ is the smaller of the two numbers $\delta_1$ and $\delta_2/k_1$. For this range of $\mathbf{v}$, we have

$$\|\psi_2 \circ \psi_1(\mathbf{v})\| = \|\psi_2(\psi_1(\mathbf{v}))\| \leqslant k_2 \|\psi_1(\mathbf{v})\| \leqslant k_2 k_1 \|\mathbf{v}\|$$

proving (5.6). If $\psi_2 \in o(W, X)$ we can make $k_2$ as small as we like by choosing $\delta_2$ (and hence $\delta$) small. This proves (5.7). If $\psi_1 \in o(V, W)$, then we can choose $k_1$ as small as we like by choosing $\delta_1$ (and hence $\delta$) sufficiently small. This proves (5.8).

If $\phi$ is a function from $V$ to $W$ and $g: V \to \mathbb{R}$ a real-valued function, the product $g(\mathbf{v})\phi(\mathbf{v})$ makes sense for any $\mathbf{v}$ that lies in the domain of both $\phi$ and $g$. So we can form the function $g\phi$ which is a map from a subset of $V$ to $W$.

If $\psi \in O(V, W)$ and $g \in I(V, \mathbb{R})$, then $g\psi \in o(V, W)$. (5.9)

*Proof*. We are told that there is a $k$ such that $\|\psi(\mathbf{v})\| \leqslant k\|\mathbf{v}\|$ in some neighborhood of the origin. Given any $\varepsilon > 0$, choose $\delta$ so small that $\|g(\mathbf{v})\| \leqslant \varepsilon/k$ for all $\mathbf{v}$ with $\|\mathbf{v}\| \leqslant \delta$. Then, for such $\mathbf{v}$,

$$\|g(\mathbf{v})\psi(\mathbf{v})\| \leqslant (\varepsilon/k)\|\psi(\mathbf{v})\| \leqslant \varepsilon \|\mathbf{v}\|$$

proving (5.9). Similar arguments prove

If $\psi \in I(V, W)$ and $g \in O(V, \mathbb{R})$, then $g\psi \in o(V, W)$. (5.10)

If $\psi \in o(V, W)$ and $g$ is a bounded function from $V \to \mathbb{R}$, then $g\psi \in o(V, W)$. (5.11)

If $\psi$ is a bounded map from $V$ to $W$ defined in some neighborhood of the origin, and $g \in o(V, \mathbb{R})$, then $g\psi \in o(V, W)$. (5.12)

(To say that a map $\psi: V \to W$ is *bounded* means there is some positive real number $k$ such that $\|\psi(\mathbf{v})\|_W \leqslant k$ for all $\mathbf{v}$ in $V$.)

We have collected all the necessary lemmas to begin the study of the differential calculus.

## 5.2. The differential calculus

Let $f: V \to W$ be defined in some neighborhood of a point $\mathbf{x} \in V$. Define the function $\nabla_{\mathbf{x}} f$ by the formula

$$\nabla_{\mathbf{x}} f(\mathbf{h}) = f(\mathbf{x} + \mathbf{h}) - f(\mathbf{x}).$$

It is defined for all $h$ in some neighborhood of $\mathbf{0}$ and measures the change in $f$ relative to its value at $\mathbf{x}$. The function $f$ is *continuous at* $\mathbf{x}$ if $\nabla_{\mathbf{x}} f \in I(V, W)$. (This means that $\nabla_{\mathbf{x}} f(\mathbf{h})$ tends to $\mathbf{0}$ as $\mathbf{h} \to \mathbf{0}$, so $f(\mathbf{x} + \mathbf{h}) \to f(\mathbf{x})$.) Recall that the function $f$ is said to be differentiable at $\mathbf{x}$ if there is a linear transformation $\mathrm{d}f_{\mathbf{x}}: V \to W$ such that

$$\nabla f_{\mathbf{x}}(\mathbf{h}) = \mathrm{d}f_{\mathbf{x}}[\mathbf{h}] + o(\mathbf{h}).$$

The linear transformation $\mathrm{d}f_{\mathbf{x}}$ is uniquely determined by this equation and is called the *differential* of $f$ at $\mathbf{x}$. Any linear function belongs to $O(V, W)$, and the sum of a function in $O(V, W)$ and a function in $o(V, W)$ lies in $O(V, W)$. From this we conclude that

If $f$ is differentiable at $\mathbf{x}$, then $\nabla f_{\mathbf{x}} \in O(V, W)$. (5.13)

(In particular, since $O(V, W) \subset I(V, W)$, we conclude that, if $f$ is differentiable at $\mathbf{x}$, then it is certainly continuous at $\mathbf{x}$.) If $f$ is a linear function, $f(\mathbf{x}) = A\mathbf{x}$, then $\nabla f_{\mathbf{x}}[\mathbf{h}] = A(\mathbf{x} + \mathbf{h}) - A\mathbf{x} = A\mathbf{h}$, so,

A linear function $f(\mathbf{x}) = A\mathbf{x}$ is differentiable at all points, and its differential is given by $\mathrm{d}f_{\mathbf{x}} = A$, independent of $\mathbf{x}$.

If $f$ is a constant function, then $\nabla f_{\mathbf{x}} \equiv 0$, and (5.4) holds with $A = 0$, so

A constant function is differentiable everywhere and its differential is identically zero.

We now state and prove the rule about the differential of a sum:

If $f$ and $g$ are two functions from $V$ to $W$ and both are differentiable at $\mathbf{x}$, then so is their sum and

$$\mathrm{d}(f + g)_{\mathbf{x}} = \mathrm{d}f_{\mathbf{x}} + \mathrm{d}g_{\mathbf{x}}. \tag{5.14}$$

*Proof.* It is clear that $\nabla(f + g)_{\mathbf{x}} = \nabla f_{\mathbf{x}} + \nabla g_{\mathbf{x}}$. Since

$$\nabla f_{\mathbf{x}} = \mathrm{d}f_{\mathbf{x}} + \phi_1$$

and

$$\nabla g_{\mathbf{x}} = \mathrm{d}g_{\mathbf{x}} + \phi_2$$

where $\phi_1$ and $\phi_2$ are in $o(V, W)$, we conclude that

$$\nabla(f + g)_{\mathbf{x}} = \mathrm{d}f_{\mathbf{x}} + \mathrm{d}g_{\mathbf{x}} + \phi_1 + \phi_2.$$

Since $(\phi_1 + \phi_2) \in o(V, W)$, this proves (5.14).

We can multiply an $\mathbb{R}$-valued function $g$ with a $W$-valued function to get a $W$-valued function. For this combination we can state the usual rule for the derivative of a product:

Suppose that $f: V \to W$ and $g: V \to \mathbb{R}$ are both differentiable at $\mathbf{x}$. Then their product, $gf$, is also differentiable at $\mathbf{x}$ and

$$\mathrm{d}(gf)_{\mathbf{x}}[\mathbf{h}] = g(\mathbf{x})\mathrm{d}f_{\mathbf{x}}[\mathbf{h}] + (\mathrm{d}g_{\mathbf{x}}[\mathbf{h}])f(\mathbf{x}).$$

*Proof.*

$$\begin{aligned}
\nabla(gf)_{\mathbf{x}}[\mathbf{h}] &= g(\mathbf{x}+\mathbf{h})f(\mathbf{x}+\mathbf{h}) - g(\mathbf{x})f(\mathbf{x}) \\
&= g(\mathbf{x}+\mathbf{h})(f(\mathbf{x}+\mathbf{h}) - f(\mathbf{x})) + (g(\mathbf{x}+\mathbf{h}) - g(\mathbf{x}))f(\mathbf{x}) \\
&= g(\mathbf{x})(f(\mathbf{x}+\mathbf{h}) - f(\mathbf{x})) + (g(\mathbf{x}+\mathbf{h}) - g(\mathbf{x}))f(\mathbf{x}) \\
&\quad + (g(\mathbf{x}+\mathbf{h}) - g(\mathbf{x}))(f(\mathbf{x}+\mathbf{h}) - f(\mathbf{x})) \\
&= g(\mathbf{x})\nabla f_{\mathbf{x}}[\mathbf{h}] + (\nabla g_{\mathbf{x}}[\mathbf{h}])f(\mathbf{x}) + (\nabla g_{\mathbf{x}}[\mathbf{h}])(\nabla f_{\mathbf{x}}[\mathbf{h}]) \\
&= g(\mathbf{x})(\mathrm{d}f_{\mathbf{x}}[\mathbf{h}] + o(\mathbf{h})) + (\mathrm{d}g_{\mathbf{x}}[\mathbf{h}] + o(\mathbf{h}))f(\mathbf{x}) + O(\mathbf{h})\cdot O(\mathbf{h}),
\end{aligned}$$

since $f$ and $g$ are both differentiable at $\mathbf{x}$ and hence both $\nabla f_{\mathbf{x}}$ and $\nabla g_{\mathbf{x}}$ are $O(\mathbf{h})$ by (5.13). Now the product of two functions which are $O(\mathbf{h})$ is $o(\mathbf{h})$ by (5.9). Both $f$ and $g$ are bounded near $\mathbf{x}$ since, in fact, $g(\mathbf{x}+\mathbf{h}) - g(\mathbf{x})$ and $f(\mathbf{x}+\mathbf{h}) - f(\mathbf{x})$ both tend to zero. The product of a bounded function and one which is $o(\mathbf{h})$ is again $o(\mathbf{h})$. Putting these facts into the last expression above gives

$$\nabla(gf)_{\mathbf{x}}[\mathbf{h}] = g(\mathbf{x})\mathrm{d}f_{\mathbf{x}}[\mathbf{h}] + (\mathrm{d}g_{\mathbf{x}}[\mathbf{h}])f(\mathbf{x}) + o(\mathbf{h})$$

which was to be proved.

We now come to the very important:

*Chain rule.* Suppose that $f: V \to W$ is differentiable at $\mathbf{x} \in V$ and that $g: W \to X$ is differentiable at $\mathbf{y} = f(\mathbf{x}) \in W$. Then $g \circ f: V \to X$ is differentiable at $\mathbf{x}$ and its differential is given by

$$\mathrm{d}(g \circ f)_{\mathbf{x}} = (\mathrm{d}g_{f(\mathbf{x})}) \cdot (\mathrm{d}f_{\mathbf{x}}). \tag{5.15}$$

(On the right-hand side of this equation we have the composition of two linear transformations, $\mathrm{d}g_{f(\mathbf{x})}: W \to X$ and $\mathrm{d}f_{\mathbf{x}}: V \to W$. On the left-hand side we have the composition of $g$ and $f$.)

*Proof.*

$$\begin{aligned}
\nabla(g \circ f)_{\mathbf{x}}[\mathbf{h}] &= g(f(\mathbf{x}+\mathbf{h})) - g(f(\mathbf{x})) \\
&= g(f(\mathbf{x}) + \nabla f_{\mathbf{x}}[\mathbf{h}]) - g(f(\mathbf{x}))
\end{aligned}$$

$$= \nabla g_{f(\mathbf{x})}[\nabla f_{\mathbf{x}}[\mathbf{h}]]$$
$$= \mathrm{d}g_{f(\mathbf{x})}[\mathrm{d}f_{\mathbf{x}}[\mathbf{h}]] + \mathrm{d}g_{f(\mathbf{x})}[o[\mathbf{h}]] + (\phi \circ \psi)(\mathbf{h}),$$

where $\phi \in o(V, X)$ (coming from the error term in $\nabla g_{f(\mathbf{x})}$) and $\psi = \nabla f_{\mathbf{x}} \in O(V, W)$ by (5.13). By (5.8) this composite function is in $o(V, X)$. Also $\mathrm{d}g_{f(\mathbf{x})}$ is linear, and hence in $O(W, X)$, and thus the second term is a composite of an element in $O(W, X)$ with an element of $o(V, W)$ and so is $o(V, X)$ by (5.7). Thus

$$\nabla(g \circ f)_{\mathbf{x}}[\mathbf{h}] = (\mathrm{d}g_{f(\mathbf{x})} \circ \mathrm{d}f_{\mathbf{x}})[\mathbf{h}] + o(\mathbf{h})$$

as was to be proved.

**Examples**

We now give some examples of differentials and the chain rule. For functions $\alpha: \mathbb{R}^1 \to \mathbb{R}^1$, the differential $\mathrm{d}\alpha_x$ when evaluated on some $h \in \mathbb{R}$ is given by multiplication by the derivative $\alpha'(x)$. Thus

$$\mathrm{d}\alpha_x[h] = \alpha'(x)h.$$

This is just the definition of the derivative $\alpha'(x)$. For example, let $\alpha: \mathbb{R}^1 \to \mathbb{R}^1$ and $\beta: \mathbb{R}^1 \to \mathbb{R}^1$ be given by

$$\alpha(y) = y^2, \quad \beta(x) = 5x^3 + 1$$

so that

$$\alpha \circ \beta(x) = (5x^3 + 1)^2.$$

Then

$\mathrm{d}\alpha_y$ is multiplication by $2y$,
$\mathrm{d}\beta_x$ is multiplication by $15x^2$,
$\mathrm{d}(\alpha \circ \beta)_x$ is multiplication by $2(5x^3 + 1)(15x^2)$

so

$\mathrm{d}\alpha_{\beta(x)}$ is multiplication by $2(5x^3 + 1)$

and

$\mathrm{d}\alpha_{\beta(x)} \circ \mathrm{d}\beta_x$ is multiplication by $15x^2$ followed by multiplication by $2(5x^3 + 1)$ or
$\mathrm{d}\alpha_{\beta(x)} \circ \mathrm{d}\beta_x$ is multiplication by $2(5x^3 + 1)\,(15x^2)$

or

$\mathrm{d}\alpha_{\beta(x)} \circ \mathrm{d}\beta_x = \mathrm{d}(\alpha \circ \beta)_x$ – the chain rule.

It is clear that the notation here is cumbersome. Leibniz's notation for functions of one variable is better:

If $\alpha$ is a function of $y$ write

$$\alpha' = \frac{\mathrm{d}\alpha}{\mathrm{d}y}$$

or rather

$$\mathrm{d}\alpha = \alpha' \mathrm{d}y.$$

This last equation is taken to *mean* that at any value of $y$

$$d\alpha_y(h) = \alpha'(y)h.$$

In other words, $dy$ is a dummy symbol into which we substitute the value of $h$. Thus

$$d(y^2) = 2y\,dy,$$

and, similarly,

$$d(5x^3 + 1) = 15x^2 dx.$$

The chain rule now *says* substitute

$$y = (5x^3 + 1)$$

and

$$dy = 15x^2 dx$$

into the formula for $d(y^2)$ to get the formula for $d[(5x^3 + 1)^2]$. The chain rule *becomes* mechanical substitution in the Leibniz notation.

We will continue to do some examples in our more cumbersome notation where, we hope, the meaning of the operations is clear.

Let $f: \mathbb{R}^1 \to \mathbb{R}^2$ and $g: \mathbb{R}^2 \to \mathbb{R}^1$ be given by

$$f(x) = \begin{pmatrix} x^2 + 1 \\ 2x - 1 \end{pmatrix}, \quad g\left(\begin{pmatrix} x \\ y \end{pmatrix}\right) = x^2 y.$$

To evaluate $df_x$, we note that

$$\begin{aligned} \nabla f_x[s] &= f(x + s) - f(x) \\ &= \begin{pmatrix} 2xs \\ 2s \end{pmatrix} + \begin{pmatrix} s^2 \\ 0 \end{pmatrix} \\ &= \begin{pmatrix} 2x \\ 2 \end{pmatrix} s + o(s) \end{aligned}$$

so that $df_x$ is represented by the matrix

$$df_x = \begin{pmatrix} 2x \\ 2 \end{pmatrix}.$$

Similarly,

$$\begin{aligned} \nabla g_{\binom{x}{y}}\left[\begin{pmatrix} s \\ t \end{pmatrix}\right] &= (x + s)^2(y + t) - x^2 y \\ &= x^2 t + 2sxy + 2sxt + s^2 y + s^2 t \\ &= (2xy, x^2)\begin{pmatrix} s \\ t \end{pmatrix} + o\left(\begin{pmatrix} s \\ t \end{pmatrix}\right) \end{aligned}$$

so that $dg_{\binom{x}{y}}$ is the matrix

$$dg_{\binom{x}{y}} = (2xy, x^2).$$

The composite function $g \circ f: \mathbb{R}^1 \to \mathbb{R}^1$ is given by

$$g \circ f(x) = (x^2 + 1)^2(2x - 1)$$

so that

$$d(g\circ f)_x = 2(x^2+1)(2x)(2x-1) + 2(x^2+1)^2.$$

The chain rule says this must equal the matrix product $dg_{f(x)}\circ df_x$ which is given by

$$dg_{f(x)}\circ df_x = (2(x^2+1)(2x-1), (x^2+1)^2)\begin{pmatrix}2x\\2\end{pmatrix}$$
$$= 2(x^2+1)(2x-1)(2x) + 2(x^2+1)^2$$

which equals $d(g\circ f)_x$.

We can also form the composite function $f\circ g\colon \mathbb{R}^2\to\mathbb{R}^2$ given by

$$f\circ g\left(\begin{pmatrix}x\\y\end{pmatrix}\right) = \begin{pmatrix}(x^2y)^2+1\\2(x^2y)-1\end{pmatrix}.$$

To compute $d(f\circ g)_{\binom{x}{y}}$, we expand

$$\nabla(f\circ g)_{\binom{x}{y}}\left(\begin{pmatrix}s\\t\end{pmatrix}\right) = \begin{pmatrix}(x+s)^4(y+t)^2+1\\2(x+s)^2(y+t)-1\end{pmatrix} - \begin{pmatrix}x^4y^2+1\\2x^2y-1\end{pmatrix}$$
$$= \begin{pmatrix}(x^4+4x^3s+6x^2s^2+4xs^3+s^4)(y^2+2yt+t^2)-x^4y^2\\2(x^2+2xs+s^2)(y+t)-2x^2y\end{pmatrix}$$
$$= \begin{pmatrix}2x^4yt+4x^3y^2s\\2x^2t+4xys\end{pmatrix}$$
$$+\begin{pmatrix}x^4t^2+4x^3s(2yt+t^2)+(6x^2s^2+4xs^3+s^4)(y+t)^2\\4xst+s^2y+s^2t\end{pmatrix}$$
$$= \begin{pmatrix}4x^3y^2 & 2x^4y\\4xy & 2x^2\end{pmatrix}\begin{pmatrix}s\\t\end{pmatrix} + o\left(\begin{pmatrix}s\\t\end{pmatrix}\right)$$

so that

$$d(f\circ g)_{\binom{x}{y}} = \begin{pmatrix}4x^3y^2 & 2x^4y\\4xy & 2x^2\end{pmatrix}.$$

The chain rule says that this must equal $df_{g(\binom{x}{y})}\circ dg_{\binom{x}{y}}$ which is given by

$$df_{g(\binom{x}{y})}\circ dg_{\binom{x}{y}} = \begin{pmatrix}2x^2y\\2\end{pmatrix}(2xy, x^2)$$
$$= \begin{pmatrix}4x^3y^2 & 2x^4y\\4xy & 2x^2\end{pmatrix}$$

which equals $d(f\circ g)_{\binom{x}{y}}$.

As another example of the chain rule, let $F\colon\mathbb{R}^2\to\mathbb{R}^2$ and $G\colon\mathbb{R}^2\to\mathbb{R}^2$ be given by

$$F\left(\begin{pmatrix}x\\y\end{pmatrix}\right) = \begin{pmatrix}x^2+y\\xy\end{pmatrix},\quad G\left(\begin{pmatrix}x\\y\end{pmatrix}\right) = \begin{pmatrix}3xy^2\\x^2\end{pmatrix}.$$

We then have

$$\nabla F_{\binom{x}{y}}\left(\binom{s}{t}\right)=\binom{(x+s)^2+(y+t)}{(x+s)(y+t)}-\binom{x^2+y}{xy}$$
$$=\binom{2xs+t}{xt+sy}+\binom{s^2}{st}$$
$$=\begin{pmatrix}2x & 1\\ y & x\end{pmatrix}\binom{s}{t}+o\left(\binom{s}{t}\right)$$

so that

$$\mathrm{d}F_{\binom{x}{y}}=\begin{pmatrix}2x & 1\\ y & x\end{pmatrix}.$$

Similarly

$$\nabla G_{\binom{x}{y}}\left(\binom{s}{t}\right)=\binom{3(x+s)(y+t)^2}{(x+s)^2}-\binom{3xy^2}{x^2}$$
$$=\binom{3sy^2+3xy(2t)}{2xs}+\binom{3xt^2+3s(2yt+t^2)}{s^2}$$
$$=\begin{pmatrix}3y^2 & 6xy\\ 2x & 0\end{pmatrix}\binom{s}{t}+o\left(\binom{s}{t}\right)$$

so that

$$\mathrm{d}G_{\binom{x}{y}}=\begin{pmatrix}3y^2 & 6xy\\ 2x & 0\end{pmatrix}.$$

The composite function $F\circ G:\mathbb{R}^2\to\mathbb{R}^2$ is given by

$$F\circ G\left(\binom{x}{y}\right)=F\left(\binom{3xy^2}{x^2}\right)=\binom{(3xy^2)^2+x^2}{3xy^2x^2}$$
$$=\binom{9x^2y^4+x^2}{3x^3y^2}.$$

We then have

$$\nabla(F\circ G)_{\binom{x}{y}}\left(\binom{s}{t}\right)$$
$$=\binom{9(x+s)^2(y+t)^4+(x+s)^2}{3(x+s)^3(y+t)^2}-\binom{9x^2y^4+x^2}{3x^3y^2}$$
$$=\binom{9(2xs)y^4+9x^2(4y^3t)+2xs}{3(3x^2s)y^2+3x^3(2yt)}$$
$$+\binom{9x^2(6y^2t^2+4yt^3+4t)+18xs[(y+1)^4-y^4]+9s^2(y+t)^4}{3x^3t^2+3(3xs^2+s^3)(y+t)^2+9x^2s(2yt+t^2)}$$
$$=\begin{pmatrix}18xy^4+2x & 36x^2y^3\\ 9x^2y^2 & 6x^3y\end{pmatrix}\binom{s}{t}+o\left(\binom{s}{t}\right)$$

so that

$$d(F\circ G)_{\binom{x}{y}} = \begin{pmatrix} 18xy^4+2x & 36x^2y^3 \\ 9x^2y^2 & 6x^3y \end{pmatrix}.$$

By the chain rule, this must equal $dF_{G(\binom{x}{y})}\circ dG_{\binom{x}{y}}$ which is given by

$$\begin{aligned} dF_{G(\binom{x}{y})}\circ dG_{\binom{x}{y}} &= \begin{pmatrix} 2(3xy^2) & 1 \\ x^2 & 3xy^2 \end{pmatrix}\begin{pmatrix} 3y^2 & 6xy \\ 2x & 0 \end{pmatrix} \\ &= \begin{pmatrix} 18xy^4+2x & 36x^2y^3 \\ 3x^2y^2+6x^2y^2 & 6x^3y \end{pmatrix} \\ &= d(F\circ G)_{\binom{x}{y}}. \end{aligned}$$

In the next few sections we will spend some time extracting important consequences of the chain rule.

We first give some more 'abstract' examples of the chain rule and introduce some notation.

## 5.3. More examples of the chain rule

Let us consider the multiplication map $g\colon \mathbb{R}^2 \to \mathbb{R}^1$ defined by

$$g\left(\begin{pmatrix} x \\ y \end{pmatrix}\right) = xy.$$

If $\mathbf{v} = \begin{pmatrix} x \\ y \end{pmatrix}$ and $\mathbf{h} = \begin{pmatrix} r \\ s \end{pmatrix}$, then

$$\begin{aligned} g(\mathbf{v}+\mathbf{h}) = (x+r)(y+s) &= xy + xs + yr + rs \\ &= g(\mathbf{v}) + xs + yr + o(\mathbf{h}) \end{aligned}$$

so

$$d_{\mathbf{v}}g(\mathbf{h}) = xs + yr,$$

and its matrix (with one row and two columns) is

$$(y, x).$$

Let $f\colon \mathbb{R}^1 \to \mathbb{R}^2$ be given. We can think of $f$ as describing a curve in the plane, or, more simply, as giving a pair of real-valued functions of one real variable,

$$f(t) = \begin{pmatrix} x(t) \\ y(t) \end{pmatrix}.$$

Then

$$f(t+h) = \begin{pmatrix} x(t+h) \\ y(t+h) \end{pmatrix} = \begin{pmatrix} x(t) + x'(t)h + o(h) \\ y(t) + y'(t)h + o(h) \end{pmatrix}$$

so

$$f(t+h) - f(t) = h\begin{pmatrix} x'(t) \\ y'(t) \end{pmatrix} + o(h)$$

or

$$f'(t) = \begin{pmatrix} x'(t) \\ y'(t) \end{pmatrix}.$$

Multiplying the matrices

$$\mathrm{d}g_{f(t)} = (y(t), x(t))$$

and

$$\mathrm{d}f_t = f'(t) = \begin{pmatrix} x'(t) \\ y'(t) \end{pmatrix}$$

gives

$$\mathrm{d}(g \circ f)_t = (g \circ f)'(t) = x'(t)y(t) + x(t)y'(t).$$

But $(g \circ f)(t) = x(t)y(t)$. Thus the chain rule implies Leibniz's formula for the derivative of the product of two functions.

Before proceeding, it will be convenient to introduce and explain some further notation. Instead of writing

$$\mathrm{d}g_{\mathbf{v}}(\mathbf{h}) = yr + xs \quad \text{where} \quad \mathbf{v} = \begin{pmatrix} x \\ y \end{pmatrix}, \quad \mathbf{h} = \begin{pmatrix} r \\ s \end{pmatrix} \quad \text{and} \quad g\left(\begin{pmatrix} x \\ y \end{pmatrix}\right) = xy,$$

it is more convenient to write all of this information as

$$\mathrm{d}(xy) = y\mathrm{d}x + x\mathrm{d}y.$$

In this equation, the symbol $\mathrm{d}x$ occurring on the right-hand side is understood as a linear map from $\mathbb{R}^2 \to \mathbb{R}^1$: the map which assigns to each vector its first coordinate. Thus

$$\mathrm{d}x(\mathbf{h}) = r \quad \text{if} \quad \mathbf{h} = \begin{pmatrix} r \\ s \end{pmatrix},$$

and similarly,

$$\mathrm{d}y(\mathbf{h}) = s.$$

In the expression $y\mathrm{d}x$, the $y$ is a function of $\mathbf{v}$, that function which assigns to $\mathbf{v}$ its second coordinate, where $\mathbf{v} = \begin{pmatrix} x \\ y \end{pmatrix}$. So the terms like $y\mathrm{d}x$ really depend on two kinds of variables, the variable $\mathbf{v}$ which tells us where we are computing the derivative and the $\mathbf{h}$ which is the measure of the small displacement. The $\mathrm{d}(xy)$ that occurs on the left-hand side is a shorthand way of writing '$\mathrm{d}g_{[\mathbf{v}]}$ where $g$ is that function defined by $g(\mathbf{v}) = xy$ when $\mathbf{v} = \begin{pmatrix} x \\ y \end{pmatrix}$'. In applying the chain rule as in the above example, we would say

> Consider $x$ as the function* on $\mathbb{R}^2$ which assigns to each vector its first coordinate. Then $(x \circ f)(t) = x(t)$ by the definition of the map $f$. By the chain rule,

* It might be instructive here to reread the lengthy discussion in section 1.3 where we discuss how a coordinate, such as $x$, is to be viewed as a function.

$d(x \circ f)_t = x'(t)dt$, where, in this equation, $x'(t)$ is a function evaluated at the point $t$ where we are computing the derivative, and $dt$ is the part which measures the small increment. So when we think of $x$ as a function of $t$ given to us by the map $f$, we make the 'substitution' $dx = x'dt$ where now $x'$ is a function of $t$. Similarly, the chain rule tells us that if we consider $y$ as a function of $t$ given to us by the map $f$, then we must 'substitute' $dy = y'dt$.

We would then write

$$d(g \circ f) = yx'dt + xy'dt$$

with $x, y, x'$ and $y'$ substituted on the right-hand side as explicit functions of $t$.

For example, suppose $x(t) = t + \sin t$, $y(t) = e^{2t}$. Then we would write

$$dg = d(xy) = ydx + xdy,$$

$$df = d\begin{pmatrix} t + \sin t \\ e^{2t} \end{pmatrix} = \begin{pmatrix} 1 + \cos t \\ 2e^{2t} \end{pmatrix} dt$$

and

$$d(g \circ f) = d((t + \sin t)(e^{2t})) = (e^{2t}(1 + \cos t) + 2e^{2t}(t + \sin t))dt.$$

Let us state the chain rule once more in diagrammatic form: We are given two differentiable maps $f: V \to W$ and $g: W \to Z$, so we can form their composite $g \circ f: V \to Z$. At some point $\mathbf{v}$ in $V$ we can apply $f$ to get to $f(\mathbf{v})$ and then $g$ to get to $g(f(\mathbf{v}))$. In computing $d(g \circ f)_{\mathbf{v}}(\mathbf{h})$ we can follow the maps along, by first applying $df_{\mathbf{v}}$ to $\mathbf{h}$ and then $dg_{f(\mathbf{v})}$ to the image.

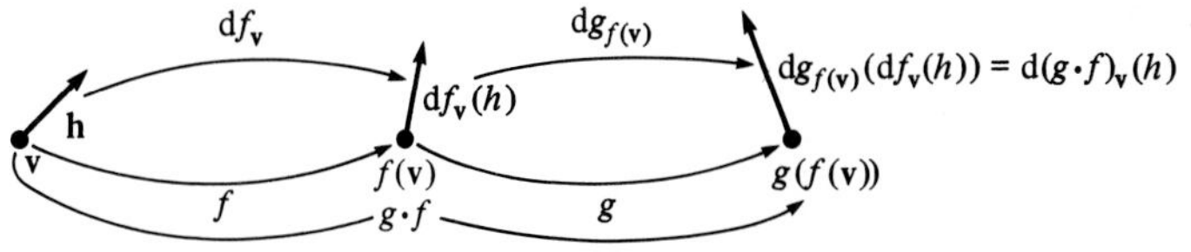


**Figure 5.5**

Let us now do some slightly more sophisticated computations with the chain rule. In these computations we will take $V$, $W$ etc. to be higher-dimensional vector spaces, so the logical purist might want to postpone studying them until after reading the chapter on linear algebra. Nevertheless, we recommend having a look at them here. We begin with a computation of the derivative of a product of two matrices. Let $V$ be the vector space consisting of pairs of $n \times n$ matrices, so a typical vector in $V$ is of the form

$$\mathbf{v} = \begin{pmatrix} A \\ B \end{pmatrix}$$

where $A$ and $B$ are $n \times n$ matrices. (This becomes a vector space by componentwise addition and scalar multiplication:

$$\text{If } \mathbf{v} = \begin{pmatrix} A \\ B \end{pmatrix} \text{ and } \mathbf{v}' = \begin{pmatrix} A' \\ B' \end{pmatrix} \text{ then } \mathbf{v} + \mathbf{v}' = \begin{pmatrix} A + A' \\ B + B' \end{pmatrix} \text{ and } a\mathbf{v} = \begin{pmatrix} aA \\ aB \end{pmatrix}.$$

This obviously makes $V$ into a vector space of dimension $2n^2$.) Let $W$ denote the vector space of all $n \times n$ matrices, and define the map $g: V \to W$ by $g\left(\begin{pmatrix} A \\ B \end{pmatrix}\right) = AB$. If $\mathbf{v} = \begin{pmatrix} A \\ B \end{pmatrix}$ and $\mathbf{h} = \begin{pmatrix} X \\ Y \end{pmatrix}$ then

$$g(\mathbf{v} + \mathbf{h}) - g(\mathbf{v}) = (A + X)(B + Y) - AB = XB + AY + XY = XB + AY + o(\mathbf{h})$$

so

$$\mathrm{d}g_{\mathbf{v}}(\mathbf{h}) = XB + AY. \tag{5.16}$$

In doing computations, we might want to use our more convenient notation which drops the subscript $\mathbf{v}$ and the values at a particular $\mathbf{h}$. We could write (5.16) as

$$\mathrm{d}(AB) = (\mathrm{d}A)B + A\mathrm{d}B. \tag{5.17}$$

In this notation, the $AB$ occurring on the left is a sloppy but convenient way of writing the function $g$. The $\mathrm{d}A$ occurring on the right is the derivative of the function which assigns to $\begin{pmatrix} A \\ B \end{pmatrix}$ the matrix $A$. This derivative when evaluated at the point $\begin{pmatrix} A \\ B \end{pmatrix}$ on the vector $\begin{pmatrix} X \\ Y \end{pmatrix}$ yields the value $X$. Thus $\mathrm{d}A$ is the linear map which assigns to each $\begin{pmatrix} X \\ Y \end{pmatrix}$ the value $X$. So, for example, $(\mathrm{d}A)B$ is the linear map which assigns to $\begin{pmatrix} X \\ Y \end{pmatrix}$ the value $XB$. In this sense, (5.17) is a shorthand form of writing (5.16).

As another example of this notation, let $f$ denote the map from $W$ to $V$ given by

$$f(A) = \begin{pmatrix} A \\ A \end{pmatrix}.$$

Since $f$ is linear, we know that its derivative is independent of $A$ and is just the same map again, evaluated on vectors, i.e.

$$\mathrm{d}f_A(X) = \begin{pmatrix} X \\ X \end{pmatrix}.$$

In the differential notation we would write this as

$$\mathrm{d}\begin{pmatrix} A \\ A \end{pmatrix} = \begin{pmatrix} \mathrm{d}A \\ \mathrm{d}A \end{pmatrix}$$

(where again, $\mathrm{d}A$ is the linear function which assigns the value $X$ to any element $X$). Now let us consider the map $h$ of $W \to W$ defined by

$$h(A) = A^2.$$

We clearly have $h(A) = g(f(A))$ or $h = g \circ f$. So the chain rule applies:

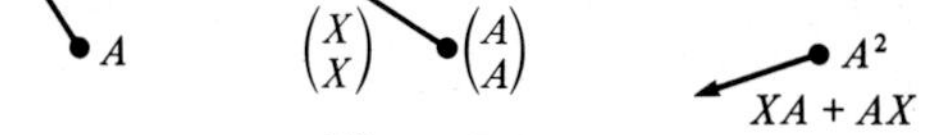


**Figure 5.6**

It says:

$$\mathrm{d}h_A(Z) = \mathrm{d}(g \circ f)_A(Z) = \mathrm{d}g_{f(A)}(\mathrm{d}f_A(Z)) = \mathrm{d}g_{f(A)}\begin{pmatrix} Z \\ Z \end{pmatrix} = ZA + AZ.$$

We would write this computation in the 'differential notation' as follows: Make the 'substitutions' $A = A$ and $B = A$ in (5.17) to obtain

$$\mathrm{d}(A^2) = (\mathrm{d}A)A + A(\mathrm{d}A).$$

(Notice once again, that on account of the non-commutative nature of matrix multiplication *this* is the correct generalization of the formula $\mathrm{d}(x^2) = 2x\mathrm{d}x$ of functions of one variable. It is *not* true that $\mathrm{d}(A^2) = 2A\mathrm{d}A$.)

**The Born expansion**

Let us now consider the map (inv) which assigns to each invertible matrix its inverse, so

$$(\mathrm{inv})(A) = A^{-1}.$$

The map (inv) is not defined on all of $W$, but only on that subset of $W$ consisting of all matrices which are invertible. *Assuming* that inv is differentiable where defined, we shall show how to compute the derivative of the map (inv) using the chain rule: Define the map $f$ by

$$f(A) = \begin{pmatrix} A \\ A^{-1} \end{pmatrix}$$

or, more symbolically,

$$f = \begin{pmatrix} (\mathrm{id}) \\ (\mathrm{inv}) \end{pmatrix}.$$

Recall that $g$ is the map defined by

$$g\begin{pmatrix} A \\ B \end{pmatrix} = AB.$$

Then $(g \circ f)(A) = AA^{-1} = I$ where $I$ is the unit matrix. In other words, $g \circ f$ is a constant, and hence $\mathrm{d}(g \circ f) = 0$. By the chain rule,

$$\mathrm{d}f_A(X) = \begin{pmatrix} \mathrm{d}_A(\mathrm{id})(X) \\ \mathrm{d}_A(\mathrm{inv})(X) \end{pmatrix} = \begin{pmatrix} X \\ \mathrm{d}(\mathrm{inv})_A(X) \end{pmatrix}$$

and, by the chain rule again,

$$0 = [\mathrm{d}_{f(A)}g](\mathrm{d}_A f(X)) = XA^{-1} + A(\mathrm{d}_A(\mathrm{inv})(X)).$$

Multiplying this equation on the left by $A^{-1}$ and solving for $\mathrm{d}_A(\mathrm{inv})(X)$ gives

$$\mathrm{d}_A(\mathrm{inv})(X) = -A^{-1}XA^{-1}.$$

In 'differential notation' we would write the preceding argument as follows: Since $AA^{-1} = I$, we know that $\mathrm{d}(AA^{-1}) = 0$. 'Substituting' $A$ and $A^{-1}$ for $A$ and $B$ in the formula $\mathrm{d}(AB) = (\mathrm{d}A)B + A(\mathrm{d}B)$ gives

$$0 = \mathrm{d}(AA^{-1}) = (\mathrm{d}A)A^{-1} + A\mathrm{d}(A^{-1})$$

and solving this equation for $\mathrm{d}(A^{-1})$ gives the formula

$$\mathrm{d}(A^{-1}) = -A^{-1}(\mathrm{d}A)A^{-1}. \tag{5.18}$$

(This is the correct generalization to matrices of the formula $\mathrm{d}(1/x) = -(1/x^2)\mathrm{d}x$ of one-variable calculus.) We pause to give a slightly different explanation of the preceding formula. Suppose that $A$ is an invertible matrix, i.e. that $\operatorname{Det} A \neq 0$. Then if $X$ is a matrix whose entries are sufficiently small, $\operatorname{Det}(A+X) \neq 0$ so that $A+X$ is also invertible. We can write

$$A + X = (I + XA^{-1})A.$$

If $X$ is sufficiently small the matrix $XA^{-1}$ will also be small and the series

$$(I + XA^{-1})^{-1} = I - (XA^{-1}) + (XA^{-1})^2 - (XA^{-1})^3 + \cdots$$

will converge. Then we have

$$\begin{aligned}(A+X)^{-1} = [(I + XA^{-1})A]^{-1} &= A^{-1}(I + XA^{-1})^{-1} \\ &= A^{-1}(I - (XA^{-1}) + (XA^{-1})^2 - \cdots)\end{aligned}$$

or

$$\begin{aligned}(A+X)^{-1} &= A^{-1} - A^{-1}XA^{-1} + A^{-1}XA^{-1}XA^{-1} \\ &\quad - A^{-1}XA^{-1}XA^{-1}XA^{-1} + \cdots.\end{aligned}$$

In the physics literature this series is known as the *Born expansion* after the famous theoretical physicist Max Born. The formula (5.18) follows from the Born expansion when we drop all terms which are of higher order in $X$. In the physics literature the approximation given by (5.18) is known as the *first Born approximation.* It is of basic importance in scattering theory. As we have seen, we did not have to know the entire Born expansion in order to derive the first Born approximation; we got it straight from the chain rule.

On the other hand, a moment's reflection shows that the Born expansion implies that

$$(A+X)^{-1} - A^{-1} = -A^{-1}XA^{-1} + o(X).$$

This *proves* that the function (inv) is differentiable – a fact that we had to *assume* in applying the chain rule.

Let $B$ be a constant matrix, and consider the map $f(A) = ABA^{-1}$. Then

$$\begin{aligned}\mathrm{d}(ABA^{-1}) &= (\mathrm{d}A)BA^{-1} + AB(\mathrm{d}A^{-1}) \\ &= (\mathrm{d}A)BA^{-1} - ABA^{-1}(\mathrm{d}A)A^{-1}.\end{aligned}$$

In other words,

$$\mathrm{d}_A f(X) = XBA^{-1} - ABA^{-1}XA^{-1}.$$

Suppose that $t \to A(t)$ is some differentiable curve of matrices, and let

$$C(t) = A(t)BA(t)^{-1}$$

where $B$ is a constant matrix and we assume that $A(t)$ is invertible for all $t$. Applying the chain rule and the preceding formula we see that

$$C'(t) = A'(t)BA(t)^{-1} - A(t)BA(t)^{-1}A'(t)A(t)^{-1}.$$

Suppose that $A(0) = I$ and $A'(0) = X$. Then setting $t = 0$ into the preceding formula gives

$$C'(0) = XB - BX.$$

This formula is one of the most basic in mathematics and physics. The right-hand side of this formula is called the *commutator of $X$ and $B$* and is denoted by $[X, B]$, so

$$[X, B] = XB - BX.$$

For example, suppose that $A(t) = \exp tX$ so

$$A(t) = I + tX + \tfrac{1}{2}t^2X^2 + \cdots.$$

Then clearly $A(0) = I$ and $A'(0) = X$ so the above formula applies. Let us verify it directly. We have $A(t)^{-1} = (\exp tX)^{-1} = \exp(-tX) = I - tX + \frac{1}{2}t^2X^2 + \cdots$ so

$$\begin{aligned} A(t)BA(t)^{-1} &= (I + tX + \tfrac{1}{2}t^2X^2 + \cdots)B(I - tX + \tfrac{1}{2}t^2X^2 - \cdots) \\ &= B + t(XB - BX) + \tfrac{1}{2}t^2(X^2B - 2XBX + BX^2) + \cdots. \end{aligned}$$

Collecting the terms which are of degree two or higher in $t$ gives

$$A(t)BA(t)^{-1} = B + t[X, B] + o(t).$$

**Kepler motion**

We have seen that the chain rule implies Leibniz's rule for the derivative of a product – even for the product of matrices where the multiplication is not commutative. We now want to apply this same reasoning to the so-called vector product in $\mathbb{R}^3$. (We will remind you of its definition in a moment.) As a consequence, we will derive Kepler's second law for planetary motion.

In three-dimensional space there is a vector product defined as follows:

$$\text{If } \mathbf{v} = \begin{pmatrix} x \\ y \\ z \end{pmatrix} \text{ and } \mathbf{w} = \begin{pmatrix} p \\ q \\ r \end{pmatrix} \text{ then } \mathbf{v} \times \mathbf{w} = \begin{pmatrix} yr - zq \\ zp - xr \\ xq - yp \end{pmatrix}.$$

It follows immediately from the definition that

$$(\mathbf{v}_1 + \mathbf{v}_2) \times \mathbf{w} = \mathbf{v}_1 \times \mathbf{w} + \mathbf{v}_2 \times \mathbf{w}, \quad \mathbf{v} \times (\mathbf{w}_1 + \mathbf{w}_2) = \mathbf{v} \times \mathbf{w}_1 + \mathbf{v} \times \mathbf{w}_2$$

$$(a\mathbf{v}) \times \mathbf{w} = \mathbf{v} \times (a\mathbf{w}) = a(\mathbf{v} \times \mathbf{w})$$

and

$$\mathbf{v} \times \mathbf{v} = \mathbf{0}.$$

It follows from the first three equations that $\times$ acts like a multiplication and hence that

$$\mathrm{d}(\mathbf{v} \times \mathbf{w}) = \mathrm{d}\mathbf{v} \times \mathbf{w} + \mathbf{v} \times \mathrm{d}\mathbf{w}.$$

In particular, if $\mathbf{v}(t)$ and $\mathbf{w}(t)$ are curves in $\mathbb{R}^3$ and if we set

$$\mathbf{u}(t) = \mathbf{v}(t) \times \mathbf{w}(t)$$

then

$$\mathbf{u}'(t) = \mathbf{v}'(t) \times \mathbf{w}(t) + \mathbf{v}(t) \times \mathbf{w}'(t).$$

Suppose that $\mathbf{r}(t)$ denotes the position at time $t$ of a particle moving in space, and suppose that $\mathbf{p}(t)$ denotes the momentum at time $t$ of the particle. The vector

$$\boldsymbol{\mu}(t) = \mathbf{p}(t) \times \mathbf{r}(t)$$

is called the *angular momentum* of the particle relative to the origin (at time $t$). Suppose that the particle has mass $m$ and that it is subject to a force $\mathbf{F}(t)$ pointing along the line from the origin to the particle, so that $\mathbf{F}(t) = c(t)\mathbf{r}(t)$. Then

$$\mathbf{r}'(t) = (1/m)\mathbf{p}(t) \quad \text{and} \quad \mathbf{p}'(t) = \mathbf{F}(t) = c(t)\mathbf{r}(t)$$

and hence

$$\begin{aligned}\boldsymbol{\mu}'(t) &= \mathbf{p}'(t) \times \mathbf{r}(t) + \mathbf{p}(t) \times \mathbf{r}'(t)\\ &= c(t)\mathbf{r}(t) \times \mathbf{r}(t) + (1/m)\mathbf{p}(t) \times \mathbf{p}(t) = 0.\end{aligned}$$

In other words, $\boldsymbol{\mu}$ must be a constant. This law is known as the *conservation of angular momentum*. Let us suppose (for simplicity) that $\boldsymbol{\mu} \neq \mathbf{0}$. It follows easily from the definition of vector multiplication that for any vectors $\mathbf{v}$ and $\mathbf{w}$ we always have $(\mathbf{v} \times \mathbf{w})\cdot\mathbf{w} = 0$. Since $\boldsymbol{\mu} = \mathbf{p}(t) \times \mathbf{r}(t)$ we conclude that $\boldsymbol{\mu}\cdot\mathbf{r}(t) = 0$ for all $t$. In other words the particle always moves in a fixed plane, the plane perpendicular to $\boldsymbol{\mu}$. Let us rotate our coordinate system in $\mathbb{R}^3$ so that $\boldsymbol{\mu}$ lies along the $z$-axis, and hence the particle lies in the $xy$-plane. Thus

$$\mathbf{r}(t) = \begin{pmatrix} x(t) \\ y(t) \\ 0 \end{pmatrix} \text{ and therefore } \mathbf{p}(t) = m\begin{pmatrix} x'(t) \\ y'(t) \\ 0 \end{pmatrix} \text{ and } \boldsymbol{\mu} = m\begin{pmatrix} 0 \\ 0 \\ x'(t)y(t) - x(t)y'(t) \end{pmatrix}.$$

Thus the condition that $\boldsymbol{\mu}$ be constant implies that the expression $x'(t)y(t) - x(t)y'(t)$ is constant. To understand the meaning of this condition, let us draw the trajectory of the particle in the $xy$-plane. Up to terms which are $o(h)$, the area bounded by the vector $\begin{pmatrix} x(t) \\ y(t) \end{pmatrix}$, the trajectory, and the vector $\begin{pmatrix} x(t+h) \\ y(t+h) \end{pmatrix}$ is the same as the area of the triangle determined by the two vectors $\begin{pmatrix} x(t) \\ y(t) \end{pmatrix}$ and $\begin{pmatrix} x(t+h) \\ y(t+h) \end{pmatrix}$; i.e. we can ignore the hatched region in figure 5.7. The area of the triangle is (up to sign) given by

$$\tfrac{1}{2}(x(t+h)y(t) - x(t)y(t+h)).$$

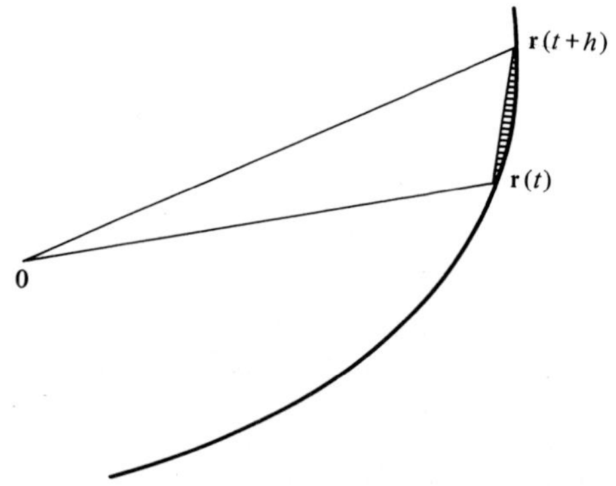


**Figure 5.7**

But

$$x(t+h) = x(t) + hx'(t) + o(h)$$

and

$$y(t+h) = y(t) + hy'(t) + o(h)$$

so the area of the triangle is given by

$$\tfrac{1}{2}(x'(t)y(t) - x(t)y'(t))h + o(h).$$

We conclude that the rate at which 'area is swept out by the radius vector' is a constant, Kepler's second law. Thus, by use of the chain rule, we see that Kepler's second law, and the fact that the particle moves in a fixed plane, follow whenever there is a central force law. The fact that the planets move in a fixed plane and sweep out equal areas in equal times is a consequence of the fact that their motion is determined by a force directed toward the sun. The preceding derivation of Kepler's second law is due to Newton.

## 5.4. Partial derivatives and differential forms

In this section we will introduce some concepts and some notation that are convenient for the chain rule. Let us consider a differentiable function $f: \mathbb{R}^k \to \mathbb{R}$. For example, take $k = 3$ and suppose that

$$f\left(\begin{pmatrix} x \\ y \\ z \end{pmatrix}\right) = x^2y^3z^4.$$

Then $df_{\mathbf{v}}$ is a linear map from $\mathbb{R}^3 \to \mathbb{R}$. So $df_{\mathbf{v}}$ can be represented as a row vector. We claim that, at any point $\mathbf{v} = \begin{pmatrix} x \\ y \\ z \end{pmatrix}$, the row vector is given by

$$df_{\mathbf{v}} = (2xy^3z^4, 3x^2y^2z^4, 4x^2y^3z^3).$$

To check this, we need only to evaluate on each of the vectors $\begin{pmatrix} 1 \\ 0 \\ 0 \end{pmatrix}, \begin{pmatrix} 0 \\ 1 \\ 0 \end{pmatrix}$ and $\begin{pmatrix} 0 \\ 0 \\ 1 \end{pmatrix}$.

For example,

$$\begin{aligned} f\left(\mathbf{v} + s\begin{pmatrix} 1 \\ 0 \\ 0 \end{pmatrix}\right) - f(\mathbf{v}) &= df_{\mathbf{v}}\begin{pmatrix} s \\ 0 \\ 0 \end{pmatrix} + o(s) \\ &= s\,df_{\mathbf{v}}\begin{pmatrix} 1 \\ 0 \\ 0 \end{pmatrix} + o(s) \end{aligned}$$

by the definition of $df_{\mathbf{v}}$ and the fact that $df_{\mathbf{v}}[\mathbf{h}]$ is linear in $\mathbf{h}$. Now

$$f\left(\mathbf{v} + s\begin{pmatrix} 1 \\ 0 \\ 0 \end{pmatrix}\right) - f\left(\begin{pmatrix} x \\ y \\ z \end{pmatrix}\right) = \frac{1}{s}\left[f\left(\begin{pmatrix} x+s \\ y \\ z \end{pmatrix}\right) - f\left(\begin{pmatrix} x \\ y \\ z \end{pmatrix}\right)\right]$$

and the limit of this expansion as $s \to 0$ is just the derivative of $f$ with respect to the variable $x$ when $y$ and $z$ are kept fixed. This is called the *partial derivative* of $f$ with respect to $x$ and is denoted by $\partial f/\partial x$. If

$$f\left(\begin{pmatrix} x \\ y \\ z \end{pmatrix}\right) = x^2y^3z^4$$

then

$$\frac{\partial f}{\partial x}\left(\begin{pmatrix} x \\ y \\ z \end{pmatrix}\right) = 2xy^3z^4$$

from elementary calculus. Thus

$$\mathrm{d}f_{\mathbf{v}}\left(\begin{pmatrix} 1 \\ 0 \\ 0 \end{pmatrix}\right) = \frac{\partial f}{\partial x}.$$

Similarly,

$$\mathrm{d}f_{\mathbf{v}}\left(\begin{pmatrix} 0 \\ 1 \\ 0 \end{pmatrix}\right) = \frac{\partial f}{\partial y}.$$

and

$$\mathrm{d}f_{\mathbf{v}}\left(\begin{pmatrix} 0 \\ 0 \\ 1 \end{pmatrix}\right) = \frac{\partial f}{\partial z}.$$

Thus

$$\mathrm{d}f_{\mathbf{v}} = \left(\frac{\partial f}{\partial x}(\mathbf{v}), \frac{\partial f}{\partial y}(\mathbf{v}), \frac{\partial f}{\partial z}(\mathbf{v})\right).$$

A direct check for $f = x^2y^3z^4$ now verifies the computation claimed above. There is a more convenient way of organizing this information. Recall that we have written $\mathrm{d}x$ for the linear function which assigns to each vector its first component. Thus

$$\mathrm{d}x = (1, 0, 0)$$

and similarly

$$\mathrm{d}y = (0, 1, 0), \quad \mathrm{d}z = (0, 0, 1).$$

Then we can write the equation

$$\mathrm{d}f_{\mathbf{v}} = \left(\frac{\partial f}{\partial x}(\mathbf{v}), \frac{\partial f}{\partial y}(\mathbf{v}), \frac{\partial f}{\partial z}(\mathbf{v})\right) = \frac{\partial f}{\partial x}(\mathbf{v})(1, 0, 0) + \frac{\partial f}{\partial y}(\mathbf{v})(0, 1, 0) + \frac{\partial f}{\partial z}(\mathbf{z})(0, 0, 1)$$

as

$$\mathrm{d}f = \frac{\partial f}{\partial x}\mathrm{d}x + \frac{\partial f}{\partial y}\mathrm{d}y + \frac{\partial f}{\partial z}\mathrm{d}z.$$

Thus

$$\mathrm{d}(x^2y^3z^4) = 2xy^3z^4\mathrm{d}x + 3x^2y^2z^4\mathrm{d}y + 4x^2y^3z^3\,\mathrm{d}z.$$

The expression on the right is a sum of three terms, each a function times a $\mathrm{d}x$

or a $\mathrm{d}y$ or a $\mathrm{d}z$. Such a sum is called a *linear differential form*. Its meaning is that it is a rule which assigns to each point of $\mathbb{R}^3$ a row vector.

The formula is consistent in the sense that if we substitute the function

$$f\left(\begin{pmatrix} x \\ y \\ z \end{pmatrix}\right) = x$$

then

$$\frac{\partial f}{\partial x} \equiv 1, \quad \frac{\partial f}{\partial y} \equiv 0, \quad \frac{\partial f}{\partial z} \equiv 0$$

to get

$$\mathrm{d}f = \mathrm{d}x.$$

With this notation, the chain rule reduces to substitution. Let us illustrate what we mean. Consider the map $\phi: \mathbb{R}^2 \to \mathbb{R}^2$ given by

$$\phi\left(\begin{pmatrix} r \\ \theta \end{pmatrix}\right) = \begin{pmatrix} r\cos\theta \\ r\sin\theta \end{pmatrix}.$$

Let $f: \mathbb{R}^2 \to \mathbb{R}$ be some function, say

$$f\left(\begin{pmatrix} x \\ y \end{pmatrix}\right) = x^3 + y^2 x.$$

Then

$$f \circ \phi\left(\begin{pmatrix} r \\ \theta \end{pmatrix}\right) = r^3 \cos\theta.$$

The map $\mathrm{d}\phi_{\binom{r}{\theta}}$ will be some $2 \times 2$ matrix: say

$$\mathrm{d}\phi_{\binom{r}{\theta}} = \begin{pmatrix} a & b \\ c & d \end{pmatrix}.$$

To find the top row of this matrix, we need only to multiply it on the left by the row vector $(1, 0)$ so

$$(1,0)\begin{pmatrix} a & b \\ c & d \end{pmatrix} = (a, b).$$

Now $(1, 0)$ is just $\mathrm{d}x$. The chain rule says that

$$\mathrm{d}x_{\phi(\binom{r}{\theta})} \circ \mathrm{d}\phi_{\binom{r}{\theta}} = \mathrm{d}(x \circ \phi)_{\binom{r}{\theta}}.$$

But $x \circ \phi\left(\begin{pmatrix} r \\ \theta \end{pmatrix}\right) = r\cos\theta$, so

$$\begin{aligned} \mathrm{d}(x \circ \phi)_{\binom{r}{\theta}} &= \cos\theta\,\mathrm{d}r - r\sin\theta\,\mathrm{d}\theta \\ &= (\cos\theta, -r\sin\theta) \end{aligned}$$

as a row vector. So $a = \cos\theta$, $b = -r\sin\theta$. Similarly

$$\begin{aligned} (c,d) = (0,1)\begin{pmatrix} a & b \\ c & d \end{pmatrix} = \mathrm{d}y_{\phi(\binom{r}{\theta})} \circ \mathrm{d}\phi_{\binom{r}{\theta}} = \mathrm{d}(y \circ \phi)_{\binom{r}{\theta}} &= \sin\theta\,\mathrm{d}r + r\cos\theta\,\mathrm{d}\theta \\ &= (\sin\theta, r\cos\theta). \end{aligned}$$

So $d\phi_{\binom{r}{\theta}}$ is the matrix

$$\begin{pmatrix} \cos\theta & -r\sin\theta \\ \sin\theta & r\cos\theta \end{pmatrix}.$$

Now

$$df = (3x^2 + y^2)\,dx + 2yx\,dy,$$

and

$$d(f\circ\phi) = 3r^2\cos\theta\,dr - r^3\sin\theta\,d\theta.$$

In principle, the chain rule says

$$(3r^2\cos^2\theta + r^2\sin^2\theta, 2r^2\sin\theta\cos\theta)\begin{pmatrix} \cos\theta & -r\sin\theta \\ \sin\theta & r\cos\theta \end{pmatrix} = (3r^2\cos\theta, -r^3\sin\theta).$$

This is, of course, correct. But in effect, the chain rule says substitute

$$x = r\cos\theta, \quad dx = \cos\theta\,dr - r\sin\theta\,d\theta,$$
$$y = r\sin\theta, \quad dy = \sin\theta\,dr + r\cos\theta\,d\theta$$

into the expression

$$df = (3x^2 + y^2)dx + (2xy)dy$$

then multiply, collect coefficients and you will get

$$d(f\circ\phi).$$

In other words, think of $x$ as a function of $r$ and $\theta$, which it *becomes* by the map $\phi$, i.e., $x$ is replaced by the function $x\circ\phi = r\cos\theta$, and then take d of this function.

In doing these computations it is convenient to remember that

$$d(gh) = g\,dh + h\,dg.$$

(Here, for example, in $\mathbb{R}^2$

$$dg = \frac{\partial g}{\partial x}dx + \frac{\partial g}{\partial y}dy$$

so

$$h\,dg = h\frac{\partial g}{\partial x}dx + h\frac{\partial g}{\partial y}dy.)$$

Then

$$d[(gh)\circ\phi] = (g\circ\phi)d(h\circ\phi) + (h\circ\phi)d(g\circ\phi).$$

Thus, in our example

$$F\begin{pmatrix} x \\ y \end{pmatrix} = x^3 + y^2x = (x^2 + y^2)x = gh\begin{pmatrix} x \\ y \end{pmatrix}$$

with

$$g = x^2 + y^2 \quad \text{and} \quad h = x.$$

Thus

$$f\circ\phi = r^2\cdot r\cos\theta$$

so

$$d(f\circ\phi) = (r\cos\theta)2r\,dr + r^2(\cos\theta\,dr - r\sin\theta\,d\theta)$$
$$= 3r^2\cos\theta\,dr - r^3\sin\theta\,d\theta.$$

This procedure is completely general: let $y_1, \ldots, y_l$ denote the coordinate functions on $\mathbb{R}^l$ – so a typical point of $\mathbb{R}^l$ is

$$\begin{pmatrix} y_1 \\ \vdots \\ y_l \end{pmatrix}.$$

Let $f: \mathbb{R}^l \to \mathbb{R}$ be a differentiable function. Then

$$\mathrm{d}f = \frac{\partial f}{\partial y_1}\mathrm{d}y_1 + \cdots + \frac{\partial f}{\partial y_l}\mathrm{d}y_l.$$

Suppose that $x_1, \ldots, x_k$ are coordinates on $\mathbb{R}^k$. Let $\phi: \mathbb{R}^k \to \mathbb{R}^l$ be a differentiable map. Define $\phi_1 = y_1 \circ \phi$, $\phi_2 = y_2 \circ \phi$, etc., so

$$\phi(\mathbf{v}) = \begin{pmatrix} \phi_1(\mathbf{v}) \\ \vdots \\ \phi_l(\mathbf{v}) \end{pmatrix} \quad \text{where} \quad \mathbf{v} = \begin{pmatrix} x_1 \\ \vdots \\ x_k \end{pmatrix}.$$

Then

$$\mathrm{d}\phi_1 = \frac{\partial \phi_1}{\partial x_1}\mathrm{d}x_1 + \cdots + \frac{\partial \phi_1}{\partial x_k}\mathrm{d}x_k$$

$$\vdots$$

$$\mathrm{d}\phi_l = \frac{\partial \phi_l}{\partial x_1}\mathrm{d}x_1 + \cdots + \frac{\partial \phi_l}{\partial x_k}\mathrm{d}x_k$$

and the linear map $\mathrm{d}\phi_{\mathbf{v}}$ is given by the matrix

$$\begin{pmatrix} \frac{\partial \phi_1}{\partial x_1}(\mathbf{v}) & \cdots & \frac{\partial \phi_1}{\partial x_k}(\mathbf{v}) \\ \vdots & & \vdots \\ \frac{\partial \phi_l}{\partial x_1}(\mathbf{v}) & \cdots & \frac{\partial \phi_l}{\partial x_k}(\mathbf{v}) \end{pmatrix}$$

or, put more simply,

$$\mathrm{d}\phi = \begin{pmatrix} \frac{\partial \phi_1}{\partial x_1} & \cdots & \frac{\partial \phi_1}{\partial x_k} \\ \vdots & & \\ \frac{\partial \phi_l}{\partial x_1} & \cdots & \frac{\partial \phi_l}{\partial x_k} \end{pmatrix}.$$

The chain rule says that

$$\mathrm{d}(f \circ \phi) = \frac{\partial f}{\partial y_1} \circ \phi \, \mathrm{d}\phi_1 + \cdots + \frac{\partial f}{\partial y_l} \circ \phi \, \mathrm{d}\phi_l$$

where the expressions $\mathrm{d}\phi_1 = \frac{\partial \phi_1}{\partial x_1}\mathrm{d}x_1 + \cdots + \frac{\partial \phi_1}{\partial x_k}\mathrm{d}x_k$ are used in this formula.

We close this section with two theoretical points.

The differentiability of $f$ at $p$ was used to define the partial derivative. The existence of the partial derivatives with respect to $x$ and with respect to $y$ does not necessarily imply the differentiability of $f$ at $p$ as can be shown by some pathological examples. Sufficient conditions for the differentiability of $f$ at $p$ are given by the following theorem.

**Theorem.** *Let $f:\mathbb{R}^2\to\mathbb{R}^1$ have continuous partial derivatives $\dfrac{\partial f}{\partial x}$ and $\dfrac{\partial f}{\partial y}$ at* $\mathbf{p}$. *Then $f$ is differentiable at* $\mathbf{p}$.

*Proof.* If $f$ is differentiable at $\mathbf{p}$, then the linear map $\mathrm{d}f_{\mathbf{p}}$ must be given by

$$\mathrm{d}f_{\mathbf{p}}\left(\begin{pmatrix} s\\ t\end{pmatrix}\right)=s\left.\frac{\partial f}{\partial x}\right|_{\mathbf{p}}+t\left.\frac{\partial f}{\partial y}\right|_{\mathbf{p}}.$$

Thus, if $\left.\dfrac{\partial f}{\partial x}\right|_{\mathbf{p}}$ and $\left.\dfrac{\partial f}{\partial y}\right|_{\mathbf{p}}$ exist, then $f$ is differentiable at $\mathbf{p}$ if and only if

$$\nabla f_{\mathbf{p}}\left(\begin{pmatrix} s\\ t\end{pmatrix}\right)=s\left.\frac{\partial f}{\partial x}\right|_{\mathbf{p}}+t\left.\frac{\partial f}{\partial y}\right|_{\mathbf{p}}+o(\mathbb{R}^2,\mathbb{R}^1).$$

Letting $\mathbf{p}=\begin{pmatrix} x_{\mathbf{p}}\\ y_{\mathbf{p}}\end{pmatrix}$, we can expand $\nabla f_{\mathbf{p}}\left(\begin{pmatrix} s\\ t\end{pmatrix}\right)$ as

$$\begin{aligned}\nabla f_{\mathbf{p}}\left(\begin{pmatrix} s\\ t\end{pmatrix}\right)&=f(x_{\mathbf{p}}+s,y_{\mathbf{p}}+t)-f(x_{\mathbf{p}},y_{\mathbf{p}})\\ &=f(x_{\mathbf{p}}+s,y_{\mathbf{p}}+t)-f(x_{\mathbf{p}},y_{\mathbf{p}}+t)+f(x_{\mathbf{p}},y_{\mathbf{p}}+t)-f(x_{\mathbf{p}},y_{\mathbf{p}}).\end{aligned}$$

The continuity of $\left.\dfrac{\partial f}{\partial x}\right|_{\mathbf{p}}$ and $\left.\dfrac{\partial f}{\partial y}\right|_{\mathbf{p}}$ implies by the mean-value theorem of one-variable calculus that

$$\begin{aligned}f(x_{\mathbf{p}}+s,y_{\mathbf{p}}+t)-f(x_{\mathbf{p}},y_{\mathbf{p}}+t)&=s\left.\frac{\partial f}{\partial x}\right|_{\begin{pmatrix} x_0\\ y_{\mathbf{p}}+t\end{pmatrix}}\\ f(x_{\mathbf{p}},y_{\mathbf{p}}+t)-f(x_{\mathbf{p}},y_{\mathbf{p}})&=t\left.\frac{\partial f}{\partial y}\right|_{\begin{pmatrix} x_{\mathbf{p}}\\ y_0\end{pmatrix}}\end{aligned}$$

for some $x_0$ and $y_0$ satisfying $x_{\mathbf{p}}<x_0<x_{\mathbf{p}}+s$ and $y_{\mathbf{p}}<y_0<y_{\mathbf{p}}+t$. Therefore

$$\nabla f_{\mathbf{p}}\left(\begin{pmatrix} s\\ t\end{pmatrix}\right)=s\left.\frac{\partial f}{\partial x}\right|_{\begin{pmatrix} x_0\\ y_{\mathbf{p}}+t\end{pmatrix}}+t\left.\frac{\partial f}{\partial y}\right|_{\begin{pmatrix} x_{\mathbf{p}}\\ y_0\end{pmatrix}}$$

so that

$$\begin{aligned}&\nabla f_{\mathbf{p}}\left(\begin{pmatrix} s\\ t\end{pmatrix}\right)-s\left.\frac{\partial f}{\partial x}\right|_{\mathbf{p}}-t\left.\frac{\partial f}{\partial y}\right|_{\mathbf{p}}\\ &=s\left\{\left.\frac{\partial f}{\partial x}\right|_{\begin{pmatrix} x_0\\ y_{\mathbf{p}}+t\end{pmatrix}}-\left.\frac{\partial f}{\partial x}\right|_{\begin{pmatrix} x_{\mathbf{p}}\\ y_{\mathbf{p}}\end{pmatrix}}\right\}+t\left\{\left.\frac{\partial f}{\partial y}\right|_{\begin{pmatrix} x_{\mathbf{p}}\\ y_0\end{pmatrix}}-\left.\frac{\partial f}{\partial y}\right|_{\begin{pmatrix} x_{\mathbf{p}}\\ y_{\mathbf{p}}\end{pmatrix}}\right\}\end{aligned}$$

As $\left\|\begin{pmatrix} s\\ t\end{pmatrix}\right\|$ tends to zero, the coefficients of $s$ and $t$ each tend to zero so that these coefficients are in $I(\mathbb{R}^2, \mathbb{R}^1)$. Since $s$ and $t$ considered as functions are each in $O(\mathbb{R}^2, \mathbb{R}^1)$, the entire expression is in $o(\mathbb{R}^2, \mathbb{R}^1)$, completing the proof.

Another important property of partial derivatives is given by

**Theorem:** *Let $f:\mathbb{R}^2 \to \mathbb{R}^1$ be differentiable at $\mathbf{p}$ with continuous second partial derivatives $\frac{\partial}{\partial y}\left(\frac{\partial f}{\partial x}\right)$ and $\frac{\partial}{\partial x}\left(\frac{\partial f}{\partial y}\right)$ at $\mathbf{p}$. Then*

$$\frac{\partial}{\partial y}\left(\frac{\partial f}{\partial x}\right)\bigg|_{\mathbf{p}} = \frac{\partial}{\partial x}\left(\frac{\partial f}{\partial y}\right)\bigg|_{\mathbf{p}}.$$

The intuitive idea behind the proof of this important theorem is very simple. Consider the four points $\begin{pmatrix} x \\ y \end{pmatrix}$, $\begin{pmatrix} x+s \\ y \end{pmatrix}$, $\begin{pmatrix} x \\ y+t \end{pmatrix}$ and $\begin{pmatrix} x+s \\ y+t \end{pmatrix}$.

$$\begin{array}{cc} \bullet\begin{pmatrix} x \\ y+t \end{pmatrix} & \bullet\begin{pmatrix} x+s \\ y+t \end{pmatrix} \\ \\ \bullet\begin{pmatrix} x \\ y \end{pmatrix} & \bullet\begin{pmatrix} x+s \\ y \end{pmatrix} \end{array}$$

We can break up the sum

$$f\left(\begin{pmatrix} x \\ y \end{pmatrix}\right) - f\left(\begin{pmatrix} x+s \\ y \end{pmatrix}\right) + f\left(\begin{pmatrix} x+s \\ y+t \end{pmatrix}\right) - f\left(\begin{pmatrix} x \\ y+t \end{pmatrix}\right)$$

in two different ways as

$$f\left(\begin{pmatrix} x+s \\ y+t \end{pmatrix}\right) - f\left(\begin{pmatrix} x+s \\ y \end{pmatrix}\right) - \left(f\left(\begin{pmatrix} x \\ y+t \end{pmatrix}\right) - f\left(\begin{pmatrix} x \\ y \end{pmatrix}\right)\right) \tag{5.19}$$

or as

$$f\left(\begin{pmatrix} x+s \\ y+t \end{pmatrix}\right) - f\left(\begin{pmatrix} x \\ y+t \end{pmatrix}\right) - \left(f\left(\begin{pmatrix} x+s \\ y \end{pmatrix}\right) - f\left(\begin{pmatrix} x \\ y \end{pmatrix}\right)\right) \tag{5.20}$$

So both of these expressions are equal. The first is

$$\nabla f_{\left(\begin{smallmatrix} x \\ y+t \end{smallmatrix}\right)}\left(\begin{pmatrix} s \\ 0 \end{pmatrix}\right) - \nabla f_{\left(\begin{smallmatrix} x \\ y \end{smallmatrix}\right)}\left(\begin{pmatrix} s \\ 0 \end{pmatrix}\right).$$

Now

$$\nabla f_{\left(\begin{smallmatrix} x \\ y+t \end{smallmatrix}\right)}\left(\begin{pmatrix} s \\ 0 \end{pmatrix}\right) = s\frac{\partial f}{\partial x}\bigg|_{\left(\begin{smallmatrix} x \\ y+t \end{smallmatrix}\right)} + o(s)$$

and

$$\nabla f_{\left(\begin{smallmatrix} x \\ y \end{smallmatrix}\right)}\left(\begin{pmatrix} s \\ 0 \end{pmatrix}\right) = s\frac{\partial f}{\partial x}\bigg|_{\left(\begin{smallmatrix} x \\ y \end{smallmatrix}\right)} + o(s).$$

If the error terms implied by the expressions $o(s)$ were actually zero, then we would have

$$\nabla f_{\left(\begin{smallmatrix} x \\ y+t \end{smallmatrix}\right)}\left(\begin{pmatrix} s \\ 0 \end{pmatrix}\right) - \nabla f_{\left(\begin{smallmatrix} x \\ t \end{smallmatrix}\right)}\left(\begin{pmatrix} s \\ 0 \end{pmatrix}\right) = s\left(\frac{\partial f}{\partial x}\left(\begin{pmatrix} x+t \\ y+t \end{pmatrix}\right) - \frac{\partial f}{\partial x}\left(\begin{pmatrix} x \\ y \end{pmatrix}\right)\right)$$

$$= s\nabla\left(\frac{\partial f}{\partial x}\right)_{\binom{x}{y}}\begin{pmatrix}0\\ t\end{pmatrix}$$

$$= st\frac{\partial}{\partial y}\left(\frac{\partial f}{\partial x}\right) + so(t).$$

Apply the same argument to (5.20) and the sum will give

$$ts\frac{\partial}{\partial x}\left(\frac{\partial f}{\partial y}\right)\begin{pmatrix}x\\ y\end{pmatrix} + to(s)$$

and hence, dividing by $ts$

$$\frac{\partial}{\partial x}\frac{\partial f}{\partial y} = \frac{\partial}{\partial y}\frac{\partial f}{\partial x}.$$

In order to make this argument work, we just need to take care in examining the error terms. We can do this by appealing to the mean value theorem in the calculus of one variable. Set

$$g(y) = f\left(\begin{pmatrix}x+s\\ y\end{pmatrix}\right) - f\left(\begin{pmatrix}x\\ y\end{pmatrix}\right).$$

The function $g$ is differentiable in $y$ and our sum (5.20) is just

$$g(y+t) - g(y).$$

By the mean-value theorem

$$g(y+t) - g(y) = tg'(\bar{y})$$

where $\bar{y}$ is some point between $y$ and $y+t$. But

$$g'(\bar{y}) = \lim_{\varepsilon\to 0}\frac{1}{\varepsilon}(g(\bar{y}+\varepsilon) - g(\bar{y}))$$

$$= \lim_{\varepsilon\to 0}\frac{1}{\varepsilon}\left(f\left(\begin{pmatrix}x+s\\ \bar{y}+\varepsilon\end{pmatrix}\right) - f\left(\begin{pmatrix}x+s\\ \bar{y}\end{pmatrix}\right) - \left(f\left(\begin{pmatrix}x\\ \bar{y}+\varepsilon\end{pmatrix}\right)\right) - f\left(\begin{pmatrix}x\\ \bar{y}\end{pmatrix}\right)\right)$$

$$= \frac{\partial}{\partial y}f\left(\begin{pmatrix}x+s\\ y\end{pmatrix}\right) - \frac{\partial}{\partial y}f\left(\begin{pmatrix}x\\ y\end{pmatrix}\right).$$

By assumption, the function $\partial f/\partial y$ is differentiable. So applying the mean-value theorem once more we get

$$\frac{\partial}{\partial y}f\left(\begin{pmatrix}x+s\\ \bar{y}\end{pmatrix}\right) - \frac{\partial}{\partial y}f\left(\begin{pmatrix}x\\ \bar{y}\end{pmatrix}\right) = s\frac{\partial}{\partial x}\left(\frac{\partial f}{\partial y}\right)\left(\begin{pmatrix}\bar{x}\\ \bar{y}\end{pmatrix}\right).$$

Thus the sum given by (5.20) equals

$$st\frac{\partial}{\partial x}\frac{\partial}{\partial y}f\left(\begin{pmatrix}\bar{x}\\ \bar{y}\end{pmatrix}\right).$$

By assumption, the function $\frac{\partial}{\partial x}\left(\frac{\partial f}{\partial y}\right)$ is continuous. So for any $\varepsilon > 0$ we can find a

$\delta > 0$ such that

$$\left|\frac{\partial}{\partial x}\frac{\partial}{\partial y}f\left(\begin{pmatrix}\bar{x}\\ \bar{y}\end{pmatrix}\right) - \frac{\partial}{\partial x}\frac{\partial}{\partial y}f\left(\begin{pmatrix}x\\ y\end{pmatrix}\right)\right| < \varepsilon \quad \text{if } |s| + |t| < \delta.$$

Thus (5.20) becomes

$$st\left(\frac{\partial}{\partial x}\left(\frac{\partial f}{\partial y}\right)\left(\begin{pmatrix}x\\ y\end{pmatrix}\right) + r_1\right)$$

where $|r_1| < \varepsilon$ if $|s| + |t| < \delta$. Similarly, (5.19) is

$$= st\left(\frac{\partial}{\partial y}\frac{\partial f}{\partial x}\left(\begin{pmatrix}x\\ y\end{pmatrix}\right) + r_2\right)$$

Assume $|st| > 0$. Dividing by $st$ we see that

$$\left|\frac{\partial}{\partial x}\frac{\partial f}{\partial y}\left(\begin{pmatrix}x\\ y\end{pmatrix}\right) - \frac{\partial}{\partial y}\frac{\partial f}{\partial x}\left(\begin{pmatrix}x\\ y\end{pmatrix}\right)\right| < r_3$$

where

$$|r_3| \leqslant |r_1| + |r_2| \leqslant 2\varepsilon \quad \text{if} \quad |s| + |t| < \delta.$$

Since the left-hand side of this inequality does not depend on $\delta$ – it is just the difference between two numbers – and $r_3$ can be made as well as we like, we conclude the equality of the crossed derivatives $\left(\text{i.e., } \frac{\partial}{\partial x}\frac{\partial f}{\partial y} \text{ and } \frac{\partial}{\partial y}\frac{\partial f}{\partial x}\right)$.

## 5.5. Directional derivatives

Let $I$ be some interval containing 0 in $\mathbb{R}^1$ and let $\gamma: I \to V$ be differentiable at 0. (As usual, $V$ can be any of our choices of vector spaces, but let us visualize the case where $V = \mathbb{R}^2$.) Suppose that $\gamma(0) = \mathbf{x}$. We will use the notation $\gamma'(0)$ to denote the vector $\mathrm{d}\gamma_0(1)$ so that

$$\gamma'(0) = \mathrm{d}\gamma_0(1) = \lim\left[(1/t)(\gamma(t) - \gamma(0))\right].$$

The vector $\gamma'(0)$ is called the *tangent vector* to the curve $\gamma$ at $t = 0$. If $\gamma_1$ is a second curve with $\gamma_1(0) = \gamma(0)$ and $\gamma_1'(0) = \gamma'(0)$, then we say that $\gamma$ and $\gamma_1$ are tangent at 0, or agree to first order at 0. If $\gamma$ is tangent to $\gamma_1$ at zero and $\gamma_1$ is tangent to $\gamma_2$ at zero, then

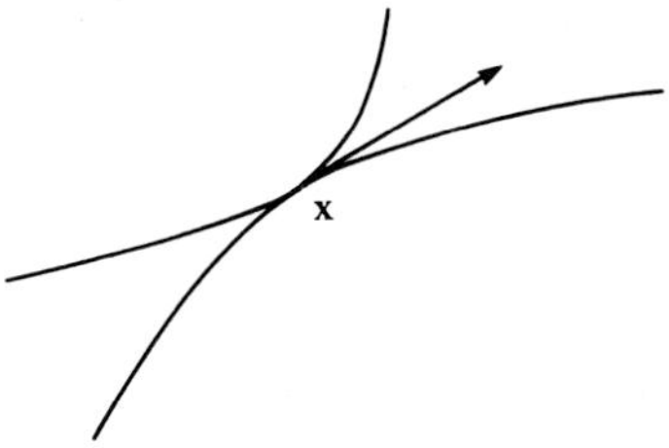


**Figure 5.8**

clearly $\gamma$ is tangent to $\gamma_2$ at zero. In other words, we have defined an equivalence relation on differentiable curves; two curves are equivalent if they agree to first order at zero. If $\gamma'(0) = \mathbf{v}$, then the pair $\{\mathbf{x}, \mathbf{v}\}$ determine the equivalence class. We visualize this equivalence class as a (little) vector $\mathbf{v}$ whose tail starts at $\mathbf{x}$, and we call it a *tangent vector at* $\mathbf{x}$. Any $\mathbf{x}$ and $\mathbf{v}$ comes from an equivalence class, because we can always consider the straight line curve

$$\gamma(t) = \mathbf{x} + t\mathbf{v}$$

which satisfies $\gamma(0) = \mathbf{x}$ and $\gamma'(0) = \mathbf{v}$. We will sometimes use a single Greek letter such as $\xi$ for a tangent vector at $\mathbf{x}$. So $\xi$ specifies both $\mathbf{x}$ and $\mathbf{v}$.

Suppose that $V = \mathbb{R}^2$. The curve $\gamma$ is then specified by giving the two functions $x\circ\gamma$ and $y\circ\gamma$, usually written as $x(t)$ and $y(t)$. Thus, for example

$$x(t) = t\sin t + 1$$
$$y(t) = \mathrm{e}^t$$

specifies the curve

$$\gamma(t) = \begin{pmatrix} t\sin t + 1 \\ \mathrm{e}^t \end{pmatrix}$$

with

$$\gamma(0) = \begin{pmatrix} 1 \\ 1 \end{pmatrix}$$

and

$$\gamma'(0) = \begin{pmatrix} 0 \\ 1 \end{pmatrix}.$$

Notice that

$$\mathrm{d}(x\circ\gamma) = \mathrm{d}(t\sin t + 1) = (\sin t + t\cos t)\mathrm{d}t$$
$$\mathrm{d}(y\circ\gamma) = \mathrm{d}(\mathrm{e}^t) = \mathrm{e}^t\mathrm{d}t$$

so that the first and second coordinates of

$$\gamma'(t) = \begin{pmatrix} \sin t + t\cos t \\ \mathrm{e}^t \end{pmatrix}$$

can be recovered as the coefficients of $\mathrm{d}t$ in $\mathrm{d}(x\circ\gamma)$ and $\mathrm{d}(y\circ\gamma)$.

Let $f\colon V \to \mathbb{R}$ be a function defined in some neighborhood of $\mathbf{p}$. For each curve $\gamma$ with $\gamma(0) = \mathbf{p}$, the function $f\circ\gamma$ is defined near 0 in $\mathbb{R}$. If $f$ is differentiable at $\mathbf{p}$ and $\gamma$ is differentiable at 0, then, by the chain rule, $f\circ\gamma$ is a (real-valued) function which is differentiable at 0 and its derivative is given by

$$(f\circ\gamma)'(0) = \mathrm{d}f_{\mathbf{p}}(\gamma'(0))$$

according to the chain rule.

In terms of our differential form notation in $\mathbb{R}^2$, we would substitute $\mathrm{d}(x\circ\gamma)$ for $\mathrm{d}x$, $\mathrm{d}(y\circ\gamma)$ for $\mathrm{d}y$ and $\partial f/\partial x, \partial f/\partial y$ for $\partial f/\partial x$ and $\partial f/\partial y$ in the expression for $\mathrm{d}f$. Thus, in our preceding example, if we took

$$f\begin{pmatrix} x \\ y \end{pmatrix} = x^2 + y^2,$$
$$df = 2x\,dx + 2y\,dy,$$
$$d(f\circ\gamma) = 2(t\sin t + 1)(\sin t + t\cos t)\,dt + 2e^t\cdot e^t\,dt$$
$$= (2(t\sin t + 1)(\sin t + t\cos t) + 2e^{2t})dt.$$

The coefficient of $dt$ is $(f\circ\gamma)'(t)$. Setting $t = 0$ gives $(f\circ\gamma)'(0)$.

Notice that $(f\circ\gamma)'(0)$ depends on $\mathbf{p}$ and $\gamma'(0)$ but on no further information about the curve $\gamma$. In short, it depends on the tangent vector $\boldsymbol{\xi}$. We shall write this value as $D_\xi f$. We call $D_\xi f$ the *directional* derivative of $f$ with respect to $\xi$. Thus

$$D_\xi f = df_{\mathbf{p}}(\mathbf{v}) \quad \text{if } \xi = \{\mathbf{p}, \mathbf{v}\}.$$

For example, if $\mathbf{v} = \begin{pmatrix} 1 \\ 0 \end{pmatrix}$ then $D_\xi f = \dfrac{\partial f}{\partial x}(\mathbf{p})$. Let $f_1$ and $f_2$ be two functions which are differentiable at $\mathbf{x}$, and let $f = f_1 + f_2$. Let $\gamma$ be a curve passing through $\mathbf{x}$ whose tangent vector at $\mathbf{p}$ is $\xi$. From the calculus of functions of one variable we know that

$$(f\circ\gamma)'(0) = (f_1\circ\gamma)'(0) + (f_2\circ\gamma)'(0)$$

and so we conclude that

$$\boxed{D_\xi(f_1 + f_2) = D_\xi f_1 + D_\xi f_2.}$$

Similarly, if we set $h = f_1 f_2$, we know from elementary calculus that

$$(h\circ\gamma)'(0) = (f_1\circ\gamma)'(0)(f_2\circ\gamma)(0) + (f_1\circ\gamma)(0)(f_2\circ\gamma)'(0)$$
$$= (f_1\circ\gamma)'(0)f_2(\mathbf{x}) + f_1(\mathbf{x})(f_2\circ\gamma)'(0),$$

since $(f_1\circ\gamma)(0) = f_1(\gamma(0)) = f_1(\mathbf{x})$ and similarly for $f_2$. Thus we can write

$$\boxed{D_\xi(f_1 f_2) = (D_\xi f_1)f_2 + f_1 D_\xi f_2.}$$

Another example of the directional derivative follows. Let $\gamma\colon \mathbb{R} \to \mathbb{R}^2$ with

$$\gamma(t) = \begin{pmatrix} t-1 \\ t^2 + 2t + 2 \end{pmatrix}, \quad f\left(\begin{pmatrix} x \\ y \end{pmatrix}\right) = x^2 y + y^3.$$

Then $df_{\left(\begin{smallmatrix} x \\ y \end{smallmatrix}\right)}$ and $\xi = \{\gamma(0), \gamma'(0)\}$ are given by

$$df_{\left(\begin{smallmatrix} x \\ y \end{smallmatrix}\right)} = (2xy, x^2 + 3y^2), \quad \xi = \left\{\begin{pmatrix} -1 \\ 2 \end{pmatrix}, \begin{pmatrix} 1 \\ 2 \end{pmatrix}\right\}$$

so that

$$D_\xi(f) = df_{\left(\begin{smallmatrix} -1 \\ 2 \end{smallmatrix}\right)}\left(\begin{pmatrix} 1 \\ 2 \end{pmatrix}\right) = (-4, 13)\begin{pmatrix} 1 \\ 2 \end{pmatrix} = -4 + 26 = 22.$$

To verify that this equals $(f\circ\gamma')(0)$, we note that

$$f\circ\gamma(t) = (t-1)^2(t^2 + 2t + 2) + (t^2 + 2t + 2)^3,$$
$$(f\circ\gamma)'(t) = 2(t-1)(t^2 + 2t + 2) + (t-1)^2(2t + 2)$$
$$+ 3(t^2 + 2t + 2)^2(2t + 2)$$

so that

$$(f\circ\gamma)'(0) = 2(-1)(2) + (-1)^2(2) + 3(2)^2(2) = 22.$$

As an example of the formula for the directional derivative of a product, let $g\colon \mathbb{R}^2 \to \mathbb{R}$ be given by $g\left(\begin{pmatrix} x \\ y \end{pmatrix}\right) = x^2 - y^2$. Then the product mapping $fg\colon \mathbb{R}^2 \to \mathbb{R}$ is

$$fg\left(\begin{pmatrix} x \\ y \end{pmatrix}\right) = f\left(\begin{pmatrix} x \\ y \end{pmatrix}\right) g\left(\begin{pmatrix} x \\ y \end{pmatrix}\right) = (x^2y + y^3)(x^2 - y^2).$$

The differentials $\mathrm{d}g$ and $d(fg)$ are given by

$$\begin{aligned} \mathrm{d}g_{\binom{x}{y}} &= (2x, -2y) \\ \mathrm{d}(fg)_{\binom{x}{y}} &= (2xy(x^2 - y^2) + (x^2y + y^3)(2x), (x^2 + 3y^2)(x^2 - y^2) \\ &\quad + (x^2y + y^3)(-2y)). \end{aligned}$$

We then have

$$\begin{aligned} \mathrm{D}_\xi(fg) &= \mathrm{d}(fg)_{\binom{-1}{2}}\left(\begin{pmatrix} 1 \\ 2 \end{pmatrix}\right) \\ &= (-4(-3) + 10(-2), 13(-3) + 10(-4))\begin{pmatrix} 1 \\ 2 \end{pmatrix} \\ &= -8 - 158 = -166. \end{aligned}$$

By the product formula, $\mathrm{D}_\xi(fg)$ must also be given by

$$\mathrm{D}_\xi(fg) = \mathrm{D}_\xi(f)g\begin{pmatrix} -1 \\ 2 \end{pmatrix} + f\begin{pmatrix} -1 \\ 2 \end{pmatrix}\mathrm{D}_\xi(g),$$

with

$$g\begin{pmatrix} -1 \\ 2 \end{pmatrix} = (-1)^2 - 2^2 = -3, \quad f\begin{pmatrix} -1 \\ 2 \end{pmatrix} = (-1)^2 2 + 2^3 = 10$$

$$\mathrm{D}_\xi(g) = \mathrm{d}g_{\binom{-1}{2}}\left(\begin{pmatrix} 1 \\ 2 \end{pmatrix}\right) = (-2, -4)\begin{pmatrix} 1 \\ 2 \end{pmatrix} = -10$$

so that

$$\mathrm{D}_\xi(fg) = 22(-3) + 10(-10) = -166$$

which agrees with the previous calculation. It will be convenient for us to think of the set of all tangent vectors at $\mathbf{x}$ as constituting a vector space, called the *tangent space* at $\mathbf{x}$ and denoted by $TV_{\mathbf{x}}$. Thus, if $\boldsymbol{\xi} = \{\mathbf{x}, \mathbf{v}\}$ and $\boldsymbol{\eta} = \{\mathbf{x}, \mathbf{w}\}$ are two tangent vectors at $\mathbf{x}$, then their sum is defined as $\boldsymbol{\xi} + \boldsymbol{\eta} = \{\mathbf{x}, \mathbf{v} + \mathbf{w}\}$. Similarly, if $\boldsymbol{\xi} = \{\mathbf{x}, \mathbf{v}\}$ and $a$ is any real number, then $a\boldsymbol{\xi} = \{\mathbf{x}, a\mathbf{v}\}$. In short, $TV_{\mathbf{x}}$ looks just like $V$ except that it has the extra dummy label $\mathbf{x}$ attached to everything. At present this seems like a cumbersome piece of excess notational baggage, but its value will become clear later.

If $\boldsymbol{\xi} = \{\mathbf{x}, \mathbf{v}\}$ and $\boldsymbol{\eta} = \{\mathbf{x}, \mathbf{w}\}$, then

$$\begin{aligned} \mathrm{D}_{\xi+\eta}f &= \mathrm{d}f_{\mathbf{x}}[\mathbf{v} + \mathbf{w}] \\ &= \mathrm{d}f_{\mathbf{x}}x[\mathbf{v}] + \mathrm{d}f_{\mathbf{x}}[\mathbf{w}] \\ &= \mathrm{D}_\xi f + \mathrm{D}_\eta f, \end{aligned}$$

so

$$\boxed{\mathrm{D}_{\xi+\eta}f = \mathrm{D}_\xi f + \mathrm{D}_\eta f.}$$

Similarly,

$$\boxed{\mathrm{D}_{a\xi}f = a\mathrm{D}_\xi f.}$$

## 5.6. The pullback notation

Let $\phi: V \to W$ be a differentiable function with $\phi(\mathbf{x}) = \mathbf{y}$. If $f: W \to \mathbb{R}$ is a function defined near $\mathbf{y}$, then $f \circ \phi$ is a function defined near $\mathbf{x}$. In order to emphasize a point of view which will be central in this book, we will denote this function by $\phi^*\mathbf{f}$ and call it the *pullback* of $f$ under $\phi$. So

$$\phi^* f = f \circ \phi,$$
$$(\phi^* f)(\mathbf{x}) = f(\phi(\mathbf{x})).$$

We think of $\phi$ as fixed and $f$ as varying, so that $\phi^*$ pulls all functions on $W$ back to $V$. Notice that

$$\phi^*(f_1 + f_2) = \phi^* f_1 + \phi^* f_2$$

and

$$\phi^*(f_1 f_2) = (\phi^* f_1)(\phi^* f_2)$$

so algebraic operations are preserved by $\phi^*$.

Thus, if $\phi: \mathbb{R}^2 \to \mathbb{R}^2$ is the map giving the transition from polar to rectangular coordinates,

$$\phi\left(\begin{pmatrix} r \\ \theta \end{pmatrix}\right) = \begin{pmatrix} r\cos\theta \\ r\sin\theta \end{pmatrix},$$

we would write

$$\phi^* x = r\cos\theta,$$
$$\phi^* y = r\sin\theta.$$

We should pause to explain our point of view about these equations. We have an $r\theta$-plane and an $xy$-plane. We are thinking of $\phi$ as the map which assigns to each point of the $r\theta$-plane a point in the $xy$-plane. We are considering $x$ as a function on the $xy$-plane: that function which assigns to each point its $x$-coordinate. Then $\phi^* x$

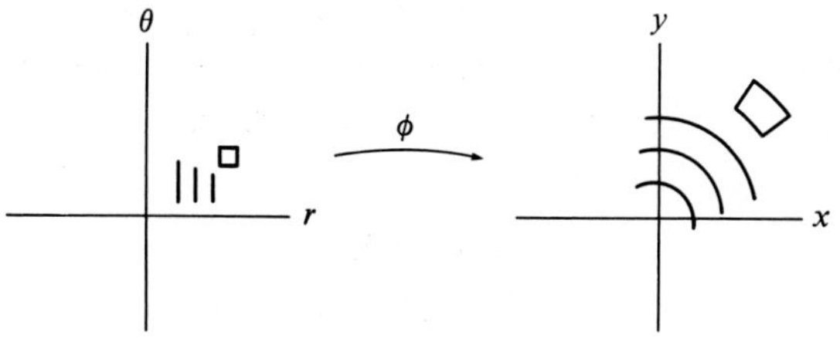


**Figure 5.9**

becomes a function on the $r\theta$-plane, and in fact,

$$\phi^*x = r\cos\theta$$

and similarly

$$\phi^*y = r\sin\theta.$$

We now want to define the pullback under $\phi$ of differential forms. We begin by defining the pullback of the basic forms $\mathrm{d}x$ and $\mathrm{d}y$. We define

$$\phi^*\mathrm{d}x = \mathrm{d}(\phi^*x) = \cos\theta\,\mathrm{d}r - r\sin\theta\,\mathrm{d}\theta,$$
$$\phi^*\mathrm{d}y = \mathrm{d}(\phi^*y) = \sin\theta\,\mathrm{d}r + r\cos\theta\,\mathrm{d}\theta$$

and for any linear differential form such as

$$x^2\mathrm{d}x + y^2\mathrm{d}y$$

we define

$$\begin{aligned}\phi^*(x^2\mathrm{d}x + y^2\mathrm{d}y) &= \phi^*(x^2)\cdot\phi^*\mathrm{d}x + \phi^*(y^2)\phi^*\mathrm{d}y\\ &= (r^2\cos^2\theta)(\cos\theta\,\mathrm{d}r - r\sin\theta\,\mathrm{d}\theta) + (r^2\sin^2\theta)(\sin\theta\,\mathrm{d}r + r\cos\theta\,\mathrm{d}\theta)\\ &= r^2(\cos^3\theta + \sin^3\theta)\mathrm{d}r + r^3(\sin^2\theta\cos\theta - \cos^2\theta\sin\theta)\mathrm{d}\theta.\end{aligned}$$

In other words, for a differential form

$$a\mathrm{d}x + b\mathrm{d}y$$

where $a$ and $b$ are functions, we define

$$\boxed{\phi^*(a\mathrm{d}x + b\mathrm{d}y) = \phi^*(a)\phi^*(\mathrm{d}x) + \phi^*(b)\phi^*(\mathrm{d}y),}$$

where we then collect coefficients of $\mathrm{d}r$ and $\mathrm{d}\theta$. Notice that if $f$ is any function in the $xy$-plane

$$\mathrm{d}f = \frac{\partial f}{\partial x}\mathrm{d}x + \frac{\partial f}{\partial y}\mathrm{d}y$$

so

$$\begin{aligned}\phi^*\mathrm{d}f &= \phi^*\left(\frac{\partial f}{\partial x}\right)\phi^*\mathrm{d}x + \phi^*\left(\frac{\partial f}{\partial y}\right)\phi^*\mathrm{d}y\\ &= \left(\frac{\partial f}{\partial x}\circ\phi\right)\mathrm{d}(x\circ\phi) + \left(\frac{\partial f}{\partial y}\circ\phi\right)\mathrm{d}(y\circ\phi)\\ &= \mathrm{d}(f\circ\phi)\end{aligned}$$

by the chain rule. Thus

$$\boxed{\phi^*\mathrm{d}f = \mathrm{d}(\phi^*f).}$$

This notation works in complete generality for differentiable maps from $\mathbb{R}^k \to \mathbb{R}^l$.

For example, consider the map $\phi$ of $\mathbb{R}^2 \to \mathbb{R}^3$ given by

$$\phi\left(\begin{pmatrix} r \\ s \end{pmatrix}\right) = \begin{pmatrix} r^2 \\ sr \\ s^2 \end{pmatrix}.$$

Then, if $x, y, z$ denote the three coordinates on $\mathbb{R}$, then

$$\begin{aligned}\phi^*x &= r^2 \quad \text{so} \quad \phi^*\mathrm{d}x = 2r\mathrm{d}r,\\ \phi^*y &= sr \quad \text{so} \quad \phi^*\mathrm{d}y = s\mathrm{d}r + r\mathrm{d}s,\\ \phi^*z &= s^2 \quad \text{so} \quad \phi^*\mathrm{d}z = 2s\mathrm{d}s.\end{aligned}$$

For any function $f$ on $\mathbb{R}^3$

$$\mathrm{d}f = \frac{\partial f}{\partial x}\mathrm{d}x + \frac{\partial f}{\partial y}\mathrm{d}y + \frac{\partial f}{\partial z}\mathrm{d}z$$

and we can compute $\phi^*(\mathrm{d}f)$ in either of two ways; either as

$$\mathrm{d}(\phi^*f)$$

or as

$$\phi^*(\mathrm{d}f) = \phi^*\left(\frac{\partial f}{\partial x}\right)\phi^*\mathrm{d}x + \phi^*\left(\frac{\partial f}{\partial y}\right)\phi^*\mathrm{d}y + \phi^*\left(\frac{\partial f}{\partial z}\right)\phi^*\mathrm{d}z.$$

For example, suppose that

$$f\left(\begin{pmatrix} x \\ y \\ z \end{pmatrix}\right) = y^2 - xz$$

so

$$\mathrm{d}f = -z\mathrm{d}x + 2y\mathrm{d}y - x\mathrm{d}z.$$

Then

$$\phi^*f = 0$$

so

$$\phi^*\mathrm{d}f = \mathrm{d}(\phi^*f) = 0.$$

Computing $\phi^*\mathrm{d}f$ directly, we get

$$\begin{aligned}&-s^2\cdot 2r\mathrm{d}r + 2sr\cdot(s\mathrm{d}r + r\mathrm{d}s) - r^2\cdot 2s\mathrm{d}s\\ &= [-s^2\cdot 2r + 2s^2r]\mathrm{d}r + [-r^2\cdot 2s + 2r^2s]\mathrm{d}s\\ &= 0.\end{aligned}$$

The general situation is now clear. If $x_1, \ldots, x_k$ are the coordinates on $\mathbb{R}^k$ and $y_1, \ldots, y_l$ are the coordinates on $\mathbb{R}^l$, then a differentiable map $\phi: \mathbb{R}^k \to \mathbb{R}^l$ is given by

$$\phi(\mathbf{v}) = \begin{pmatrix} \phi_1(\mathbf{v}) \\ \vdots \\ \phi_l(\mathbf{v}) \end{pmatrix}, \quad \mathbf{v} = \begin{pmatrix} x_1 \\ \vdots \\ x_k \end{pmatrix}.$$

Then

$$\begin{aligned}\phi^*y_1 &= \phi_1, \qquad \phi^*\mathrm{d}y_1 = \mathrm{d}\phi_1 = \frac{\partial \phi_1}{\partial x_1}\mathrm{d}x_1 + \cdots + \frac{\partial \phi_1}{\partial x_k}\mathrm{d}x_k,\\ &\vdots \qquad\qquad\qquad \vdots\\ \phi^*y_l &= \phi_l, \qquad \phi^*\mathrm{d}y_l = \mathrm{d}\phi_l.\end{aligned}$$

The operation $\phi^*$ carries functions on $\mathbb{R}^l$ to functions on $\mathbb{R}^k$ and linear differential forms on $\mathbb{R}^l$ to linear differential forms on $\mathbb{R}^k$. All algebraic operations, such as adding or multiplying two functions, adding two forms, multiplying a function by a form, are preserved by $\phi^*$. Furthermore, the chain rule says that

$$\phi^*\mathrm{d}f = \mathrm{d}\phi^*f,$$

for any function $f$ on $\mathbb{R}^l$.

Suppose that we have

$$\phi: V \to W \quad \text{and} \quad \psi: W \to Z.$$

We can compose the two maps to obtain

$$\psi \circ \phi: V \to Z.$$

If $g$ is any function on $Z$, we can form

$$\psi^*g = g \circ \psi$$

which is a function on $W$ and then

$$\phi^*(\psi^*g) = (g \circ \psi) \circ \phi$$

which is a function on $V$. By the associative law for composition, we know that

$$(g \circ \psi) \circ \phi = g \circ (\psi \circ \phi).$$

Thus

$$(\psi \circ \phi)^*g = \phi^*(\psi^*g),$$

so that on functions

$$\boxed{(\psi \circ \phi)^* = \phi^* \circ \psi^*.}$$

Notice the reversal of the order.

Suppose $V = \mathbb{R}^k$, $W = \mathbb{R}^l$ and $Z = \mathbb{R}^m$ with coordinates $x^1, \ldots, x^k$; $y^1, \ldots, y^l$; $z^1, \ldots, z^m$; then if

$$\omega = a_1 \mathrm{d}z^1 + \cdots + a_m \mathrm{d}z^m$$

is a differential form on $\mathbb{R}^m$, then

$$\psi^*\omega$$

is a differential form on $\mathbb{R}^l$ and

$$\phi^*\psi^*\omega$$

is a differential form on $\mathbb{R}^k$. If follows from the chain rule that

$$\begin{aligned} \phi^*\psi^*\mathrm{d}f &= \phi^*(\mathrm{d}\psi^*f) \\ &= \mathrm{d}\phi^*\psi^*f \\ &= \mathrm{d}(\psi \circ \phi)^*f. \end{aligned}$$

Since $\phi^*$ and $\psi^*$ preserve all algebraic operations, so does $(\psi \circ \phi)^*$,

$$\phi^*\psi^*(g\mathrm{d}f) = (\psi \circ \phi)^*(g\mathrm{d}f),$$

and since the most general linear differential form is a sum of terms like $g\mathrm{d}f$, i.e. is of the form

$$a_1 dz^1 + \cdots + a_m dz^m,$$

it follows that

$$(\psi \circ \phi)^*(\omega) = \phi^* \psi^* \omega \qquad (5.21)$$

for all linear differential forms.

Let $\gamma: \mathbb{R} \to V$ be a curve passing through $\mathbf{x}$ i.e., $\gamma(0) = \mathbf{x}$. Then $\phi \circ \gamma$ is a curve passing through $\mathbf{y}$. If $\gamma$ is differentiable at 0, then so is $\phi \circ \gamma$ and, by the chain rule,

$$(\phi \circ \gamma)'(0) = d\phi_{\mathbf{x}}(\gamma'(0)).$$

The right-hand side of this equation depends only on the tangent vector $\xi$ associated to $\gamma$. Thus $d\phi_{\mathbf{x}}$ maps tangent vectors at $\mathbf{x}$ to tangent vectors at $\phi(\mathbf{x})$:

$$d\phi_{\mathbf{x}}: TV_{\mathbf{x}} \to TW_{\phi(\mathbf{x})}$$

where we define

$$d\phi_{\mathbf{x}}\xi = \{\phi(\mathbf{x}), d\phi_{\mathbf{x}}(\mathbf{v})\} \quad \text{if} \quad \xi = \{\mathbf{x}, \mathbf{v}\}.$$

We can thus visualize the differential $d\phi_{\mathbf{x}}$ as taking infinitesimal curves through $\mathbf{x}$ into infinitesimal curves through $\phi(\mathbf{x})$.

Now let $f: W \to \mathbb{R}$ be a function which is differentiable at $\phi(\mathbf{x})$. Then by the associative law for composition

$$\begin{aligned}(\phi^* f) \circ \gamma &= f \circ \phi \circ \gamma \\ &= f \circ (\phi \circ \gamma).\end{aligned}$$

Differentiating this equation at $t = 0$ gives

$$D_\xi(\phi^* f) = D_{d\phi_{\mathbf{x}}\xi} f. \qquad (5.22)$$

This means we can pull $f$ back by $\phi$ and then take the directional derivative with respect to $\xi$, or, we can push forward by $d\phi_{\mathbf{x}}$ and then take the directional derivative of $f$ with respect to $d\phi_{\mathbf{x}}\xi$. Both procedures yield the same answer.

Letting $\xi = \{\mathbf{x}, \mathbf{v}\}$, the above identity is given explicitly by

$$D_{\{\mathbf{x},\mathbf{v}\}}(\phi^* f) = D_{\phi(\mathbf{x}), d\phi_{\mathbf{x}}[\mathbf{v}]}(f)$$

or equivalently,

$$d(f \circ \phi)_{\mathbf{x}}[\mathbf{v}] = df_{\phi(\mathbf{x})}[d\phi_{\mathbf{x}}[\mathbf{v}]]$$

which states that

$$d(f \circ \phi)_{\mathbf{x}} = df_{\phi(\mathbf{x})} \circ d\phi_{\mathbf{x}}$$

which is a special case of the chain rule.

As an example of this identity, let $f: W \to \mathbb{R}$ and $\phi: V \to W$ be given by

$$f\left(\begin{pmatrix} x \\ y \end{pmatrix}\right) = x^2 y, \quad \phi\left(\begin{pmatrix} r \\ \theta \end{pmatrix}\right) = \begin{pmatrix} r\cos\theta \\ r\sin\theta \end{pmatrix}.$$

Then

$$\mathrm{d}f_{\binom{x}{y}} = (2xy, x^2)$$

$$\mathrm{d}\phi_{\binom{r}{\theta}} = \begin{pmatrix} \cos\theta & -r\sin\theta \\ \sin\theta & r\cos\theta \end{pmatrix}$$

so that

$$\mathrm{D}_{\left\{\phi\left(\binom{r}{\theta}\right), \mathrm{d}\phi\left(\binom{r}{\theta}\right)[\mathrm{v}]\right\}}(f) = \mathrm{d}f_{\phi\binom{r}{\theta}} \circ \mathrm{d}\phi_{\binom{r}{\theta}}[v]$$

$$= (2(r\cos\theta)(r\sin\theta), (r\cos\theta)^2)\begin{pmatrix} \cos\theta & -r\sin\theta \\ \sin\theta & r\cos\theta \end{pmatrix}\begin{pmatrix} v_r \\ v_\theta \end{pmatrix}$$

$$= (2r^2\cos\theta\sin\theta, r^2\cos^2\theta)\begin{pmatrix} v_r\cos\theta - rv_\theta\sin\theta \\ v_r\sin\theta + rv_\theta\cos\theta \end{pmatrix}$$

$$= 2r^2\cos\theta\sin\theta(v_r\cos\theta - rv_\theta\sin\theta) + r^2\cos^2\theta(v_r\sin\theta + rv_\theta\cos\theta)$$

$$= 3r^2\cos^2\theta\sin\theta\, v_r + r^3(-2\cos\theta\sin^2\theta + \cos^3\theta)v_\theta.$$

To verify that this equals $\mathrm{D}_{\left\{\binom{r}{\theta}, \mathrm{v}\right\}}(\phi^*f)$, we note that

$$\phi^*f\left(\begin{pmatrix} r \\ \theta \end{pmatrix}\right) = f\left(\begin{pmatrix} r\cos\theta \\ r\sin\theta \end{pmatrix}\right) = (r\cos\theta)^2 r\sin\theta = r^3\cos^2\theta\sin\theta$$

so that

$$\mathrm{d}(\phi^*f)_{\binom{r}{\theta}} = (3r^2\cos^2\theta\sin\theta, r^3(2\cos\theta(-\sin\theta)\sin\theta + \cos^3\theta)).$$

We then have

$$\mathrm{D}_{\left\{\binom{r}{\theta}, \mathrm{v}\right\}}(\phi^*f) = \mathrm{d}(\phi^*f)_{\binom{r}{\theta}}\left(\begin{pmatrix} v_r \\ v_\theta \end{pmatrix}\right)$$

$$= 3r^2\cos^2\theta\sin\theta\, v_r + r^3(-2\cos\theta\sin^2\theta + \cos^3\theta)v_\theta.$$

---

## Summary

### A Differentials and partial derivatives

You should be able to state the definition of the differential $\mathrm{d}f$ of a function $f$ in terms of '$o$' and '$O$' notation.

You should be able to state and apply the rules for differentiating the sum, product, or composition of functions.

You should be able to express the differential of a function in terms of partial derivatives and to construct the matrix that represents the differential of a function $f = \mathbb{R}^2 \to \mathbb{R}^2$.

### B Coordinate transformations

Given a transformation that can be used to introduce new coordinates on the plane, you should be able to use the chain rule to express differentials and partial derivatives in terms of these new coordinates.

C Applications of differentials

You should know how to determine the equation of the line or plane tangent to the graph of a function at a given point.

You should be able to use the chain rule to solve 'related rate' problems that involve functions on the plane.

---

## Exercises

5.1. Show that if $f: V \to W$ is differentiable at $\boldsymbol{\alpha}$ and if $T: W \to Z$ is linear, then $T \circ f$ is differentiable at $\boldsymbol{\alpha}$ and

$$\mathrm{d}(T \circ f)_{\boldsymbol{\alpha}} = T \circ \mathrm{d}f_{\boldsymbol{\alpha}}.$$

5.2. Let $F: V \to \mathbb{R}$ be differentiable at $\boldsymbol{\alpha}$ and let $f: \mathbb{R} \to \mathbb{R}$ be a function whose derivative exists at $a = F(\boldsymbol{\alpha})$. Prove that $f \circ F$ is differentiable at $\boldsymbol{\alpha}$ and that

$$\mathrm{d}(f \circ F)_{\boldsymbol{\alpha}} = f'(a)\mathrm{d}F_{\boldsymbol{\alpha}}.$$

5.3. Let $F: V \to W$ and $G: W \to V$ be continuous maps with $G \circ F(\mathbf{v}) = \mathbf{v}$ and $F \circ G(\mathbf{w}) = \mathbf{w}$ for all $\mathbf{v}$ in $V$ and $\mathbf{w}$ in $W$. Suppose that $F$ is differentiable at $\boldsymbol{\alpha}$ and $G$ is differentiable at $\boldsymbol{\beta} = F(\boldsymbol{\alpha})$. Prove that

$$\mathrm{d}G_{\boldsymbol{\beta}} = (\mathrm{d}F_{\boldsymbol{\alpha}})^{-1}.$$

5.4. Let $f: V \to \mathbb{R}$ be differentiable at $\boldsymbol{\alpha}$. Show that $g = f^n$ is differentiable at $\boldsymbol{\alpha}$ and that

$$\mathrm{d}g_{\boldsymbol{\alpha}} = nf^{n-1}\mathrm{d}f_{\boldsymbol{\alpha}}.$$

5.5. Let $\gamma: \mathbb{R} \to \mathbb{R}^2$ denote the curve $\gamma(t) = \begin{pmatrix} e^t \\ \sin t \end{pmatrix}$, and let $F: \mathbb{R}^2 \to \mathbb{R}^2$ be the mapping

$$F\left(\begin{pmatrix} x \\ y \end{pmatrix}\right) = \begin{pmatrix} 3x^2y \\ x^2y^3 \end{pmatrix}.$$

(a) Compute the tangent vector for $\gamma$ at $t = 0$ and $t = \pi/2$

(b) Find the directional derivative of $F$ with respect to each of these tangent vectors.

5.6. Let $f: \mathbb{R}^2 \to \mathbb{R}^2$ and $g: \mathbb{R}^2 \to \mathbb{R}^2$ be given by

$$f\left(\begin{pmatrix} x \\ y \end{pmatrix}\right) = \begin{pmatrix} x^2y \\ y^3 \end{pmatrix}, \quad g\left(\begin{pmatrix} x \\ y \end{pmatrix}\right) = \begin{pmatrix} \cos xy \\ \sin xy \end{pmatrix}.$$

Verify the chain rule for the mapping $g \circ f: \mathbb{R}^2 \to \mathbb{R}^2$.

5.7. Let $g: V \to W$ be the mapping $g\left(\begin{pmatrix} x \\ y \end{pmatrix}\right) = \begin{pmatrix} x^2e^y \\ \cos xy \end{pmatrix}$, and let $\boldsymbol{\lambda}: \mathbb{R} \to V$ be the straight line

$$\boldsymbol{\lambda}(t) = \begin{pmatrix} 3 \\ -1 \end{pmatrix} + t\begin{pmatrix} 1 \\ 2 \end{pmatrix}$$

(a) Find the tangent vector at $\begin{pmatrix} 3 \\ -1 \end{pmatrix}$ to the curve $g \circ \boldsymbol{\lambda}$.

(b) Compute the directional derivative $D_{\left(\begin{smallmatrix} 3 \\ -1 \end{smallmatrix}\right),\left(\begin{smallmatrix} 1 \\ 2 \end{smallmatrix}\right)}(g)$ in two ways.

5.8. Let $\phi: V \to W$ be the mapping $\phi: \begin{pmatrix} r \\ \theta \end{pmatrix} \to \begin{pmatrix} r\cos\theta \\ r\sin\theta \end{pmatrix}$ and let $f: W \to \mathbb{R}$ be given by $f: \begin{pmatrix} x \\ y \end{pmatrix} \to x^3 y^4$. Verify that

$$D_\xi(\phi^* f) = D_{d\phi_\alpha(\xi)} f$$

for all tangent vectors $\xi = (\boldsymbol{\alpha}, \mathbf{v})$.

5.9. Define mappings $F: \mathbb{R}^2 \to \mathbb{R}^2$, $G: \mathbb{R}^2 \to \mathbb{R}^2$, $f: \mathbb{R}^1 \to \mathbb{R}^2$, and $g: \mathbb{R}^2 \to \mathbb{R}^1$ by

$$F\left(\begin{pmatrix} x \\ y \end{pmatrix}\right) = \begin{pmatrix} x^3 + y^2 \\ xy \end{pmatrix}, \quad G\left(\begin{pmatrix} x \\ y \end{pmatrix}\right) = \begin{pmatrix} 2x^2 y \\ y^2 \end{pmatrix},$$

$$f(t) = \begin{pmatrix} t^2 + 1 + \cos t \\ 3t + 2 \end{pmatrix}, \quad g\left(\begin{pmatrix} x \\ y \end{pmatrix}\right) = (x^3 y)$$

Verify that

(a) $d(F \circ G)_{\left(\begin{smallmatrix} x \\ y \end{smallmatrix}\right)} = dF_{G\left(\left(\begin{smallmatrix} x \\ y \end{smallmatrix}\right)\right)} \circ dG_{\left(\begin{smallmatrix} x \\ y \end{smallmatrix}\right)}$

(b) $d(G \circ F)_{\left(\begin{smallmatrix} x \\ y \end{smallmatrix}\right)} = dG_{F\left(\left(\begin{smallmatrix} x \\ y \end{smallmatrix}\right)\right)} \circ dF_{\left(\begin{smallmatrix} x \\ y \end{smallmatrix}\right)}$

(c) $d(g \circ f)_{(t)} = dg_{f(t)} \circ df_{(t)}$

(d) $d(f \circ g)_{\left(\begin{smallmatrix} x \\ y \end{smallmatrix}\right)} = df_{g\left(\left(\begin{smallmatrix} x \\ y \end{smallmatrix}\right)\right)} \circ dg_{\left(\begin{smallmatrix} x \\ y \end{smallmatrix}\right)}$

5.10. Let $f: \mathbb{R}^2 \to \mathbb{R}$ be differentiable in some neighborhood of $\begin{pmatrix} x_0 \\ y_0 \end{pmatrix}$ and satisfy $f\left(\begin{pmatrix} x_0 \\ y_0 \end{pmatrix}\right) \neq 0$. Show that in some neighborhood of $\begin{pmatrix} x_0 \\ y_0 \end{pmatrix}$ the mapping $g$, given by $g\left(\begin{pmatrix} x \\ y \end{pmatrix}\right) = 1/f\left(\begin{pmatrix} x \\ y \end{pmatrix}\right)$ is differentiable and that

$$dg_{\left(\begin{smallmatrix} x_0 \\ y_0 \end{smallmatrix}\right)} = -df_{\left(\begin{smallmatrix} x_0 \\ y_0 \end{smallmatrix}\right)} \Big/ \left(f\left(\begin{pmatrix} x_0 \\ y_0 \end{pmatrix}\right)\right)^2.$$

5.11. A function $f$ on the plane is defined in terms of affine coordinates $x$ and $y$ by

$$f(P) = \sqrt{(|x(P)y(P)|)}$$

(a) Is $f$ *continuous* at the origin $P_0(x = 0, y = 0)$? Justify your answer carefully in terms of the definition of continuity.

(b) Is $f$ *differentiable* at the origin? Justify your answer carefully in terms of the definition of differentiability.

5.12. So-called parabolic coordinates on the plane are defined in terms of Cartesian coordinates $x$ and $y$ by

$$\begin{pmatrix} u \\ v \end{pmatrix} = \begin{pmatrix} \sqrt{(x^2 + y^2)} - x \\ \sqrt{(x^2 + y^2)} + x \end{pmatrix}$$

(a) Express $\begin{pmatrix} du \\ dv \end{pmatrix}$ in terms of $\begin{pmatrix} dx \\ dy \end{pmatrix}$ by means of a $2 \times 2$ matrix, then invert this matrix to express $\begin{pmatrix} dx \\ dy \end{pmatrix}$ in terms of $\begin{pmatrix} du \\ dv \end{pmatrix}$.

(b) Invert the coordinate transformation by solving for $\begin{pmatrix} x \\ y \end{pmatrix}$ in terms of $u$ and $v$. Differentiate to express $\begin{pmatrix} dx \\ dy \end{pmatrix}$ in terms of $\begin{pmatrix} du \\ dv \end{pmatrix}$.

(c) Show that the curves $u = \text{constant}$ and $v = \text{constant}$ are parabolas which are perpendicular where they cross. Sketch these families of curves.

(d) Consider the function $f$ on the plane defined by $f(p) = 1/(u(p) + v(p))$. Express $d_{\mathbf{q}}f$, where $\mathbf{q}$ is the point with coordinates $u(\mathbf{q}) = 4$, $v(\mathbf{q}) = 16$, in terms of $du$ and $dv$, then in terms of $dx$ and $dy$.

(e) Suppose that a particle moves along the path defined by the function $\boldsymbol{\alpha}: \mathbb{R} \to \mathbb{R}^2$ such that

$$\begin{pmatrix} x \\ y \end{pmatrix} \circ \boldsymbol{\alpha}(t) = \begin{pmatrix} t^2 + t \\ t^3 \end{pmatrix}.$$

Calculate the derivatives of $x \circ \boldsymbol{\alpha}$, $y \circ \boldsymbol{\alpha}$, $u \circ \boldsymbol{\alpha}$, and $v \circ \boldsymbol{\alpha}$ at $t = 2$.

5.13. Consider coordinates $u$ and $v$ in the plane which are related to $x$ and $y$ by the equations

$$\begin{pmatrix} x \\ y \end{pmatrix} = \begin{pmatrix} 2uv \\ u^2 - v^2 \end{pmatrix}.$$

(a) Calculate the derivative (the *Jacobian matrix*) of this transformation at the point $u = 2$, $v = 1$ (equivalently, express $\begin{pmatrix} dx \\ dy \end{pmatrix}$ in terms of $\begin{pmatrix} du \\ dv \end{pmatrix}$ at this point).

(b) Consider the function $f(u, v) = u^2 v^3$. Find the equation, in terms of $x$ and $y$, of the line tangent to the curve $f(u, v) = 4$ at the point $u = 2$, $v = 1$ (i.e., at $x = 4$, $y = 3$). (Do not try to solve for $u$ and $v$ as functions of $x$ and $y$; just use the chain rule.)

(c) Suppose that a particle moves along the path

$$\begin{pmatrix} x \\ y \end{pmatrix} = \begin{pmatrix} \frac{1}{2}t^3 \\ \frac{3}{4}t^2 \end{pmatrix}.$$

At the instant $t = 2$, when the particle is passing through the point $\begin{pmatrix} u \\ v \end{pmatrix} = \begin{pmatrix} 2 \\ 1 \end{pmatrix}$, at what rate are its $u$ and $v$ coordinates changing; i.e., what are $du/dt$ and $dv/dt$ at this instant?

5.14. Let $\mathbb{A}$ denote an affine plane, let $P_0$ be a point in this plane. Invent a function $f: \mathbb{A} \to \mathbb{R}$, satisfying $f(P_0) = 0$, which has the property that for *any* affine coordinates $s(P)$, $t(P)$ on the plane, $\left(\frac{\partial f}{\partial s}\right)_t$ and $\left(\frac{\partial f}{\partial t}\right)_s$ are defined and equal to zero at $P_0$, yet $f$ is *not* differentiable at $P_0$. (Hint: replace 'differentiable' by 'continuous', 'partial derivative' by 'limit as a function of one coordinate', and an answer would be

$$f(P) = \begin{cases} 0 & \text{at } P_0, \\ \dfrac{x^2 y}{x^4 + y^2} & \text{otherwise} \end{cases}$$

where $x(P_0)$ and $y(P_0)$ are both zero.)

5.15. In Quadratic Crater National Monument, the altitude above sea level is described by the function

$$z(x, y) = \sqrt{(x^2 + 4y^2)}, \quad (x, y, z \text{ in kilometers}).$$

The Fahrenheit temperature is described by

$$T(x, y) = 100 + 2x - \tfrac{1}{4}x^2 y^2.$$

(a) Express $\mathrm{d}z$ and $\mathrm{d}T$ in terms of $\mathrm{d}x$ and $\mathrm{d}y$ at the point $x = 3$, $y = 2$.

(b) Find the equation of the tangent plane to the crater at the point $x = 3$, $y = 2$.

(c) At the point $x = 3$, $y = 2$, along what direction is the temperature changing most rapidly? If one follows a path along this direction, what is the rate of change of temperature with respect to altitude (accurate to the nearest degree per kilometer)?

# 6

# Theorems of the differential calculus



In Chapter 6 we continue the study of the differential calculus. We present the vector versions of the mean-value theorem, of Taylor's formula and of the inverse function theorem. We discuss critical point behavior and Lagrange multipliers. You might want to read the chapter quickly without concentrating on details of the proofs. But do the exercises.

## 6.1. The mean-value theorem

This is one of the few theorems that we will not be able to state, in the higher-dimensional calculus, with the same degree of precision as in the one-variable case. We first recall the statement in one variable. It says that, if $f$ is continuously differentiable on some interval $[a, b]$, then

$$f(b) - f(a) = f'(z)(b - a) \tag{6.1}$$

where $z$ is some interior point of the interval. The point $z$ is in fact difficult to determine explicitly, and the mean-value theorem is usually applied as an inequality:

$$\text{If } f'(x) \leqslant m \text{ for } \textit{all } x \in [a, b], \text{ then } f(b) - f(a) \leqslant m(b - a). \tag{6.2}$$

This inequality is of course an immediate consequence of the mean-value theorem as stated above, since $f'(z) \leqslant m$. But it is easy to give a direct proof of this inequality using the fundamental theorem of the calculus:

$$f(b) - f(a) = \int_a^b f'(s)\mathrm{d}s \leqslant \int_a^b m\mathrm{d}s \leqslant m(b - a). \tag{6.3}$$

For purposes of generalization to the higher-dimensional case, it is convenient to rewrite argument (6.3) slightly:

$$f(b)-f(a)=\int_0^1 \frac{\mathrm{d}}{\mathrm{d}t} f(a+t(b-a))\mathrm{d}t=\int_0^1 f'(a+t(b-a))(b-a)\,\mathrm{d}t$$

$$\leqslant (b-a)\int_0^1 f'(a+t(b-a))\mathrm{d}t \leqslant m(b-a). \tag{6.4}$$

(Notice that the second equality involved a use of the chain rule.) One advantage of (6.2) or (6.4) over the original mean-value theorem (6.1) is that it extends immediately to the case where $f$ is a mapping from $\mathbb{R}$ to $\mathbb{R}^k$. Suppose that $f$ is such a map, so $f$ is given as a function:

$$f(t)=\begin{pmatrix} f_1(t) \\ \vdots \\ f_k(t) \end{pmatrix}.$$

Then $f'(t)=\lim_{t\to 0}(1/h)(f(t+h)-f(t))=\mathrm{d}f_t[1]$ (where we think of 1 as a vector in $\mathbb{R}^1$ in the notation of the preceding sections). Clearly the vector $f'(t)$ is given as

$$f'(t)=\begin{pmatrix} f'_1(t) \\ \vdots \\ f'_k(t) \end{pmatrix}.$$

We can now write, as in the first equations of (6.4),

$$f(b)-f(a)=\int_0^1 \frac{\mathrm{d}}{\mathrm{d}t} f(a+t(b-a))\,\mathrm{d}t=\left(\int_0^1 f'(a+t(b-a))\mathrm{d}t\right)(b-a). \tag{6.5}$$

By the integral of a vector-valued function $g$ we simply mean the vector whose components are the integrals of the components:

$$\text{If}\quad g=\begin{pmatrix} g_1 \\ \vdots \\ g_k \end{pmatrix}\quad \text{then}\quad \int g(t)\mathrm{d}t=\begin{pmatrix} \int g_1(t)\mathrm{d}t \\ \vdots \\ \int g_k(t)\mathrm{d}t \end{pmatrix}.$$

Of course, we have the direct definition of the integral from approximating sums:

$$\int g(t)\mathrm{d}t=\lim_{n\to\infty}(1/n)\sum_{i=1}^{n} g(i/n).$$

Since each of the components of this approximating sum of vectors is an approximating sum for the integral of the corresponding component function, the two definitions of integral for a vector-valued function of one variable coincide. Since $\|\mathbf{v}_1+\mathbf{v}_2\| \leqslant \|\mathbf{v}_1\|+\|\mathbf{v}_2\|$, it follows for the approximating sum and hence, passing to the limit, for the integral that

$$\left\|\int g(t)\mathrm{d}t\right\| \leqslant \int \|g(t)\|\,\mathrm{d}t.$$

Substituting into (6.5) with $g=f'$, we get

$$\|f(b)-f(a)\| \leqslant m(b-a) \quad \text{if} \quad \|f'(t)\| \leqslant m \quad \text{for all} \quad t\in(a,b). \tag{6.6}$$

At this point we can see the trouble involved in trying to generalize (6.1) to an

$\mathbb{R}^k$-valued function. We could apply (6.1) to each component $f_j$ of $f$. For each such component we would get $f_j(b)-f_j(a)=f'_j(z_j)(b-a)$, but the $z_j$ would vary from one $j$ to another. There will, in general, be no point $z$ that can work for all the $f_j$s, and so the analogue of (6.1) need not be true. Nevertheless, (6.5) is true.

We now want to generalize (6.6) to the case where $f$ is a map from $V \to W$ and where $V$ is not necessarily one-dimensional. We have already observed that the generalization of $f'(x)$ is $\mathrm{d}f_x$. Now $\mathrm{d}f_x$ is a linear transformation, and we have to understand what we mean by $\|A\|$ when $A$ is a linear transformation from $V$ to $W$. We define

$$\|A\| = \max_{\|\mathbf{u}\|=1} \|A\mathbf{u}\|$$

or, equivalently,

$$\|A\| = \max_{\mathbf{v}\neq \mathbf{0}} \frac{\|A\mathbf{v}\|}{\|\mathbf{v}\|}.$$

Thus

$$\|A\mathbf{v}\| \leqslant \|A\|\,\|\mathbf{v}\| \quad \text{for all} \quad \mathbf{v}$$

and $\|A\|$ is the smallest number with this property, i.e., if

$$\|A\mathbf{v}\| \leqslant k\|\mathbf{v}\| \quad \text{for all} \quad \mathbf{v},$$

then

$$\|A\| \leqslant k.$$

If $A_1$ and $A_2$ are two linear transformations from $V$ to $W$, then

$$\|(A_1+A_2)\mathbf{v}\| = \|A_1\mathbf{v}+A_2\mathbf{v}\| \leqslant \|A_1\mathbf{v}\| + \|A_2\mathbf{v}\| \leqslant (\|A_1\|+\|A_2\|)\|\mathbf{v}\|$$

so

$$\|A_1+A_2\| \leqslant \|A_1\| + \|A_2\|. \tag{6.7}$$

For any points $\mathbf{a}$ and $\mathbf{b}$ in $V$ we shall let $[\mathbf{a},\mathbf{b}]$ denote the line segment joining $\mathbf{a}$ to $\mathbf{b}$, so $[\mathbf{a},\mathbf{b}]$ consists of all points of the form $\mathbf{a}+t(\mathbf{b}-\mathbf{a})$ for $0 \leqslant t \leqslant 1$. (This is a natural generalization of the one-dimensional notation.)

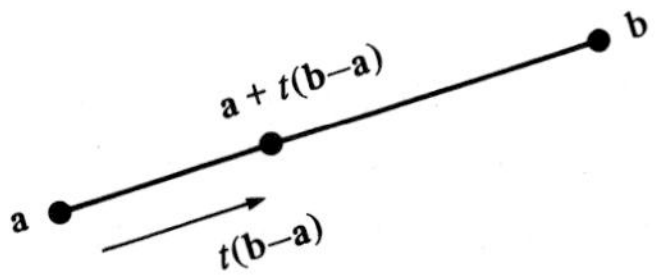


**Figure 6.1**

We wish to prove the following:

Suppose that $f: V \to W$ is differentiable at all points of $[\mathbf{a},\mathbf{b}]$ and its differential, $\mathrm{d}f_\mathbf{x}$, is a continuous function of $\mathbf{x}$ on this segment. Suppose further that

$$\|\mathrm{d}f_\mathbf{x}\| \leqslant m \quad \text{for all} \quad \mathbf{x}\in[\mathbf{a},\mathbf{b}].$$

Then

$$\|f(\mathbf{b})-f(\mathbf{a})\| \leqslant m\|\mathbf{b}-\mathbf{a}\|. \tag{6.8}$$

*Proof.* Let $h$: $[0,1]\to[\mathbf{a},\mathbf{b}]$ be the map given by

$$h(t)=\mathbf{a}+t(\mathbf{b}-\mathbf{a}).$$

Let $F=f\circ h$, so $F(t)=f(\mathbf{a}+t(\mathbf{b}-\mathbf{a}))$. Then $F$ is a differentiable map from $[0,1]\to W$ and, by the chain rule,

$$\mathrm{d}F_t=\mathrm{d}f_{h(t)}\circ\mathrm{d}h_t.$$

Now $\mathrm{d}h_t[1]=\mathbf{b}-\mathbf{a}$ so

$$F'(t)=\mathrm{d}F_t[1]=\mathrm{d}f_{h(t)}[\mathbf{b}-\mathbf{a}].$$

Also

$$\begin{aligned} f(\mathbf{b})-f(\mathbf{a})=F(1)-F(0)&=\int_0^1 F'(t)\mathrm{d}t\\ &=\int_0^1 \mathrm{d}f_{h(t)}[\mathbf{b}-\mathbf{a}]\mathrm{d}t\\ &=A(\mathbf{b}-\mathbf{a}) \end{aligned} \tag{6.9}$$

where $A$ is the linear transformation

$$A=\int_0^1 \mathrm{d}f_{h(t)}\mathrm{d}t. \tag{6.10}$$

In (6.10) we are integrating a linear-transformation-valued function, the function which assigns to each $t$ the linear transformation $\mathrm{d}f_{h(t)}$. We can treat such integrals just as we dealt with vector-valued integrals, for instance, since $V$ and $W$ have standard bases, we can identify every linear transformation with a matrix. The integral of a matrix-valued function $g$, where

$$g(t)=(g_{ij}(t))$$

is given as the matrix whose $ij$th entry is the integral of the numerical valued function $g_{ij}$. Or, as before, the integral can be given as a limit of approximating sums. It follows then from (6.7) that

$$\left\|\int g(t)\mathrm{d}t\right\| \leqslant \int \|g(t)\|\,\mathrm{d}t.$$

In particular, substituting into (6.10) and using the hypothesis that $\|\mathrm{d}f_{\mathbf{x}}\|\leqslant m$ for all $\mathbf{x}\in[\mathbf{a},\mathbf{b}]$, we conclude that

$$\|A\|\leqslant m$$

and hence, from (6.9), that (6.8) holds.

## 6.2. Higher derivatives and Taylor's formula

Let $f$: $\mathbb{R}^2\to\mathbb{R}$ be a function which is differentiable. If $\partial f/\partial x$ and $\partial f/\partial y$ are both continuously differentiable, we can form

$$\frac{\partial}{\partial x}\left(\frac{\partial f}{\partial x}\right) \quad \text{which we denote by} \quad \frac{\partial^2 f}{\partial x^2},$$

$$\frac{\partial}{\partial y}\left(\frac{\partial f}{\partial y}\right) \quad \text{which we denote by} \quad \frac{\partial^2 f}{\partial y^2},$$

and

$$\frac{\partial}{\partial x}\left(\frac{\partial f}{\partial y}\right) \quad \text{which we denote by} \quad \frac{\partial^2 f}{\partial x\, \partial y}.$$

We have already seen that

$$\frac{\partial^2 f}{\partial x\, \partial y} = \frac{\partial}{\partial y}\left(\frac{\partial f}{\partial x}\right).$$

Similarly we can define higher-order partial derivatives when they exist, and have the appropriate equality among mixed partials. For example,

$$\frac{\partial}{\partial x}\left(\frac{\partial}{\partial y}\left(\frac{\partial f}{\partial x}\right)\right) = \frac{\partial}{\partial y}\left(\frac{\partial}{\partial x}\left(\frac{\partial f}{\partial x}\right)\right) = \frac{\partial}{\partial x}\left(\frac{\partial}{\partial x}\left(\frac{\partial f}{\partial y}\right)\right)$$

etc. The significance of the second (and similarly higher) derivatives is given by Taylor's formula which we will now state and prove.

For notational simplicity, we first state and prove it at the origin. Suppose that $f$ has continuous first- and second-order derivatives near $\begin{pmatrix}0\\0\end{pmatrix}$. $\Bigg($For convenience, we will write $f(x, y)$ for $f\left(\begin{pmatrix}x\\y\end{pmatrix}\right)$.$\Bigg)$ Then

$$\begin{aligned} f(x, y) &= f(0, 0) + \int_0^1 \frac{\mathrm{d}}{\mathrm{d}t} f(tx, ty)\,\mathrm{d}t \\ &= f(0, 0) + \int_0^1 \left(\frac{\partial f}{\partial x}(tx, ty)x + \frac{\partial f}{\partial y}(tx, ty)y\right)\mathrm{d}t. \end{aligned}$$

Let

$$f_1(x, y) = \int_0^1 \frac{\partial f}{\partial x}(tx, ty)\,\mathrm{d}t \quad \text{and} \quad f_2(x, y) = \int_0^1 \frac{\partial f}{\partial y}(tx, ty)\,\mathrm{d}t.$$

Then $f_1$ and $f_2$ are differentiable functions, and

$$f(x, y) = f(0,0) + x f_1(x, y) + y f_2(x, y)$$

or, more succinctly.

$$f = f(0,0) + x f_1 + y f_2.$$

Furthermore,

$$f_1(0,0) = \frac{\partial f}{\partial x}(0,0) \quad \text{and} \quad f_2(0,0) = \frac{\partial f}{\partial y}(0,0).$$

Now apply the same argument to $f_1$ and $f_2$:

$$f_1(x, y) = f_1(0,0) + x f_{11}(x, y) + y f_{12}(x, y)$$

where

$$f_{11}(x, y) = \int_0^1 \frac{\partial f_1}{\partial x}(tx, ty)dt \quad \text{and} \quad f_{12}(x, y) = \int_0^1 \frac{\partial f_1}{\partial y}(tx, ty)dt$$

and similarly,

$$f_2 = f_2(0, 0) + xf_{21} + xf_{22}.$$

Thus

$$f = f(0, 0) + xf_1(0, 0) + yf_2(0, 0) + x^2 f_{11} + xy(f_{12} + f_{21}).$$

If $f$ has continuous derivatives up to third order, we can repeat the process once again to get

$$\begin{aligned} f = f(0, 0) &+ xf_1(0, 0) + yf_2(0, 0) + x^2 f_{11}(0, 0) + xy(f_{12}(0, 0) + f_{21}(0, 0)) \\ &+ y^2 f_{22}(0, 0) + x^3 f_{111} + x^2 y(f_{112} + f_{121} + f_{211}) \\ &+ xy^2(f_{122} + f_{212} + f_{221}) + y^3 f_{222} \end{aligned}$$

where all the functions $f_{111}, f_{112}$, etc., are continuous. If we compute the second derivatives of both sides of this equation at the origin, we conclude that

$$2f_{11}(0, 0) = \frac{\partial^2 f}{\partial x^2}(0, 0)$$

$$f_{12}(0, 0) + f_{21}(0, 0) = \frac{\partial f}{\partial x\, \partial y}(0, 0)$$

and

$$2f_{22}(0, 0) = \frac{\partial f^2}{\partial y^2}(0, 0).$$

Thus we have proved

$$\begin{aligned} f(x, y) = f(0, 0) &+ x\frac{\partial f}{\partial x}(0, 0) + y\frac{\partial f}{\partial y}(0, 0) + \tfrac{1}{2}x^2 \frac{\partial^2 f}{\partial x^2}(0, 0) \\ &+ xy\frac{\partial^2 f}{\partial x\, \partial y}(0, 0) + \tfrac{1}{2}y^2 \frac{\partial^2 f}{\partial y^2}(0, 0) + O\left\| \begin{pmatrix} x \\ y \end{pmatrix} \right\|^3. \end{aligned} \tag{6.11}$$

It is clear that, if $f$ has still higher-order continuous derivatives, we can keep on going. It is also clear that the same argument works in $\mathbb{R}^k$ as well as in $\mathbb{R}^2$. Finally, we may replace the origin $\begin{pmatrix} 0 \\ 0 \end{pmatrix}$ by any vector $\mathbf{u}$ and $\begin{pmatrix} x \\ y \end{pmatrix}$ by $\mathbf{u} + \mathbf{v}$:

*Let $f: \mathbb{R}^k \to \mathbb{R}$ have continuous derivatives up to order $n + 1$. Then there is a polynomial $P_n$ in the coordinates of $\mathbf{v}$ such that*

$$f(\mathbf{u} + \mathbf{v}) = P_n(\mathbf{v}) + O(\|\mathbf{v}\|^{n+1}).$$

The coefficients of $P_n(\mathbf{v})$ can be determined by successive differentiations and evaluation at $\mathbf{v} = \mathbf{0}$.

If $f: \mathbb{R}^2 \to \mathbb{R}^1$, the matrix of second partial derivatives

$$H_{\mathbf{p}}(f) = \begin{pmatrix} \dfrac{\partial^2 f}{\partial x^2}(\mathbf{p}) & \dfrac{\partial^2 f}{\partial x\, \partial y}(\mathbf{p}) \\ \dfrac{\partial^2 f}{\partial x\, \partial y}(\mathbf{p}) & \dfrac{\partial^2 f}{\partial y^2}(\mathbf{p}) \end{pmatrix}$$

is called the *Hessian matrix* and the corresponding quadratic form is denoted by

$$d^2 f_{\mathbf{p}}.$$

We use simply $H$ when $f$ and $\mathbf{p}$ are understood. Thus we can rewrite (6.11) as

$$f(\mathbf{p}+\mathbf{v}) = f(\mathbf{p}) + df_{\mathbf{p}}(\mathbf{v}) + \tfrac{1}{2}d^2 f_{\mathbf{p}}(\mathbf{v}) + o(\|\mathbf{v}\|^2) \tag{6.11}$$

where

$$d^2 f_{\mathbf{p}}(\mathbf{v}) = \mathbf{v}^{\mathrm{T}} H \mathbf{v} = (v_1, v_2) H \begin{pmatrix} v_1 \\ v_2 \end{pmatrix}. \tag{6.12}$$

The Hessian $d^2 f_{\mathbf{p}}$, as a quadratic form, is subject to the analysis we presented in Chapter 4. For example, if $f(P) = [x(P)]^2 y(P)$, then, at the point where $x = 2$, $y = 3$, we have

$$\frac{\partial f}{\partial x} = 2xy = 12, \quad \frac{\partial f}{\partial y} = x^2 = 4, \quad \frac{\partial^2 f}{\partial x^2} = 2y = 6, \quad \frac{\partial^2 f}{\partial x\,\partial y} = 2x = 4, \quad \frac{\partial^2 f}{\partial y^2} = 0,$$

so that at $\mathbf{p} = \begin{pmatrix} 2 \\ 3 \end{pmatrix}$ $df$ is represented by $(12, 4)$ and $H$ by $\begin{pmatrix} 6 & 4 \\ 4 & 0 \end{pmatrix}$.

## Maxima and minima

The Hessian is especially useful in analyzing the behavior of a function near a critical point where its differential $df$ is zero. If $\mathbf{P}_0 = \begin{pmatrix} x_0 \\ y_0 \end{pmatrix}$ is such a point, then

$$f(\mathbf{P}_0 + \mathbf{v}) = f(\mathbf{P}_0) + \tfrac{1}{2}d^2 f_{\mathbf{p}}(\mathbf{v}).$$

If the quadratic form $d^2 f_{\mathbf{p}}$ is *positive definite* ($H$ has two positive eigenvalues), then it follows from Taylor's formula that $f(\mathbf{P}_0 + \mathbf{v}) > f(\mathbf{P}_0)$ for small $\mathbf{v}$ and $f$ achieves a *minimum* at $\mathbf{P}_0$. If $d^2 f$ is *negative definite* ($H$ has two negative eigenvalues), then $f(\mathbf{P}_0 + \mathbf{v}) < f(\mathbf{P}_0)$ and $f$ achieves a *maximum* at $\mathbf{P}_0$. Finally, if $H$ has one positive and one negative eigenvalue, then $d^2 f(\mathbf{v})$ achieves both positive and negative values for small $\mathbf{v}$, so that $f$ achieves neither a maximum nor a minimum at $\mathbf{P}_0$; what it has there is a *saddle point*. If $H$ has one or more zero eigenvalues, and is therefore singular, we have to inspect higher derivatives to determine whether $f$ has a maximum or a minimum at $\mathbf{P}_0$.

As an example of using the Hessian, we find and classify the critical points of the function

$$f = 3x^2 + 2y^3 - 6xy.$$

To locate the critical points, we set the partial derivatives with respect to $x$ and $y$ equal to zero:

$$\frac{\partial f}{\partial x} = f'_1(x, y) = 6x - 6y = 0 \qquad \text{so } x = y$$

$$\frac{\partial f}{\partial y} = f'_2(x, y) = 6y^2 - 6x = 0 \qquad \text{so } x = y^2$$

The critical points are therefore at $x = 0$, $y = 0$ and $x = 1$, $y = 1$.

Next we calculate the second partial derivatives in order to form the Hessian:

$$\frac{\partial^2 f}{\partial x^2} = 6, \quad \frac{\partial^2 f}{\partial x\,\partial y} = -6, \quad \frac{\partial^2 f}{\partial y^2} = 12y.$$

At $x = 0$, $y = 0$, the Hessian is therefore

$$H = \begin{pmatrix} 6 & -6 \\ -6 & 0 \end{pmatrix}.$$

This has a negative determinant, hence its eigenvalues are of opposite sign and the critical point at the origin is a saddle point. To confirm this conclusion, we note that $f(x, y)$ is positive for points near the origin along the $x$-axis, while along the line $x = y$ the function is negative near the origin.

At $x = 1$, $y = 1$ the Hessian is

$$H = \begin{pmatrix} 6 & -6 \\ -6 & 12 \end{pmatrix}.$$

This has positive determinant, so its eigenvalues are of the same sign; since the trace is positive, both eigenvalues are positive. Hence $F(x, y)$ has a relative minimum at $x = 1$, $y = 1$.

On an affine plane, the only property of the second differential $\mathrm{d}^2 f$ which is independent of choice of coordinates is the number of positive, negative, and zero eigenvalues of the Hessian. On a Euclidean plane, we can inquire about another coordinate-independent property of a function $f$: namely, how its average value on a small circle surrounding a point $\mathbf{P}_0$ compares with its value at $\mathbf{P}_0$. We write, from (6.11),

$$f(\mathbf{P}_0 + \mathbf{v}) = f(P_0) + \mathrm{d}f[\mathbf{v}] + \tfrac{1}{2}\mathrm{d}^2 f(\mathbf{v}) + \text{error}.$$

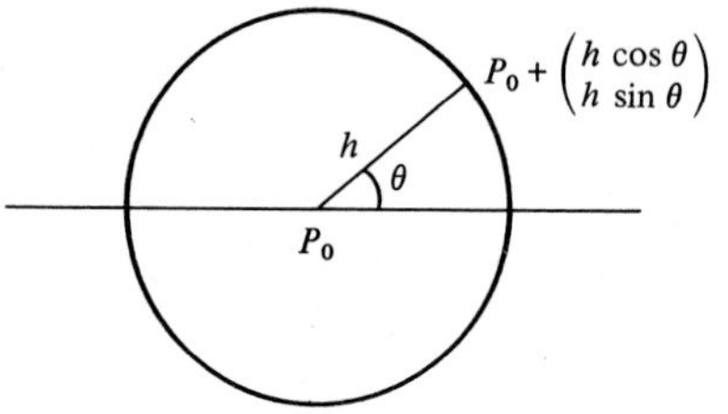


**Figure 6.2**

Since $\mathrm{d}f[-\mathbf{v}] = -\mathrm{d}f[\mathbf{v}]$, the average value of $\mathrm{d}f[\mathbf{v}]$ for any circle centered at $\mathbf{P}_0$ is clearly zero. To find the average value of $\tfrac{1}{2}\mathrm{d}^2 f(\mathbf{v})$, we set $\mathbf{v} = \begin{pmatrix} h\cos\theta \\ h\sin\theta \end{pmatrix}$, so that using (6.12)

$$\tfrac{1}{2}\mathrm{d}^2 f(\mathbf{v}) = \tfrac{1}{2}(h\cos\theta, h\sin\theta)\begin{pmatrix} \partial^2 f/\partial x^2 & \partial^2 f/\partial x\,\partial y \\ \partial^2 f/\partial x\,\partial y & \partial^2 f/\partial y^2 \end{pmatrix}\begin{pmatrix} h\cos\theta \\ h\sin\theta \end{pmatrix}$$

or

$$\tfrac{1}{2}\mathrm{d}^2 f(\mathbf{v}) = \tfrac{1}{2}h^2\left[\frac{\partial^2 f}{\partial x^2}\cos^2\theta + 2\frac{\partial^2 f}{\partial x\,\partial y}\cos\theta\sin\theta + \frac{\partial^2 f}{\partial y^2}\sin^2\theta\right].$$

Since the average value of $\cos^2\theta$ or $\sin^2\theta$ on $[0, 2\pi]$ is $\frac{1}{2}$, while the average value of $\sin\theta\cos\theta$ is zero, we see that

$$\langle \tfrac{1}{2}\mathrm{d}^2 f(\mathbf{v})\rangle_{\text{average}} = \tfrac{1}{4}h^2\left[\frac{\partial^2 f}{\partial x^2} + \frac{\partial^2 f}{\partial y^2}\right]$$

and that

$$\langle f(\mathbf{P}_0 + \mathbf{v})\rangle_{\text{average}} = f(\mathbf{P}_0) + \tfrac{1}{4}h^2\left(\frac{\partial^2 f}{\partial x^2} + \frac{\partial^2 f}{\partial y^2}\right) + \text{error}.$$

The quantity $\partial^2 f/\partial x^2 + \partial^2 f/\partial y^2$, which determines whether $f$ increases or decreases 'on the average' as we move away from $\mathbf{P}_0$, is called the *Laplacian* of $f$. By virtue of its definition in terms of an average over a circle, for any coordinates obtained from $x$ and $y$ by a rotation, the Laplacian will have the same value. For this reason the equation

$$\frac{\partial^2 f}{\partial x^2} + \frac{\partial^2 f}{\partial y^2} = 0,$$

called *Laplace's equation*, describes a property of a function on a Euclidean plane which does not depend on the choice of Euclidean coordinates. Not surprisingly, this equation arises frequently in conjunction with functions on a plane which have a physical significance: electric potential, for example, or temperature.

Before leaving the subject of maxima and minima, we shall consider the *constrained* extremum problem on the plane: where, along the curve defined by $g(\mathbf{P}) =$ constant, does the function $f(\mathbf{P})$ achieve a maximum or minimum? The condition

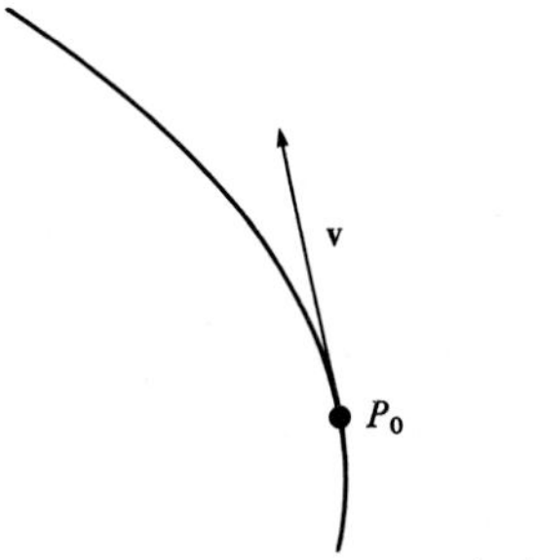


**Figure 6.3**

for such an extremum to occur at $\mathbf{P}_0$ is that $\mathrm{d}f_{\mathbf{P}_0}(\mathbf{v}) = 0$ for any vector $\mathbf{v}$ which lies tangent to the curve $g(\mathbf{P}) =$ constant. But such a vector satisfies $\mathrm{d}g(\mathbf{v}) = 0$. It follows that, at the point $\mathbf{P}_0$ where the maximum or minimum is achieved, $\mathrm{d}f_{\mathbf{P}_0}$ must be a *multiple* of $\mathrm{d}g_{\mathbf{P}_0}$: say $\mathrm{d}f = \lambda\mathrm{d}g$. Thus we are led to the *Lagrange multiplier* method for the constrained extremum problem: to maximize or minimize $f(\mathbf{P})$ along the curve $g(\mathbf{P}) = \text{constant}$, consider the function $f - \lambda g$, where $\lambda$ is an undetermined scalar called the *Lagrange multiplier*, and set $\mathrm{d}(f - \lambda g) = 0$. The resulting two equations, along with $g = \text{constant}$, determine the unknown quantities $x, y$, and $\lambda$. To determine whether the extremum thus found is a maximum, a minimum, or neither, we first find a vector $\mathbf{v}$ for which $\mathrm{d}g(\mathbf{v}) = 0$. This vector is tangent to the level curves $f = \text{constant}$ and $g = \text{constant}$ at the critical point $\mathbf{P}$. We then consider the function $h = f - \lambda g$, which, by construction, has a critical point at $\mathbf{P}$ as a function on the plane. We calculate the best quadratic approximation to $h$ near this critical point and evaluate it on our vector $\mathbf{v}$ for which $\mathrm{d}f(\mathbf{v}) = \mathrm{d}g(\mathbf{v}) = 0$. If this quantity, $\frac{1}{2}\mathrm{d}^2h(\mathbf{v})$, is positive, we claim that $h(\mathbf{P}) > h(\mathbf{P}_0)$ at all points near $\mathbf{P}_0$ on the curve. Indeed, suppose we parameterize the curve $g = 0$ by $p = p(t)$. That is, we choose a function $p: \mathbb{R} \to \mathbb{R}^2$ such that

$$g(p(t)) \equiv 0, \quad p(0) = \mathbf{p}_0 \quad \text{and} \quad p'(0) = \mathbf{v}.$$

(That this is always possible will be proved in the next section – it is a consequence of the *implicit function theorem* to be proved there.) Then

$$h \circ p = f \circ p \quad \text{since} \quad g \circ p \equiv 0.$$

Also

$$(h \circ p)'(0) = \mathrm{d}h_{\mathbf{P}_0}(\mathbf{v}) = 0$$

and

$$(f \circ p)'' = (h \circ p)''(0) = \mathrm{d}^2h_{\mathbf{P}_0}(\mathbf{v}).$$

Thus $(h \circ p)''(0) > 0$ and hence $f$ has a minimum at $\mathbf{P}_0$ along $g = 0$. Similarly, $f$ has a maximum along the curve $g = 0$ if $\mathrm{d}^2h(\mathbf{v}) < 0$.

For example, suppose we wish to maximize the quadratic form

$$Q(x, y) = 8x^2 - 12xy + 17y^2$$

on the circle $G(x, y) = x^2 + y^2 = 1$. Setting the differential of $Q - \lambda G$ equal to zero, we find

$$16x - 12y - 2\lambda x = 0$$

and

$$-12x + 34y - 2\lambda y = 0.$$

On eliminating $\lambda$ between these equations, we find

$$16 - 12(y/x) = -12(x/y) + 34$$

or

$$(x/y) - (y/x) = 3/2.$$

Thus

$$x^2 - \tfrac{3}{2}xy - y^2 = 0$$

so

$$(x - 2y)(x + \tfrac{1}{2}y) = 0.$$

The extreme values of $Q$ therefore occur where the lines $x = 2y$ and $x = -\frac{1}{2}y$

intersect the circle $x^2 + y^2 = 1$, at

$$\mathbf{P}_1 = \pm\begin{pmatrix} 2/\sqrt{5} \\ 1/\sqrt{5} \end{pmatrix} \quad \text{and} \quad \mathbf{P}_2 = \pm\begin{pmatrix} 1/\sqrt{5} \\ -2/\sqrt{5} \end{pmatrix}.$$

We now investigate the nature of these critical points. The differentials of $Q$ and $G$ are

$$\mathrm{d}Q = (16x - 12y)\,\mathrm{d}x + (-12x + 34y)\,\mathrm{d}y,$$
$$\mathrm{d}G = 2x\,\mathrm{d}x + 2y\,\mathrm{d}y.$$

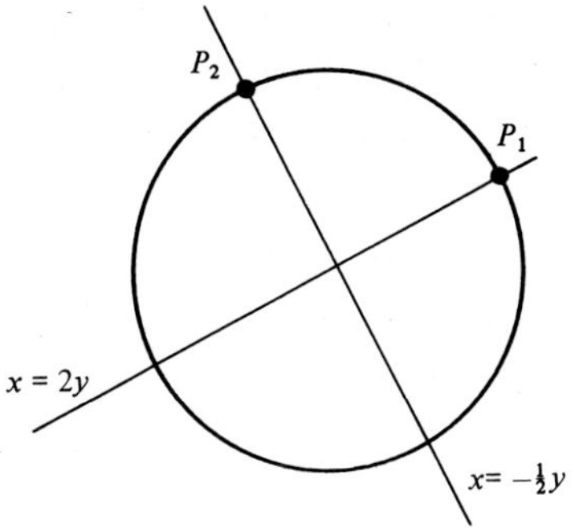


**Figure 6.4**

At $\mathbf{P}_1$, where $x = 2y$, we have

$$\mathrm{d}Q = 10x\,\mathrm{d}x + 5x\,\mathrm{d}y, \quad \mathrm{d}G = 2x\,\mathrm{d}x + x\,\mathrm{d}y.$$

As we expected, $\mathrm{d}Q$ is a multiple of $\mathrm{d}G$, with $\lambda = 5$. A vector $\mathbf{v}$ for which $\mathrm{d}Q(\mathbf{v}) = \mathrm{d}G(\mathbf{v}) = 0$ is $\begin{pmatrix} 1 \\ -2 \end{pmatrix}$. The Hessian of $Q - \lambda G$ is just the constant matrix which represents the quadratic form $Q - 5G$:

$$H = \begin{pmatrix} 3 & -6 \\ -6 & 12 \end{pmatrix}.$$

We calculate $(1, -2)\begin{pmatrix} 3 & -6 \\ -6 & 12 \end{pmatrix}\begin{pmatrix} 1 \\ -2 \end{pmatrix} = (1, -2)\begin{pmatrix} 15 \\ -30 \end{pmatrix} = 75$ and conclude that $Q$ has a *minimum* on the circle at $\mathbf{P}_1$.

At the other critical point, where $y = -2x$, we find

$$\mathrm{d}Q = 40x\,\mathrm{d}x - 80x\,\mathrm{d}y; \quad \mathrm{d}G = 2x\,\mathrm{d}x - 4x\,\mathrm{d}y, \quad \text{so} \quad \lambda = 20.$$

A vector for which $\mathrm{d}Q(\mathbf{v}) = \mathrm{d}G(\mathbf{v}) = 0$ is $\begin{pmatrix} 2 \\ 1 \end{pmatrix}$, and, on evaluating the Hessian of $Q - 20G$ for this vector, we find that

$$(2, 1)\begin{pmatrix} -12 & -6 \\ -6 & -3 \end{pmatrix}\begin{pmatrix} 2 \\ 1 \end{pmatrix} = (2, 1)\begin{pmatrix} -30 \\ -15 \end{pmatrix} = -75$$

so that $Q$ has a local maximum on the circle $x^2 + y^2 = 1$ at the point $\begin{bmatrix} 1/\sqrt{5} \\ -2/\sqrt{5} \end{bmatrix}$.

## 6.3. The inverse function theorem

Let $U$ and $V$ be vector spaces of the same dimension, and let $f: U \to V$ be a differentiable map with $f(\mathbf{p}_0) = \mathbf{q}_0$. We would like to know when there exists an inverse map $g: V \to U$ such that $g \circ f = \text{id}$. Before we formulate the appropriate theorem, we first examine some necessary limitations on the problem.

If we expect that $g$ is also to be differentiable, then the chain rule says that

$$\mathrm{d}g_{f(\mathbf{p})} \circ \mathrm{d}f_{\mathbf{p}} = \text{id}.$$

Thus the *linear* map $\mathrm{d}f_{\mathbf{p}}$ had better be invertible. If it is, then we expect the formula

$$\mathrm{d}g_{f(\mathbf{p})} = [\mathrm{d}f_{\mathbf{p}}]^{-1}$$

to hold.

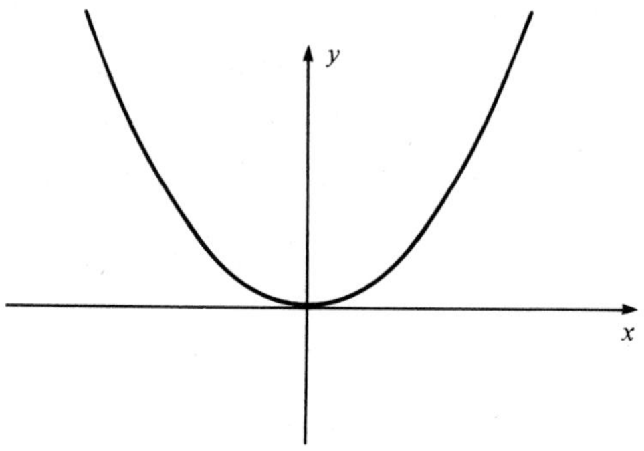


**Figure 6.5**

A familiar case where $\mathrm{d}f$ is not invertible is where $U = V = \mathbb{R}^1$ and $f(x) = x^2$. At 0, $\mathrm{d}f_0 = 0$ and there is trouble with $g(y) = \sqrt{y}$ near $y = 0$. In fact, there are three kinds of trouble. First of all, $\sqrt{y}$ is not defined (over the reals) for $y < 0$. More precisely, no point $y < 0$ is in the image of $f$. Secondly, the square root for $y$ is not uniquely specified: for a given $y > 0$ there are two values of $x$ with $x^2 = y$. Thirdly, the derivative of $\sqrt{y}$ blows up as $y \to 0$. To get around the second of these difficulties, we can proceed as follows. Suppose we choose some $x_0 \neq 0$ with $x_0^2 = y_0$. For

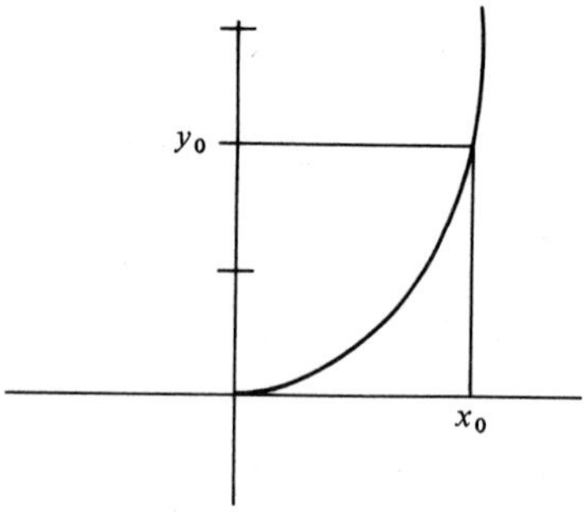


**Figure 6.6**

the sake of argument, suppose $x_0 > 0$. Then in a sufficiently small neighborhood about $y_0$ (small enough so as not to include $y = 0$), there is a unique inverse function, the square root, specified by the requirement that the values be close enough to $x_0$. (In this case 'close enough' means not to be negative – once we specify that the square root be positive, it is uniquely determined.)

We can not only assert the existence of the square root, we can give an algorithm for computing as close an approximation to the square root as we like. We recall one of these algorithms – Newton's method – but formulate it more generally.

Suppose we are given a map $f: U \to V$ with $f(\mathbf{p}_0) = \mathbf{q}_0$. We are given some $\mathbf{q}$ near $\mathbf{q}_0$ and wish to find a $\mathbf{p}$ near $\mathbf{p}_0$ such that $f(\mathbf{p}) = \mathbf{q}$. Finding $\mathbf{p}$ is the same as finding $\mathbf{p} - \mathbf{p}_0$. We wish to have

$$f(\mathbf{p}_0 + \mathbf{p} - \mathbf{p}_0) = \mathbf{q}.$$

But

$$\begin{aligned} f(\mathbf{p}_0 + \mathbf{p} - \mathbf{p}_0) &= f(\mathbf{p}_0) + \mathrm{d}f_{\mathbf{p}_0}(\mathbf{p} - \mathbf{p}_0) + o(\mathbf{p} - \mathbf{p}_0) \\ &= \mathbf{q}_0 + \mathrm{d}f_{\mathbf{p}_0}(\mathbf{p} - \mathbf{p}_0) + o(\mathbf{p} - \mathbf{p}_0). \end{aligned}$$

If we could ignore the term $o(\mathbf{p} - \mathbf{p}_0)$, we would obtain the approximate equation

$$\mathbf{q} - \mathbf{q}_0 \doteq \mathrm{d}f_{\mathbf{p}_0}(\mathbf{p} - \mathbf{p}_0)$$

or, since $\mathrm{d}f_{\mathbf{p}_0}$ has an inverse,

$$\mathbf{p} \doteq \mathbf{p}_0 + \mathrm{d}f_{\mathbf{p}_0}^{-1}(\mathbf{q} - \mathbf{q}_0).$$

This suggests defining

$$\mathbf{p}_1 = \mathbf{p}_0 + \mathrm{d}f_{\mathbf{p}_0}^{-1}(\mathbf{q} - \mathbf{q}_0)$$

as an approximate solution, then defining

and starting anew. Thus

$$\left.\begin{aligned} \mathbf{q}_1 &= f(\mathbf{p}_1) \\ \\ \mathbf{p}_2 &= \mathbf{p}_1 + \mathrm{d}f_{\mathbf{p}_1}^{-1}(\mathbf{q} - \mathbf{q}_1), \end{aligned}\right\}$$ **Newton's method**

etc.

Suppose $U = V = \mathbb{R}^1$ and $f(p) = p^2$. Then $\mathrm{d}f_p$ is multiplication by $2p$ and hence $\mathrm{d}f_p^{-1}(w) = w/2p$. Thus, in this case,

$$p_1 = p_0 + \frac{1}{2p_0}(q - q_0).$$

For example, suppose we take

$$p_0 = 3 \quad \text{so} \quad q_0 = 9$$

and take $q = 10$. Then

$$p_1 = 3 + \tfrac{1}{6}(10 - 9) = 3.166\ldots.$$

Then

$$q_1 = p_1^2 = 10.027\,77\ldots.$$

(Notice that $p_1$ is already a much better approximation to $\sqrt{10}$.) The next approximation is given by

$$p_2 = 3.1\bar{6} + \frac{1}{2 \times 3.1\bar{6}}(10 - 10.02\bar{7})$$
$$= 3.162\,281\,6\ldots.$$

Then

$$q_2 = p_2^2 = 10.000\,024$$

so $p_2$ is correct to four decimal places.

Let us give a second example, with $U = V = \mathbb{R}^2$. Suppose that the map $f$ is given by

$$f\left(\begin{pmatrix} x \\ y \end{pmatrix}\right) = \begin{pmatrix} x^3 - y^3 \\ 2xy \end{pmatrix}.$$

Then $\mathrm{d}f_{\binom{x}{y}}$ is the linear transformation whose matrix is

$$\begin{pmatrix} 3x^2 & -3y^2 \\ 2y & 2x \end{pmatrix}.$$

Suppose

$$\mathbf{p}_0 = \begin{pmatrix} 2 \\ 1 \end{pmatrix}$$

so that

$$\mathbf{q}_0 = \begin{pmatrix} 8 - 1 \\ 2 \cdot 2 \cdot 1 \end{pmatrix} = \begin{pmatrix} 7 \\ 4 \end{pmatrix},$$

and

$$\mathrm{d}f_{\mathbf{p}_0} = \begin{pmatrix} 12 & -3 \\ 2 & 4 \end{pmatrix}$$

and

$$(\mathrm{d}f_{\mathbf{p}_0})^{-1} = \frac{1}{54}\begin{pmatrix} 4 & 3 \\ -2 & 12 \end{pmatrix}.$$

Suppose we take

$$\mathbf{q} = \begin{pmatrix} 7.5 \\ 3.8 \end{pmatrix}.$$

Then

$$\mathbf{p}_1 = \mathbf{p}_0 + (\mathrm{d}f_p)^{-1}(\mathbf{q} - \mathbf{q}_0)$$
$$= \begin{pmatrix} 2 \\ 1 \end{pmatrix} + \frac{1}{54}\begin{pmatrix} 4 & 3 \\ -2 & 12 \end{pmatrix}\begin{pmatrix} 0.5 \\ -0.2 \end{pmatrix}$$
$$= \begin{pmatrix} 2.026 \\ 0.937 \end{pmatrix}.$$

we get

$$\mathbf{q}_1 = \begin{pmatrix} 7.493 \\ 3.796 \end{pmatrix}$$

which is already quite close. Notice that at each successive stage in this algorithm we have to compute a different value of $\mathrm{d}f_{\mathbf{p}_1}^{-1}$.

A mathematical theorem will be formulated which asserts that, under suitable hypotheses about $f$, Newton's method will give a sequence of points $\mathbf{p}_i$ which converge to a solution $\mathbf{p}$ of $f(\mathbf{p})=\mathbf{q}$, provided that $\mathbf{q}$ is sufficiently close to $\mathbf{q}_0$.

Another algorithm which converges much more slowly than Newton's method is to set

$$L=(\mathrm{d}f_{\mathbf{p}_0})^{-1}$$

and

$$\begin{aligned}\mathbf{p}_1 &= \mathbf{p}_0 + L(\mathbf{q}-\mathbf{q}_0),\\ \mathbf{q}_1 &= f(\mathbf{p}_1),\\ \mathbf{p}_2 &= \mathbf{p}_1 + L(\mathbf{q}-\mathbf{q}_1),\\ \mathbf{q}_2 &= f(\mathbf{p}_2), \quad \text{etc.}\end{aligned}$$

This is known as *Picard's* method. For example, with $f:\mathbb{R}\to\mathbb{R}$, $f(p)=p^2$ and $p_0=3$, we get

$$\begin{aligned}p_1 &= 3.1\bar{6}\\ q_1 &= 10.02\bar{7},\end{aligned}$$

as before, but

$$\begin{aligned}p_2 &= 3.1\bar{6}+\tfrac{1}{6}(-0.02\bar{7}) = 3.162\,036,\\ q_2 &= 9.998\,481\,7, \quad \text{etc.}\end{aligned}$$

An advantage of Picard's method is that we only need to compute $L$ once. It is easier to formulate and prove the slow convergence of Picard's or Newton's method with fewer assumptions about $f$ than it is to prove the fast convergence of Newton's method, which requires more hypotheses about $f$, as we shall see.

## Proofs of convergence

We now formulate the hypotheses we need about $f$ and prove the convergence of both methods. Recall that, if $f$ is differentiable at $\mathbf{p}$, then

$$f(\mathbf{p}^1)=f(\mathbf{p})+\mathrm{d}f_{\mathbf{p}}(\mathbf{p}^1-\mathbf{p})+o(\mathbf{p}^1-\mathbf{p})$$

which means that given any $\varepsilon$ we can find a $\delta$ such that

$$\| f(\mathbf{p}^1)-f(\mathbf{p})-\mathrm{d}f_{\mathbf{p}}(\mathbf{p}^1-\mathbf{p})\| \leqslant \varepsilon \| \mathbf{p}^1-\mathbf{p}\| \tag{6.13}$$

whenever $\|\mathbf{p}^1-\mathbf{p}\|\leqslant\delta$. The $\delta$ that is required for this inequality may depend on the point $\mathbf{p}$. Let us assume that $f$ is *uniformly differentiable* in the sense that for any $\varepsilon$ we can find a $\delta$ such that (6.13) holds for all points $\mathbf{p}$ and $\mathbf{p}^1$ in some ball centered at $\mathbf{p}_0$. So we assume that there is some $a>0$ such that given any $\varepsilon$ we can find a $\delta$ such that (6.13) holds if

$$\|\mathbf{p}^1-\mathbf{p}\|\leqslant\delta,\quad \|\mathbf{p}-\mathbf{p}_0\|\leqslant a,\quad \|\mathbf{p}^1-\mathbf{p}_0\|\leqslant a.$$

Let us also assume that $\mathrm{d}f_{\mathbf{p}}^{-1}$ is uniformly bounded, i.e., that there is some constant

$K$ such that

$$\|(\mathrm{d}f_{\mathbf{p}})^{-1}\| < K \quad \text{for all} \quad \|\mathbf{p}-\mathbf{p}_0\| \leqslant a. \tag{6.14}$$

We also assume that $\mathrm{d}f_{\mathbf{p}}$ itself is uniformly bounded, i.e., that

$$\|\mathrm{d}f_{\mathbf{p}}\| < M \quad \text{for all} \quad \|\mathbf{p}-\mathbf{p}_0\| \leqslant a.$$

Finally, for the Picard method, let us assume that $\mathrm{d}f_{\mathbf{p}}$ is continuous in the sense that for any $\varepsilon > 0$ there is a $\delta$ such that

$$\|\mathrm{d}f_{\mathbf{p}} - \mathrm{d}f_{\mathbf{p}^1}\| \leqslant \tfrac{1}{2}\varepsilon \quad \text{if} \quad \|\mathbf{p}-\mathbf{p}^1\| \leqslant \delta, \|\mathbf{p}-\mathbf{p}_0\| \leqslant a, \|\mathbf{p}^1-\mathbf{p}_0\| \leqslant a \tag{6.15}$$

and we only need assume that $(\mathrm{d}f_{\mathbf{p}_0})^{-1}$ exists.

**Convergence of Newton's method**

Now let us look at Newton's method. The step going from $\mathbf{p}_i$ to $\mathbf{p}_{i+1}$ is given by

$$\boxed{\mathbf{p}_{i+1} = \mathbf{p}_i + (\mathrm{d}f_{\mathbf{p}_i})^{-1}(\mathbf{q}-f(\mathbf{p}_i)).}$$

But

$$f(\mathbf{p}_i) = f(\mathbf{p}_{i-1} + \mathbf{p}_i - \mathbf{p}_{i-1}) = f(\mathbf{p}_{i-1}) + \mathrm{d}f_{\mathbf{p}_{i-1}}(\mathbf{p}_i - \mathbf{p}_{i-1}) + o(\mathbf{p}_i - \mathbf{p}_{i-1})$$

and

$$\mathbf{q} - f(\mathbf{p}_{i-1}) + \mathrm{d}f_{\mathbf{p}_{i-1}}(\mathbf{p}_i - \mathbf{p}_{i-1}) = \mathbf{0}.$$

Also, *if* $\|\mathbf{p}_i - \mathbf{p}_{i-1}\| < \delta$ then $\|o(\mathbf{p}_i - \mathbf{p}_{i-1})\| < \varepsilon \|\mathbf{p}_i - \mathbf{p}_{i-1}\|$ by (6.13). So

$$\|\mathbf{p}_{i+1} - \mathbf{p}_i\| \leqslant K\varepsilon \|\mathbf{p}_i - \mathbf{p}_{i-1}\|. \tag{6.16}$$

We may choose $\varepsilon$ small enough so that $K\varepsilon < \frac{1}{2}$ and also $K\varepsilon < \frac{1}{2}a$, provided $\delta$ is sufficiently small. Now

$$\mathbf{p}_1 = \mathbf{p}_0 + \mathrm{d}f_{\mathbf{p}_0}^{-1}(\mathbf{q} - \mathbf{q}_0), \ \mathbf{q}_0 = f(\mathbf{p}_0)$$

so if

$$\|\mathbf{q} - \mathbf{q}_0\| < \delta/K$$

then

$$\|\mathbf{p}_1 - \mathbf{p}_0\| \leqslant \delta.$$

If $2\delta < a$, so in particular $\delta < \frac{1}{2}a$, the point $\mathbf{p}_1$ will satisfy

$$\|\mathbf{p}_1 - \mathbf{p}_0\| < a$$

so that, in particular, $\mathbf{p}_1$ is in the domain of definition of $f$, and we can use the algorithm to define $\mathbf{p}_2$. It follows from (6.16) that

$$\|\mathbf{p}_2 - \mathbf{p}_1\| \leqslant \tfrac{1}{2}\delta \leqslant \tfrac{1}{2}(\tfrac{1}{2}a)$$

so

$$\|\mathbf{p}_2 - \mathbf{p}_0\| \leqslant \|\mathbf{p}_2 - \mathbf{p}_1\| + \|\mathbf{p}_1 - \mathbf{p}_0\| \leqslant (\tfrac{1}{2} + 1)\tfrac{1}{2}a \leqslant a.$$

Thus $\mathbf{p}_2$ is again in the ball of radius $a$ so we can apply the algorithm and (6.16) to get

$$\|\mathbf{p}_3 - \mathbf{p}_2\| \leqslant \tfrac{1}{2}\|\mathbf{p}_2 - \mathbf{p}_1\| \leqslant \tfrac{1}{4}\delta$$

and hence

$$\|\mathbf{p}_3 - \mathbf{p}_0\| \leqslant (\tfrac{1}{4} + \tfrac{1}{2} + 1)\tfrac{1}{2}a \leqslant a$$

etc. We can always continue to the next step since (by induction)

$$\|\mathbf{p}_i - \mathbf{p}_{i-1}\| \leqslant \delta/2^i$$

and

$$\|\mathbf{p}_i - \mathbf{p}_0\| \leqslant \left(\frac{1}{2^i} + \frac{1}{2^{i-1}} + \cdots + 1\right)\tfrac{1}{2}a \leqslant 2\cdot\tfrac{1}{2}a \leqslant a.$$

The sequence of points $\mathbf{p}_i$ converges to some point $\mathbf{p}$ since

$$\|\mathbf{p}_{i+k} - \mathbf{p}_i\| \leqslant \tfrac{1}{2}i(\tfrac{1}{2}N + \cdots + 1)\delta \leqslant \frac{1}{2^{i-1}}\delta \to 0 \quad \text{as} \quad i \to \infty.$$

Finally,

$$\|\mathbf{q} - f(\mathbf{p}_i)\| = \|\mathrm{d}f_{\mathbf{p}_i}(\mathbf{p}_{i+1} - \mathbf{p}_i)\| \leqslant M\|\mathbf{p}_{i+1} - \mathbf{p}_i\| \to 0$$

so, by the continuity of $f$, we see that

$$f(\mathbf{p}) = \mathbf{q}.$$

**Uniqueness of solution**

We now look at uniqueness. Notice that $K$ is determined by $f$. We are free to choose a smaller value of $a$, if we wish, without changing $K$. This is at the expense of choosing $\delta$ and hence $\delta/K$ smaller. In particular, we may assume that $a$ has been chosen so small to start with that (6.13) holds for any pair of points $\mathbf{p}$ and $\mathbf{p}^1$ where $\varepsilon K < 1$. Now for any pair of points

$$\|\mathbf{p} - \mathbf{p}^1\| = \|(\mathrm{d}f_{\mathbf{p}})^{-1}(\mathrm{d}f_{\mathbf{p}}(\mathbf{p} - \mathbf{p}^1))\| \leqslant K\|\mathrm{d}f_{\mathbf{p}}(\mathbf{p} - \mathbf{p}^1)\|.$$

If $f(\mathbf{p}) = f(\mathbf{p}^1)$, then (6.13) implies that $\|\mathrm{d}f_{\mathbf{p}}(\mathbf{p}^1 - \mathbf{p})\| \leqslant \varepsilon\|\mathbf{p}^1 - \mathbf{p}\|$ and combining these two inequalities we get

$$\|\mathbf{p} - \mathbf{p}^1\| \leqslant \varepsilon K\|\mathbf{p} - \mathbf{p}^1\|, \quad \varepsilon K < 1$$

which can only happen if $\|\mathbf{p} - \mathbf{p}^1\| = 0$, i.e., $\mathbf{p} = \mathbf{p}^1$. Thus there can be at most one solution of $f(\mathbf{p}) = \mathbf{q}$ with $\|\mathbf{p} - \mathbf{p}_0\| \leqslant a$.

**Convergence of Picard's method**

Now let us look at Picard's method. Let $L = (\mathrm{d}f_{\mathbf{p}_0})^{-1}$. Then

$$\begin{aligned}\mathbf{p}_{i+1} &= \mathbf{p}_i + L(\mathbf{q} - f(\mathbf{p}_i))\\ &= \mathbf{p}_i + L(\mathbf{q} - (f(\mathbf{p}_{i-1}) + \mathrm{d}f_{\mathbf{p}_{i-1}}(\mathbf{p}_i - \mathbf{p}_{i-1}) + o(\mathbf{p}_i - \mathbf{p}_{i-1})))\end{aligned}$$

as before. Now

$$\mathbf{q} - (f(\mathbf{p}_{i+1}) + \mathrm{d}f_{\mathbf{p}_0}(\mathbf{p}_i - \mathbf{p}_{i-1})) = \mathbf{0}$$

so

$$\begin{aligned}\|\mathbf{q} - (f(\mathbf{p}_{i-1}) + \mathrm{d}f_{\mathbf{p}_{i-1}}(\mathbf{p}_i - \mathbf{p}_{i-1}))\| &= \|(\mathrm{d}f_{\mathbf{p}_{i-1}} - \mathrm{d}f_{\mathbf{p}_0})(\mathbf{p}_i - \mathbf{p}_{i-1})\|\\ &\leqslant \frac{\varepsilon}{2}\|\mathbf{p}_i - \mathbf{p}_{i-1}\|\end{aligned}$$

provided we take $a$ small enough. Also, we can choose $\delta$ small enough so that $\varepsilon$ is replaced by $\tfrac{1}{2}\varepsilon$ in (6.13). Then

$$\|\mathbf{p}_{i+1} - \mathbf{p}_i\| \leqslant k(\tfrac{1}{2}\varepsilon\|\mathbf{p}_i - \mathbf{p}_{i-1}\| + \tfrac{1}{2}\varepsilon\|\mathbf{p}_i - \mathbf{p}_{i-1}\|) \leqslant k\varepsilon\|\mathbf{p}_i - \mathbf{p}_{i-1}\|$$

so that (6.16) holds as before, where $k = \|L\|$. Actually, we can use the mean-value theorem to rephrase the argument for the Picard method so as to avoid the unnecessary assumption of *uniform differentiability*. Indeed, consider the map

$$h(\mathbf{p}) = \mathbf{p} + L(\mathbf{q} - f(\mathbf{p}))$$

where $L = (\mathrm{d}f_{\mathbf{p}_0})^{-1}$. By the continuity of $\mathrm{d}f_{\mathbf{p}}$ we can choose $a$ small enough so that

$$\|\mathrm{d}h_{\mathbf{p}}\| = \|I - L\mathrm{d}f_{\mathbf{p}}\| \leqslant \tfrac{1}{2}.$$

Then

$$\|\mathbf{p}_{i+1} - \mathbf{p}_i\| = \|h(\mathbf{p}_i) - h(\mathbf{p}_{i-1})\| \leqslant \tfrac{1}{2}\|\mathbf{p}_i - \mathbf{p}_{i-1}\|$$

by the mean-value theorem and we can proceed as before.

We can also understand why Newton's method converges so much more rapidly, when it works. Suppose that $f$ has two continuous derivatives. Then, by Taylor's formula,

$$|f(\mathbf{p}^1) - f(\mathbf{p}) - \mathrm{d}f_{\mathbf{p}}(\mathbf{p}^1 - \mathbf{p})| \leqslant c\|\mathbf{p}^1 - \mathbf{p}\|^2$$

(where $c$ is a constant given by the maximum of $|\mathrm{d}^2 f|$), a much stronger inequality than (6.13). Going back to the proof of (6.16) and substituting this inequality, we get

$$\|\mathbf{p}_{i+1} - \mathbf{p}_1\| \leqslant k\|\mathbf{p}_i - \mathbf{p}_{i-1}\|^2.$$

If, for example, we started out with $\|\mathbf{p}_1 - \mathbf{p}_0\|$ small enough so that $k\|\mathbf{p}_1 - \mathbf{p}_0\|^{1/2} \leqslant 1$ (and $\|\mathbf{p}_1 - \mathbf{p}_0\| < 1$), the above inequality would say that

$$\|\mathbf{p}_{i+1} - \mathbf{p}_i\| \leqslant \|\mathbf{p}_i - \mathbf{p}_{i-1}\|^{3/2}$$

so

$$\|\mathbf{p}_{i+1} - \mathbf{p}_i\| \leqslant \|\mathbf{p}_1 - \mathbf{p}_0\|^{(3/2)^n},$$

an exponential rate of decrease instead of the geometrical one $\|\mathbf{p}_{i+1} - \mathbf{p}_i\| \leqslant (\varepsilon K)^n\|\mathbf{p}_1 - \mathbf{p}_0\|$ given by the Picard method.

Let us summarize what we know so far. We have shown that under suitable hypotheses there exists a ball $B$ around $\mathbf{q}_0 = f(\mathbf{p}_0)$ and a ball $C$ around $\mathbf{p}_0$ such that for each $\mathbf{q} \in B$ there is a unique $\mathbf{p} \in C$ with $f(\mathbf{p}) = \mathbf{q}$. In other words, we have defined a map

$$g: B \to C$$

such that

$$f \circ g = \mathrm{id}$$

and

$$g \circ f = \mathrm{id}.$$

**Differentiability of solution**

We now want to prove that $g$ is differentiable. Notice that the uniqueness of $g$ implies that $g$ is actually continuous. Indeed, suppose that $g(\mathbf{q}) = \mathbf{p}$. Draw a small ball around $\mathbf{p}$. Then apply the results obtained so far to $\mathbf{p}$ and $\mathbf{q}$. This means that there will be a small ball around $\mathbf{q}$ and a function inverse to $f$ mapping into a

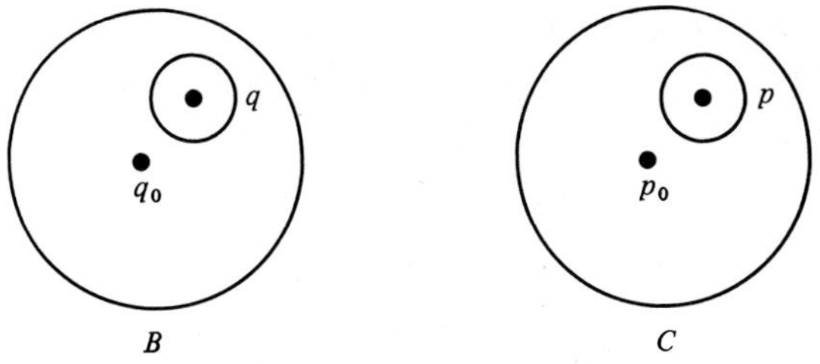


**Figure 6.7**

small ball around **p**. But, by uniqueness, this inverse must coincide with $g$. Hence $g$ maps a small ball around **q** into a small ball around **p**, i.e., is continuous. Now

$$\mathbf{v} = f(g(\mathbf{q}+\mathbf{v})) - f(g(\mathbf{q})) = \mathrm{d}f_{\mathbf{p}}(g(\mathbf{q}+\mathbf{v}) - g(\mathbf{q})) + o(g(\mathbf{q}+\mathbf{v}) - g(\mathbf{q})).$$

Applying $(\mathrm{d}f_{\mathbf{p}})^{-1}$ to both sides we get

$$(\mathrm{d}f_{\mathbf{p}})^{-1}(\mathbf{v}) = g(\mathbf{q}+\mathbf{v}) - g(\mathbf{q}) + o(g(\mathbf{q}+\mathbf{v}) - g(\mathbf{q})).$$

Since $g$ is continuous we can choose $\mathbf{v}$ small enough so that $\| o(g(\mathbf{q}+\mathbf{v}) - g(\mathbf{q}))\|$ is smaller than $\frac{1}{2}\| g(\mathbf{q}+\mathbf{v}) - g(\mathbf{q})\|$. The preceding equation implies that

$$\|(\mathrm{d}f_{\mathbf{p}})^{-1}(\mathbf{v})\| + \| o(g(\mathbf{q}+\mathbf{v}) - g(\mathbf{q}))\| \geqslant \| g(\mathbf{q}+\mathbf{v}) - g(\mathbf{q})\|$$

so

$$\| g(\mathbf{q}+\mathbf{v}) - g(\mathbf{q})\| \leqslant 2\|(\mathrm{d}f_{\mathbf{p}})^{-1}(\mathbf{v})\|$$

i.e., $g(\mathbf{q}+\mathbf{v}) - g(\mathbf{q}) = O(\mathbf{v})$. But then

$$o(g(\mathbf{q}+\mathbf{v}) - g(\mathbf{q})) = o(O(\mathbf{v})) = o(\mathbf{v})$$

so from the above we have

$$g(\mathbf{q}+\mathbf{v}) - g(\mathbf{p}) = (\mathrm{d}f_{\mathbf{p}})^{-1}(\mathbf{v}) + o(\mathbf{v}),$$

i.e., $g$ is differentiable at **q** with derivative

$$\mathrm{d}g_{f(\mathbf{p})} = (\mathrm{d}f_{\mathbf{p}})^{-1}.$$

We have thus proved the

**Inverse function theorem.** *Let $f: U \to V$ be continuously differentiable with $f(\mathbf{p}_0) = \mathbf{q}_0$ and $\mathrm{d}f_{\mathbf{p}_0}$ invertible. Then there exist balls $B$ and $C$ around $\mathbf{q}_0$ and $\mathbf{p}_0$ such that there is a unique map $g: B \to C$ such that $f \circ g = \mathrm{id}$. This map is continuously differentiable and*

$$\mathrm{d}g_{f(\mathbf{P})} = (\mathrm{d}f_{\mathbf{P}})^{-1}.$$

## The implicit function theorem

Let us draw some consequences of the inverse function theorem. Suppose $G: \mathbb{R}^2 \to \mathbb{R}^1$ with $G(x_0, y_0) = 0$ and

$$\frac{\partial G}{\partial y} \neq 0 \quad \text{at} \quad (x_0, y_0).$$

Let $f: \mathbb{R}^2 \to \mathbb{R}^2$ be defined by

$$f\left(\begin{pmatrix} x \\ y \end{pmatrix}\right) = \begin{pmatrix} x \\ G(x, y) \end{pmatrix}.$$

Then

$$\mathrm{d}f = \begin{pmatrix} 1 & 0 \\ \dfrac{\partial G}{\partial x} & \dfrac{\partial G}{\partial y} \end{pmatrix}$$

as a matrix, and is nonsingular at $\begin{pmatrix} x_0 \\ y_0 \end{pmatrix}$. By the inverse function theorem we can find an inverse map $g$ with $f \circ g = \mathrm{id}$. We may write

$$g\left(\begin{pmatrix} u \\ v \end{pmatrix}\right) = \begin{pmatrix} F(u, v) \\ H(u, v) \end{pmatrix}$$

so that the equation $f \circ g = \mathrm{id}$ becomes

$$F(u, v) = u,$$
$$G(F(u, v), H(u, v)) = v.$$

Substituting the first equation into the second gives

$$G(u, H(u, v)) = v$$

and setting $v = 0$, $h(u) = H(u, 0)$ gives

$$G(u, h(u)) = 0.$$

*The function $h(u)$ is differentiable and is the unique solution to this equation.* The existence of uniqueness and differentiability of $h$ is the content of the *implicit function theorem.* Thus the implicit function theorem in one variable is a *consequence* of the inverse function theorem in two variables. We state it once more as a formal theorem.

**The implicit function theorem.** *Let $G$ be a differentiable function with $G(x_0, y_0) = 0$ and $(\partial G/\partial y)(x_0, y_0) \neq 0$. Then there exists a unique function $h(x)$ defined near $x = x_0$ such that $h(x_0) = y_0$ and $G(x, h(x)) \equiv 0$. The function $h$ is differentiable and $h'(x) = -(\partial G/\partial x)/(\partial G/\partial y)$.*

We can reformulate the preceding argument. The simplest map (other than the constant map) that we can imagine from $\mathbb{R}^2 \to \mathbb{R}^1$ is projection onto one of the factors

$$\pi: \mathbb{R}^2 \to \mathbb{R}^1 \quad \pi\left(\begin{pmatrix} u \\ v \end{pmatrix}\right) = v.$$

Now let $G: \mathbb{R}^2 \to \mathbb{R}^1$ be any continuously differentiable map and suppose that $\mathrm{d}G_{\mathbf{p}_0}$ is *surjective* (which, in our case, where the range is one-dimensional, means that $\mathrm{d}G_{\mathbf{p}_0} \neq 0$). We claim that there are local *changes of coordinates*, i.e., maps

$$g: \mathbb{R}^2 \to \mathbb{R}^2$$

locally defined and having a differentiable inverse so that

$$G \circ g = \pi.$$

Indeed, since $\mathrm{d}G_{\mathbf{p}_0} \neq 0$, we can make a preliminary linear change of coordinates in the plane so that $\partial G/\partial y \neq 0$. Then the above argument applies to give a map $g$ such that $G \circ g = \pi$. Thus, if we allow arbitrary changes of coordinates, the most general continuously differentiable map with $\mathrm{d}G_{\mathbf{p}}$ surjective 'looks like' projection onto a factor. For example

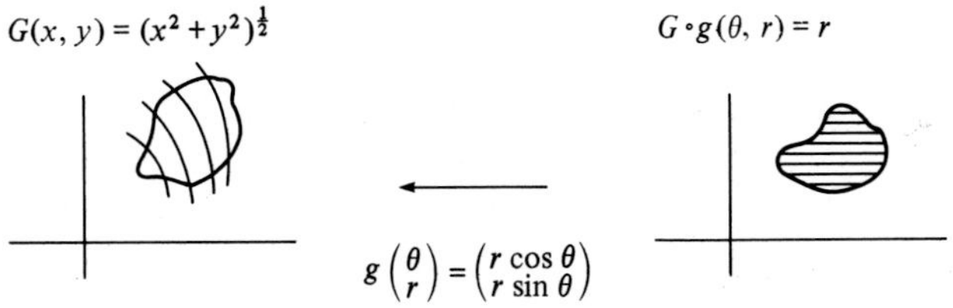


**Figure 6.8**

The simplest non-trivial map from $\mathbb{R}^1 \to \mathbb{R}^2$ is the map $i$ which simply injects $\mathbb{R}^1$ as the '$x$-axis'

$$i\colon \mathbb{R}^1 \to \mathbb{R}^2 \quad i(x) = \begin{pmatrix} x \\ 0 \end{pmatrix}.$$

We claim that if $G\colon \mathbb{R}^1 \to \mathbb{R}^2$ is any continuously differentiable map with $\mathrm{d}G_{\mathbf{p}_0} \neq 0$, we can find a change of coordinates, i.e., a continuously differentiable map $f$ with differentiable inverse such that

$$f \circ G = i.$$

**Figure 6.9**

Indeed, by a preliminary linear change of variables in the plane, we can arrange that

$$\mathrm{d}G_{\mathbf{p}_0} = \begin{pmatrix} 1 \\ 0 \end{pmatrix}.$$

By a translation we may assume that $G(\mathbf{p}_0) = \begin{pmatrix} \mathbf{p}_0 \\ 0 \end{pmatrix}$. Now define

$$F\colon \mathbb{R}^2 \to \mathbb{R}^2 \quad F\left(\begin{pmatrix} x \\ y \end{pmatrix}\right) = \begin{pmatrix} G_1(x) \\ G_2(x) + y \end{pmatrix}$$

where

$$G(t) = \begin{pmatrix} G_1(t) \\ G_2(t) \end{pmatrix}.$$

Then

$$\mathrm{d}F_{\binom{p_0}{0}} = \begin{pmatrix} 1 & 0 \\ 0 & 1 \end{pmatrix}$$

and hence $F$ has a continuously differentiable inverse. Now

$$F \circ i = \begin{pmatrix} G_1 \\ G_2 \end{pmatrix} = G$$

by the definitions of $F$ and $i$. Hence, taking $f = F^{-1}$

$$i = f \circ G.$$

Locally, we can 'straighten out any curve' by a change of variables.

## 6.4. Behavior near a critical point

Suppose that $f$ has a critical point at $\mathbf{p}_0$. Let us assume that we have made a preliminary choice of coordinates so that $\mathbf{p}_0 = \mathbf{0}$ in $\mathbb{R}^2$. Suppose $\mathrm{d}^2 f_0$ is non-degenerate i.e. that $\mathrm{Det}\,(\mathrm{d}^2 f_0) \neq 0$. In other words, we assume that the symmetric matrix-valued function

$$H = \begin{pmatrix} \dfrac{\partial^2 f}{\partial x_1^2} & \dfrac{\partial^2 f}{\partial x_1 \partial x_2} \\ \dfrac{\partial^2 f}{\partial x_1 \partial x_2} & \dfrac{\partial^2 f}{\partial x_2^2} \end{pmatrix}$$

is non-degenerate at $\mathbf{0}$. By the results of section 4.2, we know that we can make a linear change of coordinates $L$ so that $LH(\mathbf{0})L^{\mathrm{T}}$ has one of the three forms

$$LH(\mathbf{0})L^{\mathrm{T}} = \begin{cases} \begin{pmatrix} 1 & 0 \\ 0 & 1 \end{pmatrix}, \\ \begin{pmatrix} -1 & 0 \\ 0 & 1 \end{pmatrix}, \\ \text{or} \begin{pmatrix} -1 & 0 \\ 0 & -1 \end{pmatrix}. \end{cases}$$

Let us assume that we have made this preliminary linear change of coordinates, so that $\mathrm{d}^2 f_0$ already has one of these three standard forms. Now by our proof of the Taylor expansion, we know that

$$\begin{aligned} f(x, y) &= f(\mathbf{0}) + b_{11}(x, y)x^2 + 2b_{12}(x, y)xy + b_{22}(x, y)y^2 \\ &= f(\mathbf{0}) + (x, y)\begin{pmatrix} b_{11} & b_{12} \\ b_{12} & b_{22} \end{pmatrix}\begin{pmatrix} x \\ y \end{pmatrix} \\ &= f(\mathbf{0}) + (x, y)B\begin{pmatrix} x \\ y \end{pmatrix} \end{aligned}$$

where the $b_{ij}$ are continuous functions of $x$ and $y$ and the matrix valued function $B$ when evaluated at the origin is just $\mathrm{d}^2 f_0$, i.e.,

$$B(\mathbf{0}) = H(\mathbf{0})$$

or

$$b_{11}(\mathbf{0}) = \frac{\partial^2 f}{\partial x}(\mathbf{0}),$$

$$b_{12}(\mathbf{0}) = \frac{\partial^2 f}{\partial x \partial y}(\mathbf{0}),$$

$$b_{22}(\mathbf{0}) = \frac{\partial^2 f}{\partial y^2}(\mathbf{0}).$$

Now $B$ is a symmetric matrix. Let us apply the Gram–Schmidt procedure to $B(\mathbf{x})\begin{pmatrix}1\\0\end{pmatrix}$ and $B(\mathbf{x})\begin{pmatrix}0\\1\end{pmatrix}$. By the continuity of $B$ we know that the scalar products

$$(1,0)B(\mathbf{x})\begin{pmatrix}1\\0\end{pmatrix},$$

$$(1,0)B(\mathbf{x})\begin{pmatrix}0\\1\end{pmatrix}$$

and

$$(0,1)B(\mathbf{x})\begin{pmatrix}0\\1\end{pmatrix}$$

depend continuously on $x$. Hence for $\mathbf{x}$ close enough to zero we can find an invertible matrix $Q(\mathbf{x})$ (given by the Gram–Schmidt procedure) such that

$$B(\mathbf{x}) = Q(\mathbf{x})^{\mathrm{T}} B(\mathbf{0}) Q(\mathbf{x}).$$

The Gram–Schmidt algorithm guarantees that $Q$ is a differentiable function of $\mathbf{x}$. Thus

$$f(\mathbf{x}) = f(\mathbf{0}) + \mathbf{x}^{\mathrm{T}} Q(x)^{\mathrm{T}} H(\mathbf{0}) Q(\mathbf{x})\mathbf{x}$$

or

$$f(\mathbf{x}) = f(\mathbf{0}) + \mathbf{y}^{\mathrm{T}} H(\mathbf{0})\mathbf{y},$$

where

$$\mathbf{y} = Q(\mathbf{x})\mathbf{x}.$$

Now the map $\mathbf{x} \mapsto \mathbf{y}$ given by this formula is invertible by the inverse function theorem! In more detail: let $\phi\colon \mathbb{R}^2 \mapsto \mathbb{R}^2$ be defined by

$$\phi(\mathbf{x}) = Q(\mathbf{x})\mathbf{x}.$$

Then, by the product formula,

$$\mathrm{d}\phi_0(\mathbf{x}) = (\mathrm{d}Q_0(\mathbf{x}))\mathbf{0} + Q(\mathbf{0})\mathbf{x}$$

or

$$\mathrm{d}\phi_0 = \mathrm{id}.$$

Thus the inverse function theorem guarantees that $\phi$ has a differentiable inverse

$$\mathbf{x} = \psi(\mathbf{y}).$$

But then $(\psi^* f)(\mathbf{y}) = f(\mathbf{x})$ so

$$\psi^* f(\mathbf{y}) = f(\mathbf{0}) + \mathbf{y}^{\mathrm{T}} H(\mathbf{0})\mathbf{y}.$$

In other words, $\psi^*(f - f(0,0))$ is quadratic! We have proved that, near any non-degenerate critical point, it is possible to introduce coordinates $y_1$ and $y_2$ such that

$$f(y_1, y_2) = f(\mathbf{0}) + Q(y_1, y_2)$$

where

$$Q(y_1, y_2) = \pm(y_1^2 + y_2^2), \quad \text{or} \quad -y_1^2 + y_2^2.$$

Which of the three alternatives holds is determined by the normal form (the number of negative eigenvalues) of $\mathrm{d}^2 f_{\mathbf{0}}$.

This proof is completely general – it works in $n$ dimensions: So, if $\mathbf{0}$ is a non-degenerate critical point of $f$, it is possible to find coordinates in terms of which

$$f(\mathbf{y}) = f(\mathbf{0}) + Q(\mathbf{y})$$

where

$$Q(\mathbf{y}) = \pm y_1^2 \pm y_2^2 + \cdots.$$

The number of – signs (called the *index of* $Q$) is the same as the number of negative eigenvalues of the matrix $\mathrm{d}^2 f_0$. This result is known as *Morse's lemma.* We will make use of this lemma in our study of asymptotic integrals in Chapter 21.

---

## Summary

### A — Higher derivatives

You should be able to write down the Taylor expansion of a function on the plane through terms involving second partial derivatives.

You should be able to apply the chain rule in order to express second partial derivatives or second-order partial differential equations in terms of new coordinates.

### B — Critical points

You should be able to locate the critical points of a function on the plane and to classify each critical point as a maximum, minimum, or saddle point.

You should know how to use the method of Lagrange multipliers to find the critical values of a function of several variables subject to constraints.

### C — Inverse functions

You should be able to state and apply the inverse function theorem.

You should know how to use Newton's method to find an approximate solution to $f(\mathbf{p}) = \mathbf{q}$ where $f$ is a function from $\mathbb{R}^2$ to $\mathbb{R}^2$.

---

## Exercises

6.1. Let $F(x, y) = \begin{cases} 0 & \text{if } x = y = 0 \\ \dfrac{x^3 y}{x^2 + y^2} & \text{otherwise.} \end{cases}$

(a) Calculate $\partial F/\partial x$ and $\partial F/\partial y$. Are they continuous at $(0,0)$?
(b) Calculate $\partial^2 F/\partial x \partial y$ and $\partial^2 F/\partial y \partial x$. Are they continuous at $(0,0)$? (Note: If they are not, you may *not* compute their values at the origin by finding a general formula and trying to let $x$ and $y$ both approach zero!)
(c) Show that $(\partial^2 F/\partial x \partial y)(0,0) \neq (\partial^2 F/\partial y \partial x)(0,0)$.
(d) Invent a smooth curve through the origin described by $x = X(t)$, $y = Y(t)$ with $X(0) = Y(0) = 0$, such that the function $G(t) = F(X(t), Y(t))$ is *not* differentiable at the origin.

6.2. Find and classify all the critical points of the function $F: \mathbb{R}^2 \to \mathbb{R}$ given by

$$F(x, y) = x^3 + y^3 - 3xy.$$

6.3. Let $F(x, y) = x^2 y - 3xy + \frac{1}{2}x^2 + y^2$.
(a) Find the equation of the tangent plane to the graph of $z = F(x, y)$ at the point $x = 2$, $y = 2$, $z = 2$.
(b) The function $F(x, y)$ has three critical points, two of which lie on the line $x = y$. Locate these critical points and classify each as maximum, minimum or saddle point.

6.4. Consider the function $F$ on $\mathbb{R}^2$ given by

$$F(x, y) = x^2 - 4xy + y^2 - 6x^{-1}.$$

(a) Find the equation of the plane tangent to the graph $z = F(x, y)$ at the point corresponding to $x = -1$, $y = -2$.
(b) Locate the critical point of this function and determine its nature.

6.5. Find and classify all critical points of the function $F(x, y) = y^2 + (x^2 - 3x)\log y$, defined in the upper half-plane $y > 0$.

6.6. Show that the function $F(x, y) = y(e^{4x} - 1) + 9x^2 + 6y^2$ has a critical point at the origin, and determine the nature of this critical point. Describe the level curves of $F(x, y)$ in the neighborhood of the origin. Sketch a couple of typical curves. Describe the level curves of $F(x, y)$ in the neighborhood of the point $x = 0$, $y = 1$, and sketch typical curves.

6.7. Find the critical points of the following function:

$$F(x, y) = 5x^3 - 3x^2 y + 6xy^2 - 4y^3 - 27x + 27y$$

and determine their nature. (At a suitable point in the calculation add two equations. The resulting homogeneous polynomial factors. The critical points have integer coordinates.)

6.8.(a) Find the critical points of the function $F(x, y) = xy^2 e^{-(x+y)}$.
(b) Determine the nature of the critical point which is not at the origin. Sketch, as accurately as you can, some level curves near the point.
(c) For the critical point at the origin, the Hessian vanishes and is no help. Figure out whether the critical point is a maximum, minimum, or saddle point. Sketch some level curves near the origin.

6.9. Let $x$ and $y$ be the usual affine coordinate functions on a plane. Another pair of coordinate function on the right half-plane ($x > 0$) is defined by the equations.

$$u = x^2 - y^2, \quad v = 2xy$$

(a) Express $\mathrm{d}u$ and $\mathrm{d}v$ in terms of $\mathrm{d}x$ and $\mathrm{d}y$, and write the matrix which expresses $\begin{pmatrix} \mathrm{d}u \\ \mathrm{d}v \end{pmatrix}$ in terms of $\begin{pmatrix} \mathrm{d}x \\ \mathrm{d}y \end{pmatrix}$ at the point $P$ with coordinates $x = 2, y = 1, u = 3, v = 4$.

(b) Find the approximate $x$ and $y$ coordinates of a point $Q$ such that $u(Q) = 3.5, v(Q) = 4$.

(c) Let $\phi$ denote the electric potential function on the plane. Given that at the point $P(x = 2,\ y = 1,\ u = 3,\ v = 4)(\partial\phi/\partial u) = 2$ and $\partial\phi/\partial v = -1$, calculate $\partial\phi/\partial x$ and $\partial\phi/\partial y$ at this point. Describe the direction along which $\phi$ increases most rapidly.

(d) At the same point, express $\partial^2\phi/\partial y\partial x$ in terms of partial derivatives of $\phi$ with respect to $u$ and $v$. Your answer may also involve explicit functions of $x$ and $y$, of course.

6.10. Suppose that coordinates $u$ and $v$ on the plane are expressed in terms of $x$ and $y$ by

$$\begin{pmatrix} u \\ v \end{pmatrix} = \begin{pmatrix} \cos\alpha & -\sin\alpha \\ \sin\alpha & \cos\alpha \end{pmatrix}\begin{pmatrix} x \\ y \end{pmatrix}.$$

Let $f$ be a twice-differentiable function on the plane. Show that

$$\frac{\partial^2 f}{\partial u^2} + \frac{\partial^2 f}{\partial v^2} = \frac{\partial^2 f}{\partial x^2} + \frac{\partial^2 f}{\partial y^2}$$

6.11. Polar coordinates $r, \theta$ on the plane are related to Cartesian coordinates by the equations

$$\begin{pmatrix} x \\ y \end{pmatrix} = \begin{pmatrix} r\cos\theta \\ r\sin\theta \end{pmatrix}.$$

Suppose $f: \mathbb{R}^2 \to \mathbb{R}$ is a function satisfying Laplace's equation,

$$\frac{\partial^2 f}{\partial x^2} + \frac{\partial^2 f}{\partial y^2} = 0.$$

Express this equation entirely in terms of derivatives of $f$ with respect to $r$ and $\theta$.

6.12. Let $f: \mathbb{R}^2 \to \mathbb{R}$ be a twice-differentiable function. If $\begin{pmatrix} x \\ y \end{pmatrix} = \begin{pmatrix} r\cos\theta \\ r\sin\theta \end{pmatrix}$ express $\partial^2 f/\partial\theta^2$ in terms of partial derivatives of $f$ with respect to $x$ and $y$.

6.13. Let $f: \mathbb{R}^2 \to \mathbb{R}$ be a twice-differentiable function. If $x \neq 0$ and

$$\begin{pmatrix} r \\ \theta \end{pmatrix} = \begin{pmatrix} \sqrt{(x^2 + y^2)} \\ \arctan(y/x) \end{pmatrix},$$

express $\partial^2 f/\partial x^2$ in terms of partial derivatives of $f$ with respect to $r$ and $\theta$.

6.14. Given that $\partial f/\partial x = f + \partial f/\partial y$, show that $\partial^2 f/\partial x^2 - \partial^2 f/\partial y^2 = f + 2(\partial f/\partial y)$.

6.15. With polar coordinates as in exercise 6.11:

(a) Let $f$ be a function on the plane. Suppose that at the point whose coordinates are $x = 3$, $y = 4$, $\partial f/\partial x = 2$ and $\partial f/\partial y = 1$. Calculate $\partial f/\partial r$ and $\partial f/\partial\theta$ at this point.

(b) Suppose that $f$ satisfies the partial differential equation

$$\frac{\partial^2 f}{\partial r^2} + \frac{1}{r}\frac{\partial f}{\partial r} + \frac{1}{r^2}\frac{\partial^2 f}{\partial\theta^2} = 0.$$

Express this equation entirely in terms of partial derivatives of $f$ with respect to $x$ and $y$.

6.16. Suppose that $f$ is a function on the plane which satisfies Laplace's equation $\partial^2 f/\partial x^2 + \partial^2 f/\partial y^2 = 0$. Express this equation in terms of the parabolic coordinates of exercise 5.13. It may involve $\partial^2 f/\partial u^2$, $\partial^2 f/\partial v^2$, $\partial^2 f/\partial u\partial v$, $\partial f/\partial u$, $\partial f/\partial v$, $u$, and $v$, but not $x$, $y$, or any partial derivatives with respect to $x$ or $y$.

6.17. Let $\phi: \mathbb{R}^2 \to \mathbb{R}^2$ be the mapping

$$\phi\left(\begin{pmatrix} x \\ y \end{pmatrix}\right) = \begin{pmatrix} e^x + e^y \\ e^x - e^y \end{pmatrix}.$$

Show that $\phi$ can be inverted in the neighborhood of any point and compute the Jacobian of the inverse map.

6.18. Consider the surface in $\mathbb{R}^3$ defined by $z = F(x, y)$, where $F(x, y) = x^2 - 2xy + 2y^2 + 3x + 4y$.

(a) Find the *best affine approximation* to $F$ near the point $\begin{pmatrix} x \\ y \end{pmatrix} = \begin{pmatrix} 1 \\ 1 \end{pmatrix}$.

(b) Write the equation of the plane tangent to the surface at $\begin{pmatrix} 1 \\ 1 \end{pmatrix}$.

(c) Find the equation of the line normal to the surface at the same point.

(d) The equation $F(x, y) = 8$ defines a function $y = g(x)$. Evaluate $g'(1)$.

6.19. An important problem of statistical mechanics is the following: Consider a physical system which can have energy $+E$, 0 or $-E$. Let $x$ denote the probability that the energy is $+E$; let $y$ denote the probability that the energy is $-E$. Then $1 - x - y$ is the probability that the energy is zero. Maximize the *entropy* $S$, defined as

$$S(x, y) = -x \log x - y \log y - (1 - x - y) \log (1 - x - y)$$

subject to the constraint that the average energy is $E_0$; i.e.,

$$F(x, y) = xE - yE = E_0.$$

Solve this problem using a Lagrange multiplier $\beta$, and show that $x = e^{-\beta E}$, $y = e^{+\beta E}$, where $E_0 = -2E \sinh \beta E$. (The Lagrange multiplier in this case turns out to equal $1/T$, where $T$ is absolute temperature.)

6.20. Consider the function $\alpha: \mathbb{R}^2 \to \mathbb{R}^2$ defined by the formula

$$\alpha(P) = F\left(\begin{pmatrix} x \\ y \end{pmatrix}\right) = \begin{pmatrix} xy^{3/2} \\ x^{3/2} - y^2 \end{pmatrix}$$

(a) Calculate the $2 \times 2$ Jacobian matrix which represents the linear part of the best affine approximation to $\alpha$ near the point $\begin{pmatrix} x \\ y \end{pmatrix} = \begin{pmatrix} 4 \\ 1 \end{pmatrix}$. Use this matrix to determine the approximate value of $F\left(\begin{pmatrix} 4.2 \\ 1.1 \end{pmatrix}\right)$.

(b) Use the matrix to obtain an approximate solution of $F\left(\begin{pmatrix} x \\ y \end{pmatrix}\right) = \begin{pmatrix} 4.2 \\ 6.6 \end{pmatrix}$.

6.21. Let $f(x, y) = \sqrt{(x^2 y^4 + 9x^2)}$.

(a) Find the *best affine approximation* to this function near the point $x = 1$, $y = 2$.

(b) At the point $x = 1$, $y = 2$, along what direction is the rate of change of the function $f(x, y)$ greatest?

(c) A solution of the equations

$$f(x, y) = 5,$$
$$x + y = 3$$

is $x = 1$, $y = 2$. Construct an approximate solution to the equations

$$f(x, y) = 5.34,$$
$$x + y = 3.05$$

by using the approximation from part (a).

6.22 Functions $s$ and $t$ are defined in terms of the affine coordinate functions $x$ and $y$ on the region $x > 0$, $y > 0$ of the plane by

$$s = xy, \quad t = \log y - \log x.$$

(a) Express the differentials $\mathrm{d}s$ and $\mathrm{d}t$, at the point whose coordinates are $x = 1$, $y = 2$, in terms of $\mathrm{d}x$ and $\mathrm{d}y$.

(b) At the point $x = 1$, $y = 2$, the values of $s$ and $t$ are $s = 2$, $t = \log 2 \approx 0.693$. Use the Jacobian matrix at this point to find the approximate $x$ and $y$ coordinates of a point where $s = 2.02$, $t = 0.723$.

(c) Let $f$ be a twice differentiable function on the plane. Express $\partial f/\partial x$, $\partial f/\partial y$ and $\partial^2 f/(\partial y\,\partial x)$ in terms of $x$, $y$, and partial derivatives of $f$ with respect to $s$ and $t$.

# 7

# Differential forms and line integrals



Chapters 7 and 8 are meant as a first introduction to the integral calculus. Chapter 7 is devoted to the study of linear differential forms and their line integrals. Particular attention is paid to the behavior under change of variables. Other one-dimensional integrals such as arc length are also discussed.

## Introduction

In this chapter we shall disclose the true geometric meaning of linear differential forms: they are objects which are to be integrated over oriented paths to yield numbers. We begin with some examples. Consider the one-form

$$\omega = \tfrac{1}{2}(x\,dy - y\,dx).$$

By its definition it is the rule which assigns to every point $\begin{pmatrix} x \\ y \end{pmatrix}$ the row vector $\frac{1}{2}(-y, x)$. Now a row vector is a linear function on vectors. The row vector $\frac{1}{2}(-y, x)$ is the linear function that assigns to the vector $\mathbf{h} = \begin{pmatrix} r \\ s \end{pmatrix}$ the number

$$\omega_{\binom{x}{y}}[\mathbf{h}] = \tfrac{1}{2}(xs - yr)$$

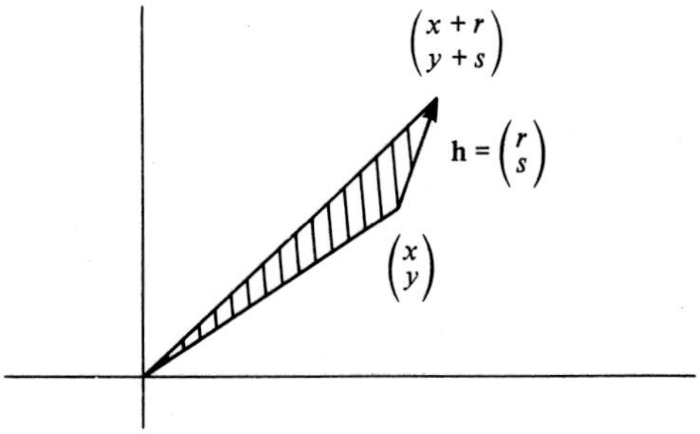


**Figure 7.1**

which is just the oriented area of the triangle from the origin to $\begin{pmatrix} x \\ y \end{pmatrix}$ to $\begin{pmatrix} x \\ y \end{pmatrix} + \begin{pmatrix} r \\ s \end{pmatrix}$.

Suppose we had a curve $\alpha(t) = \begin{pmatrix} x(t) \\ y(t) \end{pmatrix}$. We can choose a number of points $\mathbf{p}_i$ on the curve and let $\mathbf{h}_i = \mathbf{p}_{i+1} - \mathbf{p}_i$. Then the sum

$$\Sigma \omega_{\mathbf{p}_i}[\mathbf{h}_i]$$

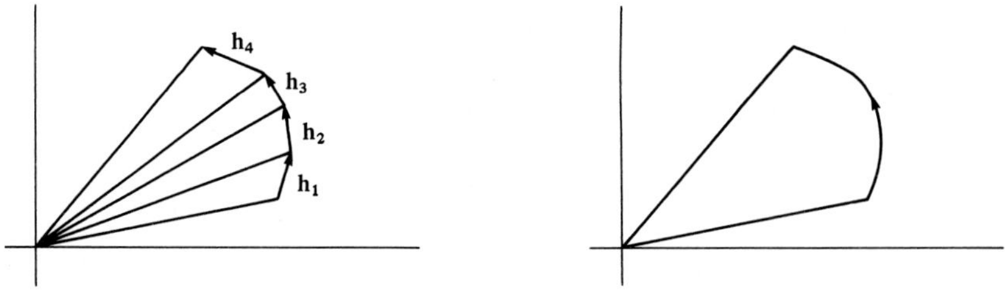


**Figure 7.2**

is the total oriented area of the various triangles. In the limit, as the polygon joining the $\mathbf{p}_i$ approximates the curve, we expect this sum will tend to a limit – the (oriented) area swept out by the radius vector moving along the curve. We have already encountered this notion in our study of Kepler's second law.

A second example to keep in mind is the notion of a *force field*. In three dimensions, a force field

$$\omega = F\,dx + G\,dy + H\,dz$$

gives a linear function

$$(F, G, H)$$

at each point of space. This linear function measures the resistance or impetus to any *infinitesimal motion* – it assigns

$$(F, G, H)\begin{pmatrix} v_x \\ v_y \\ v_z \end{pmatrix} = Fv_x + Gv_y + Hv_z$$

to any displacement vector $\mathbf{v}$ at the point $\mathbf{p}$. Along any path, $\Gamma$, we expect to be able to integrate and get

$$\int_\Gamma \omega = \text{the work done by moving along } \Gamma.$$

Notice that we wish to be able to assign work to all paths. We can imagine a two-dimensional universe in which a force field would be

$$\omega = F\,dx + G\,dy.$$

For example, see figure 7.3, we can imagine feeling the influence of gravity while being constrained to move on a surface

$$z = f(x, y).$$

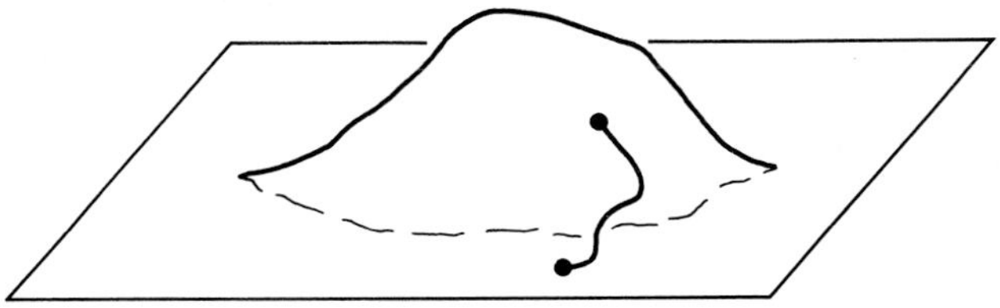

**Figure 7.3**

The force field would then be proportional to

$$\omega = \mathrm{d}f = \frac{\partial f}{\partial x}\mathrm{d}x + \frac{\partial f}{\partial y}\mathrm{d}y.$$

Suppose we had a perfectly reversible electric car. (By perfectly reversible we mean that all the energy of braking is returned to the battery – no air or other kind of resistance.) We could keep track, using a meter, of the total energy flow into and out of the battery. Let us call this $B_F - B_I$ (the difference between the final and initial readings of the battery). We can also consider the kinetic energy at the beginning and end of the trip, $KE_I$ and $KE_F$. The principle of conservation of energy says that

$$KE_F - KE_I + B_F - B_I = \int_\Gamma \omega = \text{the work done along the path.}$$

(Throughout this discussion we are assuming that the forces are *not* velocity dependent: that there is a definite force field where the force depends only on the location in space.)

Notice a subtle difference in viewpoint from the use of force in Newton's laws. In Newton's laws, we are interested in *predicting* how a particle will move – if we set a pebble rolling on our surface, how will it continue to move? Newton says that the motion is determined by the equations

$$\frac{\mathrm{d}\mathbf{p}}{\mathrm{d}t} = \mathbf{F} \quad \mathbf{p} = m\mathbf{v} = \text{momentum} \quad \mathbf{v} = \text{velocity vector.}$$

In our present discussion we are interested in how much energy is used in *driving* along a given curve. The force field $\omega$ assigns energies to paths $\Gamma$. If $\omega = \mathrm{d}f$, then we expect that the total work done along the path is just the potential difference

$$f(Q) - f(P)$$

**Figure 7.4**

where $P$ and $Q$ are the initial and final points of the path.

With this motivation in mind, we now turn to the mathematical discussion.

## 7.1. Paths and line integrals

By an *oriented path* in the plane (or in $\mathbb{R}^k$) we shall mean a curve which is to be traversed in a specified sense. A path like $\Gamma$, whose endpoints do not coincide, has a well-defined 'beginning' ($P_a$ in figure 7.5) and 'end' ($P_b$); interchanging 'beginning' and 'end' reverses the orientation. A closed path like $\Gamma_2$ has no well-defined endpoints; any point $P$ can function as both 'beginning' and 'end'. For this sort of closed path, it is still possible to assign an orientation, which then determines a 'beginning' and 'end' for any piece of the path.

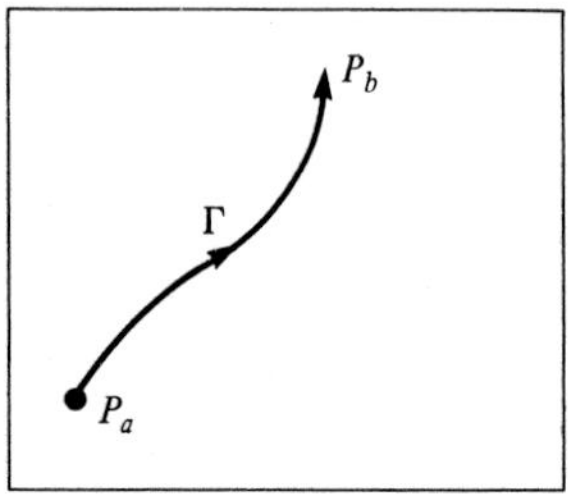


**Figure 7.5**

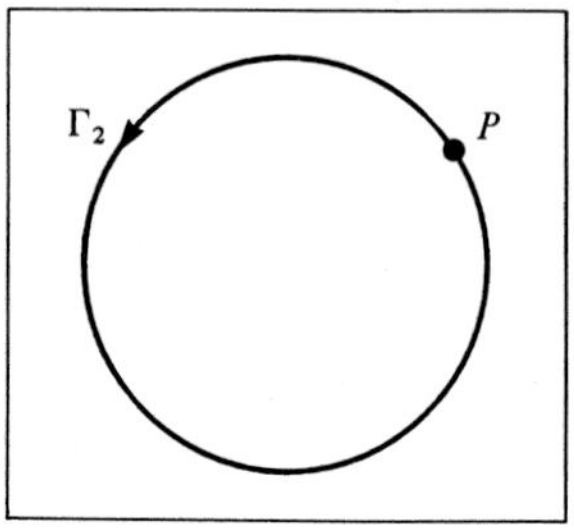


**Figure 7.6**

Physically, such a path is appropriate to represent the trajectory of a particle in circumstances where we know through what points the particle moved, and in what order, but not the speed with which the particle traveled. It is permissible for a segment of a path to be traversed two or more times; for example, a particle might move twice counterclockwise around the unit circle, or it might move from $P_a$ to $Q$, then back to $R$, then forward again to $P_b$. Such paths may be difficult to represent unambiguously by drawing curves with arrows attached, but they make good physical sense, and as we shall see, they are easy to describe in terms of functions.

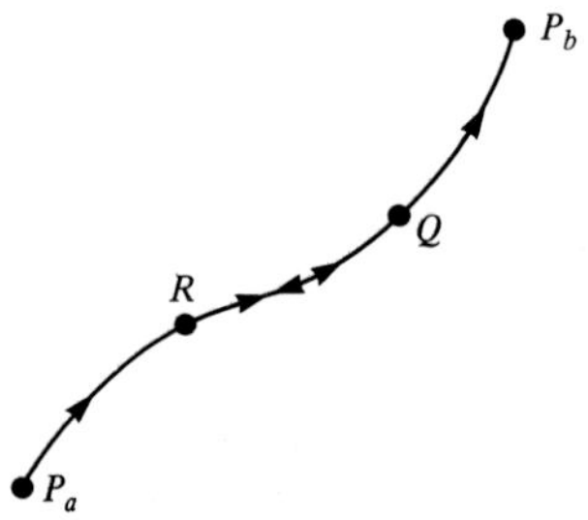


**Figure 7.7**

We shall restrict our attention exclusively to *piecewise differentiable paths*– continuous paths for which a well-defined tangent exists at all except possibly a finite number of points. Such a path can be described as the image of an interval of the real line under a continuous map

$$\alpha: \mathbb{R} \to \mathbb{R}^2$$

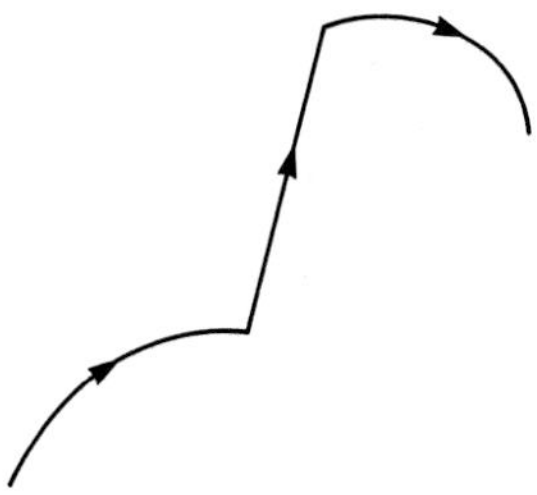

**Figure 7.8**

which is differentiable except at finitely many points where $\alpha$ may not be differentiable. The function $\alpha$ is called a *parameterization* of the path. Physically we may think of $\alpha$ as the function which assigns to each instant of time the position of the particle at that instant. We usually describe $\alpha$ by specifying the pullback of the

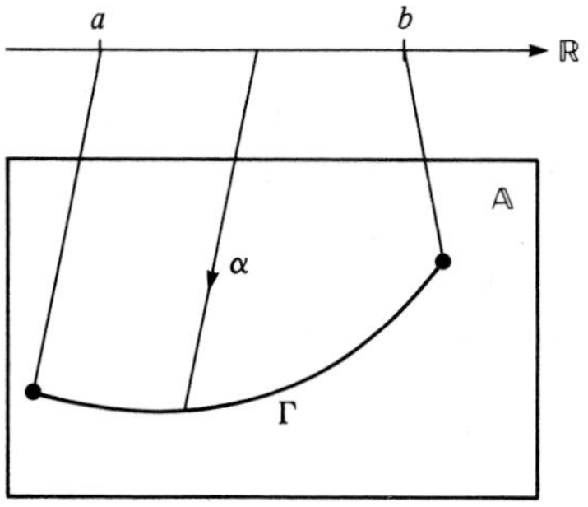


**Figure 7.9**

coordinate functions; that is, by writing formulas which give the numerical values of the $x$- and $y$-coordinates as functions of time. For example, if a particle moves along a circular arc of radius $R$ from $\begin{pmatrix} R \\ 0 \end{pmatrix}$ to $\begin{pmatrix} 0 \\ R \end{pmatrix}$ we may describe its path by

$$\alpha: t \to \begin{pmatrix} R\cos t \\ R\sin t \end{pmatrix} \quad 0 \leqslant 2t \leqslant \pi$$

or by

$$\alpha^* x = R\cos t, \quad \alpha^* y = R\sin t.$$

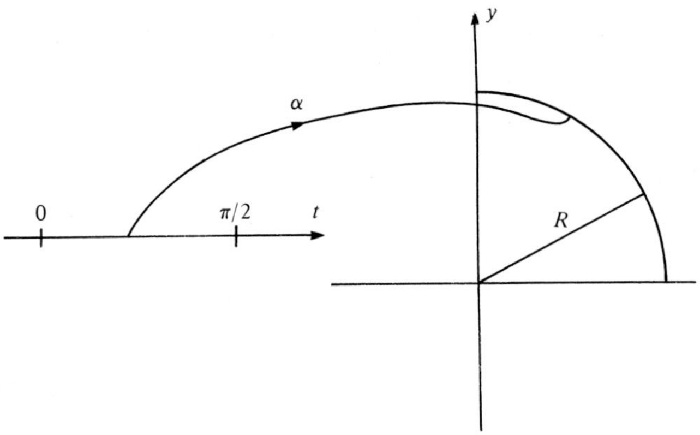


**Figure 7.10**

A given path may have many different parameterizations, which correspond physically to traversing the path at different rates. For example, the segment of the hyperbola $y^2 - x^2 = 1$ from $\begin{pmatrix} 0 \\ 1 \end{pmatrix}$ to $\begin{pmatrix} 1 \\ \sqrt{2} \end{pmatrix}$ may be parameterized in any of the following ways:

$$\begin{array}{ll} \alpha^*x = t, \alpha^*y = \sqrt{(t^2+1)} & 0 \leqslant t \leqslant 1 \\ \alpha^*x = \sinh t, \alpha^*y = \cosh t & 0 \leqslant t \leqslant \operatorname{arcsinh} 1 \\ \alpha^*x = \tan t, \alpha^*y = \sec t & 0 \leqslant 4t \leqslant \pi \end{array}$$

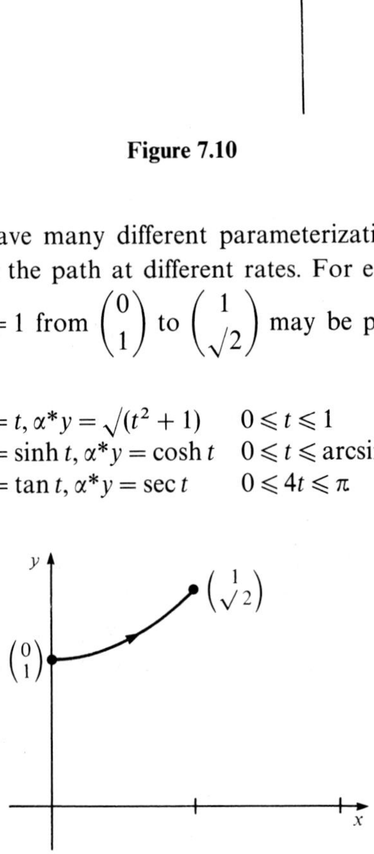


**Figure 7.11**

Paths that involve traversing portions of the same curve more than once are frequently easy to describe in parameterized form. For example, the parameterization

$$\alpha^*x = \cos t, \alpha^*y = \sin t \quad 0 \leqslant t \leqslant 4\pi$$

describes the unit circle traversed twice counterclockwise, while

$$\alpha^*x = \sin^2 t, \alpha^*y = \sin^2 t \quad 0 < 2t < 3\pi$$

describes the line segment from $\begin{pmatrix}0\\0\end{pmatrix}$ to $\begin{pmatrix}1\\1\end{pmatrix}$ traversed forward from $\begin{pmatrix}0\\0\end{pmatrix}$ to $\begin{pmatrix}1\\1\end{pmatrix}$, then backward, then forward again.

As a practical matter, each differentiable segment of a path is usually parameterized separately. For example, the path shown in figure 7.12 might be parameterized as

$$\alpha^*x = t, \alpha^*y = t \quad 0 \leqslant t \leqslant 1$$

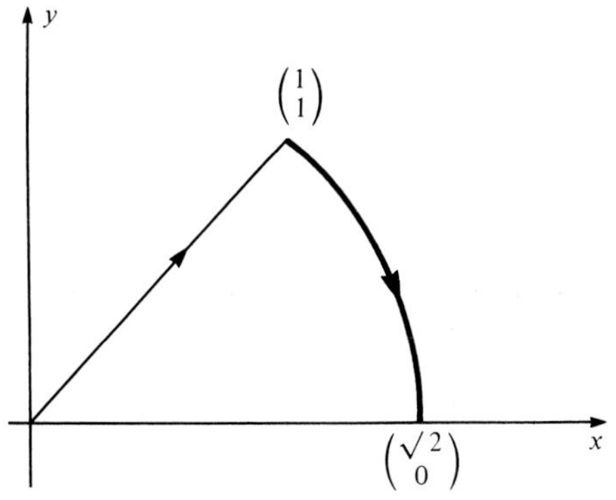


**Figure 7.12**

followed by

$$\alpha^*x = \sqrt{2}\cos(\pi/4 - t), \alpha^*y = \sqrt{2}\sin(\pi/4 - t) \quad 0 \leqslant 4t \leqslant \pi.$$

A piecewise differentiable path can be well-approximated by a sequence of vectors laid 'head to tail'. We simply introduce subdivision points, $P_0, P_1, P_2, \ldots, P_N$, where $P_0$ is the beginning of the path, $P_N$ its end, being sure to include as subdivision points any points where the curve fails to have a tangent. Then we introduce the displacement vectors

$$\mathbf{v}_0 = P_0P_1, \mathbf{v}_1 = P_1P_2, \mathbf{v}_{N-1} = P_{N-1}P_N.$$

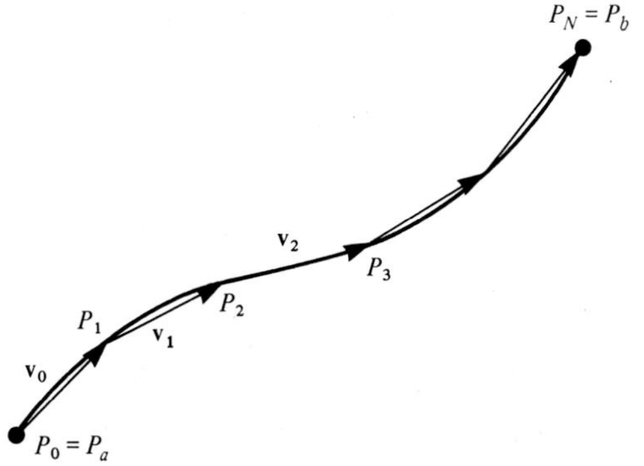


**Figure 7.13**

By choosing the subdivision points close enough together, we can in this manner approximate any piecewise smooth path as accurately as we like by a polygonal path given by a sequence of vectors. Incidentally, it is easy to construct paths that cannot be well-approximated in this manner, but such paths cannot be parameterized by differentiable functions, and we shall not concern ourselves with them.

We now describe a natural way in which a differential form assigns a real number to any path. Each segment of the path is specified by a point $P_i$ with a vector $\mathbf{v}_i$ attached. The differential form $\omega$ assigns to such a segment the real number $\omega(P_i)[\mathbf{v}_i]$. We form the sum over all segments:

$$I_N = \sum_{i=0}^{N-1} \omega(P_i)[\mathbf{v}_i]$$

which is very much like a Riemann sum for an ordinary integral. We now take the limit as the number of subdivision points increases in such a way that all vectors $\mathbf{v}_i$ approach zero. If this limit exists, independent of the precise manner in which the subdivision points are chosen, it defines the *line integral* of the one-form $\omega$ over the path $\Gamma$, which we denote by $\int_\Gamma \omega$. That is,

$$\int_\Gamma \omega = \lim_{N\to\infty} \sum_{i=0}^{N-1} \omega(P_i)[\mathbf{v}_i].$$

We shall soon prove that for piecewise differentiable paths, the limit exists, and is independent of the subdivision, and shall give a formula for $\int_\Gamma \omega$ using pullback.

Three properties of the line integral $\int_\Gamma \omega$ are apparent from the definition:

(1) It is linear in $\omega$: that is, if $\omega = \omega_1 + \omega_2$, then $\int_\Gamma \omega = \int_\Gamma \omega_1 + \int_\Gamma \omega_2$ and if $\omega = \lambda\omega'$ for a real number $\lambda$, then $\int_\Gamma \omega = \lambda \int_\Gamma \omega'$. Both these results follow immediately from the definition of the sum of differential forms and the product of a differential form and a real number.

(2) If $\Gamma$ consists of $\Gamma_1$ followed by $\Gamma_2$, then

$$\int_\Gamma \omega = \int_{\Gamma_1} \omega + \int_{\Gamma_2} \omega.$$

This implies that we can always subdivide a piecewise differentiable path into differentiable portions, as suggested by figure 7.14, and calculate the line integral over each portion.

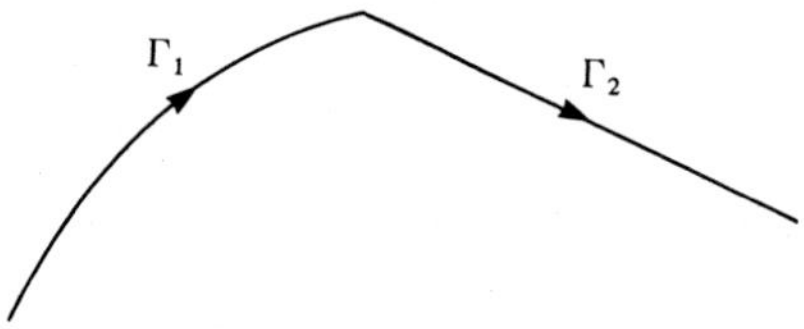


**Figure 7.14**

(3) If $\Gamma$ and $\Gamma'$ differ only in their orientation, then

$$\int_{\Gamma'} \omega = -\int_\Gamma \omega.$$

This is true because reversing the orientation of $\Gamma$ just changes the sign of each of the vectors $\mathbf{v}_i$, and, since $\omega$ is linear,

$$\omega(P)[-\mathbf{v}] = -\omega(P)[\mathbf{v}].$$

We turn now to the problem of computing the numerical value of a line integral. The strategy is to reduce the problem to calculation of an ordinary integral over the parameter for the path of integration. The parameterization

$$\alpha: \mathbb{R} \to \mathbb{R}^2$$

maps an interval $[a, b]$ of the real line into the path $\Gamma$. We assume that $a$, the lower bound of the interval $[a, b]$, is mapped onto the beginning point $P_a$ of the path, while $b$ is mapped into the end point $P_b$. By looking separately at the smooth pieces of our path, we may assume that $\alpha$ can be described by a pair of differentiable functions,

$$\alpha^*x = X(t), \alpha^*y = Y(t),$$

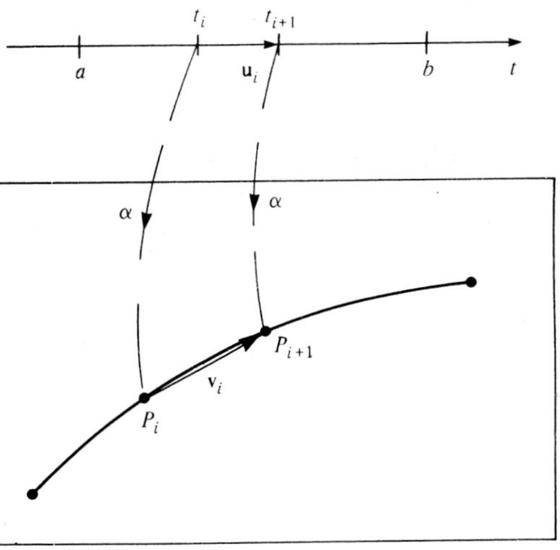


**Figure 7.15**

so that, by the chain rule,

$$\alpha^*\mathrm{d}x = X'(t)\,\mathrm{d}t, \alpha^*\mathrm{d}y = Y'(t)\,\mathrm{d}t.$$

We may assume that our subdivision of the path corresponds to a subdivision

$$a = t_0 < t_1 < \cdots < t_n = b$$

so

$$P_i = \alpha(t_i) = \begin{pmatrix} X(t_i) \\ Y(t_i) \end{pmatrix}.$$

Suppose that

$$\omega = g\mathrm{d}x + h\mathrm{d}y.$$

Thus our approximating expression to the line integral can be written as

$$\begin{aligned}&\sum_i \omega(P_i)[\alpha(t_{i+1})-\alpha(t_i)]\\&\quad=\sum_i\{g(P_i)\mathrm{d}x[\alpha(t_{i+1})-\alpha(t_i)]+h(P_i)\mathrm{d}y[\alpha(t_{i+1})-\alpha(t_i)]\}\\&\quad=\sum_i\{g(\alpha(t_i))(X(t_{i+1})-X(t_i))+h(\alpha(t_i))(Y(t_{i+1})-Y(t_i))\}.\end{aligned}\tag{7.1}$$

Recall that

$$\alpha^*\omega=f\,\mathrm{d}t$$

where

$$f(t)=g(\alpha(t))X'(t)+h(\alpha(t))Y'(t).$$

We will show (under appropriate hypotheses on $\alpha$, $f$ and $g$) that the approximating expression (7.1) converges to

$$\int_a^b f\,\mathrm{d}t\tag{7.2}$$

as the subdivision (and hence the polygonal approximation to our path) gets more and more refined. This will prove that the limit is independent of the choice of subdivisions. So we wish to compare (7.1) and (7.2) for a fixed subdivision and show that their difference tends to zero as the mesh size, $\max_i(t_{i+1}-t_i)$ goes to zero. Now we can write

$$X(t_{i+1})-X(t_i)=\int_{t_i}^{t_{i+1}}X'(s)\,\mathrm{d}s$$

and

$$Y(t_{i+1})-Y(t_i)=\int_{t_i}^{t_{i+1}}Y'(s)\,\mathrm{d}s$$

so (7.1) can be written as

$$\sum\left\{g(\alpha(t_i))\int_{t_i}^{t_{i+1}}X'(s)\,\mathrm{d}s+h(\alpha(t_i))\int_{t_i}^{t_{i+1}}Y'(s)\,\mathrm{d}s\right\}$$

while (7.2) can be written as

$$\sum_i\int_{t_i}^{t_{i+1}}f(s)\,\mathrm{d}s=\sum_i\left\{\int_{t_i}^{t_{i+1}}g(\alpha(s))X'(s)\mathrm{d}s+\int_{t_i}^{t_{i+1}}h(\alpha(s))Y'(s)\,\mathrm{d}s\right\}.$$

The difference between these two expressions is that for (7.1) we have $g(\alpha(t_i))$ or $h(\alpha(t_i))$ occurring outside the integrals in each summand, while in (7.2) we have $g(\alpha(s))$ and $h(\alpha(s))$ occurring under the integral sign. It is intuitively clear that, for $f$ and $g$ continuous and $\alpha$ smooth, the sum of these differences is negligible for a fine enough subdivision. Here are the assumptions we shall make in order to get a precise estimate on the difference between (7.1) and (7.2). Weaker assumptions would suffice, but require more careful argument.

(i) We assume that $g$ and $h$ are uniformly continuous, i.e., that for any $\varepsilon>0$ there is an $\eta>0$ such that $\|P-Q\|<\eta$ implies that $|g(P)-g(Q)|<\varepsilon$ and $|h(P)-g(Q)|<\varepsilon$. This is an assumption about $\omega$. By the mean-value theorem it will hold (with $\eta N = \varepsilon$) if the derivatives $\mathrm{d}g$ and $\mathrm{d}h$ satisfy $\|\mathrm{d}g\| < N$ and $\|\mathrm{d}h\| < N$ at all points.

(ii) We assume that there is a constant $M$ such that

$$|X'(t)| < M$$

and

$$|Y'(t)| < M$$

for all $t$. This is an assumption about the path $\alpha$.

By the mean-value theorem, we can find a $\delta > 0$ such that for any $t'$ and $t''$ with

$$|t' - t''| < \delta$$

we have

$$|\alpha(t') - \alpha(t'')| < \eta.$$

Let us choose our subdivision so that its mesh size is less than $\delta$, i.e., $|t_{i+1} - t_i| < \delta$ for all $i$. Thus by (i) we have

$$|g(\alpha(s)) - g(\alpha(t_i))| < \varepsilon \quad \text{for} \quad t_i \leqslant s \leqslant t_{i+1}$$

with a similar estimate for $h$. Thus

$$\begin{aligned} &\left| g(\alpha(t_i)) \int_{t_i}^{t_{i+1}} X'(t)\,\mathrm{d}t - \int_{t_i}^{t_{i+1}} g(\alpha(s))X'(s)\,\mathrm{d}s \right| \\ &\quad \leqslant \int_{t_i}^{t_{i+1}} |g(\alpha(t_i)) - g(\alpha(s))|\,|X'(s)|\,\mathrm{d}s \\ &\quad \leqslant \varepsilon M |t_{i+1} - t_i|. \end{aligned}$$

Summing up, and with a similar estimate for the $h$ term, we see that the difference between (7.1) and (7.2) is at most

$$\sum_i \varepsilon M |t_{i+1} - t_i| + \varepsilon M |t_{i+1} - t_i| = 2\varepsilon M |b - a|.$$

We can arrange to have $\varepsilon$ as small as we like by making $\delta$, i.e., the mesh size, small enough. This proves our assertion.

As $f\,\mathrm{d}t = \alpha^*\omega$ we can write

$$\int_a^b f\,\mathrm{d}t = \int_a^b \alpha^*\omega$$

where the right-hand side here is *defined* to be the left. We can thus write

$$\int_\Gamma \omega = \int_a^b \alpha^*\omega$$

In this equation, the left-hand side has an obvious intuitive meaning, while we use the right-hand side for computation.

**Example**

As an example of the use of this result, we evaluate the integral $I$ of $\omega = xy\mathrm{d}x + x^2\mathrm{d}y$ along two different paths joining $\begin{pmatrix}0\\1\end{pmatrix}$ to $\begin{pmatrix}1\\0\end{pmatrix}$. Path $\Gamma_1$ lies along the parabola $y = 1 - x^2$, while $\Gamma_2$ is a straight line segment. See figure 7.16.

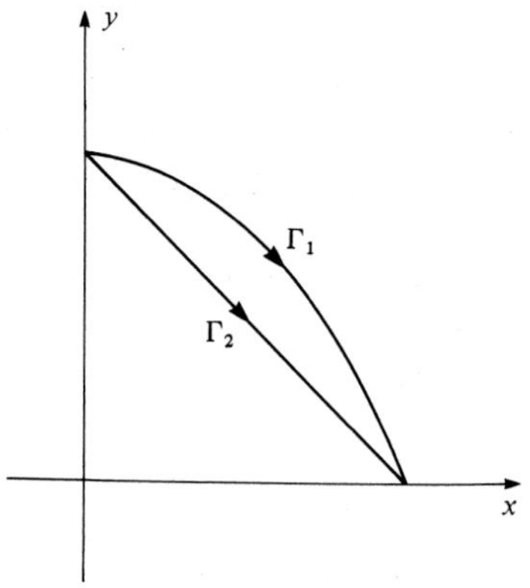


**Figure 7.16**

To parameterize $\Gamma_1$ we introduce

$$\alpha: t \mapsto \begin{pmatrix} t \\ 1-t^2 \end{pmatrix} \quad 0 \leqslant t \leqslant 1,$$

i.e.,

$$\alpha^* x = t, \qquad \alpha^* y = 1 - t^2$$

so that

$$\alpha^* \mathrm{d}x = \mathrm{d}t, \ \alpha^* \mathrm{d}y = -2t\mathrm{d}t.$$

The orientation is correct: $\alpha(0) = \begin{pmatrix} 0 \\ 1 \end{pmatrix}$, while $\alpha(1) = \begin{pmatrix} 1 \\ 0 \end{pmatrix}$.

Then

$$\alpha^* \omega = t(1-t^2)\,\mathrm{d}t + t^2(-2t\,\mathrm{d}t) = (t - 3t^3)\,\mathrm{d}t$$

and

$$\int_{\Gamma_1} \omega = \int_0^1 (t - 3t^3)\,\mathrm{d}t = [\tfrac{1}{2}t^2 - \tfrac{3}{4}t^4]_0^1 = -\tfrac{1}{4}.$$

To parameterize $\Gamma_2$ we introduce

$$\beta: t \rightarrow \begin{pmatrix} t \\ 1-t \end{pmatrix},$$

i.e.,

$$\beta^* x = t, \qquad \beta^* y = 1 - t$$

so that

$$\beta^* \mathrm{d}x = \mathrm{d}t, \ \beta^* \mathrm{d}y = -\mathrm{d}t$$

and

$$\beta^* \omega = t(1-t)\,\mathrm{d}t + t^2(-\mathrm{d}t) = (t - 2t^2)\,\mathrm{d}t.$$

For this path, then,

$$\int_{\Gamma_2} \omega = \int_0^1 (t - 2t^2)\,\mathrm{d}t = [\tfrac{1}{2}t^2 - \tfrac{2}{3}t^3]_0^1 = -\tfrac{1}{6}.$$

**Exact forms**

In this example, you will note, the value of the line integral depends on the path, not just upon the endpoints. This is true in general, but there is one important exception. Suppose that the one-form $\omega = \mathrm{d}f$. In this case

$$\int_\Gamma \mathrm{d}f = \int_a^b \alpha^*(\mathrm{d}f) = \int_a^b \mathrm{d}(\alpha^* f).$$

By the fundamental theorem of calculus,

$$\int_a^b \mathrm{d}(\alpha^* f) = \alpha^* f(b) - \alpha^* f(a).$$

But $\alpha^* f(b) = f(\alpha(b)) = f(P_b)$, where $P_b$ is the endpoint of $\Gamma$, and similarly, $\alpha^* f(a) = f(P_a)$. We conclude that, if $\Gamma$ extends from $P_a$ to $P_b$,

$$\int_\Gamma \mathrm{d}f = f(P_b) - f(P_a)$$

independent of the choice of $\Gamma$.

Notice that this result, combined with the preceding calculation, shows that not every differential form $\omega$ can be written as $\omega = \mathrm{d}f$. Indeed, it is easy to write down a necessary condition: suppose

$$\omega = G\,\mathrm{d}x + H\,\mathrm{d}y = \mathrm{d}f = \frac{\partial f}{\partial x}\mathrm{d}x + \frac{\partial f}{\partial y}\mathrm{d}y.$$

By the equality of cross derivatives, i.e. since $\partial^2 f/\partial x\partial y = \partial^2 f/\partial y\partial x$, we must have

$$\frac{\partial G}{\partial y} = \frac{\partial H}{\partial x}.$$

In the example above, $G = xy$ and $H = x^2$ so

$$\frac{\partial G}{\partial y} = x \neq \frac{\partial H}{\partial x} = 2x.$$

A differential form $\omega$ which can be written as $\omega = \mathrm{d}f$ is called *exact*.

We will now show that *locally* (we shall explain what this entails) the condition

$$\frac{\partial G}{\partial y} = \frac{\partial H}{\partial x}$$

is enough to guarantee that $\omega = \mathrm{d}f$ for some $f$, determined up to a constant.

We first choose a convenient point $P_a$ and declare that $f(P_a) = 0$. We then define $f$ by the rule $f(P) = \int_\Gamma \omega$, where $\Gamma$ is a convenient path extending from $P_a$ to $P$. Of course, we could add a constant to $f$ without changing its differential $\mathrm{d}f$. The choice of $P_a$ in effect chooses this *constant of integration.*

Let us describe this procedure in terms of coordinates. Suppose that

$$\omega = G(x, y)\,\mathrm{d}x + H(x, y)\,\mathrm{d}y.$$

For simplicity we may assume $P_a$ to be the origin; so that $f\left(\begin{pmatrix}0\\0\end{pmatrix}\right) = 0$. The most convenient path $\Gamma$ joining the origin to the point $\begin{pmatrix} x \\ y \end{pmatrix}$ is a straight line segment, easily parameterized by

$$\alpha: t \to \begin{pmatrix} xt \\ yt \end{pmatrix} \quad 0 \leqslant t \leqslant 1.$$

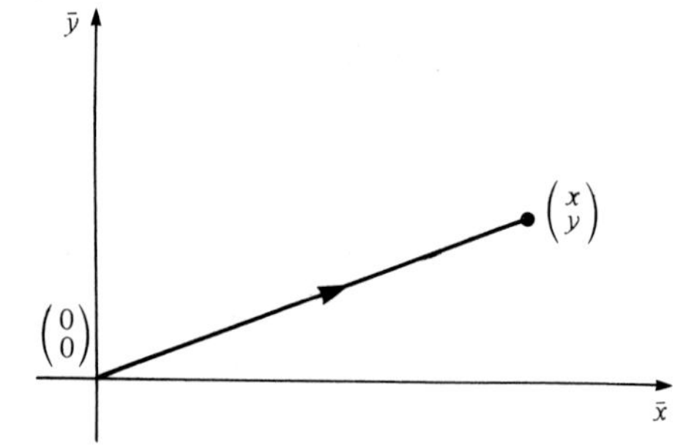


**Figure 7.17**

Since we are using $x$ and $y$ to describe the *endpoints* of the path $\Gamma$, we will use $\bar{x}$ and $\bar{y}$ as names for the dummy variables of integration. Thus

$$\alpha^*\bar{x} = xt, \qquad \alpha^*\bar{y} = yt,$$
$$\alpha^*\mathrm{d}\bar{x} = x\,\mathrm{d}t, \quad \alpha^*\mathrm{d}\bar{y} = y\,\mathrm{d}t$$

and

$$\alpha^*\omega = G(xt, yt)x\,\mathrm{d}t + H(xt, yt)y\,\mathrm{d}t.$$

Then

$$f(P) = \int_0^1 \alpha^*\omega$$

so that

$$f(x, y) = \int_0^1 [xG(xt, yt) + yH(xt, yt)]\,\mathrm{d}t \tag{7.3}$$

is a formula by which we reconstruct a function $f$ from $\omega$. Notice that this construction will succeed only if the functions $G$ and $H$ are defined everywhere on the path $\Gamma$.

So far we have not used any hypothesis on $\omega$, other than that it be defined along the paths of integration. So we do not expect, in general, that $\mathrm{d}f = \omega$. Here is where our hypothesis will come in. Let us compute $\partial f/\partial x$. By differentiating with respect to $x$ under the integral sign in the definition of $f$, we see that

$$\frac{\partial f}{\partial x} = \int_0^1 G(xt, yt)\,\mathrm{d}t + \int_0^1 \left( xt\frac{\partial G}{\partial x}(xt, yt) + ty\frac{\partial H}{\partial x}(xt, yt) \right)\mathrm{d}t.$$

Now

$$\frac{\mathrm{d}}{\mathrm{d}t}(tG(xt, yt)) = G(xt, yt) + tx\frac{\partial G}{\partial x}(xt, yt) + ty\frac{\partial G}{\partial y}(xt, yt)$$

so

$$G(x, y) = [tG(xt, yt)]_0^1 = \int_0^1 \frac{d}{dt}(tG(xt, yt))\,dt$$
$$= \int_0^1 \left( G(xt, yt) + tx\frac{\partial G}{\partial x}(xt, yt) + ty\frac{\partial G}{\partial y}(xt, yt) \right) dt.$$

Substituting this into the expression for $\partial f/\partial x$, we see that

$$\frac{\partial f}{\partial x} = G(x, y) + \int_0^1 ty\left( \frac{\partial H}{\partial x}(xt, yt) - \frac{\partial G}{\partial y}(xt, yt) \right) dt.$$

Under our assumption

$$\frac{\partial G}{\partial y} = \frac{\partial H}{\partial x}$$

everywhere, so

$$\frac{\partial f}{\partial x} = G.$$

Similarly

$$\frac{\partial f}{\partial y} = H$$

or

$$\omega = df.$$

As an explicit example of reconstructing a function from its differential, we consider

$$\omega = 2xy^3 dx + 3x^2y^2 dy.$$

Since $\omega$ is defined everywhere, and

$$\frac{\partial}{\partial y}(2xy^3) = 6xy^2 = \frac{\partial}{\partial x}(3x^2y^2),$$

$\omega$ is a differential, $df$. We find $f$ by calculating

$$f(x, y) = \int_0^1 [xG(xt, yt) + yH(xt, yt)]\,dt$$
$$f(x, y) = \int_0^1 (2x^2y^3 + 3x^2y^3)t^4\,dt = x^2y^3.$$

Indeed,

$$d(x^2y^3) = 2xy^3\,dx + 3x^2y^2\,dy.$$

**Closed forms that are not exact**

Our construction shows that if $\omega$ is defined and bounded in a *star-shaped region*, a region $R$, whose points may all be joined to some interior point, $O$, by straight line segments lying in $R$, then the condition $\partial G/\partial y = \partial H/\partial x$ is sufficient to show that $\omega$ is exact. If $G$ or $H$ fails to be defined at one point in the region, then this conclusion no longer holds.

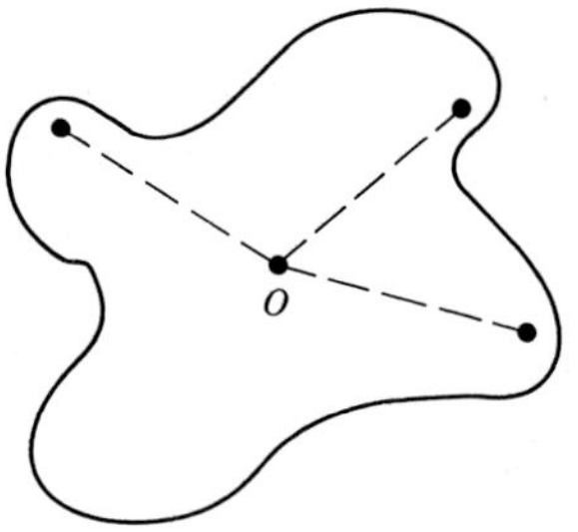


**Figure 7.18**

For example

$$\omega = -\frac{y}{x^2+y^2}\,\mathrm{d}x + \frac{x}{x^2+y^2}\,\mathrm{d}y$$

satisfies $\partial G/\partial y = \partial H/\partial x$, but $\omega$ is not defined at $x=0$, $y=0$. In this case there exists no function $f$ for which $\omega = \mathrm{d}f$. Indeed, the integral of $\omega$ around the unit circle is easily shown to be different from zero. Take

$$\alpha^*x = \cos t, \qquad \alpha^*y = \sin t,$$
$$\alpha^*\mathrm{d}x = -\sin t\,\mathrm{d}t, \quad \alpha^*\mathrm{d}y = \cos t$$

so that

$$\alpha^*\omega = \sin^2 t\,\mathrm{d}t + \cos^2 t\,\mathrm{d}t = \mathrm{d}t.$$

Then

$$\int_{\substack{\text{unit}\\ \text{circle}}} \omega = \int_0^{2\pi} \mathrm{d}t = 2\pi.$$

If $\omega$ were exact, its integral around this closed path would have to be zero. In a later chapters we shall consider this and related ideas, which are of great significance for electromagnetic theory, in detail.

A form $\omega = G\,\mathrm{d}x + H\,\mathrm{d}y$ defined in some region of $\mathbb{R}^2$ is called *closed* if $\partial H/\partial x = \partial G/\partial y$. If the region is star-shaped we have proved that a closed form is exact. In general this is not true.

**Pullback and integration**

Since the definition of the line integral $\int_\Gamma \omega$ is independent of any specific choice of parameterization for $\Gamma$, it is clear that the calculated value of the integral cannot

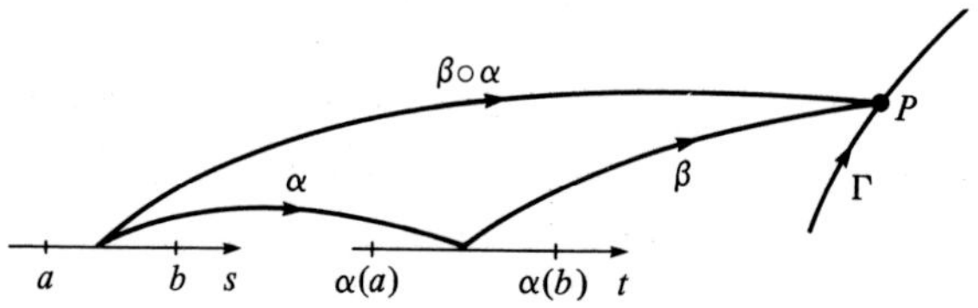


**Figure 7.19**

depend on the parameterization. Still, it is worth demonstrating this independence explicitly. Suppose that we have two alternative parameterizations of $\Gamma$ in terms of parameters $s$ and $t$.

Then there exists a one-to-one mapping $\alpha$ of the $s$-line into the $t$-line, as shown in figure 7.19, so that we may write the parameterizations of $\Gamma$ as

$$P = \beta(t) \quad \text{and} \quad P = \beta(\alpha(s)) = \beta \circ \alpha(s).$$

Using the parameter $s$, we calculate

$$\int_\Gamma \omega = \int_a^b (\beta \circ \alpha)^* \omega = \int_a^b \alpha^*(\beta^* \omega) = \int_{\alpha(a)}^{\alpha(b)} \beta^* \omega$$

by the chain rule, and by the change-of-variables formula for ordinary integrals.

This is exactly what we would have obtained by using the parameter $t$. As a practical matter, this means that using a different parameterization is equivalent computationally to making a change of variable in the integral set up by using the original parameterization.

It is also possible to transform a line integral from one *plane* to another, as suggested by figure 7.20. Here $\beta$ is a differentiable one-to-one mapping of the path $\Gamma$ in plane $\mathbb{A}$ into a path $\beta(\Gamma)$ in plane $\mathbb{B}$. Given a one-form $\omega$ on plane $\mathbb{B}$, we have defined its pullback by

$$\beta^*\omega(P)[\mathbf{v}] = \omega(\beta(P))[d\beta_P[\mathbf{v}]].$$

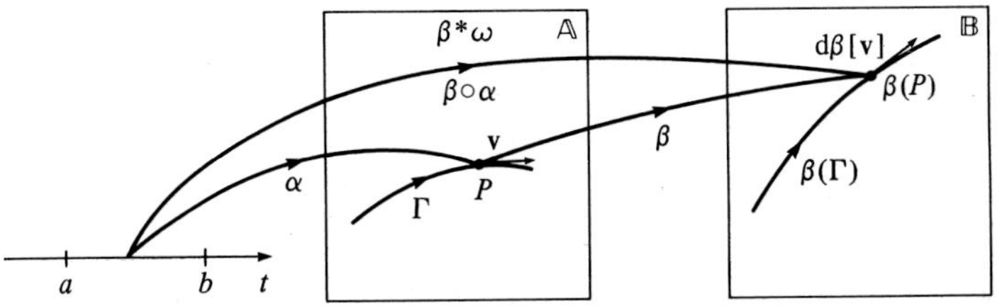


**Figure 7.20**

We claim that

$$\int_{\beta(\Gamma)} \omega = \int_\Gamma \beta^* \omega.$$

Indeed, by definition,

$$\int_\Gamma \beta^*\omega = \lim_{N\to\infty} \sum_{i=0}^{N-1} \beta^*\omega(P_i)[\mathbf{v}_i]$$

$$= \lim_{N\to\infty} \sum_{i=0}^{N-1} \omega(\beta(P_i))[d\beta_{P_i}[\mathbf{v}_i]].$$

But, as $N \to \infty$, the vectors $d\beta[\mathbf{v}_i]$ lie along the path $\beta(\Gamma)$ more and more closely, so that this last sum, in the limit $N \to \infty$, equals the integral $\int_{\beta(\Gamma)} \omega$. Indeed if we parameterize $\Gamma$ by the mapping $\alpha$, we have

$$\int_\Gamma \beta^*\omega = \int_a^b \alpha^*(\beta^*\omega) = \int_a^b (\beta \circ \alpha)^*\omega$$

by the chain rule. But $\beta \circ \alpha$ is a parameterization of $\beta(\Gamma)$, so this last integral equals $\int_{\beta(\Gamma)} \omega$, which is what we wanted to prove.

**Example**

The implication of this last result is that we may introduce any convenient coordinate system in the plane for purposes of evaluating a line integral. For example, if we wish to evaluate the integral of $\omega = x\,dy - y\,dx$ over the unit semi-circle from $\begin{pmatrix} 1 \\ 0 \end{pmatrix}$ to $\begin{pmatrix} -1 \\ 0 \end{pmatrix}$ in the $xy$-plane, we may express the semicircle as the image of a directed line segment $\Gamma$ in the *polar coordinate plane* by means of the mapping $\beta: \begin{pmatrix} r \\ \theta \end{pmatrix} \rightarrow \begin{pmatrix} r\cos\theta \\ r\sin\theta \end{pmatrix}$. Then

$$\beta^*\omega = (r\cos\theta)\mathrm{d}(r\sin\theta) - (r\sin\theta)\mathrm{d}(r\cos\theta) = r^2\,\mathrm{d}\theta.$$

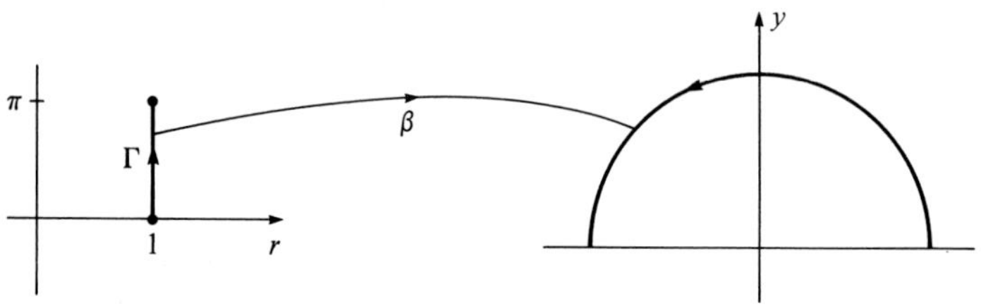


**Figure 7.21**

On the segment $\Gamma$, $r = 1$ and so

$$\int_\Gamma \beta^*\omega = \int_0^\pi \mathrm{d}\theta = \pi.$$

Of course, calculation of $\int_{\beta(\Gamma)} \omega$, using the obvious parameterization $t \rightarrow \begin{pmatrix} \cos t \\ \sin t \end{pmatrix}$, leads to exactly the same integral.

## 7.2. Arc length

So far we have considered only directed line integrals, evaluated over an *oriented* path on a plane where no scalar product is necessarily defined. Given a scalar

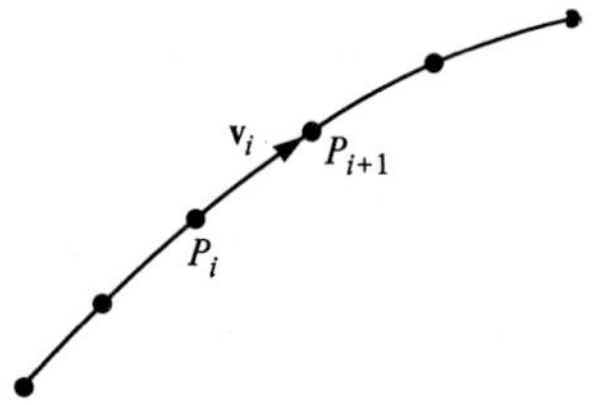


**Figure 7.22**

product, it is possible to define also an *absolute* line integral of a *function* $f$ over a path $\Gamma$. To define the integral $\int_\Gamma f\,ds$, we again break the path $\Gamma$ into short segments, with $\mathbf{v}_i = P_iP_{i+1}$, then take the *length* of each segment by using the scalar product: $s_i = \|\mathbf{v}_i\| = \sqrt{(\mathbf{v}_i, \mathbf{v}_i)}$. The integral is again defined as the limit of a sum:

$$\int f\,ds = \lim_{N\to\infty} \sum_{i=0}^{N-1} f(P_i)s_i.$$

Clearly in this case the orientation of $\Gamma$ does not matter, since the length of $\mathbf{v}_i$ is the same as the length of $-\mathbf{v}_i$.

To evaluate an absolute line integral, it is again convenient to parameterize $\Gamma$. We write

$$\alpha^*x = X(t), \quad \alpha^*y = Y(t)$$

so that

$$\alpha^*dx = X'(t)\,dt, \quad \alpha^*dy = Y'(t)\,dt.$$

Then the vector $\mathbf{u}_i = t_it_{i+1}$ is mapped into the vector

$$d\alpha[\mathbf{u}_i] = \begin{pmatrix} X'(t) \\ Y'(t) \end{pmatrix} dt[\mathbf{u}_i].$$

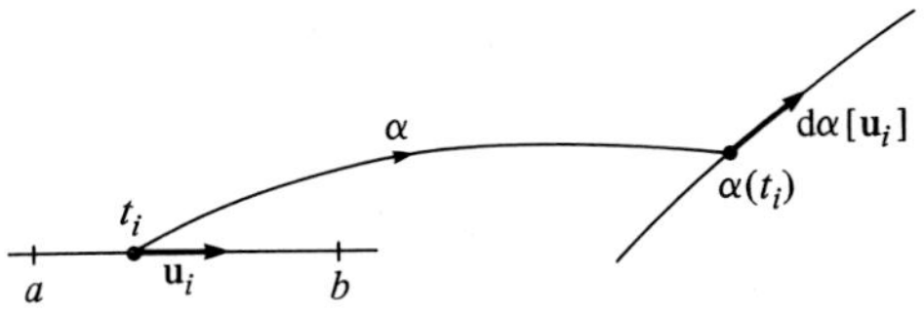


**Figure 7.23**

If the scalar product is the ordinary Euclidean one, the length of this vector is

$$\|d\alpha[\mathbf{u}_i]\| = \sqrt{(X'(t_i)^2 + Y'(t_i)^2)}\,dt[\mathbf{u}_i].$$

By definition

$$f(P_i) = f(\alpha(t_i)) = \alpha^* f(t_i)$$

so that

$$\int f\,ds = \lim_{N\to\infty} \sum_{i=0}^{N-1} \alpha^* f(t_i)\sqrt{(X'(t_i)^2 + Y'(t_i)^2)}(t_{i+1} - t_i)$$

which may be recognized as the integral

$$\int_\Gamma f\,ds = \int_a^b \alpha^* f(t)\sqrt{(X'(t)^2 + Y'(t)^2)}\,dt.$$

If $f = 1$, the integral $\int ds$ defines the length of the curve $\Gamma$. More generally, $f(P)$ might represent the linear mass density of a thin wire in the form of the curve $\Gamma$; then $\int f\,ds$ represents the total mass of the wire.

A physically important example of an absolute line integral for which the relevant scalar product is *not* Euclidean is the calculation of the *proper time* associated with the world line of a moving particle, which is the elapsed time as measured by a clock moving with the particle. In this case, since $t$ always increases along a world line, it will serve as a parameter, so we write

$$\alpha^*t = t, \qquad \alpha^*(x) = X(t)$$

and

$$\alpha^*\mathrm{d}t = \mathrm{d}t, \quad \alpha^*(\mathrm{d}x) = X'(t)\,\mathrm{d}t.$$

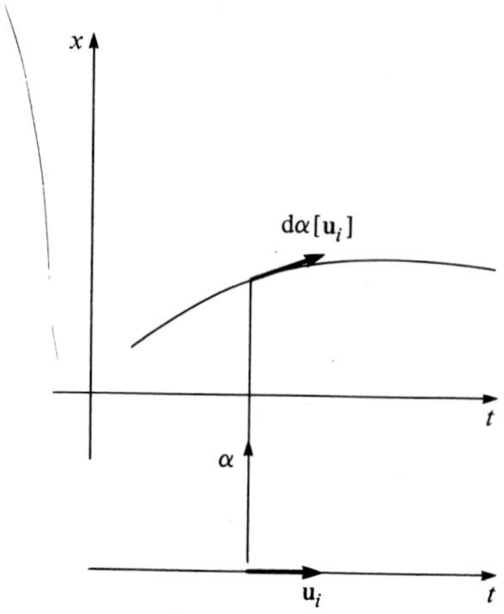


**Figure 7.24**

Now the *Lorentz* length of the vector

$$\mathrm{d}\alpha[\mathbf{u}_i] = \begin{pmatrix} 1 \\ X'(t) \end{pmatrix} \mathrm{d}t[\mathbf{u}_i]$$

is $\sqrt{(1 - X'(t)^2)}\,\mathrm{d}t[\mathbf{u}_i]$. Since $X'(t)$ is just the velocity $v$ of the particle, we may write

$$\int_\Gamma \sqrt{(1 - v^2)}\,\mathrm{d}t$$

as the integral which defines proper time.

---

## Summary

A Line integrals

You should be able to explain the meaning of a line integral of the form $\int_\Gamma \omega$ and to list and apply properties of this integral.

You should know the prescription for evaluating a line integral by pullback, and you should be able to introduce appropriate parameterizations for evaluating line integrals and specified paths.

B Differentials and differential forms

Given a one-form $\omega$ defined on a star-shaped region of the plane, you should be able to determine whether or not $\omega$ can be expressed as the differential of a function $f$, and to calculate such a function $f$ if one exists.

---

## Exercises

7.1. Let $\omega = (y\cos xy + e^x)dx + (x\cos xy + 2y)dy$.

(a) Evaluate $\int_\Gamma \omega$ along the segment of the parabola $y = x^2$ from $\begin{pmatrix}0\\0\end{pmatrix}$ to $\begin{pmatrix}1\\1\end{pmatrix}$. Use the parameterization $\phi$ described by the pullback

$$\begin{pmatrix}\phi^* x\\ \phi^* y\end{pmatrix} = \begin{pmatrix}t\\ t^2\end{pmatrix}.$$

(b) Evaluate $\int_\Gamma \omega$ for the case where $\Gamma$ is the straight line joining the origin to the point $\begin{pmatrix}\alpha\\ \beta\end{pmatrix}$. Do the same for the case where $\Gamma$ consists of the segment $0 \leqslant x \leqslant \alpha$ on the $x$-axis, followed by the segment $x = \alpha$, $0 \leqslant y \leqslant \beta$.

(c) Find a function $f(x, y)$ such that $\omega = df$.

7.2. Let $\omega = (y\cos xy + e^x)dx + (x\cos xy + 2y)dy$.

(a) Evaluate $\int_\Gamma \omega$ along the parabola $\gamma$ defined by

$$\begin{pmatrix}x\\ y\end{pmatrix} = \begin{pmatrix}t\\ t^2\end{pmatrix} \quad \text{for} \quad 0 \leqslant t \leqslant 1.$$

(b) Find $f(x, y)$ such that $\omega = df$.

7.3. Let $\omega = ydx - xdy$.

(a) Evaluate $\int_\gamma \omega$ along the semicircle $\gamma$ from $\begin{pmatrix}-1\\0\end{pmatrix}$ to $\begin{pmatrix}1\\0\end{pmatrix}$ defined by

$$\begin{pmatrix}x\\ y\end{pmatrix} = \begin{pmatrix}-\cos t\\ \sin t\end{pmatrix}$$

for $0 < t < \pi$.

(b) Show explicitly that you can obtain a different value from that in (a) by choosing a different curve joining $\begin{pmatrix}-1\\0\end{pmatrix}$ to $\begin{pmatrix}1\\0\end{pmatrix}$.

7.4. Let $\omega = (15x^2y^2 - 3y)dx + (10x^3y - 3x)dy$. Evaluate $\int_\Gamma \omega$, where $\Gamma$ is the path from $(-1, 0)$ to $(1, 0)$ along the semicircle $x^2 + y^2 = 1$, $y \geqslant 0$.

7.5.(a) Evaluate $\int_\Gamma \omega$, where

$$\omega = dx + 2xdy$$

and $\Gamma$ is the segment of the parabola $x = 1 - y^2$ between $y = -1$ and $y = +1$, as shown in figure 7.25.

(b) Find a constant $a$ and a function $f$ such that $e^{ay}\omega = df$.

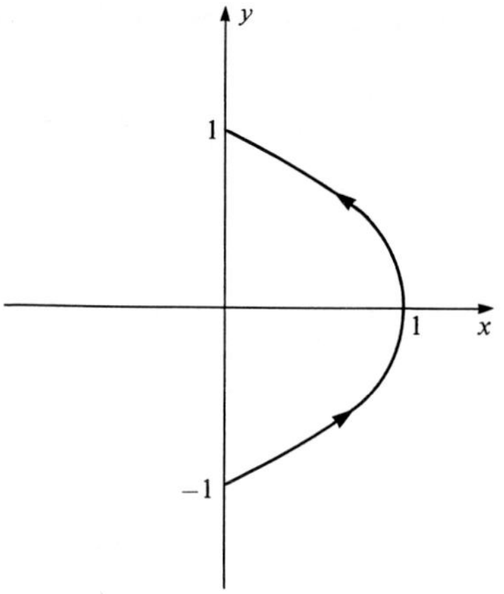


**Figure 7.25**

7.6.(a) Let $\omega = (x^2 - 2xy)\mathrm{d}x + (y^2 - 2xy)\mathrm{d}y$. Evaluate $\int_\Gamma \omega$, where $\Gamma$ is the path from $(-1, 1)$ to $(1, 1)$ along the parabola $y = x^2$.

(b) Let $\sigma = 30x^2y^5\mathrm{d}x + 40x^3y^4\mathrm{d}y$. Find an integer $n$ and a function $f$ such that $\mathrm{d}f = (xy)^n\sigma$.

7.7. Let $\omega = 10y^2\mathrm{d}x + 4xy\mathrm{d}y$.

(a) Evaluate $\int_\Gamma \omega$, where $\Gamma$ is the circular arc of radius 1 joining $\begin{pmatrix}1\\0\end{pmatrix}$ to $\begin{pmatrix}0\\1\end{pmatrix}$. (Hint: use $\theta$ as a parameter.)

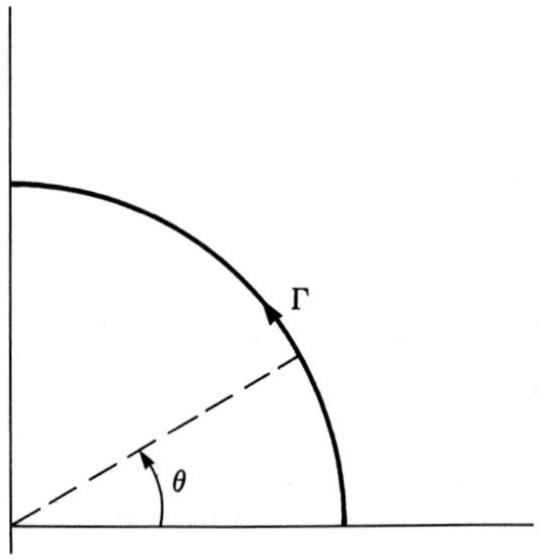


**Figure 7.26**

(b) Find an integer $n$ and a function $f$ such that $x^n\omega = \mathrm{d}f$.

7.8.(a) Show that the differential form

$$\omega = 3xy\mathrm{d}x + 2x\mathrm{d}y$$

is not the differential of any function $f$.

(b) Find the equation for the one-parameter family of curves with the property that $\omega(\mathbf{v}) = 0$ for any vector $\mathbf{v}$ which is tangent to one of the curves.

$\left(\text{Hint: If } y = F(x),\ \mathbf{v} = \begin{pmatrix} 1 \\ F'(x) \end{pmatrix}, \text{ and you have a differential equation for } F(x).\right)$

(c) Find functions $f(x, y)$ and $g(x, y)$ such that $\mathrm{d}f = g\omega$.
(Hint: $f$ must be constant along the curves which you found in part (b).)

7.9.(a) Sketch the semi-ellipse described by the polar equation

$$r = \frac{9}{5 - 4\cos\theta} \quad \text{for} \quad 0 \leqslant \theta \leqslant \pi.$$

Recalling that $x = r\cos\theta$, $y = r\sin\theta$, show that this semi-ellipse is part of the graph of

$$\frac{(x-4)^2}{25} + \frac{y^2}{9} = 1.$$

(b) Express the differential form

$$\omega = \frac{xy\mathrm{d}y - y^2\mathrm{d}x}{x^2 + y^2}$$

in terms of polar coordinates (in terms of $r, \theta, \mathrm{d}r$, and $\mathrm{d}\theta$).

(c) Evaluate $\int_\Gamma \omega$, where $\Gamma$ is the semi-ellipse of part (a), using polar coordinates. The coordinate $\theta$ makes a convenient parameter.

(d) Evaluate $\int_\Gamma \omega$ by using $x$ and $y$ as coordinates. A convenient parameterization is the one defined by the mapping

$$t \mapsto \begin{pmatrix} 4 + 5\cos t \\ 3\sin t \end{pmatrix} \quad 0 \leqslant t \leqslant \pi.$$

7.10.(a) Suppose that $u$ and $v$ are curvilinear coordinates on a region $D$ on the plane, with the Jacobian

$$\mathrm{Det}\begin{pmatrix} \partial u/\partial x & \partial u/\partial y \\ \partial v/\partial x & \partial v/\partial y \end{pmatrix}$$

nowhere zero on $D$. Let $\omega$ be a smooth differential form defined on $D$, let $\Gamma$ be a curve in $D$. Show that $\int_\Gamma \omega$ has the same value whether $\omega$ and $\Gamma$ are expressed in terms of $x$ and $y$ or in terms of $u$ and $v$. (The preceding problem was an example of this result.)

(b) Let $\Gamma$ be a closed path described in polar coordinates by $\rho = F(\theta)$, with $F(\theta) > 0$ and $F(2\pi) = F(0)$. Show that the area enclosed by this closed path equals $\int_\Gamma \omega$, where $\omega = \frac{1}{2}\rho^2\mathrm{d}\theta$.
(Hint: Try expressing $\omega$ in terms of Cartesian coordinates.)

7.11. The state of a gas confined to a cylinder can be represented by a point in a plane. In terms of coordinates $P$ (pressure) and $V$ (volume) on this plane, the quantity of heat absorbed by the gas during a process represented by a path $\Gamma$ in the plane is $Q = \int_\Gamma \omega$, where

$$\omega = \tfrac{5}{2}P\mathrm{d}V + \tfrac{3}{2}V\mathrm{d}P.$$

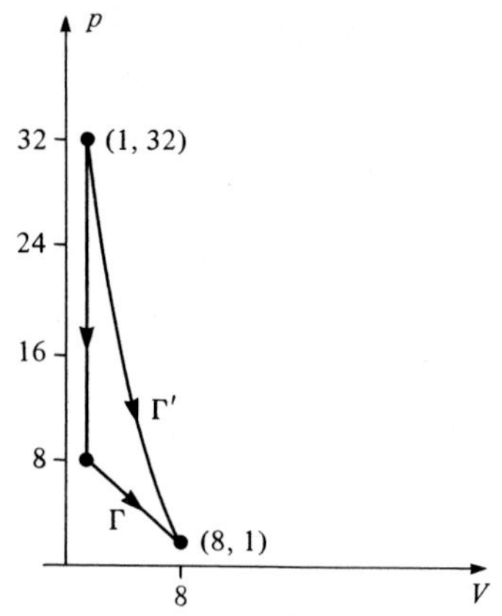


**Figure 7.27**

(a) Evaluate $\int_\Gamma \omega$ where $\Gamma$ is the broken line shown in figure 7.27, connecting $V=1$, $P=32$ to $V=1$, $P=8$, then to $V=8$, $P=1$.
(b) Evaluate $\int_{\Gamma'} \omega$ where $\Gamma'$ is the curve $PV^{5/3}=32$ joining $V=1, P=32$ to $V=8$, $P=1$.
(c) Find a function $S$ such that $\mathrm{d}S=\omega/PV$.

7.12.(a) A uniform wire of mass $M$ is bent into a semicircle of radius $R$ as shown in figure 7.28. Find the $y$-coordinate of its center of mass, and calculate its moment of inertia about the $x$-axis.

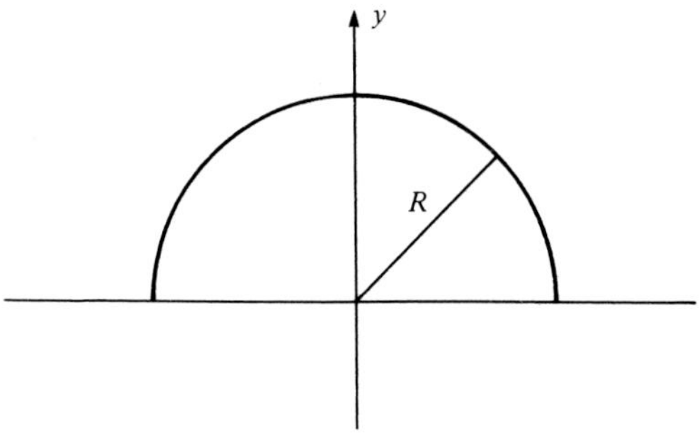


**Figure 7.28**

(b) Solve the same problem for the case where the linear density of the wire is proportional to $y$, with the mass still equal to $M$.

7.13. Express the length of the cubic curve $y=(x-2)^3$ connecting $\begin{pmatrix} 1 \\ -1 \end{pmatrix}$ and $\begin{pmatrix} 3 \\ 1 \end{pmatrix}$ in the Euclidean plane as an integral.

7.14. Consider the differential form

$$\omega = 5y\mathrm{d}x + 3x\mathrm{d}y.$$

(a) Find the integers $m$ and $n$ such that $\mathrm{d}(x^m y^n \omega)=0$.

(b) For these integers find a function $f$ such that $\mathrm{d}f = x^m y^n \omega$.

(c) If we map the $uv$-plane into the $xy$-plane so that

$$\begin{pmatrix} \alpha \\ \beta \end{pmatrix} \mapsto \begin{pmatrix} \alpha^2 + \beta \\ \beta^2 + 2\alpha \end{pmatrix}$$

what is the pullback of $\omega$?

(d) Calculate $\int \omega$ over the path $\Gamma_1$ and $\Gamma_2$ connecting $\begin{pmatrix} 0 \\ 0 \end{pmatrix}$ and $\begin{pmatrix} 1 \\ 1 \end{pmatrix}$ where $\Gamma_1$ goes in two straight segments via $\begin{pmatrix} 1 \\ 0 \end{pmatrix}$ and $\Gamma_2$ in two straight segments via $\begin{pmatrix} 0 \\ 1 \end{pmatrix}$.

(e) Evaluate the absolute arc-length integral $\int_{\Gamma_1} \|\omega\| \, \mathrm{d}s$.
(You may leave one term of your answer in the form of an ordinary definite integral.)

# 8

# Double integrals



Chapter 8 continues the study of integral calculus and is devoted to the study of exterior two-forms and their corresponding two-dimensional integrals. The exterior derivative is introduced and invariance under pullback is stressed. The two-dimensional version of Stokes' theorem, i.e. Green's theorem, is proved. Surface integrals in three-space are studied.

## 8.1. Exterior derivative

We have already seen how the differential of a function $f$ provides the best linear approximation to the change in the value of $f$ as we move from a point $\mathbf{P}$ to a nearby point $\mathbf{P}+\mathbf{v}$. To be specific,

$$\mathrm{d}f(\mathbf{P})[\mathbf{v}] = f(\mathbf{P}+\mathbf{v}) - f(\mathbf{P}) + o(\mathbf{v})$$

where the error, $o(\mathbf{v})$, goes to zero faster than the length of $\mathbf{v}$ if $\mathbf{v}$ is made small. We can think of $\mathrm{d}f$ as a linear function whose value on the vector $\mathbf{v}$ is determined by the values of the function $f$ itself on the boundary (endpoints) of the segment defined by $\mathbf{v}$, in the limit where the vector $\mathbf{v}$ becomes very small.

Using a similar approach, we can construct from a one-form $\tau$ a *two-form* $\mathrm{d}\tau$, called the *exterior derivative* of $\tau$, which is, at each point $\mathbf{P}$, a *bilinear* function of *two* vectors $\mathbf{v}$ and $\mathbf{w}$: Given a point $\mathbf{P}$, an *ordered* pair of vectors $\mathbf{v}$, $\mathbf{w}$, and a one-form $\tau$, we can obtain a number by integrating $\tau$ around the parallelogram spanned by $\mathbf{v}$ and $\mathbf{w}$, moving 'forward' from $\mathbf{P}$ along the first vector $\mathbf{v}$ of the pair, eventually backwards along the second vector $\mathbf{w}$. If the vectors $\mathbf{v}$ and $\mathbf{w}$ are small enough, and $\tau$ is reasonably well-behaved near $\mathbf{P}$, then we expect the value of this integral to be approximately *bilinear*, i.e., to depend approximately *linearly* on $\mathbf{v}$ (for fixed $\mathbf{w}$) and linearly on $\mathbf{w}$ (for fixed $\mathbf{v}$). Denoting the parallelogram spanned by $h\mathbf{v}$ and

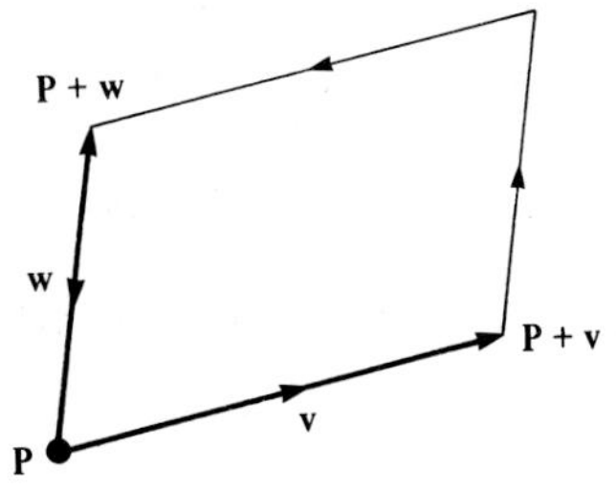


**Figure 8.1**

$k\mathbf{w}$ by $\mathrm{P}(h,k)$ we would like to define $\mathrm{d}\tau\,(\mathbf{P})$ by

$$\int_{\mathrm{P}(h,k)} \tau = hk\,\mathrm{d}\tau(\mathbf{P})[\mathbf{v},\mathbf{w}] + \text{error} \tag{8.1}$$

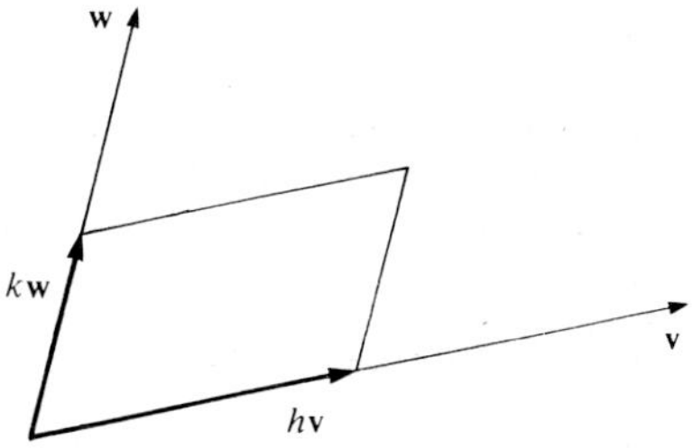


**Figure 8.2**

where (we hope) the error term goes to zero faster than $h$ as $h \to 0$ (with $k$ fixed), and also goes to zero faster than $k$ as $k \to 0$ (with $h$ fixed).

If $\mathrm{d}\tau(\mathbf{P})$ exists, it is unique. The proof is essentially the same as the proof that the differential of a function is unique. Suppose that equation (8.1) holds for two different bilinear functions $\mathrm{d}\tau$ and $\overline{\mathrm{d}\tau}$. Then, letting $\sigma$ denote the difference $\mathrm{d}\tau - \overline{\mathrm{d}\tau}$, we would have

$$0 = hk\sigma[\mathbf{v},\mathbf{w}] + \text{error}.$$

Dividing by $hk$ and letting $h$ approach zero, we find

$$0 = \sigma[\mathbf{v},\mathbf{w}] + \frac{1}{k}\lim_{h\to 0}(\text{error}/h).$$

But the error approaches zero faster than $h$, so $\sigma[\mathbf{v},\mathbf{w}] = 0$. This proves that $\sigma[\mathbf{v},\mathbf{w}]$ is the zero function, so that $\overline{\mathrm{d}\tau}$ cannot be different from $\mathrm{d}\tau$.

We turn next to the problem of calculating $\mathrm{d}\tau(\mathbf{P})$ and proving that it exists. For simplicity, we assume initially that $\tau$ is of the form $f\,\mathrm{d}x$, where $f$ is twice differentiable everywhere near $\mathbf{P}$. Here $\mathrm{d}x$ is the form which assigns to every tangent vector its $x$-component. In particular, $\mathrm{d}x[h\mathbf{v}] = h\mathrm{d}x[\mathbf{v}] = hv_x$ if $\mathbf{v} = \begin{pmatrix} v_x \\ v_y \end{pmatrix}$. A typical contribution from one side of the parallelogram, say the side from $\mathbf{P}$ to $\mathbf{P}+h\mathbf{v}$, is found by using the parameterization $t \to \mathbf{P}+th\mathbf{v}$ so that the contribution is

$$\int_0^1 f(\mathbf{P}+th\mathbf{v})x'(t)\mathrm{d}t = \int_0^1 f(\mathbf{P}+th\mathbf{v})\mathrm{d}x[h\mathbf{v}]\mathrm{d}t = h\mathrm{d}x[\mathbf{v}]\int_0^1 f(\mathbf{P}+th\mathbf{v})\mathrm{d}t.$$

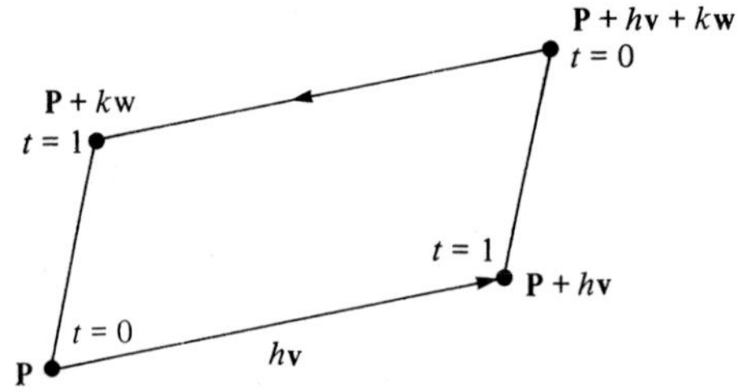


**Figure 8.3**

The contribution from the opposite side, from $\mathbf{P}+h\mathbf{v}+k\mathbf{w}$ to $\mathbf{P}+k\mathbf{w}$, is similarly

$$-h\mathrm{d}x[\mathbf{v}]\int_0^1 f(\mathbf{P}+k\mathbf{w}+th\mathbf{v})\mathrm{d}t.$$

Combining the two terms, we obtain

$$-h\mathrm{d}x[\mathbf{v}]\int_0^1 [f(\mathbf{P}+th\mathbf{v}+k\mathbf{w})-f(\mathbf{P}+th\mathbf{v})]\mathrm{d}t.$$

Since $f$ is assumed twice differentiable, we may apply Taylor's formula

$$f(\mathbf{p}+th\mathbf{v}+k\mathbf{w})-f(\mathbf{p}+k\mathbf{w}) = \mathrm{d}f_{(\mathbf{p}+th\mathbf{v})}[k\mathbf{w}]+O(k^2)$$

to write this last expression as

$$-h\mathrm{d}x[\mathbf{v}]\int_0^1 \mathrm{d}f_{(\mathbf{P}+th\mathbf{v})}[k\mathbf{w}]\,\mathrm{d}t + O(hk^2).$$

From the other two sides of the parallelogram we obtain terms which combine similarly to give

$$+k\mathrm{d}x[\mathbf{w}]\int_0^1 \mathrm{d}f_{(\mathbf{P}+tk\mathbf{w})}[h\mathbf{v}]\mathrm{d}t + O(h^2k).$$

Substituting these results into the integral around the parallelogram, we get

$$\int_{\mathbf{P}(h,k)} \tau = hk\left[-\int_0^1 \mathrm{d}f_{(\mathbf{P}+th\mathbf{v})}[\mathbf{w}]\mathrm{d}t\cdot\mathrm{d}x[\mathbf{v}] + \int_0^1 \mathrm{d}f_{(\mathbf{P}+tk\mathbf{w})}[\mathbf{v}]\mathrm{d}t\cdot\mathrm{d}x[\mathbf{w}]\right]$$
$$+O(h^2k)+O(hk^2).$$

Now $\mathrm{d}f_{(\mathbf{P}+th\mathbf{v})}$ is just the row vector

$$\left(\frac{\partial f}{\partial x}(\mathbf{P}+th\mathbf{v}),\ \frac{\partial f}{\partial y}(\mathbf{P}+th\mathbf{v})\right).$$

By assumption, the partial derivatives of $f$ are differentiable. Hence, by the mean-value theorem

$$\frac{\partial f}{\partial x}(\mathbf{P} + th\mathbf{v}) = \frac{\partial f}{\partial x}(\mathbf{P}) + O(h)$$

so

$$\mathrm{d}f_{(\mathbf{P}+th\mathbf{v})}[\mathbf{w}] = \mathrm{d}f_{\mathbf{P}}[\mathbf{w}] + O(h)$$

or, upon integration,

$$\int_0^1 \mathrm{d}f_{(\mathbf{P}+th\mathbf{v})}[\mathbf{w}]\,\mathrm{d}t = \mathrm{d}f_{\mathbf{P}}[\mathbf{w}] + O(h).$$

Substituting into our integral around the parallelogram gives

$$\int_{\mathbf{P}(h,k)} \tau = hk(\mathrm{d}f_{\mathbf{P}}[\mathbf{v}]\mathrm{d}x[\mathbf{w}] - \mathrm{d}f_{\mathbf{P}}[\mathbf{w}]\mathrm{d}x[\mathbf{v}]) + O(h^2k) + O(hk^2).$$

We thus get our desired expression (8.1) if we set

$$\boxed{\mathrm{d}\tau_{\mathbf{P}}[\mathbf{v}, \mathbf{w}] = \mathrm{d}f_{\mathbf{P}}[\mathbf{v}]\mathrm{d}x[\mathbf{w}] - \mathrm{d}f_{\mathbf{P}}[\mathbf{w}]\mathrm{d}x[\mathbf{v}].}$$

We see that $\mathrm{d}\tau$ is an antisymmetric function of its two arguments: $\mathrm{d}\tau[\mathbf{w}, \mathbf{v}] = -\mathrm{d}\tau[\mathbf{v}, \mathbf{w}]$.

We can express $\mathrm{d}\tau$ more concisely by introducting the exterior or 'wedge' product of two one-forms, defined by

$$\boxed{(\sigma \wedge \lambda)[\mathbf{v}, \mathbf{w}] = \sigma[\mathbf{v}]\lambda[\mathbf{w}] - \sigma[\mathbf{w}]\lambda[\mathbf{v}]}$$

where $\sigma$ and $\lambda$ are one-forms, $\mathbf{v}$ and $\mathbf{w}$ are vectors. Then we may write

$$\mathrm{d}\tau_{\mathbf{P}}[\mathbf{v}, \mathbf{w}] = (\mathrm{d}f_{\mathbf{P}} \wedge \mathrm{d}x)[\mathbf{v}, \mathbf{w}]$$

or, more concisely,

$$\mathrm{d}\tau = \mathrm{d}f \wedge \mathrm{d}x.$$

From the definition of the wedge product it is clear that

$$\boxed{\lambda \wedge \sigma = -\sigma \wedge \lambda,}$$

i.e., the product is antisymmetric. In particular, $\sigma \wedge \sigma = -\sigma \wedge \sigma = 0$: the wedge product of any one-form with itself is zero.

It is also apparent that

$$(\sigma + \omega) \wedge \lambda = (\sigma \wedge \lambda) + (\omega \wedge \lambda)$$

the wedge product is distributive with respect to addition.

Consider now the most general one-form $f\mathrm{d}x + g\mathrm{d}y$. The same argument applied to $g\mathrm{d}y$ will lead to

$$\mathrm{d}(g\mathrm{d}y) = \mathrm{d}g \wedge \mathrm{d}y.$$

Since the integral of $\omega$ is linear in $\omega$, we get

$$\mathrm{d}(f\mathrm{d}x + g\mathrm{d}y) = \mathrm{d}f \wedge \mathrm{d}x + \mathrm{d}g \wedge \mathrm{d}y$$

as can also be verified directly from the definition of d. But we may express $\mathrm{d}f$ and $\mathrm{d}g$ in terms of $\mathrm{d}x$ and $\mathrm{d}y$:

$$\mathrm{d}f = \frac{\partial f}{\partial x}\mathrm{d}x + \frac{\partial f}{\partial y}\mathrm{d}y; \quad \mathrm{d}g = \frac{\partial g}{\partial x}\mathrm{d}x + \frac{\partial g}{\partial y}\mathrm{d}y.$$

Since $\mathrm{d}x \wedge \mathrm{d}x = 0$ and $\mathrm{d}y \wedge \mathrm{d}y = 0$, we find

$$\mathrm{d}\tau = \frac{\partial f}{\partial y}\mathrm{d}y \wedge \mathrm{d}x + \frac{\partial g}{\partial x}\mathrm{d}x \wedge \mathrm{d}y.$$

Finally, since $\mathrm{d}y \wedge \mathrm{d}x = -\mathrm{d}x \wedge \mathrm{d}y$, we have

$$\boxed{\mathrm{d}\tau = \left(\frac{\partial g}{\partial x} - \frac{\partial f}{\partial y}\right)\mathrm{d}x \wedge \mathrm{d}y}$$

As an example, let

$$\tau = x^2y^2\mathrm{d}x + x^3y\mathrm{d}y.$$

Then

$$\begin{aligned}\mathrm{d}\tau &= \mathrm{d}(x^2y^2) \wedge \mathrm{d}x + \mathrm{d}(x^3y) \wedge \mathrm{d}y \\ &= (2xy^2\mathrm{d}x + 2x^2y\mathrm{d}y) \wedge \mathrm{d}x + (3x^2y\mathrm{d}x + x^3\mathrm{d}y) \wedge \mathrm{d}y \\ &= 2x^2y\mathrm{d}y \wedge \mathrm{d}x + 3x^2y\mathrm{d}x \wedge \mathrm{d}y = x^2y\mathrm{d}x \wedge \mathrm{d}y.\end{aligned}$$

In the special case where the one-form $\tau$ is exact, $\tau = \mathrm{d}\phi$, we find that $\mathrm{d}\tau = 0$. This is obvious from the definition of $\mathrm{d}\tau$: since $\mathrm{d}\tau[\mathbf{v}, \mathbf{w}]$ is the best linear approximation to the line integral $\int\tau$ around a parallelogram, and since the integral of a differential around any closed path is zero, clearly $\mathrm{d}(\mathrm{d}\phi) = 0$. Alternatively, we may prove the same result by direct computation:

$$\tau = \mathrm{d}\phi = \frac{\partial\phi}{\partial x}\mathrm{d}x + \frac{\partial\phi}{\partial y}\mathrm{d}y,$$

$$\mathrm{d}\tau = \frac{\partial^2\phi}{\partial y\,\partial x}\mathrm{d}y \wedge \mathrm{d}x + \frac{\partial^2\phi}{\partial x\,\partial y}\mathrm{d}x \wedge \mathrm{d}y = 0$$

because of the equality of mixed second partial derivatives. Thus we see that the condition for a form to be closed is precisely that

$$\boxed{\mathrm{d}\tau = 0}$$

We have shown that, if $f$ is differentiable and if $x$ is the coordinate function, then $\mathrm{d}(f\mathrm{d}x) = \mathrm{d}f \wedge \mathrm{d}x$. We now use this result to prove a more general product formula

$$\mathrm{d}(f\tau) = \mathrm{d}f \wedge \tau + f\mathrm{d}\tau$$

where $f$ is a differentiable function and $\tau$ a differentiable one-form. Writing $\tau = g\mathrm{d}x + h\mathrm{d}y$, we have $f\tau = (fg)\mathrm{d}x + (fh)\mathrm{d}y$, so that

$$\mathrm{d}(f\tau) = \mathrm{d}(fg) \wedge \mathrm{d}x + \mathrm{d}(fh) \wedge \mathrm{d}y.$$

Therefore

$$d(f\tau) = g\,df \wedge dx + f\,dg \wedge dx + h\,df \wedge dy + f\,dh \wedge dy$$

and we see that

$$d(f\tau) = df \wedge (g\,dx + h\,dy) + f(dg \wedge dx + dh \wedge dy),$$

i.e.,

$$d(f\tau) = df \wedge \tau + f\,d\tau.$$

## 8.2. Two-forms

Since the most general one-form in the plane has the expression $f\,dx + g\,dy$, the most general product of two one-forms will be some function multiple of $dx \wedge dy$. We call such an expression a *two-form* so a two-form looks like

$$\sigma = f\,dx \wedge dy = F(x, y)\,dx \wedge dy.$$

We want to think of the value of the two-form at $P$, i.e., $F(P)\,dx \wedge dy$, as a rule which assigns numbers to pairs of vectors.

To understand the 'constant' two-form $dx \wedge dy$, we first evaluate it on the pair of unit vectors $(\mathbf{e}_x, \mathbf{e}_y)$. By definition,

$$dx \wedge dy[\mathbf{e}_x, \mathbf{e}_y] = dx[\mathbf{e}_x]dy[\mathbf{e}_y] - dx[\mathbf{e}_y]dy[\mathbf{e}_x] = 1\cdot 1 - 0\cdot 0 = 1.$$

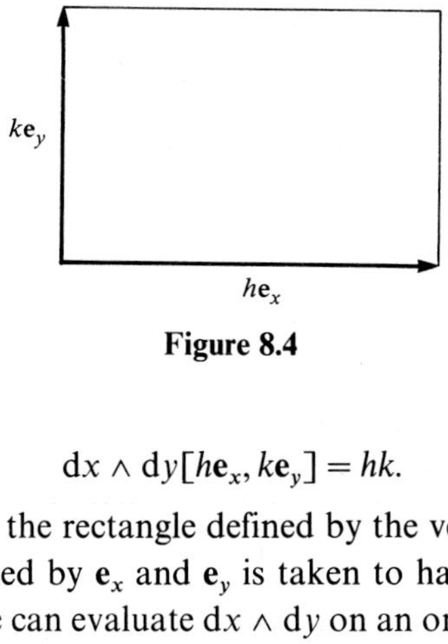


**Figure 8.4**

More generally,

$$dx \wedge dy[h\mathbf{e}_x, k\mathbf{e}_y] = hk.$$

Clearly this is the area of the rectangle defined by the vectors $h\mathbf{e}_x$ and $k\mathbf{e}_y$, in units where the rectangle defined by $\mathbf{e}_x$ and $\mathbf{e}_y$ is taken to have unit area.

More generally still, we can evaluate $dx \wedge dy$ on an ordered pair of vectors $(\mathbf{v}, \mathbf{w})$. We may write $\mathbf{v} = a\mathbf{e}_x + c\mathbf{e}_y$, $\mathbf{w} = b\mathbf{e}_x + d\mathbf{e}_y$, so that, in terms of the matrix $A = \begin{pmatrix} a & b \\ c & d \end{pmatrix}$, $\mathbf{v} = A\begin{pmatrix} 1 \\ 0 \end{pmatrix}$, $\mathbf{w} = A\begin{pmatrix} 0 \\ 1 \end{pmatrix}$. Then

$$\begin{aligned} dx \wedge dy[\mathbf{v}, \mathbf{w}] &= dx[\mathbf{v}]dy[\mathbf{w}] - dx[\mathbf{w}]dy[\mathbf{v}] \\ &= ad - bc \\ &= \text{Det } A. \end{aligned}$$

A general two-form $\sigma$ is a function of three variables $\mathbf{p}, \mathbf{v}$, and $\mathbf{w}$. For fixed $\mathbf{p}$ it is a bilinear and antisymmetric function of $\mathbf{v}$ and $\mathbf{w}$. If $\sigma = f\,\mathrm{d}x \wedge \mathrm{d}y$ then

$$\sigma(\mathbf{p})(\mathbf{v}, \mathbf{w}) = f(\mathbf{p})\,\mathrm{Det}\,A$$

where $\mathbf{v}, \mathbf{w}$ and $A$ are as above. We can think of $\sigma(\mathbf{p})$ as assigning a notion of signed area to each parallelogram based at $\mathbf{p}$. The signed area of the parallelogram spanned by $\mathbf{v}$ and $\mathbf{w}$ (in that order) is $\sigma(\mathbf{p})(\mathbf{v}, \mathbf{w})$.

It is important to remember that the value of $\mathrm{d}x \wedge \mathrm{d}y$ on a parallelogram depends on the orientation of the parallelogram, as determined by the ordering of the vectors which define the parallelogram. On the oriented rectangle which corresponds to the pair $[\mathbf{e}_x, \mathbf{e}_y]$, the value of $\mathrm{d}x \wedge \mathrm{d}y$ is $+1$, while the value of $\mathrm{d}y \wedge \mathrm{d}x$ is $-1$. On the same rectangle with opposite orientation, which corresponds to the pair $[\mathbf{e}_y, \mathbf{e}_x]$, the value of $\mathrm{d}y \wedge \mathrm{d}x$ is $+1$ but the value of $\mathrm{d}x \wedge \mathrm{d}y$ is $-1$. More generally, to evaluate a two-form $\tau$ on a parallelogram defined by $\mathbf{v}$ and $\mathbf{w}$, we look at the orientation to determine which vector, $\mathbf{v}$ or $\mathbf{w}$ is 'first', then evaluate $\tau[\mathbf{v}, \mathbf{w}]$ or $\tau[\mathbf{w}, \mathbf{v}]$ as appropriate.

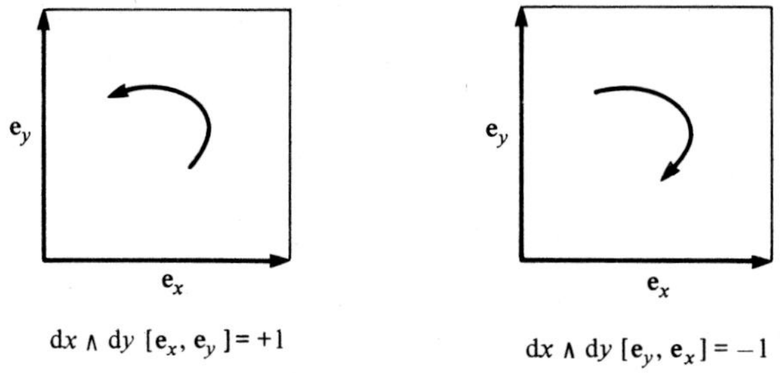


**Figure 8.5**

## 8.3. Integrating two-forms

Since a two-form $\tau$ assigns a number to each small oriented parallelogram (pair of vectors) just as a one-form assigns a number to each small directed line segment (vector), we can integrate two-forms over a region $R$ in the plane much as we integrate one-forms along paths. Given a rectangular region $R$, oriented as shown

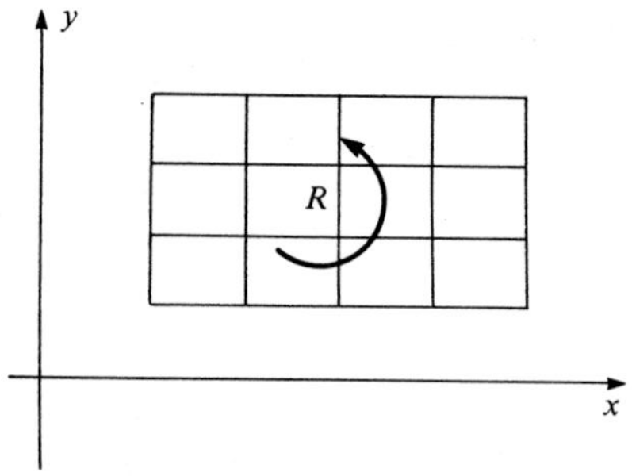


**Figure 8.6**

in figure 8.6, we break it up into $N_x N_y$ small rectangles, then form the Riemann sums

$$S_{N_x N_y} = \sum_{i=0}^{N_x-1} \sum_{j=0}^{N_y-1} \sigma(P_{ij})\left[\frac{b-a}{N_x}\mathbf{e}_x, \frac{d-c}{N_y}\mathbf{e}_y\right].$$

If

$$\sigma = F(x, y)\mathrm{d}x \wedge \mathrm{d}y,$$

then

$$S_{N_x N_y} = \frac{b-a}{N_x}\frac{d-c}{N_y}\sum_{i=0}^{N_x-1}\sum_{j=0}^{N_y-1} F(x_i, y_j)$$

where, of course,

$$x_i = a + \frac{i}{N_x}(b-a), \quad y_j = c + \frac{j}{N_y}(d-c).$$

We then define the integral of the two-form $\tau$ over the oriented region $R$ as

$$\int_R \tau = \lim_{\substack{N_x \to \infty \\ N_y \to \infty}} S_{N_x N_y}$$

provided the limit is independent of the refinements of the partition.

We may evaluate the double integral of $F(x, y)\mathrm{d}x \wedge \mathrm{d}y$ over the rectangle $R$ as an iterated integral. To evaluate the expression

$$I = \lim_{\substack{N_x \to \infty \\ N_y \to \infty}} \frac{b-a}{N_x}\frac{d-c}{N_y}\sum_{j=0}^{N_x-1}\sum_{j=0}^{N_y-1} F(x_i, y_j)$$

we may first sum over $j$ for each fixed $i$, then let $N_y \to \infty$ before summing over $i$. Since

$$\lim_{N_y \to \infty} \frac{d-c}{N_y}\sum_{j=0}^{N_y-1} F(x_i, y_j) = \int_c^d F(x_i, y)\mathrm{d}y$$

by the definition of the ordinary Riemann integral of a function of one variable, we have

$$I = \lim_{N_x \to \infty} \frac{b-a}{N_x}\sum_{i=0}^{N_x-1}\int_c^d F(x_i, y)\mathrm{d}y.$$

Again recognizing the limit of a Riemann sum as an integral, we may express this as

$$I = \int_a^b \left(\int_c^d F(x, y)\mathrm{d}y\right)\mathrm{d}x$$

an *iterated* integral which can be evaluated by techniques of single-variable calculus. We could equally well have summed first over $i$, then over $j$, to obtain

$$I = \int_c^d \left(\int_a^b F(x, y)\mathrm{d}x\right)\mathrm{d}y.$$

In evaluating the integral of a two-form $\tau$ over an oriented rectangle $R$, we must pay attention to the orientation of the rectangle. If $R$ is oriented so that $x$ is the 'first' coordinate, $y$ the 'second', as in figure 8.7(a), we write $\tau = F(x, y)\mathrm{d}x \wedge \mathrm{d}y$,

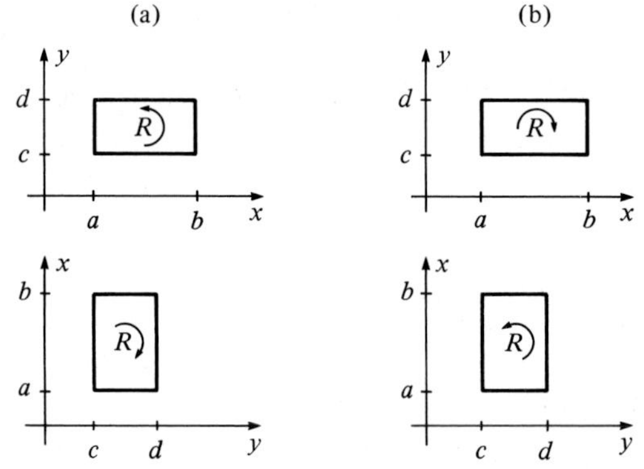


**Figure 8.7** **(a)** Oriented with $x$ first. **(b)** Oriented with $y$ first.

then evaluate the iterated integral.

$$\int_a^b\left(\int_c^d F(x,y)\,\mathrm{d}y\right)\mathrm{d}x \quad \text{or} \quad \int_c^d\left(\int_a^b F(x,y)\,\mathrm{d}x\right)\mathrm{d}y.$$

If, on the other hand, $R$ is oriented so that $y$ is the 'first' coordinate, as in figure 8.7(b), we must write

$$\tau = G(y,x)\,\mathrm{d}y \wedge \mathrm{d}x \quad (\text{where } G(y,x) = -F(x,y))$$

and evaluate the iterated integral

$$\int_a^b\left(\int_c^d G(y,x)\,\mathrm{d}y\right)\mathrm{d}x \quad \text{or} \quad \int_c^d\left(\int_a^b G(y,x)\,\mathrm{d}x\right)\mathrm{d}y.$$

Reversing the orientation of $R$ changes the *sign* of the integral.

In the case of a line integral, the notion of orientation – and why the sign changes when we reverse the orientation – is intuitively clear. (For example, in the case of the force field, the line integral gave the work along the path, a difference in energy readings, $E_B - E_A$, on some internal meter, perhaps. Changing the direction replaces $E_B - E_A$ with $E_A - E_B$.) It is important to have a similar intuitive example for our two-dimensional integrals. Here is one: One way of visualizing a change in orientation in the plane is to look at it from above and from below. That is, suppose that we imagine our $xy$-plane as being the $z = 0$ plane in three-dimensional space. Then a rotation which is clockwise when viewed from above will appear counterclockwise when viewed from below. So choosing an orientation on a surface in space is closely related to choosing a 'side' of the surface. Now imagine that material is flowing through the surface. For instance, imagine that the surface is

**Figure 8.8**

a piece of a cell membrane and we are interested in the transport of a particle ion across the membrane.

Then, of course, in using the word 'across' we must specify a definite choice of direction – a definite 'side' regarded as 'in' – for the surface. Thus in measuring the total flux across the surface, we must choose an orientation. Changing the orientation will change the sign of the total flux.

## Double integrals

Frequently one encounters *absolute* double integrals, which are to be evaluated over a region in the plane which has no orientation. If, for example, $\sigma$ represents the density (mass per unit area) of a plane lamina in the shape of a rectangle $R$, then the mass of the lamina is given by the double integral

$$M = \int_R \sigma \mathrm{d}A.$$

Clearly $M$ must be a positive number; orientation of $R$ cannot matter. We may regard $\mathrm{d}A$ in such an integral as a function which assigns to any small parallelogram its true geometrical area; that is, the absolute value of its directed area. If we are using $x$ and $y$ as coordinates, we may write $\mathrm{d}A = \mathrm{d}x\,\mathrm{d}y$ or $\mathrm{d}A = \mathrm{d}y\,\mathrm{d}x$; the

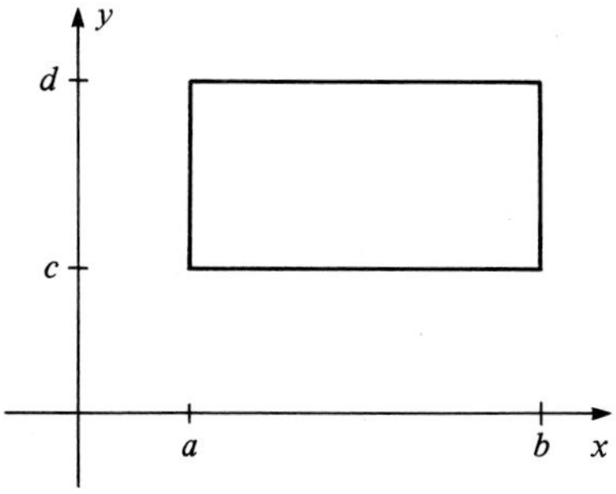


**Figure 8.9**

order of the coordinates does not matter. The absolute integral $\int_R F(x, y)\mathrm{d}x\,\mathrm{d}y$ may be evaluated as the iterated integral

$$\int_a^b \left( \int_c^d F(x, y)\mathrm{d}y \right) \mathrm{d}x$$

or as

$$\int_c^d \left( \int_a^b F(x, y)\mathrm{d}x \right) \mathrm{d}y.$$

The important point is that there are two quite distinct types of geometric objects – expressions such as $\sigma \mathrm{d}A$, which we may call *densities*, which assign numbers to regions $R$ by integration independent of any orientation – and two-forms – expressions like $\tau = F\mathrm{d}x \wedge \mathrm{d}y$, whose evaluation depends on a choice of orientation. They are each appropriate in quite different physical contexts. As we shall see, they behave differently, under change of variables or pull-back. We shall return to this important point later.

**Double integrals as iterated integrals**

Double integrals may be evaluated as iterated integrals for any region which is bounded by lines $x = \text{constant}$ and by function graphs which do not cross. For example, in the region in figure 8.10, bounded on the left by $x = a$, on the right by

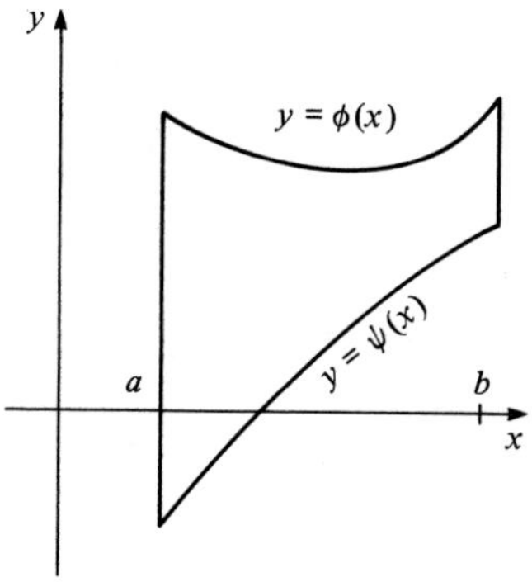


**Figure 8.10**

$x = b$, on top by the graph of $y = \phi(x)$, and on the bottom by $y = \psi(x)$, the double integral $\int_R F(x, y)\mathrm{d}x\,\mathrm{d}y$ may be evaluated as

$$\int_a^b \left( \int_{\psi(x)}^{\phi(x)} F(x, y) \right) \mathrm{d}y \mathrm{d}x.$$

As an illustration, we calculate the integral $I = \int_R 2xy\,\mathrm{d}x\,\mathrm{d}y$ over the quarter-circle bounded by $x = 0$, $x = a$, $y = 0$ and the circular arc $y = \sqrt{(a^2 - x^2)}$. Integrating

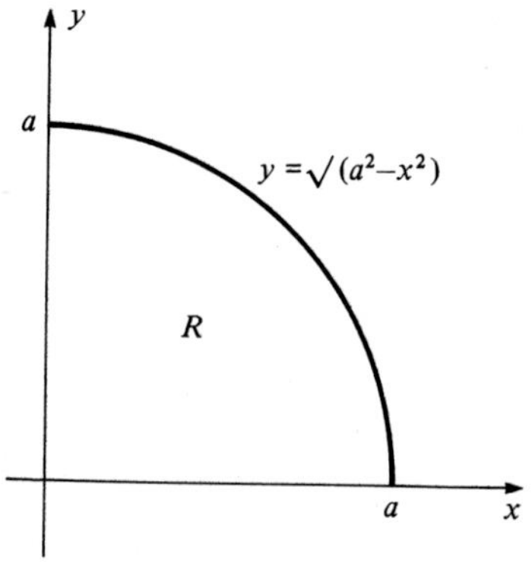


**Figure 8.11**

first over $y$, then over $x$, we find

$$I = \int_0^a \left( \int_0^{\sqrt{(a^2 - x^2)}} 2xy\,\mathrm{d}y \right) \mathrm{d}x$$

$$I = \int_0^a [xy^2]_0^{\sqrt{(a^2 - x^2)}}\,\mathrm{d}x = \int_0^a x(a^2 - x^2)\,\mathrm{d}x = \tfrac{1}{4}a^4.$$

Sometimes a double integral is more easily evaluated as an iterated integral if the integration over $x$ is performed first. For example, the integral $I = \int_S F(x, y)\,\mathrm{d}x\,\mathrm{d}y$ over the region $S$ in figure 8.12, which is *not* easily evaluated by first integrating

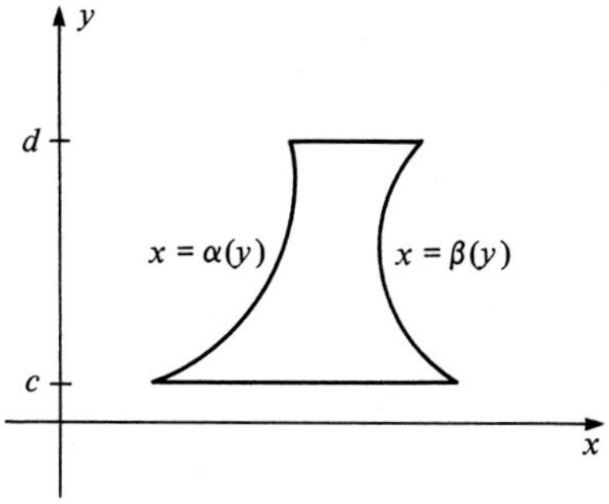


**Figure 8.12**

over $y$, may be calculated as

$$I = \int_c^d \left( \int_{\alpha(y)}^{\beta(y)} F(x, y)\,\mathrm{d}x \right) \mathrm{d}y.$$

Even for regions which are not rectangular, it is frequently possible to evaluate a double integral as an iterated integral in either order. Suppose, for example, that we wish to evaluate $I = \int_R y\,\mathrm{d}x\,\mathrm{d}y$ for the region $R$ between the parabola $y = x^2$

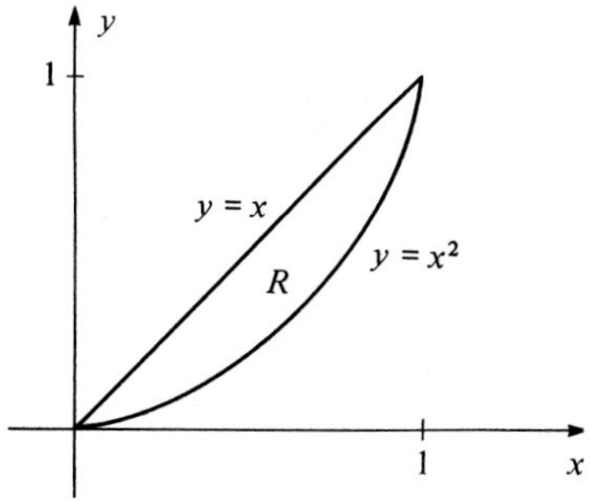


**Figure 8.13**

and the line $y = x$. This may be evaluated as

$$I = \int_0^1 \left( \int_{x^2}^x y\,\mathrm{d}y \right) \mathrm{d}x = \int_0^1 [\tfrac{1}{2}y^2]_{x^2}^x \,\mathrm{d}x = \frac{1}{2}\int_0^1 (x^2 - x^4)\,\mathrm{d}x = \tfrac{1}{15}.$$

Alternatively, we may describe the line as $x = y$, the parabola as $x = \sqrt{y}$, and integrate first over $x$:

$$I = \int_0^1 \left( \int_y^{\sqrt{y}} \mathrm{d}x \right) y\,\mathrm{d}y = \int_0^1 [\sqrt{y} - y]y\,\mathrm{d}y = \int_0^1 (y^{3/2} - y^2)\,\mathrm{d}y = \tfrac{1}{15}.$$

Sometimes it pays to regard an iterated integral as a double integral in order to reverse the order of integration. For example, the integral

$$I = \int_0^1 \left( \int_x^1 \mathrm{e}^{-y^2}\,\mathrm{d}y \right) \mathrm{d}x$$

is unattractive to evaluate as it is written. We can, however, convert it to the double integral

$$I = \int_R \mathrm{e}^{-y^2}\,\mathrm{d}x\,\mathrm{d}y$$

where $R$ is the triangular region bounded by $x = 0$, $y = 1$, and the line $y = x$. This

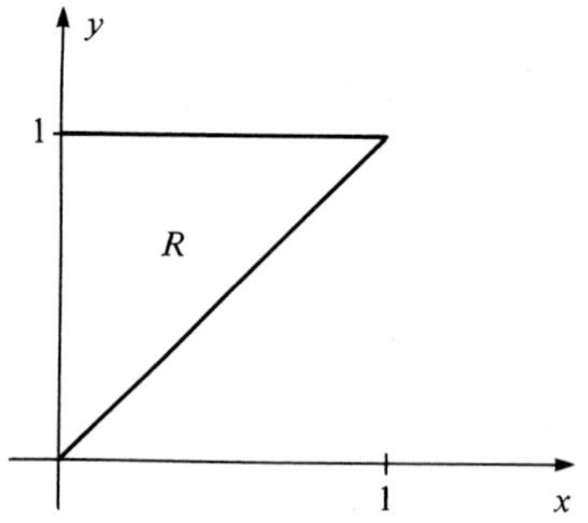


**Figure 8.14**

double integral can be evaluated by integrating first with respect to $x$, then with respect to $y$:

$$I = \int_0^1 \mathrm{e}^{-y^2} \left( \int_0^y \mathrm{d}x \right) \mathrm{d}y$$

so

$$I = \int_0^1 y\mathrm{e}^{-y^2}\,\mathrm{d}y = \int_0^1 \tfrac{1}{2}\mathrm{e}^{-u}\,\mathrm{d}u = \tfrac{1}{2}(1 - \mathrm{e}^{-1}).$$

Incidentally, the original integral can be evaluated as it is written. If we define an antiderivative of $\mathrm{e}^{-y^2}$ by

$$G(y) = \int_0^y \mathrm{e}^{-t^2}\,\mathrm{d}t,$$

so that

$$G'(y) = e^{-y^2},$$

then

$$I = \int_0^1 dx \int_x^1 e^{-y^2} dy = \int_0^1 (G(1) - G(x))\, dx.$$

Now integration by parts yields

$$I = [(G(1) - G(x))x]_0^1 - \int_0^1 x(-G'(x))\, dx.$$

The first term vanishes at both limits. Since $G'(x) = e^{-x^2}$, we find that

$$I = \int_0^1 xe^{-x^2} dx = \tfrac{1}{2}(1 - e^{-1}),$$

exactly as before.

Sometimes, in order to evaluate a double integral in terms of integrals over $x$ and $y$, it is necessary to divide up the region of integration. For example, to

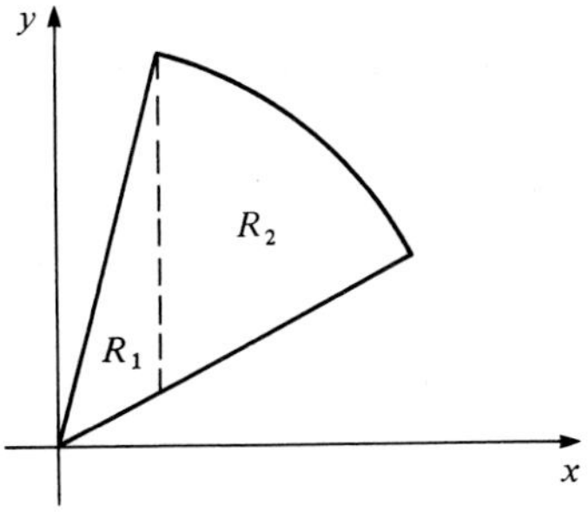


**Figure 8.15**

evaluate $\int F(x, y)\, dx\, dy$ over the circular sector shown, we first divide the sector into regions $R_1$ and $R_2$, then evaluate

$$\int_{R_1} F(x, y)\, dx\, dy + \int_{R_2} F(x, y)\, dx\, dy$$

by converting each integral to an iterated integral. A more natural way to evaluate the same integral is to introduce polar coordinates. We shall discuss this important problem of change of variables in section 8.5.

## 8.4. Orientation

We have seen that the sign of a line integral depends on the orientation of the path and that of a two-form on the orientation of the plane. We hope that you have an intuitive idea of what orientation means, but suspect that you might feel the need for a precise mathematical definition. That is our purpose in this section.

Before plunging into abstract mathematical definition, let us consider the problem. In the plane, for example, it is intuitively clear that there are two possible orientations:

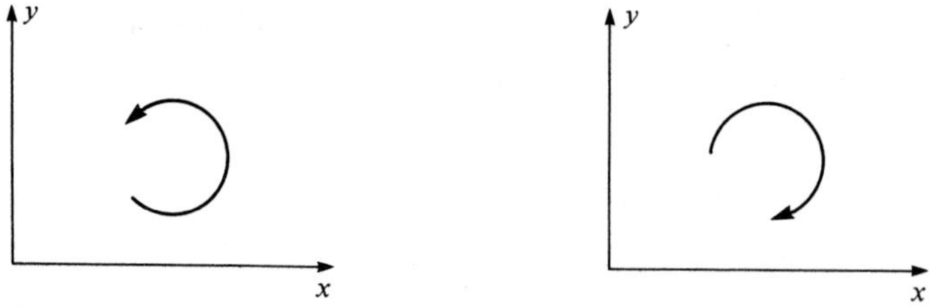


**Figure 8.16**

We cannot *intrinsically* characterize one or another but do know that they are *different* and that there are only two of them. Similarly for the line:

**Figure 8.17**

or for three-space when we try to describe right- or left-handedness:

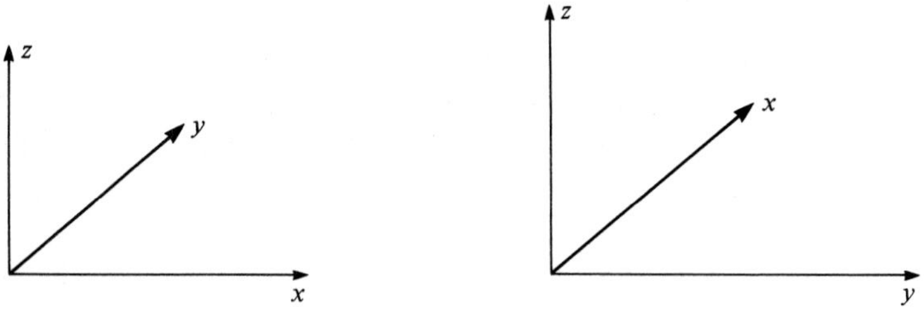


**Figure 8.18**

Getting back to the plane, we do know (see section 1.5) that a nonsingular matrix $A$ preserves or reverses orientation according as Det $A$ is positive or negative. This provides us with the clue that we need for the general definition:*

Let $V$ be an abstract two-dimensional vector space. As we saw at the end of Chapter 1, giving a basis of $V$ is the same as giving an isomorphism $L: V \to \mathbb{R}^2$. If $L$ and $L'$ are two such bases, then

$$L' = BL$$

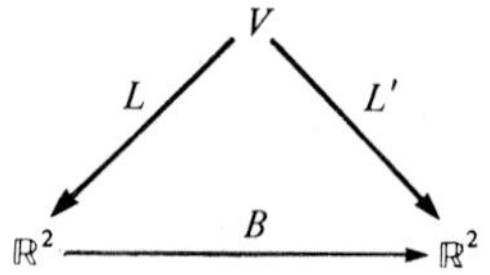


**Figure 8.19**

* As we shall see the same definition works for one-, three- or $n$-dimensional vector spaces.

where $B$ is a nonsingular $2 \times 2$ matrix. Let us call $L$ and $L'$ *similar* if $\operatorname{Det} B > 0$ and *opposite* if $\operatorname{Det} B < 0$. We claim that the set of all bases of $V$ decomposes into two collections; call them $\mathscr{F}_1$ and $\mathscr{F}_2$. All bases in the collection $\mathscr{F}_1$ are mutually similar, as are all the bases in the collection $\mathscr{F}_2$; and every basis in the collection $\mathscr{F}_1$ is opposite to every basis in the collection $\mathscr{F}_2$. Indeed, pick some basis $L_1$. Let $\mathscr{F}_1$ consist of all bases of the form

$$BL_1 \quad \operatorname{Det} B > 0$$

and let $\mathscr{F}_2$ consist of all bases of the form

$$BL_1 \quad \operatorname{Det} B < 0.$$

Every basis must belong to one or the other of these collections. If $L$ and $L'$ both belong to $\mathscr{F}_1$, then

$$L = BL_1, \quad L' = B'L_1 \qquad \operatorname{Det} B > 0, \quad \operatorname{Det} B' > 0$$

so

$$L' = B'B^{-1}L \quad \text{and} \quad \operatorname{Det} B'B^{-1} = \operatorname{Det} B' (\operatorname{Det} B)^{-1} > 0,$$

so $L$ and $L'$ are similar. If both $\operatorname{Det} B < 0$ and $\operatorname{Det} B' < 0$ in the above, we shall get that $L$ and $L'$ are similar; while, if one of the determinants is positive and the other negative, we see that $L$ and $L'$ are opposite. Thus, if we let $\mathscr{F}$ denote the collection of *all* bases of $V$, we have

$$\mathscr{F} = \mathscr{F}_1 \cup \mathscr{F}_2, \quad \mathscr{F}_1 \cap \mathscr{F}_2 = \varnothing.$$

An orientation in $V$ is *defined* to be a choice of one or the other of these two collections. In other words, an orientation on $V$ is defined to be a collection of bases of $V$ such that any two bases in the collection are similar, and any basis similar to a basis in the collection is in the collection.

Notice that giving a basis, $L$, of $V$ *determines* an orientation on $V$ – the set of all bases similar to $L$.

Once we have *chosen* an orientation on $V$, then a basis $L'$ will be called *good* or *positive* if it belongs to the collection and *bad* or *negative* if it does not. Thus once we have chosen an orientation, every basis is either good or bad. (Of course, if we had chosen the opposite orientation, these appellations would be reversed.)

Let $W$ be a second two-dimensional vector space and let $A\colon V \to W$ be a linear isomorphism. That is, $A$ has an inverse $A^{-1}\colon W \to V$. Suppose that we have chosen an orientation, $\mathscr{O}_V$ on $V$ and an orientation $\mathscr{O}_W$ on $W$. Let $M \in \mathscr{O}_W$ be a good basis

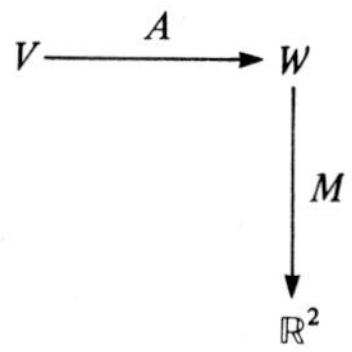


**Figure 8.20**

of $W$. Now $M: W \to \mathbb{R}^2$ so $M \circ A: V \to \mathbb{R}^2$ is an isomorphism; hence is a basis of $V$. So there are now two possibilities; $M \circ A$ is either good or bad. We can put this alternative another way. Let $L$ be a good basis of $V$.

$$C = \mathrm{Mat}_{L,M}(A) = MAL^{-1}$$

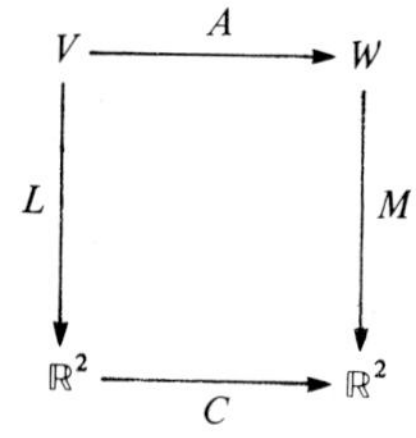


**Figure 8.21**

Then $C = MAL^{-1}$ is the change of basis matrix between $MA$ and $L$; in other words

$$MA = CL.$$

So $MA$ is good if and only if $\mathrm{Det}\, C > 0$. If we replace $L$ by $L' = B_1 L(\mathrm{Det}\, B_1 > 0)$ and $M$ by $M' = B_2 M(\mathrm{Det}\, B_2 > 0)$, then

$$C' = M'AL'^{-1} = B_2 MAL^{-1} B_1^{-1} = B_2 C B_1^{-1}$$

so

$$\mathrm{Det}\, C' = \mathrm{Det}\, B_2\, \mathrm{Det}\, C\, \mathrm{Det}\, B_1^{-1}$$

has the same sign as $\mathrm{Det}\, C$. Thus the question of whether $\mathrm{Det}\, C$ is positive or negative is independent of the particular choice of $L \in \mathcal{O}_V$ and $M \in \mathcal{O}_W$.

If $\mathrm{Det}\, C > 0$ we say that $A$ is *orientation-preserving* (or positive). If $\mathrm{Det}\, C < 0$ we say that $A$ is *orientation-reversing* (or negative). Suppose that $V \xrightarrow{A} W$ and $W \xrightarrow{A'} Z$ are two linear isomorphisms and we have chosen orientations on each of the three spaces. Then it is easy to check that

If $A$ and $A'$ are orientation-preserving, so is $A' \circ A$;

If $A$ and $A'$ are both orientation-reversing, then $A' \circ A$ is orientation-preserving;

If one is orientation preserving and the other is orientation reversing, then $A' \circ A$ is orientation reversing.

Let $\phi: V \to W$ be a differentiable map. Then, at each $\mathbf{p} \in V, \mathrm{d}\phi_{\mathbf{p}}: V \to W$ is a linear map. We say that $\phi$ is orientation preserving if $\mathrm{d}\phi_{\mathbf{p}}$ is an orientation preserving linear map for every $\mathbf{p}$. (In particular, we assume that $\mathrm{d}\phi_{\mathbf{p}}$ is a linear isomorphism for each $\mathbf{p}$.)

In our definition, we have assumed that $V$ was two-dimensional. This is irrelevant. For example, if $V$ is three-dimensional, the same definitions work. We merely need to know that a $3 \times 3$ matrix is invertible if and only if its determinant is not zero and that any three linearly independent vectors in a three-dimensional space $V$

form a basis of $V$ – hence an isomorphism with $\mathbb{R}^3$. Then the discussion at the end of Chapter 1 applies as do all the preceding discussions in this section. Similarly for an $n$-dimensional space – this requires the definition of $n$ dimensions and of the determinant of an $n \times n$ matrix. We will discuss these topics in Chapters 10 and 11. (In one dimension, a basis is just a non-zero vector, $1 \times 1$ matrix $(a)$ is just a number and this number $a$ can be regarded as the determinant of $(a)$.)

## 8.5. Pullback and integration for two-forms

The usual motivation for introducing new coordinates in a double integral is to simplify the integration. For example, the region $W$ shown in the $xy$-plane can be expressed as the image of a rectangle $R$ in the $r\theta$-plane by making the familiar polar coordinate transformation $\alpha$ defined by $\alpha^*x = r\cos\theta$, $\alpha^*y = r\sin\theta$. We can

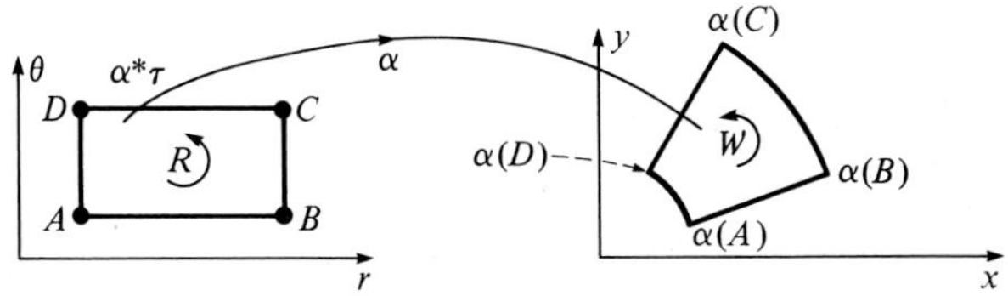


**Figure 8.22**

use this transformation to convert a directed double integral $\int_W \tau$ into the integral of a suitable two-form in the $r\theta$-plane. This is achieved by defining the pullback of a two-form, $\alpha^*\tau$, so that, if $W = \alpha(R)$, then

$$\int_W \tau = \int_R \alpha^*\tau.$$

To define pullback of a two-form, we extend the definition of the pullback of a one-form in the obvious way. Just as, for a one-form $\lambda$, we defined $\alpha^*\lambda$ by

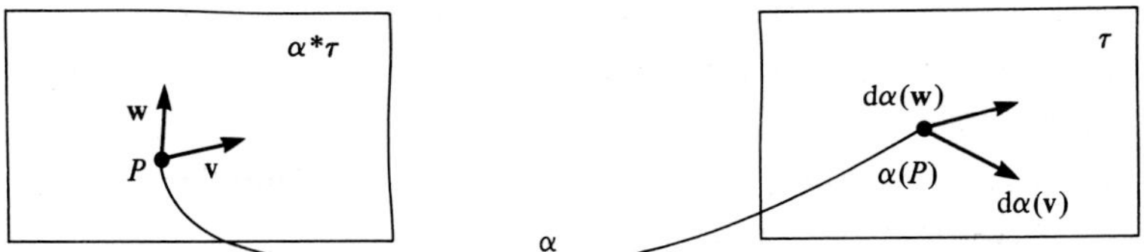


**Figure 8.23**

$\alpha^*\lambda(\mathbf{p})[\mathbf{v}] = \lambda(\alpha(\mathbf{p}))[d\alpha[\mathbf{v}]]$, we now define the pullback of a two-form by applying $d\alpha$ to both the vector arguments of $\tau$. That is,

$$\alpha^*\tau(\mathbf{p})[\mathbf{v},\mathbf{w}] = \tau(\alpha(\mathbf{p}))[d\alpha[\mathbf{v}], d\alpha[\mathbf{w}]].$$

This definition will ensure that the formula $\int_R \alpha^*\tau = \int_{\alpha(R)} \tau$ will hold for any rectangular region $R$, provided of course that $\alpha$ is differentiable and orientation preserving and that both integrals exist. We approximate $\int_R \alpha^*\tau$ as a Riemann sum over many small rectangles:

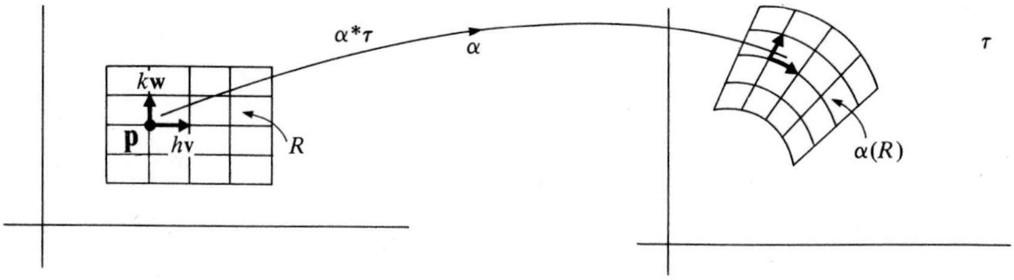


**Figure 8.24**

The contribution of the rectangle at point $\mathbf{p}$ is $\alpha^*\tau(\mathbf{p})[h\mathbf{v}, k\mathbf{w}]$. By definition, this equals $\tau(\alpha(\mathbf{p}))[\mathrm{d}\alpha[h\mathbf{v}], \mathrm{d}\alpha[k\mathbf{w}]]$, that is, the value of $\tau$ on the parallelogram which is the best linear approximation to the image under $\alpha$ of the rectangle defined by $h\mathbf{v}$ and $k\mathbf{w}$.

Of course, the image of the rectangle under $\alpha$ is not precisely a parallelogram

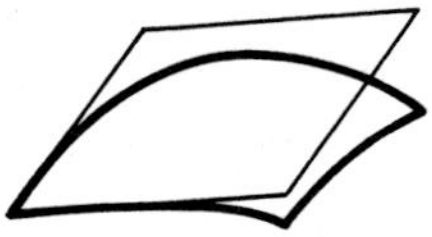

**Figure 8.25**

and the value

$$\tau(\alpha(\mathbf{p}))[\mathrm{d}\alpha(h\mathbf{v}), \mathrm{d}\alpha(k\mathbf{w})]$$

does not precisely equal the integral of $\tau$ over the image of the rectangle. So we make two types of error: replacing the image $\alpha$(rect.) by a parallelogram, call it P, and so

(i) $\int_{\alpha(\text{rect.})}$ by $\int_{\mathrm{P}}$,

then

(ii) replacing $\int_{\mathrm{P}}$ by $\tau(\alpha(\mathbf{p}))$ $[\mathrm{d}\alpha(h\mathbf{v}), \mathrm{d}\alpha(h\mathbf{w})]$.

Now, if $\tau$ is continuous, the error involved in (ii) is clearly $o(hk)$: if $\tau$ had uniformly bounded first derivatives on the entire region bounded by $\kappa$ say, then

$$|\tau(\mathbf{q}) - \tau(\alpha(\mathbf{p}))| \leqslant \kappa(h^2 + k^2)^{1/2} \quad \text{for any} \quad \mathbf{q} \text{ in P}.$$

Thus the error involved in (ii) is at most

$$\kappa(h^2 + k^2)^{1/2}hk.$$

The error involved in (i) can be estimated by Taylor's formula; for example, replacing the curved image of each side by an approximating straight line. The error here (assuming the first and second derivatives of $\alpha$ are bounded over $R$) will be a sum of terms bounded by $h^2k$ and $hk^2$ multiplied by a suitable constant, thus

$$\alpha^*(\tau(\mathbf{p}_i))[h\mathbf{v}, h\mathbf{w}] = \int_{\alpha(\text{rect.})} \tau + \text{error}$$

where

$$|\text{error}| < C(h^2 + k^2)^{1/2} \times (\text{the area of } R).$$

Summing over all rectangles, we get

$$\sum \alpha^*(\tau(\mathbf{p}_i))[h\mathbf{v}, h\mathbf{w}] = \int_{\alpha(R)} \tau + \text{error}$$

where

$$|\text{error}| < C(h^2 + k^2)^{1/2} \times (\text{the area of } R).$$

As we make the mesh finer and finer, the sum on the left approaches $\int_R \alpha^*\tau$ while the error on the right approaches zero. It follows that

$$\int_R \alpha^*\tau = \int_{\alpha(R)} \tau.$$

For a more careful proof of this important result, not requiring such stringent hypotheses on $\alpha$ and on $\tau$, and valid in $n$ dimensions, see Loomis and Sternberg, *Advanced Calculus* section 8.11. In fact, we recommend Chapter 8 of Loomis and Sternberg (which can be read independently of the rest of the book) for its treatment of the theory of integration.

We still need a procedure for computing the pullback of a two-form. Since any two-form in the plane can be expressed as a wedge product of two one-forms, we first calculate $\alpha^*(\lambda \wedge \sigma)$, where $\lambda$ and $\sigma$ are one-forms. By definition,

$$\alpha^*(\lambda \wedge \sigma)[\mathbf{v}, \mathbf{w}] = (\lambda \wedge \sigma)[d\alpha[\mathbf{v}], \quad d\alpha[\mathbf{w}]] = \lambda[d\alpha[\mathbf{v}]]\sigma[d\alpha[\mathbf{w}]] - \lambda[d\alpha[\mathbf{w}]]\sigma[d\alpha[\mathbf{v}]].$$

On the other hand,

$$(\alpha^*\lambda \wedge \alpha^*\sigma)[\mathbf{v}, \mathbf{w}] = \alpha^*\lambda[\mathbf{v}]\alpha^*\sigma[\mathbf{w}] - \alpha^*\lambda[\mathbf{w}]\alpha^*\sigma[\mathbf{v}].$$

By the definition of pullback for a one-form, $\alpha^*\lambda[\mathbf{v}] = \lambda[d\alpha[\mathbf{v}]]$. It follows then that

$$\alpha^*(\lambda \wedge \sigma) = \alpha^*\lambda \wedge \alpha^*\sigma$$

that is, pullback commutes with the wedge product.

Since the most general two-form in the $xy$-plane is of the form

$$\tau = f\,dx \wedge dy$$

we find immediately that

$$\alpha^*\tau = (\alpha^*f)d(\alpha^*x) \wedge d(\alpha^*y).$$

If, for example, $\alpha^*x = r\cos\theta$, $\alpha^*y = r\sin\theta$, then

$$d(\alpha^*x) \wedge d(\alpha^*y) = (\cos\theta\,dr - r\sin\theta\,d\theta) \wedge (\sin\theta\,dr + r\cos\theta\,d\theta) = r\,dr \wedge d\theta.$$

We have therefore established the change of variables formula for polar coordinates:

$$\int_R (\alpha^*F)(r,\theta)r\,\mathrm{d}r \wedge \mathrm{d}\theta = \int_{\alpha(R)} F(x,y)\,\mathrm{d}x \wedge \mathrm{d}y.$$

The two-form $r\,\mathrm{d}r \wedge \mathrm{d}\theta$ assigns to any small parallelogram in the $r\theta$-plane not its directed area ($\mathrm{d}r \wedge \mathrm{d}\theta$ does that) but rather the directed area *of its image* in the $xy$-plane under the transformation $\alpha$.

We can now establish the general change of variables formula for directed double integrals. Let $R$ be an oriented region in the $uv$-plane which is carried by the differentiable transformation $\alpha$ into the oriented region $\alpha(R)$ in the $xy$-plane. We

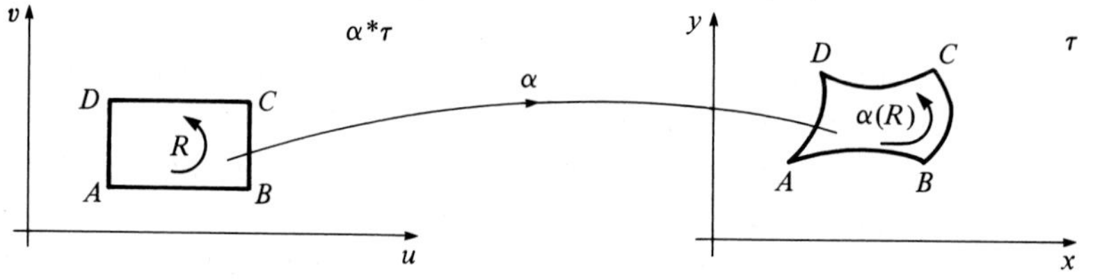


**Figure 8.26**

describe $\alpha$ by specifying the pullback of the coordinate functions $x$ and $y$:

$$\alpha^*x = X(u,v), \quad \alpha^*y = Y(u,v)$$

so that

$$\mathrm{d}(\alpha^*x) = \frac{\partial(\alpha^*x)}{\partial u}\mathrm{d}u + \frac{\partial(\alpha^*x)}{\partial v}\mathrm{d}v = \frac{\partial X}{\partial u}\mathrm{d}u + \frac{\partial X}{\partial v}\mathrm{d}v,$$

$$\mathrm{d}(\alpha^*y) = \frac{\partial(\alpha^*y)}{\partial u}\mathrm{d}u + \frac{\partial(\alpha^*y)}{\partial v}\mathrm{d}v = \frac{\partial Y}{\partial u}\mathrm{d}u + \frac{\partial Y}{\partial v}\mathrm{d}v$$

The two-form $\tau$ on the $xy$-plane may be expressed as $\tau = f\,\mathrm{d}x \wedge \mathrm{d}y = F(x,y)\,\mathrm{d}x \wedge \mathrm{d}y$. Its pullback is

$$\alpha^*\tau = \alpha^*f\,\mathrm{d}(\alpha^*x) \wedge \mathrm{d}(\alpha^*y) = \alpha^*f\left[\frac{\partial(\alpha^*x)}{\partial u}\frac{\partial(\alpha^*y)}{\partial v} - \frac{\partial(\alpha^*x)}{\partial v}\frac{\partial(\alpha^*y)}{\partial u}\right]\mathrm{d}u \wedge \mathrm{d}v$$

or, equivalently,

$$\alpha^*\tau = F(X(u,v), Y(u,v))\left[\frac{\partial X}{\partial u}\frac{\partial Y}{\partial v} - \frac{\partial X}{\partial v}\frac{\partial Y}{\partial u}\right]\mathrm{d}u \wedge \mathrm{d}v.$$

We recognize the factor in square brackets as the determinant of the *Jacobian matrix* $J$ which represents $\mathrm{d}\alpha$ relative to the given coordinates,

$$J = \begin{pmatrix} \dfrac{\partial(\alpha^*x)}{\partial u} & \dfrac{\partial(\alpha^*x)}{\partial v} \\ \dfrac{\partial(\alpha^*y)}{\partial u} & \dfrac{\partial(\alpha^*y)}{\partial v} \end{pmatrix} = \begin{pmatrix} \dfrac{\partial X}{\partial u} & \dfrac{\partial X}{\partial v} \\ \dfrac{\partial Y}{\partial u} & \dfrac{\partial Y}{\partial v} \end{pmatrix}$$

so that we may write

$$\int_{\alpha(R)} f\,\mathrm{d}x \wedge \mathrm{d}y = \int_R \alpha^* f \,\mathrm{Det}\, J\, \mathrm{d}u \wedge \mathrm{d}v.$$

This is entirely reasonable. Since Det $J$ is the 'area-transforming' factor for the linear transformation d$\alpha$, Det $J$ d$u \wedge$ d$v$ assigns to any small region in the $uv$-plane the directed area of its image in the $xy$-plane. If the ordering of $u$ and $v$ has been determined by the orientation of $R$ and the ordering of $x$ and $y$ by the orientation of $\alpha(R)$, and if $\alpha$ is orientation-preserving, then the Jacobian matrix $J$ will have a positive determinant. Reversing the order of $u$ and $v$, or of $x$ and $y$, corresponds to a change in orientation in the $uv$- or $xy$-planes. It will interchange columns or rows of $J$ and thereby change the sign of Det $J$.

As an illustration of the change of variables formula, we calculate the area of the oriented region $W$ bounded by the $x$-axis, the line $y = mx$, the hyperbola $x^2 - y^2 = 1$, and the hyperbola $x^2 - y^2 = 4$. To achieve this we write $W = \alpha(R)$ where $\alpha^*x = u \cosh v$, $\alpha^*y = u \sinh v$. Then $\alpha$ maps the oriented rectangle $R$, defined by $1 \leqslant u \leqslant 2, 0 \leqslant v \leqslant \operatorname{arctanh} m$, into the region $W$. For example, the vertical segment $u = 2$ is mapped into a portion of the hyperbola $x^2 - y^2 = 4$.

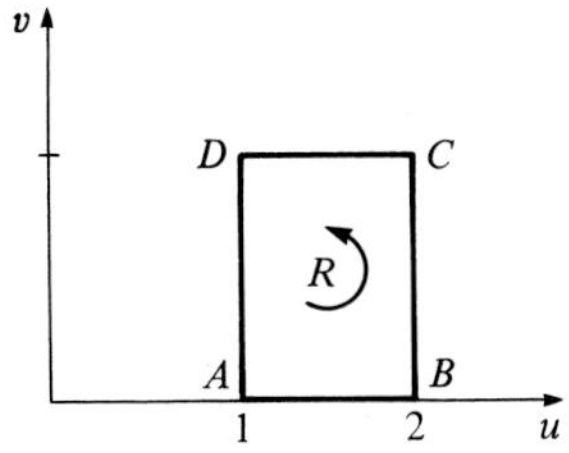


**Figure 8.27**

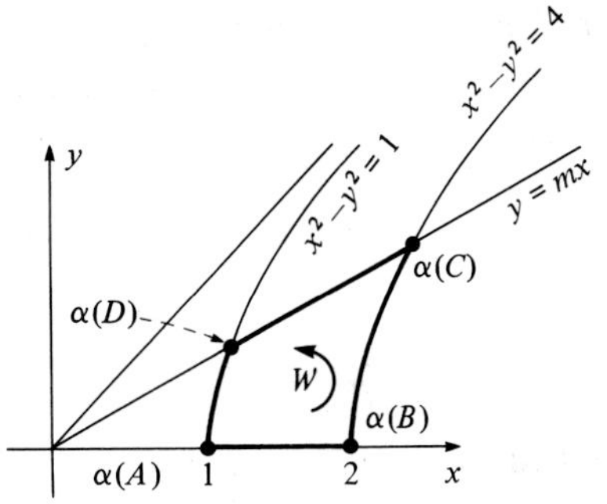


**Figure 8.28**

Since $W$ has the '$x$-first' orientation, its area is $A = \int_W \mathrm{d}x \wedge \mathrm{d}y$. Pulling back, we have $A = \int_R \alpha^*(\mathrm{d}x \wedge \mathrm{d}y) = \int_R \mathrm{Det}\, J\, \mathrm{d}u \wedge \mathrm{d}v$. Here $\alpha^*(\mathrm{d}x \wedge \mathrm{d}y) = \mathrm{d}(u \cosh v) \wedge \mathrm{d}(u \sinh v) = (\cosh v\, \mathrm{d}u + u \sinh v\, \mathrm{d}v) \wedge (\sinh v\, \mathrm{d}u + u \cosh v\, \mathrm{d}v) = u \cosh^2 v\, \mathrm{d}u \wedge \mathrm{d}v + u \sinh^2 v\, \mathrm{d}v \wedge \mathrm{d}u = u\, \mathrm{d}u \wedge \mathrm{d}v$ so that

$$A = \int_R u\,\mathrm{d}u \wedge \mathrm{d}v = \int_1^2 u\,\mathrm{d}u \int_0^{\tanh^{-1} m} \mathrm{d}v = \tfrac{3}{2} \tanh^{-1} m.$$

Equivalently, we may compute

$$\mathrm{Det}\, J = \det\begin{pmatrix} \cosh v & u \sinh v \\ \sinh v & u \cosh v \end{pmatrix} = u$$

in order to see that $A = \int_R u\,\mathrm{d}u \wedge \mathrm{d}v$.

Clearly the secret of a useful coordinate transformation is to make the boundary of the region $W$ be the image of the sides of a rectangle in the $uv$-plane. If, for example, $W$ is the triangle bounded by the coordinate axes and the line $x+y=1$, a useful

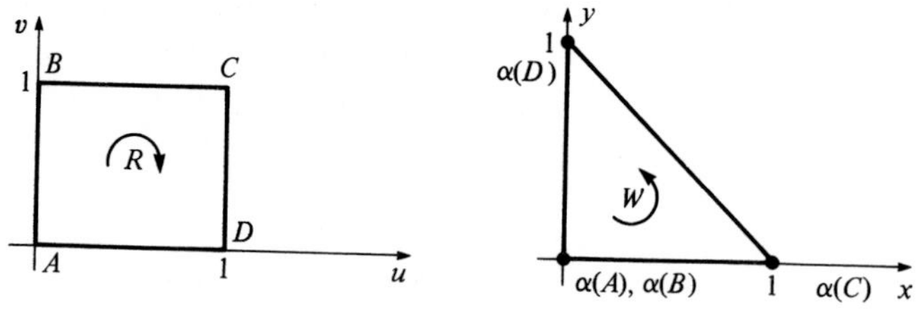


**Figure 8.29**

coordinate transformation will be one which carries the lines $u =$ constant, for a fixed interval in $v$, into segments $x + y =$ constant between the coordinate axes. Such a coordinate transformation is described by $\alpha^*x = uv$, $\alpha^*y = u(1-v)$, which has the property that $\alpha^*(x+y) = u$. You should convince yourself that $\alpha$ carries the unit square $R$ in the $uv$-plane into the region $W$, but that $R$ must be given the '$v$-first' orientation in order to make the orientation of $\alpha(R)$ agree with that of $W$. Confirmation of this fact is that, when $v$ is taken as the first coordinate, the Jacobian

$$J = \begin{pmatrix} \dfrac{\partial(\alpha^*x)}{\partial v} & \dfrac{\partial(\alpha^*x)}{\partial u} \\ \dfrac{\partial(\alpha^*y)}{\partial v} & \dfrac{\partial(\alpha^*y)}{\partial u} \end{pmatrix} = \begin{pmatrix} u & v \\ -u & 1-v \end{pmatrix}$$

has determinant $u$, which is positive. (If $u$ were the first coordinate, $\operatorname{Det} J$ would equal $-u$.)

We may use this coordinate transformation to evaluate the integral

$$I = \int_W \frac{e^{-(x+y)}}{\sqrt{(xy)}}\,dx \wedge dy$$

which would be very difficult as an iterated integral over $x$ and $y$. We find

$$\alpha^*\left(\frac{e^{-(x+y)}}{\sqrt{(xy)}}\right) = \frac{e^{-u}}{\sqrt{[u^2v(1-v)]}} = \frac{e^{-u}}{u}\frac{1}{\sqrt{[v(1-v)]}}$$

and $\alpha^*(dx \wedge dy) = (u dv + v du) \wedge (-u dv + (1-v) du) = u dv \wedge du$, so that the integral in the $uv$-plane is simply

$$I = \int_R \frac{e^{-u}}{\sqrt{[v(1-v)]}}\,dv \wedge du.$$

Note that, because $R$ has the '$v$-first' orientation, we write $\dfrac{e^{-u}}{\sqrt{[v(1-v)]}}dv \wedge du$, not its negative, before converting to an iterated integral. The final result is $I =$

$\int_0^1 e^{-u} du \int_0^1 \frac{dv}{\sqrt{[v(1-v)]}} = (1 - e^{-1})\pi$. (In evaluating the second integral we used the fact that $\int \frac{dv}{\sqrt{[v(1-v)]}} = -\arcsin(-2v+1)$.)

We turn finally to the question of changing variables in an *absolute* double integral, $I = \int_W f\, dx dy$. To make such a variable change, we may first convert $I$ to a directed double integral $I = \int_W f\, dx \wedge dy$, giving $W$ the '$x$-first' orientation. We next

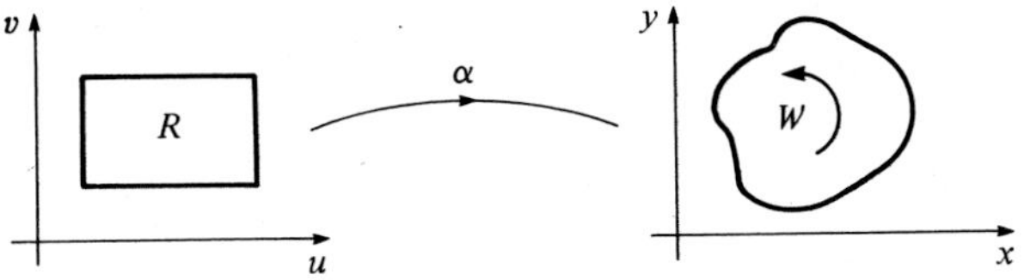


**Figure 8.30**

write $W$ as $\alpha(R)$; this procedure assigns an orientation to $R$. If $\operatorname{Det} J\left(\frac{x, y}{u, v}\right)$ is positive, $R$ has the '$u$-first' orientation, and

$$I = \int_{\substack{R \\ (u\,\text{first})}} \alpha^* f(\operatorname{Det} J) du \wedge dv = \int_{\substack{R \\ (\text{unoriented})}} \alpha^* f(\operatorname{Det} J) du dv.$$

If $\operatorname{Det} J\left(\frac{x, y}{u, v}\right)$ is negative, $R$ has the $v$-first orientation, and

$$I = \int_{\substack{R \\ (v\,\text{first})}} \alpha^* f(\operatorname{Det} J)(dv \wedge du) = -\int_{\substack{R \\ (\text{unoriented})}} \alpha^* f(\operatorname{Det} J) du dv.$$

In either case, the rule is to use the *absolute value* of $\operatorname{Det} J$:

$$I = \int_R \alpha^* f \left| \operatorname{Det} J\left(\frac{x, y}{u, v}\right) \right| du dv.$$

When this rule is used, questions of orientation, or of the order of coordinates, never arise; interchanging $x$ and $y$, or $u$ and $v$, does not affect the absolute value of $\operatorname{Det} J$.

In Chapter 15 we will discuss integration of forms in higher dimensions.

## 8.6. Two-forms in three-space

In the preceding section, we defined pullback for two-forms. The computational rule was very simple: if $\omega_1$ and $\omega_2$ are *linear* differential forms, then

$$\phi^*(\omega_1 \wedge \omega_2) = \phi^*\omega_1 \wedge \phi^*\omega_2.$$

If $f$ is a function and $\tau$ is a two-form, then

$$\phi^*(f\tau) = \phi^*(f)\phi^*\tau.$$

If $\tau_1$ and $\tau_2$ are two-forms, then

$$\phi^*(\tau_1 + \tau_2) = \phi^*\tau_1 + \phi^*\tau_2.$$

In short, all algebraic operations are preserved. We also defined the linear operator d going from one-form to two-forms,

$$\mathrm{d}(f\,\mathrm{d}g) = \mathrm{d}f \wedge \mathrm{d}g,$$

or, more generally,

$$\mathrm{d}(f\omega) = \mathrm{d}f \wedge \omega + f\,\mathrm{d}\omega,$$

and

$$\mathrm{d}(\omega_1 + \omega_2) = \mathrm{d}\omega_1 + \mathrm{d}\omega_2.$$

The pullback $\phi^*$ commutes with d in the sense that

$$\phi^*(\mathrm{d}f) = \mathrm{d}\phi^*f \quad f \text{ a function}$$

and

$$\phi^*\mathrm{d}\omega = \mathrm{d}\phi^*\omega \quad \omega \text{ a one-form.}$$

The notion of a two-form makes perfectly good sense in $\mathbb{R}^3$: a typical two-form in $\mathbb{R}^3$ (where the coordinates are $x, y, z$) is an expression of the form

$$a\,\mathrm{d}x \wedge \mathrm{d}y + b\,\mathrm{d}x \wedge \mathrm{d}z + c\,\mathrm{d}y \wedge \mathrm{d}z,$$

where $a, b$ and $c$ are functions. If

$$\omega = A\,\mathrm{d}x + B\,\mathrm{d}y + C\,\mathrm{d}z$$

is a one-form, then the rules for d and for exterior multiplication give

$$\begin{aligned}\mathrm{d}\omega &= \mathrm{d}A \wedge \mathrm{d}x + \mathrm{d}B \wedge \mathrm{d}y + \mathrm{d}C \wedge \mathrm{d}z\\ &= \left(\frac{\partial A}{\partial x}\mathrm{d}x + \frac{\partial A}{\partial y}\mathrm{d}y + \frac{\partial A}{\partial z}\mathrm{d}z\right) \wedge \mathrm{d}x + \left(\frac{\partial B}{\partial x}\mathrm{d}x + \frac{\partial B}{\partial y}\mathrm{d}y + \frac{\partial B}{\partial z}\mathrm{d}z\right) \wedge \mathrm{d}y\\ &\quad + \left(\frac{\partial C}{\partial x}\mathrm{d}x + \frac{\partial C}{\partial y}\mathrm{d}y + \frac{\partial C}{\partial z}\mathrm{d}z\right) \wedge \mathrm{d}z\\ &= \left(\frac{\partial B}{\partial x} - \frac{\partial A}{\partial y}\right)\mathrm{d}x \wedge \mathrm{d}y + \left(\frac{\partial C}{\partial x} - \frac{\partial A}{\partial z}\right)\mathrm{d}x \wedge \mathrm{d}z + \left(\frac{\partial C}{\partial y} - \frac{\partial B}{\partial z}\right)\mathrm{d}y \wedge \mathrm{d}z.\end{aligned}$$

If $\phi\colon \mathbb{R}^2 \to \mathbb{R}^3$ and $\tau$ is a two-form in $\mathbb{R}^3$, then $\phi^*\tau$ is a two-form on $\mathbb{R}^2$. If $R$ is some region in $\mathbb{R}^2$ and we have chosen an orientation on $\mathbb{R}^2$, then we can form the integral $\int_R \phi^*\tau$ which we might think of as the 'integral of $\tau$ over the oriented surface $\alpha(R)$'.

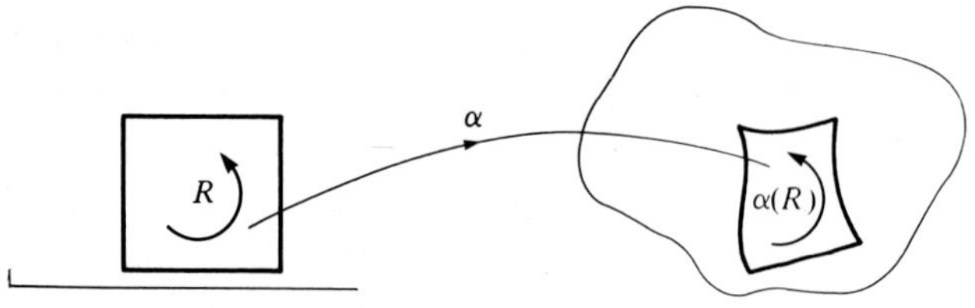


**Figure 8.31**

## 8.7. The difference between two-forms and densities

Let us return to two dimensions. We have seen that the pullback, $\alpha^*\tau$, is defined for any two-form $\tau$. Explicitly, if $\alpha$ maps the $uv$-plane into the $xy$-plane, then

$$\alpha^*(f\,dx \wedge dy) = (\alpha^*f)\cdot(\operatorname{Det} J)du \wedge dv \tag{8.2}$$

where $J$ is the Jacobian matrix

$$J = \begin{pmatrix} \dfrac{\partial\alpha_1}{\partial u} & \dfrac{\partial\alpha_1}{\partial v} \\ \dfrac{\partial\alpha_2}{\partial u} & \dfrac{\partial\alpha_2}{\partial v} \end{pmatrix}, \quad \alpha = \begin{pmatrix} \alpha_1 \\ \alpha_2 \end{pmatrix}.$$

This formula is correct for any differentiable map $\alpha$. We also proved that, *if* $\alpha$ is one-to-one and *orientation preserving*, then

$$\int_R \alpha^*(\tau) = \int_{\alpha(R)} \tau. \tag{8.3}$$

This is true when we consider the integral of a two-form, where the orientation matters. Suppose, however, we want to consider an absolute integral. Then a choice of orientation should not matter – but then the formula for change of variables or pullback for the expression $f\,dA$ or $f\,dx\,dy$ should not be the same as (8.2). In fact, if we go back to the proof of the change of variables formula on page 290 we replaced

$$\int_{\alpha(\text{rect.})} \text{by}\, \tau(P)[d\alpha(h\mathbf{v}), d\alpha(k\mathbf{w})]$$

$$= f(P) \times \text{oriented area of rect.}$$

It is clear that in the absolute integral case we must replace oriented area by absolute area. So we must replace (8.2) by

$$\alpha^*(f\,dx\,dy) = \alpha^*f\cdot|\operatorname{Det} J|\,du\,dv. \tag{8.4}$$

The reader should check – as an instructive exercise – that, if $\alpha$ is one-to-one and invertible, then

$$\int_R \alpha^*(f\,dx\,dy) = \int_{\alpha(R)} f\,dx\,dy$$

without any conditions on orientation.

So two-forms, $\tau = f\,dx \wedge dy$, and densities like $f\,dx\,dy$ are quite different objects – they transform differently under change of variables. For example, a density can be positive or negative (as in a density of electric charge): if $f \geqslant 0$ then making a change of variables replaces $f\,dx\,dy$ by $\alpha^*(f)|\operatorname{Det} J|\,du\,dv$ and $\alpha^*f\cdot|\operatorname{Det} J|$ is still positive (if $\operatorname{Det} J \neq 0$. which will hold if $\alpha^{-1}$ is differentiable). But it makes no sense to ask whether a two-form $\tau$ is positive or negative – since the factor DET $J$ which enters into (8.2) can be positive or negative. It is only when we choose an orientation (and so only allow orientation-preserving changes of variable – those for which $\operatorname{Det} J > 0$) that we can identify two-forms and densities.

## 8.8. Green's theorem in the plane

In considering line integrals we have encountered one generalization of the fundamental theorem of the calculus, namely

$$\int_\gamma \mathrm{d}f = f(B) - f(A)$$

where the path $\gamma$ runs from $A$ to $B$. This theorem relates the integral of $\mathrm{d}f$ over a one-dimensional region (the path $\gamma$) to the values of $f$ itself on the boundary of the path (the endpoints of the path).

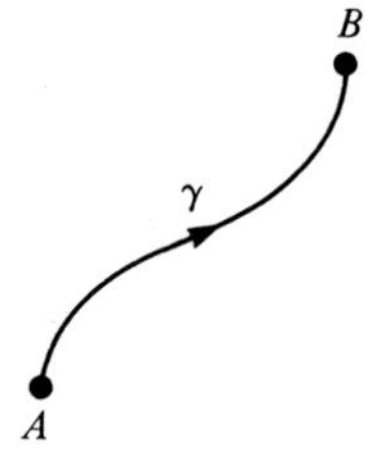


**Figure 8.32**

A similar result involving a two-dimensional region and its one-dimensional boundary is known as *Green's theorem.* This theorem states that, for any differentiable one-form $\tau$ and any oriented region $R$ in the plane,

$$\int_R \mathrm{d}\tau = \int_{\partial R} \tau.$$

Here the integral on the left is the integral of the two-form $\mathrm{d}\tau$ over the region $R$, while the integral on the right is the integral of the one-form $\tau$ over the path $\partial R$ which is the *boundary* of $R$. The sense in which the path $\partial R$ is to be traversed is determined by the orientation of $R$. For example, if $R$ is an annular region with a counter-clockwise orientation, as shown in figure 8.33, then $\partial R$ consists of the outer bounding circle traversed counterclockwise and the inner bounding circle traversed clockwise. If $R$ were given a clockwise orientation, the $\partial R$ would consist of the same two circular paths, but each traversed in the opposite sense.

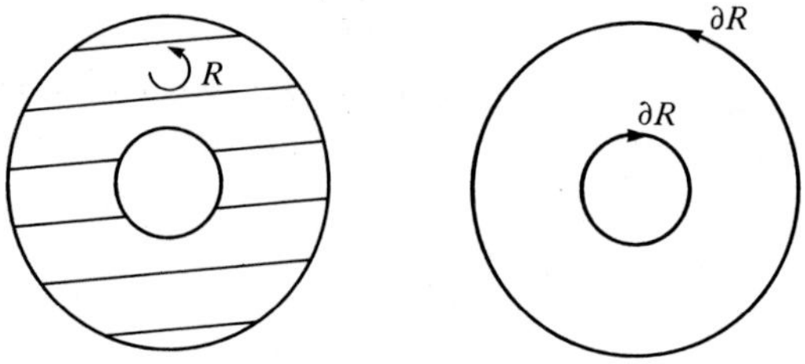


**Figure 8.33**

Before proving Green's theorem formally, it is worth reviewing the definition of the operator d acting on a one-form in order to see why such a theorem ought to hold. Recall that we defined $\mathrm{d}\tau$ as an antisymmetric bilinear function on a pair of vectors $\mathbf{v}, \mathbf{w}$ with the property that

$$\mathrm{d}\tau(P)[h\mathbf{v}, k\mathbf{w}] = \int_{\mathrm{P}(h,k)} \tau + \text{error}$$

where $\mathrm{P}(h, k)$ is the parallelogram spanned by vectors $h\mathbf{v}$ and $k\mathbf{w}$, and the error term goes to zero faster than the product $hk$ (faster than the area of the parallelogram). If

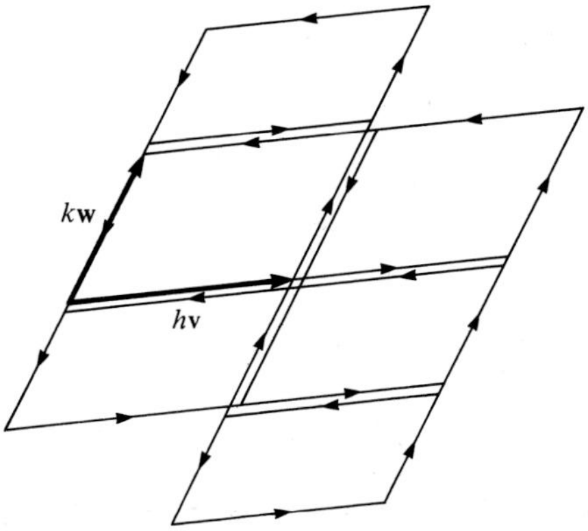


**Figure 8.34**

we now consider a region which is a union of $N$ parallelograms, each spanned by vectors $h\mathbf{v}$ and $k\mathbf{w}$, we have

$$\sum_{i=1}^{N} \mathrm{d}\tau(P_i)[h\mathbf{v}, k\mathbf{w}] = \sum_{i=1}^{N} \int_{\mathrm{P}_i} \tau + \sum_{i=1}^{N} (\text{error})_i.$$

In the sum of line integrals over the parallelograms, the contributions from the interior segments, each of which is common to two parallelograms, cancel, since each segment appears once with each orientation. Thus all that remains is a single line integral around the boundary of $R$, and we have

$$\sum_{i=1}^{N} \mathrm{d}\tau(P_i)[h\mathbf{v}, k\mathbf{w}] = \int_{\partial R} \tau + \sum_{i=1}^{N} (\text{error})_i.$$

Now, as $h$ and $k$ approach zero, the sum on the left side approaches the integral $\int_R \mathrm{d}\tau$. Since $N$ is proportional to $1/hk$, while each error term goes to zero *faster than hk*, the sum of the error terms approaches zero as $h$ and $k$ approach zero, and we have

$$\int_R \mathrm{d}\tau = \int_{\partial R} \tau$$

for any region which is a union of parallelograms.

To prove Green's theorem more rigorously, we first consider the special case where $\tau = G(x, y)\mathrm{d}y$ and evaluate the line integral $\int_{\partial R}\tau$ around a rectangle with sides parallel to the $x$- and $y$-axes. The only contributions to the integral come from the

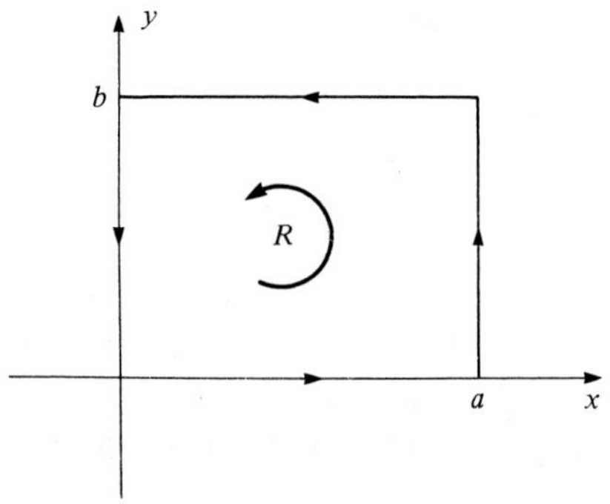


**Figure 8.35**

side $x = a$, traversed from $y = 0$ to $y = b$, and from the side $x = 0$, traversed from $y = b$ to $y = 0$. Thus

$$\int_{\partial R}\tau = \int_0^b G(a, y)\mathrm{d}y + \int_b^0 G(0, y)\mathrm{d}y$$
$$= \int_0^b G(a, y)\mathrm{d}y - \int_0^b G(0, y)\mathrm{d}y.$$

But, by the fundamental theorem of calculus,

$$G(a, y) - G(0, y) = \int_0^a \frac{\partial G}{\partial x}(x, y)\mathrm{d}x.$$

We may therefore express $\int_{\partial R}\tau$ as the iterated integral

$$\int_0^b \left(\int_0^a \frac{\partial G}{\partial x}(x, y)\mathrm{d}x\right)\mathrm{d}y$$

which in turn is equal to the directed double integral

$$\int_R \frac{\partial G}{\partial x}(x, y)\mathrm{d}x \wedge \mathrm{d}y.$$

A similar argument, applied to the one-form $\int F(x, y)\mathrm{d}x$, yields the result

$$\int_{\partial R} F(x, y)\mathrm{d}x = \int_0^a \left(\int_0^b \frac{\partial F}{\partial y}(x, y)\right)\mathrm{d}x \wedge \mathrm{d}y$$
$$= -\int_0^a (F(x, b) - F(x, 0))\mathrm{d}x.$$

Adding this to the previous result, we have

$$\int_{\partial R} F(x, y)\mathrm{d}x + G(x, y)\mathrm{d}y = \int_R \frac{\partial G}{\partial x}(x, y) - \frac{\partial F}{\partial y}(x, y)\,\mathrm{d}x \wedge \mathrm{d}y.$$

But, of course, if

$$\tau = F(x, y)\mathrm{d}x + G(x, y)\mathrm{d}y,$$

then

$$\mathrm{d}\tau = -\frac{\partial F}{\partial y}(x, y)\mathrm{d}x \wedge \mathrm{d}y + \frac{\partial G}{\partial x}(x, y)\mathrm{d}x \wedge \mathrm{d}y$$

so we have again proved that

$$\int_{\partial R} \tau = \int_R \mathrm{d}\tau,$$

which is Green's theorem.

We can now extend the proof of Green's theorem to any region in the plane which is the image of a rectangle under a smooth transformation $\alpha$. The strategy is familiar: we pull back the integrals $\int_{\partial[\alpha(R)]} \tau$ and $\int_{\alpha(R)} \mathrm{d}\tau$ to the $st$-plane, in which the region of integration is just a rectangle:

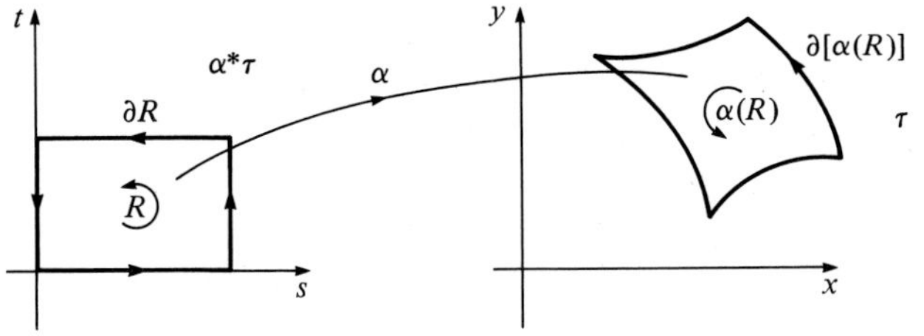


**Figure 8.36**

It is clear that, if $\alpha$ is continuous, then the boundary of $R$ is carried into the boundary of $\alpha(R)$. Therefore

$$\int_{\partial[\alpha(R)]} \tau = \int_{\partial R} \alpha^*\tau.$$

But for the rectangular region $R$ we have already proved that

$$\int_{\partial R} \alpha^*\tau = \int_R \mathrm{d}(\alpha^*\tau).$$

Furthermore,

$$\int_{\alpha(R)} \mathrm{d}\tau = \int_R \alpha^*(\mathrm{d}\tau)$$

by the definition of pullback. To prove that $\int_{\partial[\alpha(R)]}\tau = \int_{\alpha(R)}\mathrm{d}\tau$, therefore, we need only to show that

$$\alpha^*(\mathrm{d}\tau) = \mathrm{d}(\alpha^*\tau).$$

This is easily done by direct computation.

Let $\tau = f\mathrm{d}x + g\mathrm{d}y$. Using the rule $\alpha^*(\mathrm{d}x) = \mathrm{d}(\alpha^*x)$, $\alpha^*(\mathrm{d}y) = \mathrm{d}(\alpha^*y)$, we have

$$\alpha^*\tau = (\alpha^*f)\mathrm{d}(\alpha^*x) + (\alpha^*g)\mathrm{d}(\alpha^*y).$$

Then, using the rule $d(f\,dh) = df \wedge dh$, we have

$$d(\alpha^*\tau) = d(\alpha^*f) \wedge d(\alpha^*x) + d(\alpha^*g) \wedge d(\alpha^*y). \qquad (*)$$

On the other hand, we know that

$$d\tau = df \wedge dx + dg \wedge dy.$$

Using the rule $\alpha^*(\sigma \wedge \omega) = \alpha^*\sigma \wedge \alpha^*\omega$, we have

$$\alpha^*d\tau = \alpha^*(df) \wedge \alpha^*(dx) + \alpha^*(dg) \wedge \alpha^*(dy)$$

so that

$$\alpha^*d\tau = d(\alpha^*f) \wedge d(\alpha^*x) + d(\alpha^*g) \wedge d(\alpha^*y).$$

Comparing with (8.2) above, we see that

$$d(\alpha^*\tau) = \alpha^*(d\tau).$$

Thus we have

$$\int_{\alpha(\partial R)} \tau = \int_{\partial R} \alpha^*\tau = \int_R d(\alpha^*\tau) = \int_R \alpha^*(d\tau) = \int_{\alpha(R)} d\tau$$

which proves Green's theorem for a region which is the image of a rectangle.

We have proved Green's theorem for a region which is the image of a rectangle. Unfortunately, this is not general enough. We would like to consider more general polygons. Now, in the plane, every polygon can be decomposed into triangles. Indeed, we can decompose any polygon into convex polygons:

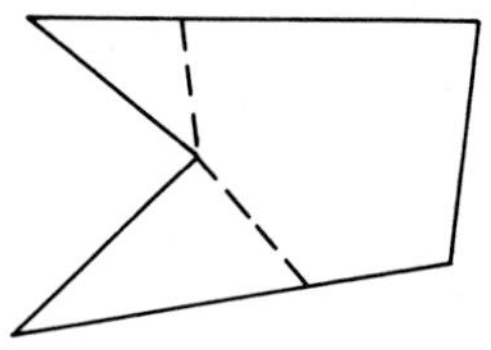

**Figure 8.37**

Any convex polygon can be decomposed into triangles by simply choosing a point in the interior and joining it to all the vertices:

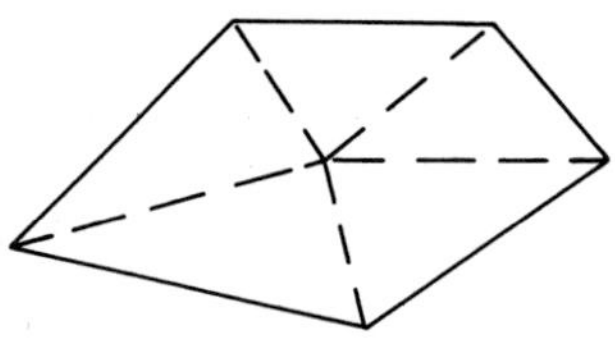

**Figure 8.38**

If we knew Green's theorem for one-forms and triangles, then we would know it for all polygons, since the integrations over the interior boundaries cancel out:

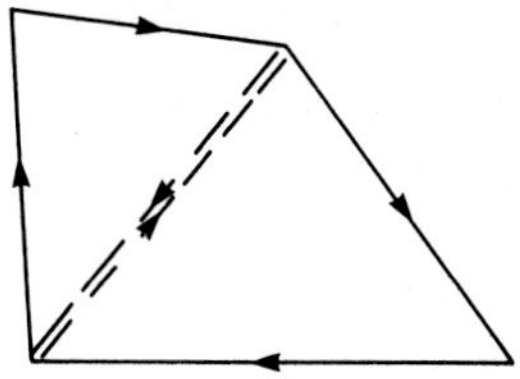

**Figure 8.39**

We thus need only to prove the theorem for triangles. Since any triangle can be mapped into any other by an affine transformation, the invariance of the integrals under smooth (in particular, affine) transformations means that it is enough to prove it for a single triangle. So consider the triangle $T_0 = \{0 \leqslant x \leqslant y \leqslant 1\}$ in the $xy$-plane.

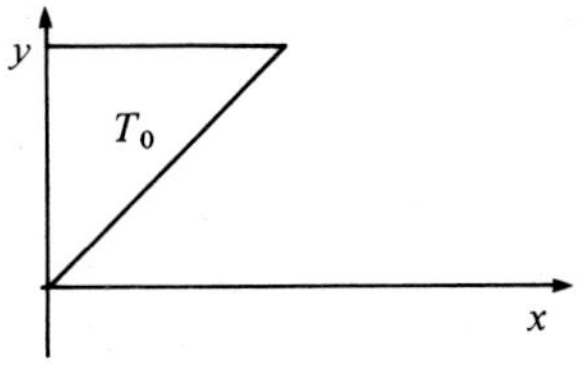


**Figure 8.40**

We may consider the quadrilateral $T_\varepsilon\{x \leqslant y, \varepsilon \leqslant y \leqslant 1\}$ for any $\varepsilon$. Clearly, by passing to the limit, if we know Green's theorem for $T_\varepsilon$, $\varepsilon > 0$, it will follow for the triangle $T_0$, since the line and area integrals around the little tip become vanishingly small. But $T$ for $\varepsilon > 0$ is the image of the rectangle $0 \leqslant u \leqslant 1$, $\varepsilon \leqslant v \leqslant 1$ in the $uv$-plane under the map

$$x = uv,$$
$$y = v.$$

And this map has a smooth inverse,

$$u = x/y,$$
$$v = y,$$

so long as $y \geqslant \varepsilon > 0$. Hence we have reduced the theorem to a case we already know – the image of a rectangle. QED

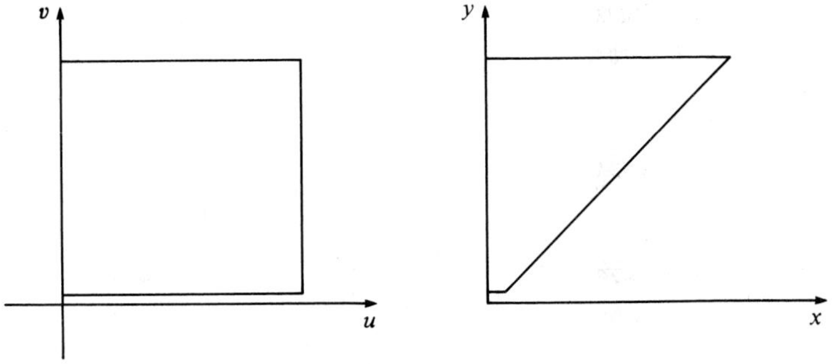


**Figure 8.41**

As an illustration of Green's theorem, we consider the integral of the one-form

$$\tau = x^2 dy$$

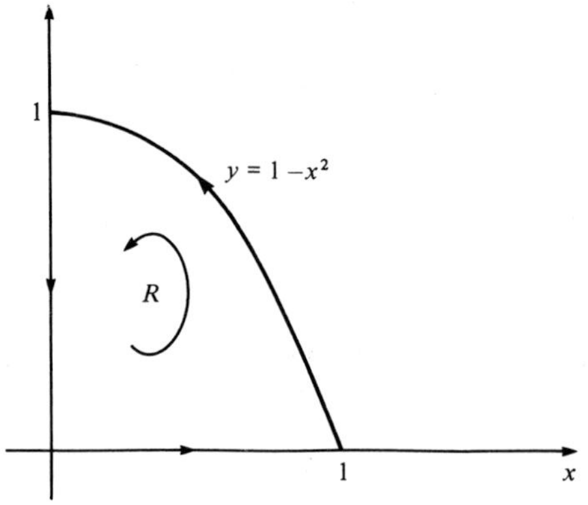


**Figure 8.42**

over the closed path $\partial R$ formed by the line segment from $\begin{pmatrix}0\\0\end{pmatrix}$ to $\begin{pmatrix}1\\0\end{pmatrix}$, the parabola $y = 1 - x^2$ from $\begin{pmatrix}1\\0\end{pmatrix}$ to $\begin{pmatrix}0\\1\end{pmatrix}$, and the line segment from $\begin{pmatrix}1\\0\end{pmatrix}$ to $\begin{pmatrix}0\\0\end{pmatrix}$. The two line segments contribute nothing to the integral. Parameterizing the parabola by

$$\beta^*x = 1 - t, \quad \beta^*y = 1 - (1-t)^2 = 2t - t^2 \quad 0 < t < 1$$

we find

$$\beta^*\tau = (1-t)^2(2-2t)dt = 2(1-t)^3 dt,$$

and

$$\int_{\partial R} \tau = \int_0^1 2(1-t)^3 dt = \tfrac{1}{2}.$$

According to Green's theorem, the integral of $d\tau = 2x dx \wedge dy$ over the region $R$ should have the same value. Evaluating this integral as an iterated integral, we obtain

$$\int_0^1 2x dx \int_0^{1-x^2} dy = \int_0^1 2(1-x^2)x dx = \tfrac{1}{2},$$

as expected.

We can use Green's theorem to obtain expressions for the area of a region in terms of line integrals. For example, if $\tau = x dy$, then $d\tau = dx \wedge dy$, and

$$\int_{\partial R} \tau = \int_R dx \wedge dy = \text{area of } R,$$

assuming that $R$ has a counterclockwise orientation. Of course, any one-form $\tau' = x dy + df$, where $f$ is an arbitrary differentiable function, has the same property, since

$$d\tau' = d(x dy) + d(df) = dx \wedge dy = d\tau.$$

Choosing $f = -\frac{1}{2}xy$, for example, we obtain

$$\tau' = x dy - \tfrac{1}{2}x dy - \tfrac{1}{2}y dx = \tfrac{1}{2}(x dy - y dx).$$

On introducing polar coordinates by the formulas

$$\alpha^*x = r\cos\theta, \quad \alpha^*y = r\sin\theta^{\dagger}$$

we obtain, after some calculation,

$$\alpha^*\tau' = \tfrac{1}{2}r^2 d\theta$$

which leads to the well-known formula

$$A = \int_0^{2\pi} \tfrac{1}{2}r^2 d\theta$$

for the area of the region bounded by a closed curve which is described in terms of polar coordinates.

The basic formulas of this chapter:

$$\lambda \wedge \sigma = -\sigma \wedge \lambda$$
$$(\omega_1 + \omega_2) \wedge \sigma = \omega_1 \wedge \sigma + \omega_2 \wedge \sigma$$
$$d(f\omega) = df \wedge \omega + f d\omega$$
$$\alpha^*(\lambda \wedge \sigma) = \alpha^*\lambda \wedge \alpha^*\sigma$$
$$\alpha^*(\tau_1 + \tau_2) = \alpha^*\tau_1 + \alpha^*\tau_2$$
$$\alpha^* d\tau = d\alpha^*\tau$$
$$\int_R \alpha^*\tau = \int_{\alpha R} \tau \text{ for orientation-preserving } \alpha$$
$$\int_{\partial R} \tau = \int_R d\tau$$

## Summary

### A Two-forms

Given a differential one-form $\tau$ on the plane, you should be able to state the definition of its exterior derivative $d\tau$ and to calculate $d\tau$ in terms of coordinates $x$ and $y$.

You should know how to define and evaluate the integral of a two-form over an oriented rectangular region of the plane.

† Frequently in applications the $\alpha^*$ is dropped, and one writes simply $x = r\cos\theta$, $y = r\sin\theta$.

B Double integrals

You should be able to evaluate double integrals over regions of the plane by carrying out iterated integrals with appropriate limits of integration, and to reverse the order of integration in an integral by converting it to a double integral.

Given a transformation from one region of the plane to another, you should be able to evaluate integrals over the second region by pullback, and you should be able to invent such transformations to simplify the evaluation of double integrals.

C Green's theorem

You should be able to state and apply Green's theorem in the plane.

---

## Exercises

8.1. In each of the following cases, $u$ and $v$ are functions on a plane where $x$ and $y$ are affine coordinates. Express $dx \wedge dy$ in terms of $du \wedge dv$. Make a sketch showing typical curves $u =$ constant and $v =$ constant in the first quadrant $(x, y > 0)$ and try to give a geometric interpretation to the relations between $dx \wedge dy$ and $du \wedge dv$ by applying both to a parallelogram whose sides are tangent to $u =$ constant and $v =$ constant respectively.

(a) $x = u \cos v$. $y = u \sin v$.

(b) $x = u \cosh v$, $y = u \sinh v$.

(c) $x = u^2 - v^2$, $y = 2uv$.

8.2. Evaluate $\iint_S x^2 y^2 \, dx dy$, where $S$ is the bounded portion of the first quadrant lying between the hyperbolas $xy = 1$ and $xy = 2$ and the straight lines $y = x$ and $y = 4x$.

8.3.(a) Show, by reversing the order of integration, that

$$\int_0^a \left( \int_0^y e^{m(a-x)} f(x) dx \right) dy = \int_0^a (a - x) e^{m(a-x)} f(x) dx$$

where $a$ and $m$ are constants, $a > 0$.

(b) Show that $\int_0^x (\int_0^v [\int_0^u f(t) dt] du) dv = \frac{1}{2} \int_0^x (x - t)^2 f(t) dt$.
If you do this in two steps, you never actually have to consider a triple integral!

8.4. Evaluate the iterated integral

$$I = \int_0^1 y \left( \int_0^{1-y^2} \frac{\sin \pi x}{x^4} dx \right) dy$$

by expressing it as a double integral over a suitable region $W$, then evaluating the integral as an iterated integral in the opposite order. Make a sketch to show the region $W$. (You may want to consult an integral table if you find the evaluation of the single integrals hard.)

8.5. Consider the mapping defined by the equations

$$x = u + v, \quad y = v - u^2.$$

(a) Compute the Jacobian determinant of this mapping as a function of $u$ and $v$.

(b) A triangle $T$ in the $uv$-plane has vertices $(0, 0)$, $(2, 0)$, $(0, 2)$. Sketch its image $S$ in the $xy$-plane.

(c) Calculate the area of $S$ by a double integral over $S$ and also by a double integral over $T$.

(d) Evaluate

$$\iint_S \frac{dx\,dy}{(x-y+1)^2}.$$

8.6.(a) Let $s$ denote the unit square in the $uv$-plane. Describe and sketch the image of $s$ under the mapping

$$\phi:\begin{pmatrix} u \\ v \end{pmatrix} \mapsto \begin{pmatrix} uv \\ v(2-u^2) \end{pmatrix}.$$

Label $\phi(A)$, $\phi(B)$, $\phi(C)$ on your sketch.

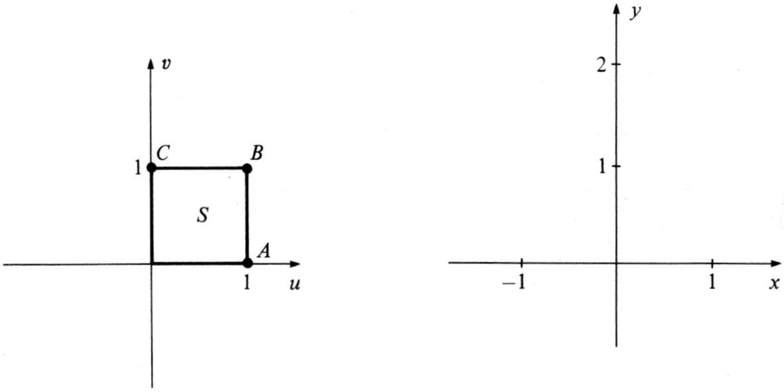


**Figure 8.43**

(b) Evaluate $\iint y\,dx\,dy$ over this region in the $xy$-plane.

(c) Evaluate this same integral by integrating the appropriate function over the square $s$ in the $uv$-plane.

8.7.(a) Evaluate the integral

$$\int_W (x+2y)\,dx\,dy$$

for the triangular region $W$ shown in figure 8.44.

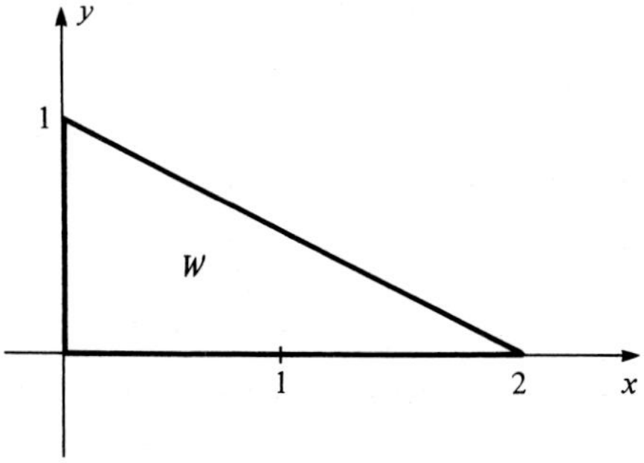


**Figure 8.44**

(b) Evaluate the same integral by using coordinates $u$ and $v$ related to $x$ and $y$ by $x = 2uv, y = u - uv$.

8.8. Evaluate

$$\int_0^a \left( \int_0^x \sqrt{(x^2 + y^2)}\mathrm{d}y \right) \mathrm{d}x$$

first as an iterated integral over $y$ and $x$, then by using polar coordinates.

8.9.(a) Evaluate the iterated integral

$$\int_0^1 \left( \int_1^{1/v} u^5 v^9 \mathrm{d}u \right) \mathrm{d}v.$$

(b) Interpret this integral as a double integral over a suitable region in the $uv$-plane. Draw a picture of this region labelling its boundary curves clearly. (Do not be concerned, here or later, by the fact that the region is unbounded.)

(c) Reinterpret the double integral as an iterated integral in the other order, and evaluate this integral.

(d) Make the substitution $u = x^2 y^{-3}, v = x^{-1} y^2$ in the double integral, obtaining a new double integral in the $xy$-plane. Draw a picture of the domain of this new integral.

(e) Convert the new double integral to an iterated integral and evaluate it.

(f) Show that $x, y$ are differentiable coordinates on the whole of the (open) first quadrant of the $uv$-plane.

8.10.(a) Evaluate

$$\iint_Q y \mathrm{d}A,$$

where $Q$ is the first quadrant of the unit disk, by converting it to an iterated integral in $x$ and $y$. ('$A$' refers to the usual area in the $xy$-plane.)

(b) Introduce polar coordinates $r$ and $\theta$ into the $xy$-plane as usual and convert the given integral into an integral over a suitable region in the $r\theta$-plane. Describe this region carefully. (By the $r\theta$-plane is meant a new copy of $\mathbb{R}^2$ in which the usual coordinates are called $r$ and $\theta$.)

(c) Convert the new integral to an iterated integral in $r$ and $\theta$ and evaluate it again.

8.11.(a) Evaluate the integral $I = \int_W 2y \mathrm{d}x \wedge \mathrm{d}y$ as the sum of two iterated integrals in the $xy$-plane. The region $W$ is bounded by the lines $y = \frac{1}{2}x$ and $y = 2x$ and the hyperbolas $xy = 2$ and $xy = 8$.

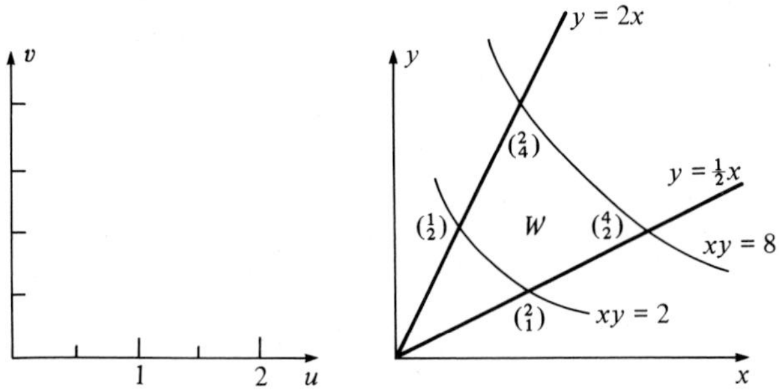


**Figure 8.45**

(b) Find a rectangle $R$ in the $uv$-plane such that $W$ is the image of $R$ under the transformation $\alpha$ described by $\alpha^*x = 2uv$, $\alpha^*y = u/v$.

(c) Calculate $\alpha^*(2y\mathrm{d}x \wedge \mathrm{d}y)$.

(d) Evaluate the integral $I$ as an integral over the region $R$.

8.12.(a) Evaluate the line integral $\int(2y^2 + 3x)\mathrm{d}x + 2xy\mathrm{d}y$ over the curve $\gamma$ shown in figure 8.46, which consists of the line segments $0 \leqslant x \leqslant 2$ and $0 \leqslant y \leqslant 2$ and the circular arc $x^2 + y^2 = 4$ for $x \geqslant 0, y \geqslant 0$.

(b) Construct a double integral over the region bounded by $\gamma$ which must be equal to the line integral in (a). Evaluate this double integral by transforming to polar coordinates.

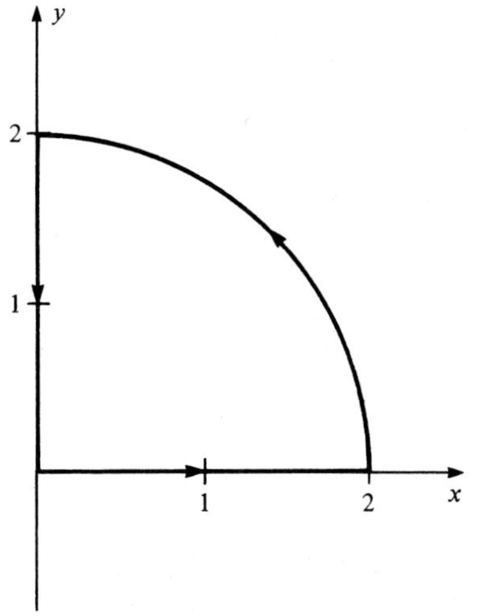


**Figure 8.46**

(c) Find a function $f(x)$ with the property that

$$\int f(x)[(2y^2 + 3x)\mathrm{d}x + 2xy\mathrm{d}y] = 0$$

when the integral is evaluated around *any* closed curve in the plane.

8.13. One way to change coordinates in a directed double integral $I = \int_W f\mathrm{d}x \wedge \mathrm{d}y$, where $W = \phi(S)$, is to use Green's theorem to express $I$ as a line integral over the closed path $\partial W$, transform the result to a line integral in the $uv$-plane, then use Green's theorem again to express $I$ as a double integral in the $uv$-plane. Use this approach to derive the change of variables formula for double integrals.

8.14. Let $u$ and $v$ be functions on the plane whose first and second partial derivatives with respect to $x$ and $y$ are continuous. Let $S$ be a connected region in the plane with boundary $\partial S$. Show that

$$\int_{\partial S}\left[\left(v\frac{\partial u}{\partial x} - u\frac{\partial v}{\partial x}\right)\mathrm{d}x + \left(u\frac{\partial v}{\partial y} - v\frac{\partial u}{\partial y}\right)\mathrm{d}y\right] = 2\iint_S\left(u\frac{\partial^2 v}{\partial x\partial y} - v\frac{\partial^2 u}{\partial x\partial y}\right)\mathrm{d}x\mathrm{d}y.$$

8.15.(a) The moment of inertia for a plane lamina $S$ of uniform density (mass per

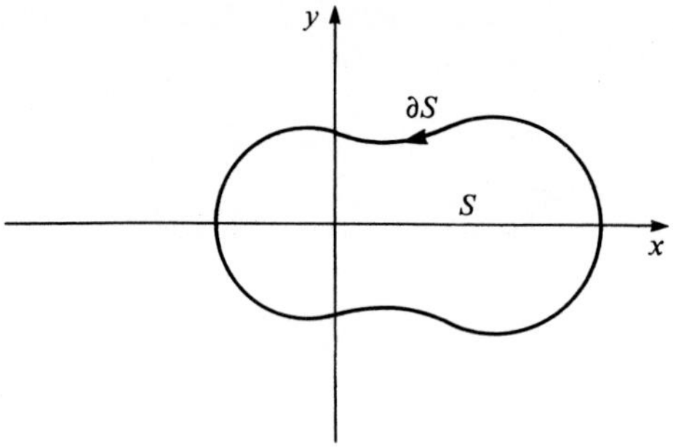


**Figure 8.47**

unit area) $\sigma$ about the $x$-axis is

$$I_x = \sigma \int_S y^2 \, \mathrm{d}x \, \mathrm{d}y.$$

Show that $I_x = \sigma \int_{\partial S} x y^2 \, \mathrm{d}y$ where $\partial S$ is the boundary of the lamina traversed counter clockwise.

(b) The moment of inertia about the $y$-axis is

$$I_y = \sigma \int x^2 \, \mathrm{d}x \, \mathrm{d}y.$$

Find two different differential forms $\omega$ and $\tau$ such that

$$I_y = \sigma \int_{\partial S} \omega = \sigma \int_{\partial S} \tau.$$

(c) The moment of inertia about the $z$-axis is

$$I_z = \sigma \int (x^2 + y^2) \, \mathrm{d}x \, \mathrm{d}y = I_x + I_y.$$

Find a differential form $\Omega$ such that $I_z = \sigma \int_{\partial S} \Omega$. Express $\Omega$ in terms of polar coordinates $r$ and $\theta$.

# 9

# Gaussian optics



Chapter 9 presents an example of how the results of the first eight chapters can be applied to a physical theory – optics. It is all in the nature of applications, and can be omitted without any effect on the understanding of what follows.

## 9.1. Theories of optics

In the history of physics it is often the case that, when an older theory is superseded by a newer one, the older theory retains its validity, either as an approximation to the newer theory, an approximation that is valid for an interesting range of circumstances, or as a special case of the newer theory. Thus Newtonian mechanics can be regarded as an approximation to relativistic mechanics, valid when the velocities that arise are very small in comparison to the velocity of light. Similarly, Newtonian mechanics can be regarded as an approximation to quantum mechanics, valid when the bodies in question are sufficiently large. Kepler's laws of planetary motion are a special case of Newton's laws, valid for the inverse square law of force between two bodies. Kepler's laws can also be regarded as an approximation to the laws of motion derived from Newtonian mechanics when we ignore the effects of the planets on each other's motion.

The currently held theory of light is known as quantum electrodynamics. It describes very successfully and very accurately the interaction of light with charged particles, explaining both the discrete character of light, as evinced in the photoelectric effect, and the wave-like character of electromagnetic radiation. The triumph of nineteenth century physics was Maxwell's electromagnetic theory, which was a self-contained theory explaining electricity, magnetism and electromagnetic radiation. Maxwell's theory can be regarded as an approximation to quantum electrodynamics, valid in that range where it is safe to ignore quantum effects. Maxwell's theory fails to explain a whole range of phenomena that occur at the atomic or subatomic level.

One of Maxwell's remarkable discoveries was that visible light is a form of electromagnetic radiation, as is *radiant heat*. In fact, since Maxwell, optics is a special chapter of the theory of electricity and magnetism which treats electromagnetic vibrations of all wavelengths, from the shortest $\gamma$ rays of radioactive substances (having a wavelength of one hundred-millionth of a millimeter) through the X-rays, the ultraviolet, visible light, the infra-red, to the longest radio waves (having a wavelength of many kilometers). In the flood of invisible light that is accessible to the mental eye of the physicist, the physiological eye is almost blind, so small is the interval of vibrations that it converts into sensations.

Maxwell's theory dealt with the source of electromagnetic radiation as well as its propagation. Before Maxwell, there was a fairly well-developed wave theory of light, due mainly to Fresnel, which dealt rather successfully with the propagation of light in various media, but had nothing to say about the production of light. Fresnel's theory did account for three physical effects which could not be explained by earlier theories – diffraction, interference, and polarization. Diffraction has to do with the behavior of light in the immediate vicinity of surfaces through which it is transmitted or reflected. A typical diffraction effect is the fact that we cannot produce an absolutely straight, arbitrarily narrow beam of light. For example, we might try to produce such a beam by lining up two opaque screens with holes in them, to collimate light arriving from the left of one of them. When the holes get very small (of the order of the wavelength of the light), we find that the region to the right of the second screen is suffused with light, instead of there being a narrow beam.

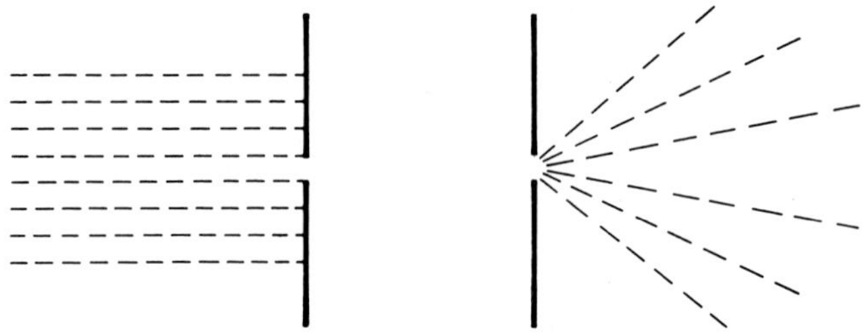

**Figure 9.1**

'Interference' refers to those phenomena where the wave character of light manifests itself by the constructive or destructive superposition of light travelling different paths. Typical is the famous Young interference experiment illustrated in figure 9.2. 'Polarization' refers to the fact that when light passes through certain materials, it appears to acquire a preferred direction in the plane perpendicular to the ray; such effects can be observed, for example, by using Polaroid filters.

Geometrical optics is the approximation to wave optics in which the wave character of light is ignored. It is valid whenever the dimensions of the various apertures are very large when compared to the wavelength of the light, and when we

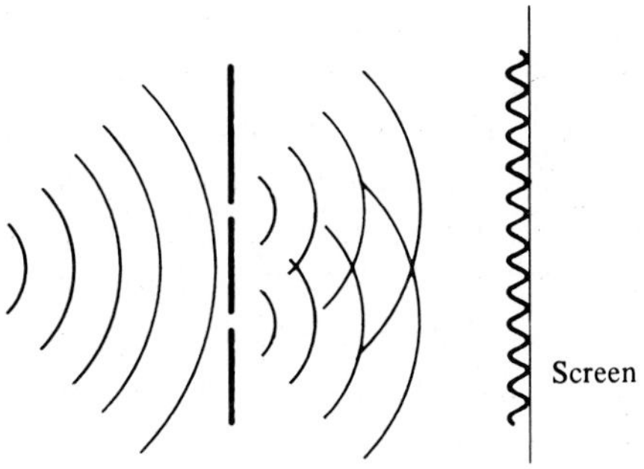


**Figure 9.2**

do not examine too closely what is happening in the neighborhood of shadows or foci. It does not account for diffraction, interference or polarization.

Linear optics is an approximation to geometrical optics that is valid when the various angles which enter into consideration are small. In linear optics one makes the approximation $\sin\theta \doteq \theta$, $\tan\theta \doteq \theta$, $\cos\theta \doteq 1$, etc.; i.e., all expressions which are quadratic (or of higher order) in the angles are ignored. For example, in geometrical optics, *Snell's law* says that if light passes from a region whose index of refraction (relative to vacuum) is $n$, into a region whose index of refraction is $n'$, then $n \sin i = n' \sin i'$ where $i$ and $i'$ are the angles that the light ray makes with the normal to the surface separating the regions. In linear optics we replace this law by the simpler law

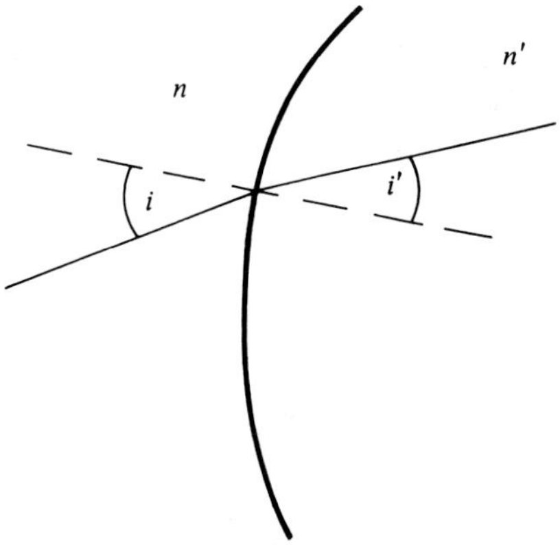


**Figure 9.3**

$ni = n'i'$, which is a good approximation if $i$ and $i'$ are small. (This approximate law was known to Ptolemy.) The deviations between geometrical optics and the linear optics approximation are known as (geometrical) aberrations. For instance, if a bundle of parallel rays is incident on a spherical mirror, a careful examination of the reflected rays shows that they do not all intersect at a common point. The rays near the diameter do intersect near a common focal point. In linear optics we restrict

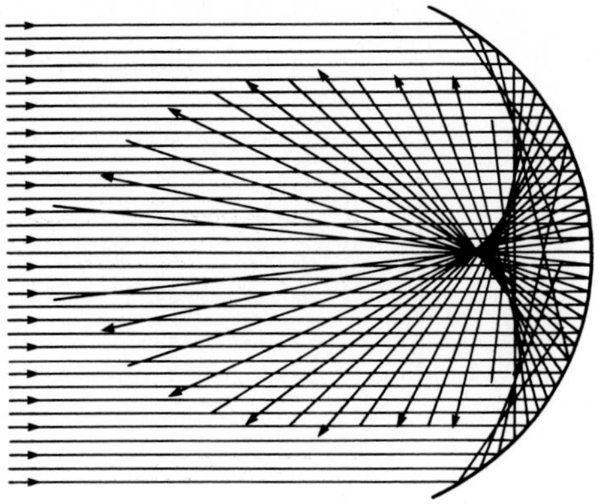

**Figure 9.4** Spherical aberration.

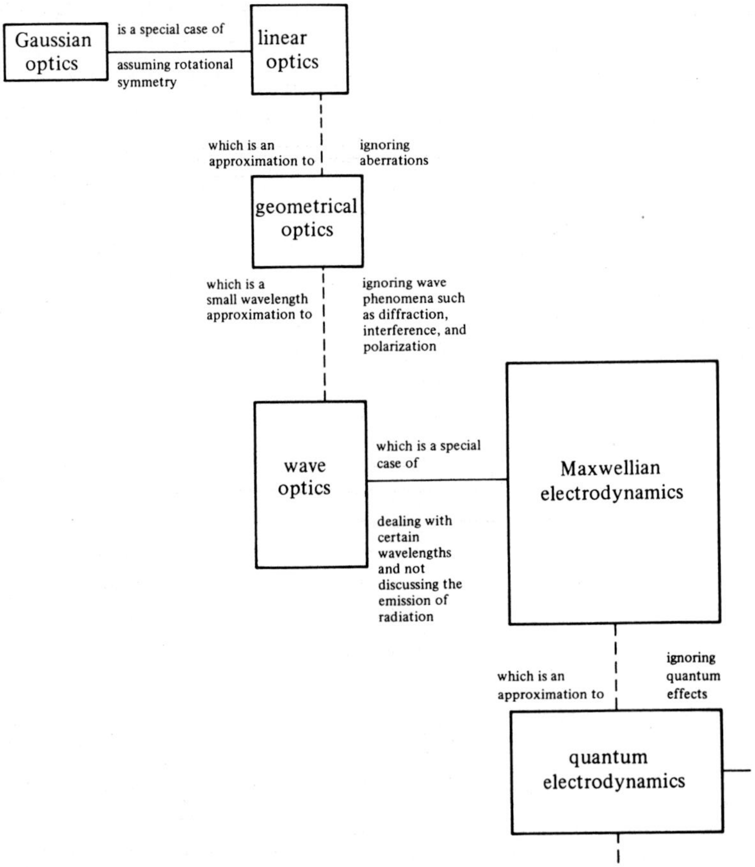


**Figure 9.5**

ourselves to rays close enough to the diameter so that we may assume that there is a common focus. (This deviation from focussing for a spherical mirror is a case of *spherical aberration.*)

Gaussian optics is a special case of linear optics in which it is assumed that all the surfaces that enter are rotationally symmetric about a central axis. This is a very important special case since all ground lenses and most polished mirrors have this property. We can summarize our discussion in figure 9.5.

## 9.2. Matrix methods

In Gaussian optics we are interested in tracing the trajectory of a light ray as it passes through the various refracting surfaces of the optical system (or is reflected by reflecting surfaces). We introduce a coordinate system so that the $z$-axis (pointing from left to right in our diagram) coincides with the optical axis (i.e., the axis of symmetry of our system). We shall restrict attention to coaxial rays – those that lie in a plane with the optical axis.*

By rotational symmetry, it is clearly sufficient to restrict attention to rays lying in one fixed plane. The trajectory of a ray, as it passes through the various refracting surfaces of the system, will consist of a series of straight lines. Our problem is to relate the straight line of the ray after it emerges from the system to the entering straight line. For this we need to have a way of specifying straight lines. We do so as follows: we choose some fixed $z$ value. This amounts to choosing a plane perpendicular to the optical axis, called the reference plane. Then a straight line is specified by two numbers, its height, $q$ above the axis at $z$, and the angle $\theta$ that the line makes with the optical axis. The angle $\theta$ will be measured in radians and considered positive if a counterclockwise rotation carries the positive $z$-direction into the direction of the ray along the straight line. It is convenient to choose new reference planes, suitably

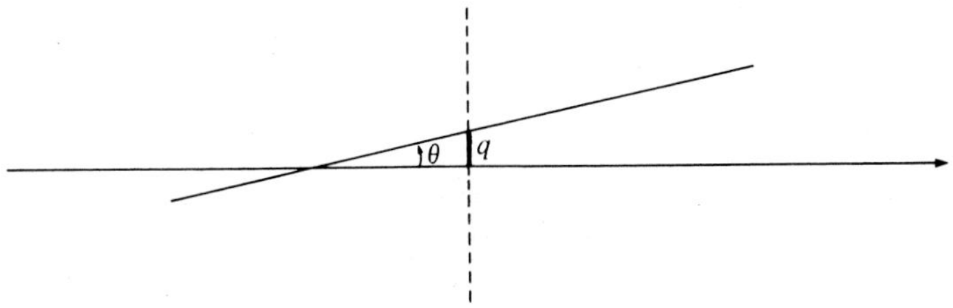


**Figure 9.6**

adjusted to each stage in the calculation. Thus, for example, if light enters our optical system from the left and emerges from the right, we would choose one reference plane $z_1$ to the left of the system of lenses and a second reference plane $z_2$ to the right.

* Although this is introduced here as a simplifying assumption, it can be proved that linearity implies that the study of the most general ray can be reduced to the study of coaxial rays by projection onto two perpendicular components.

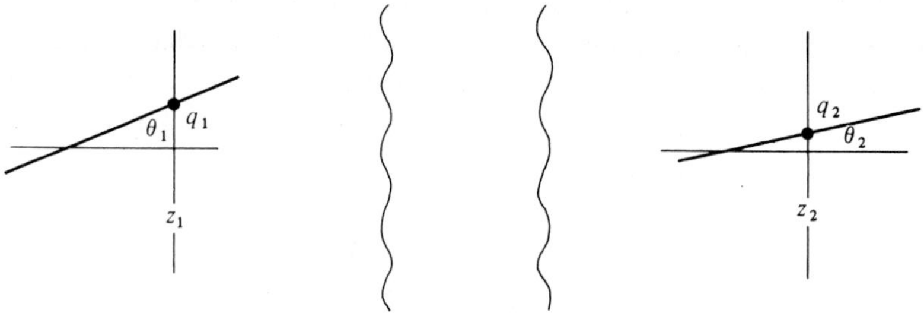


**Figure 9.7**

A ray enters the system as a straight line specified by $q_1$ and $\theta_1$ at $z_1$ and emerges as a straight line specified by $q_2$ and $\theta_2$ at $z_2$. Our problem, for any system of lenses, is to find the relation between $(q_2, \theta_2)$ and $(q_1, \theta_1)$.

Now comes a simple but crucial step, of far reaching significance, which is basic to the geometry of optics and of mechanics.

Replace the variable $\theta$ by $p = n\theta$ where $n$ is the index of refraction of the medium at the reference plane. (In mechanics, the corresponding step is to replace velocity by momentum.)

We thus describe a light ray by the vector $\begin{pmatrix} q \\ p \end{pmatrix}$ and our problem is to find $\begin{pmatrix} q_2 \\ p_2 \end{pmatrix}$ as a function of $\begin{pmatrix} q_1 \\ p_1 \end{pmatrix}$. Since we are ignoring all terms quadratic or higher, it follows from our approximation that $\begin{pmatrix} q_2 \\ p_2 \end{pmatrix}$ is a linear function of $\begin{pmatrix} q_1 \\ p_1 \end{pmatrix}$, i.e., that

$$\begin{pmatrix} q_2 \\ p_2 \end{pmatrix} = M_{21} \begin{pmatrix} q_1 \\ p_1 \end{pmatrix}$$

for some matrix $M_{21}$. The key effect of our choice of $p$ instead of $\theta$ as variable is the assertion that

$$\operatorname{Det} M_{21} = 1.$$

In other words, that the study of Gaussian optics is equivalent to the study of the group of $2 \times 2$ real matrices of determinant one, the group $\mathrm{Sl}(2, \mathbb{R})$. To prove this, observe that if we have three reference planes, $z_1, z_2$, and $z_3$, situated so that the light ray going from $z_1$ to $z_3$ passes through $z_2$, then by definition

$$M_{31} = M_{32} M_{21}.$$

Thus, if our optical system is built out of two components, we need only verify $\operatorname{Det} M = 1$ for each component separately. To simplify the exposition, assume that our system does not contain mirrors.

**The basic components**

Any refracting lens system can be considered as the composite of several systems of two basic types.

(a) A translation, in which the ray continues to travel in a straight line between two reference planes lying in the same medium. To describe such a system we must specify the gap, $t$, between the planes and the refractive index, $n$, of the medium. It is clear for such a system that $\theta$ and hence $p$ do not change and that $q_2 = q_1 + (t/n)p_1$.

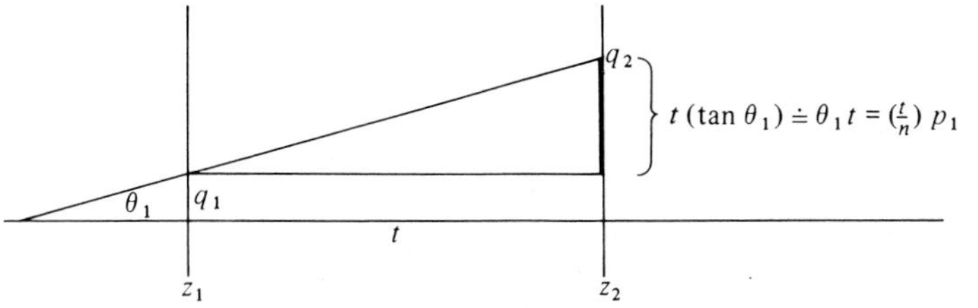


**Figure 9.8**

We write $T = t/n$ (called the reduced distance) and see that

$$\begin{pmatrix} q_2 \\ p_2 \end{pmatrix} = \begin{pmatrix} 1 & T \\ 0 & 1 \end{pmatrix} \begin{pmatrix} q_1 \\ p_1 \end{pmatrix}, \quad \text{Det} \begin{pmatrix} 1 & T \\ 0 & 1 \end{pmatrix} = 1.$$

(b) Refraction at the boundary surface between two regions of differing refractive index. We must specify the curvature of the surface and the two indices of refraction, $n_1$ and $n_2$. The two reference planes will be taken immediately to the left and immediately to the right of the surface.

At such a surface of refraction, the $q$ value does not change. The angle, and hence the $p$, changes according to (the linearized version of) Snell's law. Now Snell's law involves the slope of the tangent to the surface at the point of refraction. In our approximation, we are ignoring quadratic terms in this slope, hence terms of degree three or higher in the surface. We may thus assume that the curve giving the intersection of this surface with our plane is a parabola

$$z - z_1 = \tfrac{1}{2} k q^2.$$

Then the derivative of $z$ with respect to $q$ is $z'(q) = kq$, which is $\tan(\pi/2 - \psi)$ where $\psi$ is the angle in figure 9.9. For small angles $\theta$, i.e., for small values of $q$, $\psi$ will be close to $\pi/2$ and hence we may replace $\tan(\pi/2 - \psi)$ by $\pi/2 - \psi$, if we are willing to drop higher order terms in $q$ or $p$. Thus $\pi/2 - \psi = kq$ is our Gaussian approximation. On the other hand, if $(\pi/2 - i_1)$ denotes the angle that the incident ray makes with this

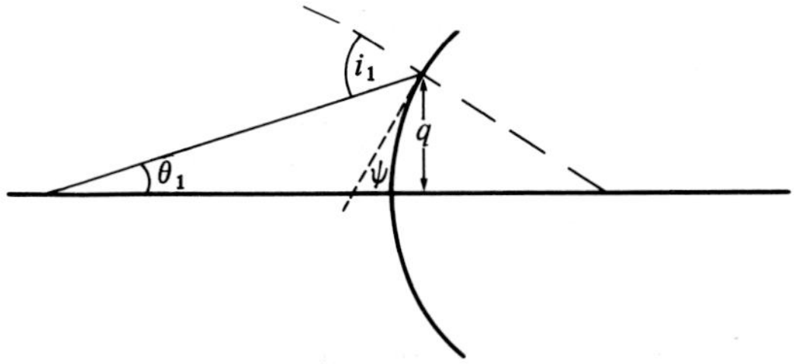


**Figure 9.9**

tangent line, then the fact that the sum of the interior angles of a triangle add up to $\pi$ shows that $(\pi - \psi) + \theta_1 + (\pi/2 - i_1) = \pi$ or

$$i_1 = \theta_1 + kq$$

and similarly

$$i_2 = \theta_2 + kq$$

where $q = q_1 = q_2$ is the point where the rays hit the refracting surface. Multiplying the first equation by $n_1$ and the second equation by $n_2$, and using Snell's law in the approximate form $n_1 i_1 = n_2 i_2$, give

$$\begin{pmatrix} q_2 \\ p_2 \end{pmatrix} = \begin{pmatrix} 1 & 0 \\ -P & 1 \end{pmatrix} \begin{pmatrix} q_1 \\ p_1 \end{pmatrix}$$

where $P = (n_2 - n_1)k$ is called the *power* of the refracting surface.

**Conjugate planes**

Thus each Gaussian optical system between two reference planes corresponds to a matrix

$$M = \begin{pmatrix} A & B \\ C & D \end{pmatrix} \quad \text{with} \quad AD - BC = 1$$

and one can set up a dictionary which translates properties of the matrix into optical properties.

For instance, the two planes are called *conjugate* (or in focus with one another) for any $q_1$ at $z_1$, if all the light rays leaving $q_1$ converge to the same point $q_2$ at $z_2$. This of course means that $q_2$ should not depend on $p_1$, i.e., that

$$B = 0.$$

**The thin lens**

Notice that the product of two matrices of the form $\begin{pmatrix} 1 & 0 \\ -P & 1 \end{pmatrix}$ again has this same form

$$\begin{pmatrix} 1 & 0 \\ -P_1 & 1 \end{pmatrix} \begin{pmatrix} 1 & 0 \\ -P_2 & 1 \end{pmatrix} = \begin{pmatrix} 1 & 0 \\ -(P_1 + P_2) & 1 \end{pmatrix}.$$

This gives the equation for the so-called *thin lens* consisting of refracting surfaces with negligible separation between them. In this case, the reference planes $z_1$ and $z_2$

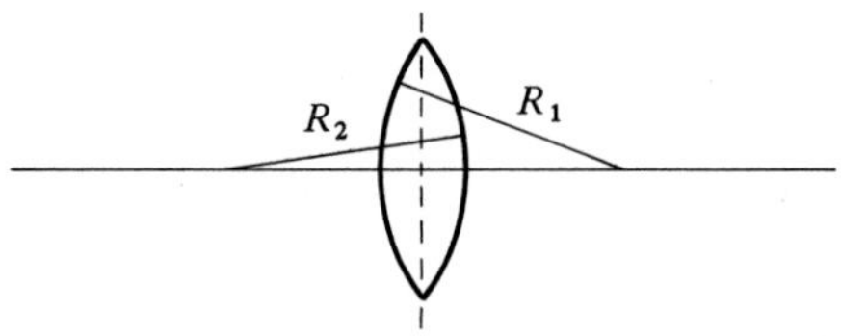


**Figure 9.10**

can conveniently both be taken to coincide with the plane of the lens. The plane $z_1$ relates, of course, to rays incident from the left, while $z_2$ relates to rays which emerge from the lens and continue to the right.

The matrix for the left refracting surface is

$$\begin{pmatrix} 1 & 0 \\ -\dfrac{n_2-n_1}{R_1} & 1 \end{pmatrix}.$$

The matrix for the right refracting surface is

$$\begin{pmatrix} 1 & 0 \\ -\dfrac{n_1-n_2}{R_2} & 1 \end{pmatrix}.$$

(Note that $R_2$ is negative in figure 9.10.) Multiplying these matrices, we find that the matrix for the thin lens is

$$\begin{pmatrix} 1 & 0 \\ -1/f & 1 \end{pmatrix}, \quad \text{where} \quad 1/f=(n_2-n_1)(1/R_1-1/R_2).$$

We shall assume that the lens is in a vacuum, so $n_1=1$ and $n_2>1$. In the case where $R_1$ is positive, $R_2$ is negative, and $n_2-n_1>0$ (a double-convex lens), the *focal length* $f$ is positive. If we calculate the matrix of the thin lens between a reference plane $F_1$

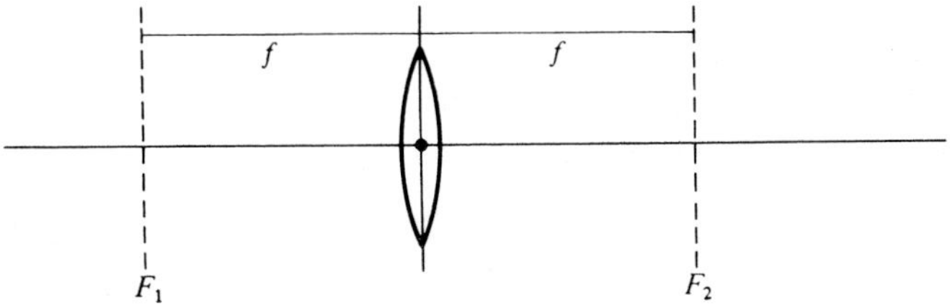


**Figure 9.11**

located a distance $f$ to the left of the lens and a reference plane $F_2$ located a distance $f$ to the right, we find

$$\begin{pmatrix} 1 & f \\ 0 & 1 \end{pmatrix}\begin{pmatrix} 1 & 0 \\ -1/f & 1 \end{pmatrix}\begin{pmatrix} 1 & f \\ 0 & 1 \end{pmatrix}=\begin{pmatrix} 0 & f \\ -1/f & 0 \end{pmatrix}.$$

The plane $F_1$ is called the *first focal plane.* If a ray, incident on the lens, passes through this plane at $q_1=0$ with slope $p_1$, then the outgoing ray has

$$\begin{pmatrix} q_2 \\ p_2 \end{pmatrix}=\begin{pmatrix} 0 & f \\ -1/f & 0 \end{pmatrix}\begin{pmatrix} 0 \\ p_1 \end{pmatrix}=\begin{pmatrix} fp_1 \\ 0 \end{pmatrix},$$

i.e., it has zero slope and so is parallel to the axis. Conversely, if the incident ray has zero slope, the outgoing ray has

$$\begin{pmatrix} q_2 \\ p_2 \end{pmatrix}=\begin{pmatrix} 0 & f \\ -1/f & 0 \end{pmatrix}\begin{pmatrix} q_1 \\ 0 \end{pmatrix}=\begin{pmatrix} 0 \\ -q_1/f \end{pmatrix},$$

i.e., it crosses the axis in the second focal plane. More generally, we can see that $p_2$ is independent of $q_1$, so that incident rays passing through a given point in the first focal plane emerge as parallel rays, all with the same slope. Furthermore, $q_2$ is independent of $p_1$, so that incident rays all emerge to pass through the same position in the second focal plane.

As a simple illustration of the use of matrix methods to locate an image, suppose that we take reference plane $z_1$ to lie a distance $s_1$ to the left of a thin lens, while $z_2$ lies a distance $s_2$ to the right of the lens. Between these planes, the matrix is

$$\begin{pmatrix} 1 & s_2 \\ 0 & 1 \end{pmatrix}\begin{pmatrix} 1 & 0 \\ -1/f & 1 \end{pmatrix}\begin{pmatrix} 1 & s_1 \\ 0 & 1 \end{pmatrix} = \begin{pmatrix} 1 - s_2/f & s_2 + s_1 - s_1 s_2/f \\ -1/f & 1 - s_1/f \end{pmatrix}.$$

The planes are conjugate if the upper right entry of this matrix is zero. Thus we obtain $1/s_1 + 1/s_2 = 1/f$, the well-known thin lens equation. We shall write this as

$$s_1 + s_2 - P s_1 s_2 = 0,$$

where $P = 1/f$.

We can solve this equation for $s_2$ so long as $s_1 + f \neq 1/P$. Thus each plane other than the one corresponding to $s_1 = f$ has a unique conjugate plane. For $s_1 = f$, i.e., at the first focal plane, all light rays entering from a single point $q$ emerge parallel, so the conjugate plane to the first focal plane is 'at infinity'. A similar discussion (with right and left interchanged) applies to the second focal plane.

For $s_1 \neq f$ and $s_2$ corresponding to the conjugate plane, the magnification is given by

$$\frac{q_2}{q_1} = 1 - \frac{s_2}{f} = 1 - s_2\left(\frac{1}{s_1} + \frac{1}{s_2}\right) = -\frac{s_2}{s_1}.$$

If $s_1$ and $s_2$ are both positive (object to left of lens, image to right), then the magnification is negative, which means that the image is inverted.

By multiplying matrices, it is straightforward to construct the matrix for any combination of thin lenses. For example, in the case of thin lenses with focal length $f_1$ and $f_2$, separated by distance $l$ in air, we find the matrix

$$\begin{pmatrix} 1 & 0 \\ -1/f_2 & 1 \end{pmatrix}\begin{pmatrix} 1 & l \\ 0 & 1 \end{pmatrix}\begin{pmatrix} 1 & 0 \\ -1/f_1 & 1 \end{pmatrix} = \begin{pmatrix} 1 - l/f_1 & l \\ l/f_1 f_2 - 1/f_2 - 1/f_1 & 1 - l/f_2 \end{pmatrix}$$

between the reference plane $z_1$ (first lens) and $z_2$ (second lens).

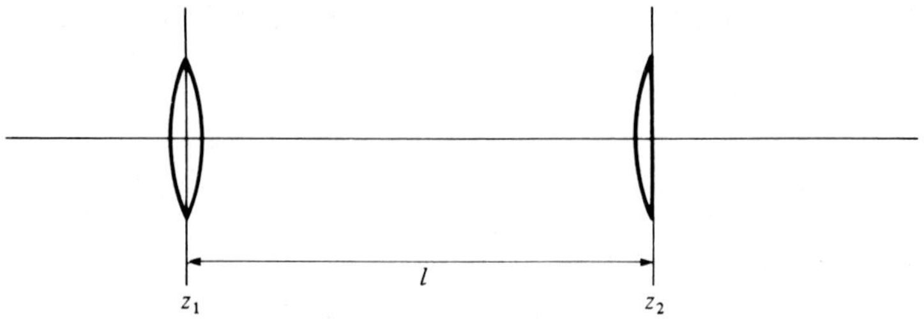


**Figure 9.12**

**The telescope**

A particularly interesting situation arises when $l = f_1 + f_2$, for then the matrix takes the form

$$\begin{pmatrix} A & B \\ 0 & D \end{pmatrix},$$

i.e., $C = 0$. This means that $p_2 = Dp_1$, i.e. that the outgoing directions depend only on the incoming directions. The condition is satisfied in the *astronomical telescope*, which consists of an objective lens of large positive focal length $f_1$ and an eyepiece of small positive focal length $f_2$, separated by a distance $f_1 + f_2$. Such a telescope converts parallel rays from a distant star into parallel rays which are presented to the eye.

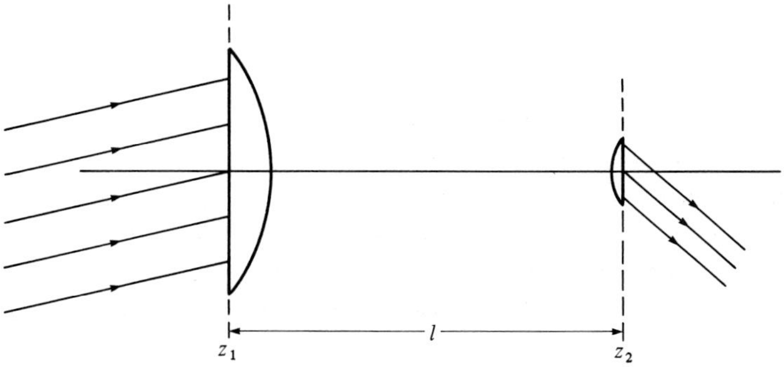


**Figure 9.13**

The *angular magnification* of such a telescope is the ratio of the slope of the *outgoing* rays to the slope of the *incoming* rays, which equals

$$D = 1 - \frac{l}{f_2} = 1 - \frac{f_1 + f_2}{f_2} = -\frac{f_1}{f_2}.$$

This magnification is negative (the image is inverted) and its magnitude is the ratio of the focal length of the objective to that of the eyepiece.

**The general system**

We now want to show that *any* $2 \times 2$ matrix with determinant 1 can arise as the matrix of some optical system. First of all, suppose that the matrix is telescopic, i.e., $C = 0$. Then $A \neq 0$, and if $P \neq 0$, then $\begin{pmatrix} 1 & 0 \\ -P & 1 \end{pmatrix}\begin{pmatrix} A & B \\ 0 & D \end{pmatrix} = \begin{pmatrix} A & B \\ -PA & D - PB \end{pmatrix}$ has $PA \neq 0$, so is not telescopic. We shall show that every matrix $\begin{pmatrix} A & B \\ C & D \end{pmatrix}$ with $C \neq 0$ can be written as

$$\begin{pmatrix} A & B \\ C & D \end{pmatrix} = \begin{pmatrix} 1 & t \\ 0 & 1 \end{pmatrix}\begin{pmatrix} 1 & 0 \\ C & 1 \end{pmatrix}\begin{pmatrix} 1 & s \\ 0 & 1 \end{pmatrix} \qquad (9.1)$$

and thus arises as an optical matrix. If $C=0$, then we need only multiply $\begin{pmatrix} A & B \\ -PA & D-PB \end{pmatrix}$ by $\begin{pmatrix} 1 & 0 \\ P & 1 \end{pmatrix}$ on the left to get $\begin{pmatrix} A & B \\ 0 & D \end{pmatrix}$, so it too is an optical matrix. To prove (9.1), consider

$$\begin{pmatrix} 1 & s \\ 0 & 1 \end{pmatrix}\begin{pmatrix} A & B \\ C & D \end{pmatrix}\begin{pmatrix} 1 & t \\ 0 & 1 \end{pmatrix}=\begin{pmatrix} A+sC & t(A+sC)+B+sD \\ C & Ct+D \end{pmatrix}.$$

Since $C \neq 0$, we can choose $s$ so that $A+sC=1$ and then choose $t=-(B+sD)$. The resulting matrix has 1 in the upper left-hand corner and zero in the upper right-hand corner. This implies that the lower right-hand corner is also 1 so that the matrix on the right has the form

$$\begin{pmatrix} 1 & 0 \\ C & 1 \end{pmatrix}$$

and this proves our assertion.

**Gauss decomposition**

Notice that $s$ and $t$ were uniquely determined. Thus, for any non-telescopic optical system, there are two unique planes such that the matrix between them has the form $\begin{pmatrix} 1 & 0 \\ C & 1 \end{pmatrix}$. These planes are conjugate to one another and have magnification one. Gauss called them the *principal planes*. If we start with the optical matrix $\begin{pmatrix} 1 & 0 \\ C & 1 \end{pmatrix}$ between the two principal planes, we can proceed exactly as for the thin lens, to find the conjugate plane to any plane. All we have to do is write $C=-P=-1/f$. For instance, the two focal planes are located $f$ units to the right and left of the principal planes:

$$\begin{pmatrix} 1 & f \\ 0 & 1 \end{pmatrix}\begin{pmatrix} 1 & 0 \\ -1/f & 1 \end{pmatrix}\begin{pmatrix} 1 & f \\ 0 & 1 \end{pmatrix}=\begin{pmatrix} 0 & f \\ -1/f & 0 \end{pmatrix}.$$

Gauss gave the following interpretation in terms of *ray tracing* of the decomposition we derived above for the more general non-telescopic system. Suppose a ray $\begin{pmatrix} q \\ 0 \end{pmatrix}$, parallel to the axis, enters the system at $z_1$. When it reaches the second principal

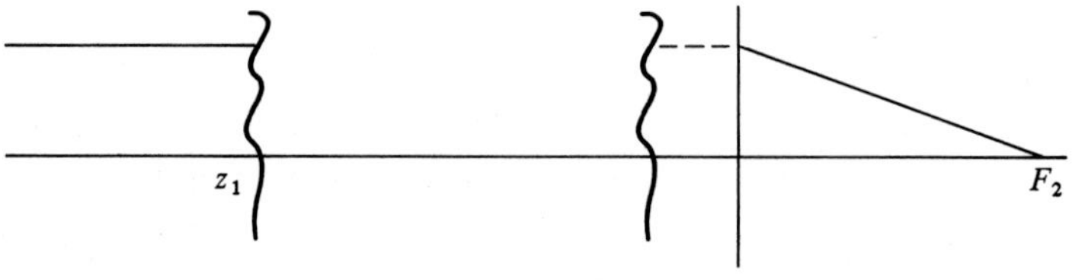


**Figure 9.14**

plane, it is at the same height but is bent into $\begin{pmatrix} q \\ -q/f \end{pmatrix}$ and is focussed on the axis at the second focal point, $F_2$. Similarly, a ray emerging from the first focal point is bent at the first principal plane into a ray parallel to the axis and arrives at $z_2$ still parallel to the axis and at the same height above the axis as it was at the first principal plane.

We see that the most general optical system which is not telescopic can be expressed simply in terms of three parameters – the location of the two principal planes and the focal length. (We know that there should be three parameters, since there are only three free parameters in the matrix, the fourth matrix coefficient being determined by the fact that the determinant must equal 1.)

Once we have located the principal planes, we have also located the focal planes by

$$H_1 - F_1 = f \quad \text{and} \quad F_2 - H_2 = f.$$

If we use the two focal planes as the reference planes for our system, then, by the very definition of focal planes, we know that the optical matrix for these two planes must have zeros in the upper left-hand corner and in the lower right-hand corner. Thus the matrix between the two focal planes is given by

$$\begin{pmatrix} 0 & f \\ -1/f & 0 \end{pmatrix}.$$

Suppose that we now consider two other planes, $y_1$ and $y_2$, related to the focal planes by

$$F_1 - y_1 = n_1 x_1 \quad \text{and} \quad y_2 - F_2 = n_2 x_2.$$

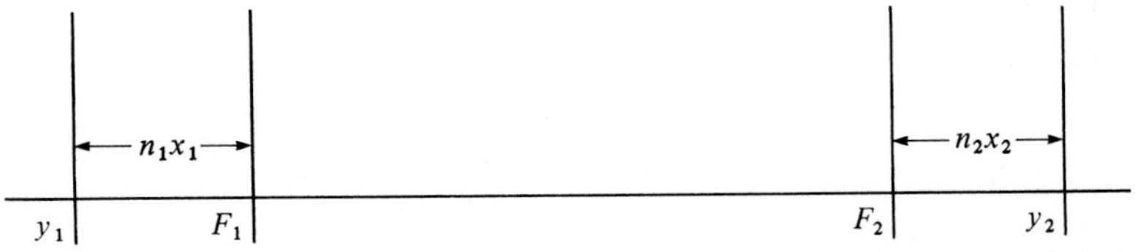


**Figure 9.15**

The matrix between these two planes will be

$$\begin{pmatrix} 1 & x_2 \\ 0 & 1 \end{pmatrix}\begin{pmatrix} 0 & f \\ -1/f & 0 \end{pmatrix}\begin{pmatrix} 1 & x_1 \\ 0 & 1 \end{pmatrix} = \begin{pmatrix} -x_2/f & f-(x_1x_2/f) \\ -1/f & -x_1/f \end{pmatrix}.$$

We see that $y_1$ and $y_2$ are conjugate if and only if $x_1 x_2 = f^2$ (this is known as *Newton's equation*), in which case the magnification is given by

$$m = -x_2/f = -f/x_1.$$

For instance, if $y_2$ lies to the right of $F_2$, so that $x_2$ is positive, and if the focal length, $f$, is positive, then $m$ is negative, i.e., the image is inverted.

We can summarize the results of this section as follows: Let Sl(2, ℝ) denote the group of all 2 × 2 matrices of determinant 1. We have shown that there is an isomorphism between Sl(2, ℝ) and Gaussian optics. Each matrix corresponds to an optical system, multiplication of matrices corresponds to composition of the corresponding system.

We next turn to Hamilton's ideas, in embryonic form.

## 9.3. Hamilton's method in Gaussian optics

Suppose that $z_1$ and $z_2$ are planes in an optical system which are *not* conjugate. This means that the $B$ term in the optical matrix is not zero. Thus, from the equations

$$q_2 = Aq_1 + Bp_1$$
$$p_2 = Cq_1 + Dp_1$$

we can solve for $p_1$ and $p_2$ in terms of $q_1$ and $q_2$ as

$$p_1 = (1/B)(q_2 - Aq_1)$$

and

$$p_2 = (1/B)(Dq_2 - q_1)$$

(where we have used the fact that $AD - BC = 1$). This has the following geometrical significance: given a point $q_1$ on the $z_1$-plane and a point $q_2$ on the $z_2$-plane, there exists a unique light ray joining these two points. (This is exactly what fails to happen if the planes are conjugate. For conjugate planes, if $q_2$ is the image of $q_1$, there will be an infinity of light rays joining $q_1$ and $q_2$; in fact, all light rays leaving $q_1$ arrive at $q_2$. If $q_2$ is not the image of $q_1$, then there will be no light ray joining $q_1$ and $q_2$.) Let $W = W(q_1, q_2)$ be the function

$$W(q_1, q_2) = (1/2B)(Aq_1^2 + Dq_2^2 - 2q_1q_2) + K$$

where $K$ is a constant. Then we can write the equations for $p_1$ and $p_2$ as

$$p_1 = -(\partial W/\partial q_1) \quad \text{and} \quad p_2 = \partial W/\partial q_2.$$

Hamilton called this function the *point characteristic* of the system. In the modern physics literature this function is sometimes called the *eikonal*. Suppose that $z_1, z_2$ and $z_3$ are planes such that no two of them are conjugate, with $z_1 < z_2 < z_3$, and such that $z_2$ does not coincide with a refracting surface. Let $W_{21}$ be the point characteristic for the $z_1$–$z_2$ system and let $W_{32}$ be the point characteristic for the $z_2$–$z_3$ system. We claim that (up to an irrelevant additive constant) the point characteristic for the $z_1$–$z_3$ system is given by

$$W_{31}(q_1, q_3) = W_{21}(q_1, q_2) + W_{32}(q_2, q_3)$$

where, in this equation, $q_2 = q_2(q_1, q_3)$ is taken to be the point where the ray from $q_1$ to $q_3$ hits the $z_2$-plane.

To see why this is so, we first observe that since the $z_2$-plane does not coincide with a refracting surface, the direction of the ray does not change at $z_2$. Thus

$$p_2 = (\partial W_{21}/\partial q_2)(q_1, q_2) = -(\partial W_{32}/\partial q_2)(q_2, q_3).$$

$$\frac{\partial W_{31}}{\partial q_1} = \frac{\partial W_{21}}{\partial q_1} + \frac{\partial W_{21}}{\partial q_2}\frac{\partial q_2}{\partial q_1} + \frac{\partial W_{32}}{\partial q_2}\frac{\partial q_2}{\partial q_1}$$
$$= \frac{\partial W_{21}}{\partial q_1} + \frac{\partial q_2}{\partial q_1}\left(\frac{\partial W_{21}}{\partial q_2} + \frac{\partial W_{32}}{\partial q_2}\right)$$
$$= \frac{\partial W_{21}}{\partial q_1} = -p_1$$ ; likewise for $\frac{\partial W_{31}}{\partial q_3} = p_3$

Now apply the chain rule to conclude that $\partial W_{31}/\partial q_1 = -p_1$ and similarly that $\partial W_{31}/\partial q_3 = p_3$ at $(q_1, q_3)$.

The function $W$ is determined by the above properties only up to an additive constant. Hamilton showed that, by an appropriate choice of the constant, we can arrange that *$W(q_1, q_2)$ is the optical length of the light ray joining $q_1$ to $q_2$* where the *optical length* is defined as follows. For a line segment of length $l$ in a medium of constant index of refraction, $n$, the optical length is $nl$. A *path*, $\gamma$, is defined to be a broken line segment, where each component segment lies in a medium of constant index of refraction. If the component segments have length $l_i$ and lie in media of refractive index $n_i$, then the optical length of $\gamma$ is

$$L(\gamma) = \sum n_i l_i.$$

Let us prove Hamilton's result within the framework of our Gaussian optics approximation. Our approximation is such that terms in $p$ and $q$ of degree higher than one are dropped from the derivatives of $W$. Thus, in computing optical length and $W$, we must retain terms up to degree two but may ignore terms higher than the second. We will prove this by establishing the following general formula for the optical length (in the Gaussian approximation) of a light ray $\gamma$ whose incoming parameters, at $z_1$, are $\begin{pmatrix} q_1 \\ p_1 \end{pmatrix}$ and whose outgoing parameters, at $z_2$, are $\begin{pmatrix} q_2 \\ p_2 \end{pmatrix}$:

$$L(\gamma) = L_{\text{axis}} + \tfrac{1}{2}(p_2 q_2 - p_1 q_1)$$

where $L_{\text{axis}}$ denotes the optical length from $z_1$ to $z_2$ of the axis $(p_1 = q_1 = 0 = p_2 = q_2)$ of the system. Notice that once this is proved, then, if we assume that $z_1$ and $z_2$ are non-conjugate, we can solve for $p_2$ and $p_1$ as functions of $q_1$ and $q_2$, i.e., substituting $p_1 = (1/B)(q_2 - Aq_1)$, $p_2 = (1/B)(Dq_2 - q_1)$ into the above formula gives our expression for $W$ with $K = L_{\text{axis}}$.

To prove the above formula for $L(\gamma)$, we observe that it behaves correctly when we combine systems: if we have $z_1, z_2$ and $z_3$, then the length along the axis certainly adds, and $\tfrac{1}{2}(p_2q_2 - p_1q_1) + \tfrac{1}{2}(p_3q_3 - p_2q_2) = \tfrac{1}{2}(p_3q_3 - p_1q_1)$. So we need only prove the formula for our two fundamental cases.

(1) If $n$ is constant,

$$\begin{aligned} L(\gamma) &= n(d^2 + (q_2 - q_1)^2)^{1/2} \\ &\doteq nd + \frac{1}{2}\frac{n}{d}(q_2 - q_1)^2 \\ &= nd + \frac{1}{2}\left[\frac{n}{d}(q_2 - q_1)\right](q_2 - q_1) \\ &= nd + \tfrac{1}{2}p(q_2 - q_1) \end{aligned}$$

(Taylor series for $\sqrt{d^2 + x^2}$ at $x = 0$)

where $p_2 = p_1 \doteq p = (n/d)(q_2 - q_1)$ is the formula which holds for this case.

(2) At a refracting surface, $z' - z = \tfrac{1}{2}kq^2$ with index of refraction $n_1$ to the left and $n_2$ to the right. Here the computation must be understood in the following

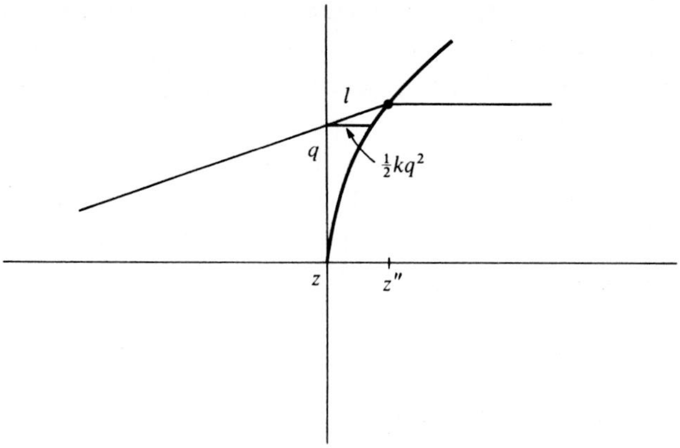


**Figure 9.16**

sense. Suppose we choose some point $z_3$ to the left and some point $z_4$ to the right of our refracting surface. If $n_1$ were equal to $n_2$, the optical length would be $n_1l_3 + n_2(l + l_4)$ where $l_3$ is the portion of the ray to the left of our plane and $l + l_4$ is the portion to the right, and where $l_4$ is the portion to the right of the surface. (We have drawn the figure with $k > 0$, but a similar argument works for $k < 0$.) If $n_2 \neq n_1$, then $n_2l_4$ will be different, but would be calculated by (1) from $z$ to $z_4$. In addition, an effect of the refracting surface is to replace $n_2l$ by $n_1l$ in the above expression, i.e., to modify the optical length by

$$(n_1 - n_2)l.$$

This is the contribution at the refracting surface. Now

$$l = (z'' - z)\operatorname{cosec}\theta_1$$

where $z''$ is determined by the pair of equations

$$z'' - z = \tfrac{1}{2}kq''^2$$
$$q'' = (\tan\theta_1)(z'' - z) + q.$$

It is clear that up to terms of higher order we may take $z'' = z' = \frac{1}{2}kq^2 + z$ and replace $\operatorname{cosec}\theta_1$ by 1 so

$$\begin{aligned}(n_1 - n_2)l &\doteq \tfrac{1}{2}k(n_1 - n_2)q^2 = -\tfrac{1}{2}pq^2\\ &= \tfrac{1}{2}[k(n_1 - n_2)q]q\\ &= \tfrac{1}{2}(p_2 - p_1)q\end{aligned}$$

since $q_2 = q_1 = q$ and $p_2 = p_1 - pq$ at a refracting surface, where $p = k(n_2 - n_1)$. This completes the proof of our formula.

## 9.4. Fermat's principle

Let us consider a refracting surface with power $p = (n_1 - n_2)k$ located at $z$. Here $P$ might be zero. Consider planes $z_1$ to the left and $z_2$ to the right of $z$. We assume constant index of refraction between $z_1$ and $z$ and between $z$ and $z_2$. Let $q_1$ be a point on the $z_1$-plane, $q_2$ a point on the $z_2$-plane and $q$ a point on the $z$-plane. Consider the path consisting of three pieces: the light ray joining $q_1$ to $q$, across

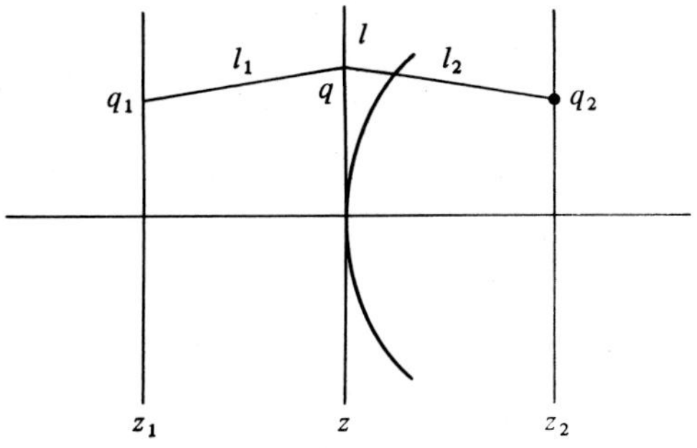


**Figure 9.17**

the surface of refraction at $q$ and then the light ray joining $q$ to $q_2$. This path will not, in general, be an optical path, since $q$ can be arbitrary. However, its optical length

$$n_1 l_1 + nl + n_2 l_2$$

is given, in the Gaussian approximation, by the sum of three terms, as we saw in the last section:

$$L(q_1, q_1 q_2) = L_{\text{axis}} + \tfrac{1}{2}(p_1(q - q_1) + p_2(q_2 - q) - pq^2).$$

In this expression,

$$p_1 = \frac{n_1}{d_1}(q - q_1)$$

and

$$p_2 = \frac{n_2}{d_2}(q_2 - q)$$

so we can write

$$L = L_{\text{axis}} + \frac{1}{2}\frac{d_1}{n_1}p_1^2 + \frac{d_2}{n_2}p_2^2 - pq^2.$$

Suppose that we hold $q_1$ and $q_2$ fixed, and look for that value of $q$ which extremizes $L$: in other words, we wish to solve the equation $\partial L/\partial q = 0$ for fixed values of $q_1$ and $q_2$. Substituting into the last expression for $L$, together with facts that $\partial p_1/\partial q = (n_1/d_1)(\partial q_2/\partial q) = -n_2/d_2$, we obtain the equation

$$p_1 - p_2 - pq = 0.$$

In other words:

$$p_2 = pq + p_1.$$

But this is precisely the relation between $p_1$ and $p_2$ given by the refraction matrix at $z$.

We have thus proved the following fact. *Let us fix $q_1$ and $q_2$ and consider the set of paths joining $q_1$ to $q_2$ which consists of two segments, from $q'$ to $q_2$. Among all such paths, the actual light ray can be characterized as that path for which L takes on an extreme value*, i.e., *for which*

$$\frac{\partial L}{\partial q'} = 0.$$

This is (our Gaussian approximation to) the famous Fermat principle of *least time*. Let us substitute $p_1 = (n_1/d_1)(q - q_1)$ and $p_2 = (n_2/d_2)(q_2 - q)$ into our formula for $L$ to obtain a third expression for $L$:

$$L = n_1 d_1 + n_2 d_2 + \tfrac{1}{2}[(n_1/d_1)(q - q_1)^2 + (n_2/d_2)(q_2 - q)^2 - pq^2].$$

The coefficient of $q^2$ is $n_1/d_1 + n_2/d_2 - P$. Thus the extremum is a *minimum* if

$$(n_1/d_1) + (n_2/d_2) - P > 0$$

and a *maximum* if

$$n_1/d_1 + n_2/d_2 - P < 0.$$

If $P > 0$ we see that we get a minimum for small values of $d_1$ and $d_2$ but a maximum for large values of $d_1$ and $d_2$. The situation is indeterminate (and we cannot, in general, solve for $q'$) when

$$n_1/d_1 + n_2/d_2 = P$$

which is precisely the condition that the planes be conjugate. Thus, we get a minimum if the conjugate plane to $z_1$ does not lie between $z_1$ and $z_2$ and a maximum otherwise. The fact that $L$ is minimized only up to the first conjugate point is true in a more general setting, where it is known as the *Morse index theorem*.

To see an intuitively obvious example of this phenomenon, let us consider light being reflected from a concave spherical mirror. We take a point $Q$ inside the sphere and let the light shine along a diameter so that it bounces back to $Q$. Then it is clear that the distance to the mirror is a local minimum if $Q$ is closer than the center, and a maximum otherwise.

## 9.5. From Gaussian optics to linear optics

What happens if we drop the assumption of rotational symmetry but retain the approximation that all terms higher than the first order in the angles and distances to one can be ignored? First of all, in specifying a ray, we now need four variables: $q_x$ and $q_y$, which specify where the ray intersects a plane transverse to the $z$-axis, and two angles, $\theta_x$ and $\theta_y$, which specify the direction of the ray. A direction in three-dimensional space is specified by a unit vector, $\mathbf{v} = (v_x, v_y, v_z)$. If $\mathbf{v}$ is close to pointing in the positive $z$-direction, it will have the form $\mathbf{v} = (\theta_x, \theta_y, v_z)$, where $v_z \doteq 1 - \frac{1}{2}(\theta_x^2 + \theta_y^2) \doteq 1$, provided $\theta_x$ and $\theta_y$ are small. Again, we replace the $\theta$ variables by $p$ variables, where $p_x = n\theta_x$ and $p_y = n\theta_y$. (If the medium is anisotropic, as is the case in certain kinds of crystals, the relation between the $\theta$ variables and the $p$ variables can be more complicated, but we will not concern ourselves with that here.) All of this, of course, is taking place at some fixed plane. If we consider two planes $z_1$ and $z_2$, the ray will correspond to vectors

$$\mathbf{u}_1 = \begin{pmatrix} q_{x1} \\ q_{y1} \\ p_{x1} \\ p_{y1} \end{pmatrix} \quad \text{and} \quad \mathbf{u}_2 = \begin{pmatrix} q_{x2} \\ q_{y2} \\ p_{x2} \\ p_{y2} \end{pmatrix}$$

at the respective planes.

Our problem is to find the form of the relationship between $\mathbf{u}_1$ and $\mathbf{u}_2$. Since we are ignoring all higher-order terms, we know that

$$\mathbf{u}_2 = M\mathbf{u}_1,$$

where $M$ is some $4 \times 4$ matrix. Our problem is to ascertain what kind of $4 \times 4$ matrices can actually arise in linear optics. The most obvious guess is that $M$ must

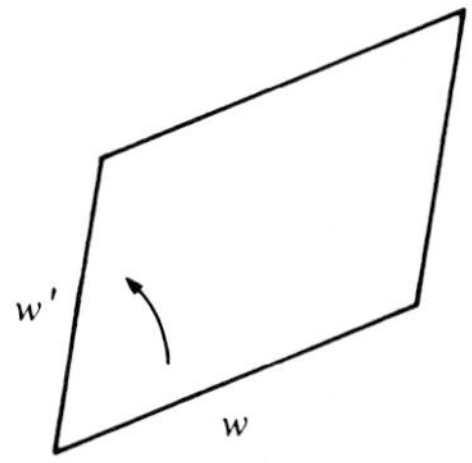


**Figure 9.18**

satisfy the requirement Det $M = 1$. This is not the right answer, however. It is true that all optical matrices must have unit determinant, but it is not true that all $4 \times 4$ matrices of determinant 1 can actually arise as transformation matrices in linear optics. There is a stronger condition that must be imposed. In order to explain what this stronger condition is, we first go back and reformulate the condition that a $2 \times 2$ matrix has determinant 1. We then formulate this condition in four variables. Let

$$\mathbf{w} = \begin{pmatrix} q \\ p \end{pmatrix} \quad \text{and} \quad \mathbf{w}' = \begin{pmatrix} q' \\ p' \end{pmatrix}$$

be two vectors in the plane. We defined in section 4.9 an antisymmetric 'product', $\omega(\mathbf{w}, \mathbf{w}')$, between these two vectors by the formula

$$\omega(\mathbf{w}, \mathbf{w}') = qp' - q'p.$$

The geometric meaning of $\omega(\mathbf{w}, \mathbf{w}')$ is that it represents the oriented area of the parallelogram spanned by the vectors $\mathbf{w}$ and $\mathbf{w}'$ (see figure 9.18). It is clear from both the definition and the geometry that $\omega$ is antisymmetric:

$$\omega(\mathbf{w}, \mathbf{w}') = -\omega(\mathbf{w}', \mathbf{w}).$$

A $2 \times 2$ matrix preserves area and orientation if and only if its determinant equals 1. Thus a $2 \times 2$ matrix $M$ has determinant 1 if and only if

$$\omega(M\mathbf{w}, M\mathbf{w}') = \omega(\mathbf{w}, \mathbf{w}')$$

for all $\mathbf{w}$ and $\mathbf{w}'$. Now suppose that

$$\mathbf{u} = \begin{pmatrix} q_x \\ q_y \\ p_x \\ p_y \end{pmatrix} \quad \text{and} \quad \mathbf{u}' = \begin{pmatrix} q'_x \\ q'_y \\ p'_x \\ p'_y \end{pmatrix}$$

are two vectors in four-dimensional space. We define

$$\omega(\mathbf{u}, \mathbf{u}') = q_x p'_x - q'_x p_x + q_y p'_y - q'_y p_y.$$

The product $\omega$ is still antisymmetric,

$$\omega(\mathbf{u}', \mathbf{u}) = -\omega(\mathbf{u}, \mathbf{u}')$$

but the geometric significance of $\omega$ is not so transparent.

It turns out that a $4 \times 4$ matrix $M$ can arise as the transformation matrix of a linear optical system if and only if

$$\omega(M\mathbf{u}, M\mathbf{u}') = \omega(\mathbf{u}, \mathbf{u}'),$$

for all vectors $\mathbf{u}$ and $\mathbf{u}'$. These kinds of matrices are called (linear) *canonical transformations* in the physics literature, and are called (linear) *symplectic transformations* in the mathematics literature. They (and their higher-dimensional generalizations) play a crucial role in theoretical mechanics and geometry.

After developing some of the basic facts about the group of linear symplectic transformations in four variables, we shall see that our arguments showing that Gaussian optics is equivalent to $\mathrm{Sl}(2, \mathbb{R})$ can be used to show that linear optics is equivalent to $\mathrm{Sp}(4, \mathbb{R})$, the group of linear symplectic transformations in 4 variables.

In general, let $V$ be any (finite-dimensional, real) vector space. A bilinear form $\Omega$ on $V$ is any function $\Omega: V \times V \to \mathbb{R}$ that is linear in each variable when the other variable is held fixed; that is, $\Omega(\mathbf{u}, \mathbf{v})$ is a linear function of $\mathbf{v}$ for each fixed $\mathbf{u}$ and a linear function of $\mathbf{u}$ for each fixed $\mathbf{v}$. We say that $\Omega$ is antisymmetric if $\Omega(\mathbf{u}, \mathbf{v}) = -\Omega(\mathbf{v}, \mathbf{u})$ for all $\mathbf{u}$ and $\mathbf{v}$ in $V$. We say that $\Omega$ is nondegenerate if the linear function $\Omega(\mathbf{u}, \cdot)$ is not identically zero unless $\mathbf{u}$ itself is zero. An antisymmetric, nondegenerate bilinear form on $V$ is called a *symplectic form.* A vector space possessing a given symplectic form is called a *symplectic vector space*, or is said to have a *symplectic structure.* If $V$ is a symplectic vector space with symplectic form $\Omega$, and if $A$ is a linear transformation of $V$ into itself, we say that $A$ is a *symplectic transformation* if $\Omega(A\mathbf{u}, A\mathbf{v}) = \Omega(\mathbf{u}, \mathbf{v})$ for all $\mathbf{u}$ and $\mathbf{v}$ in $V$. It is a theorem (cf Guillemin & Sternberg, *Symplectic Techniques in Physics* Chapter II) that every symplectic vector space must be even-dimensional and that every symplectic linear transformation must have determinant 1 and, hence, be invertible. It is clear that the inverse of any symplectic transformation must be symplectic and that the product of any two symplectic transformations must be symplectic. The collection of all symplectic linear transformations is known as the symplectic group (of $V$), and is denoted by $\mathrm{Sp}(V)$.

Now let us assume that $V = \mathbb{R}^n + \mathbb{R}^n$ and write the typical vector in $V$ as

$$\mathbf{u} = \begin{pmatrix} \mathbf{q} \\ \mathbf{p} \end{pmatrix}, \quad \text{where} \quad \mathbf{q} = \begin{pmatrix} q_1 \\ \vdots \\ q_n \end{pmatrix} \quad \text{and} \quad \mathbf{p} = \begin{pmatrix} p_1 \\ \vdots \\ p_n \end{pmatrix}$$

on $V$ there is the symplectic form $\Omega$ given by

$$\Omega(\mathbf{u}, \mathbf{u}') = \mathbf{p} \cdot \mathbf{q}' - \mathbf{p}' \cdot \mathbf{q},$$

where $\cdot$ denotes ordinary scalar product in $\mathbb{R}^n$. In terms of the scalar product $\mathbf{u} \cdot \mathbf{u}' = \mathbf{q} \cdot \mathbf{q}' + \mathbf{p} \cdot \mathbf{p}'$ we can write this as

$$\Omega(\mathbf{u}, \mathbf{u}') = \mathbf{u}' \cdot J\mathbf{u},$$

where $J$ is the $2n \times 2n$ matrix $\begin{pmatrix} 0 & I \\ -I & 0 \end{pmatrix}$ and $I$ is the $n \times n$ identity matrix. A linear transformation $T$ on $V$ is symplectic if, for all $\mathbf{u}$ and $\mathbf{u}'$,

$$\Omega(T\mathbf{u}, T\mathbf{u}') = \Omega(\mathbf{u}, \mathbf{u}').$$

We can write this as

$$T^{\mathrm{T}} J T \mathbf{u} \cdot \mathbf{u}' = J\mathbf{u} \cdot \mathbf{u}',$$

where $T^{\mathrm{T}}$ denotes the transpose of $T$ relative to the scalar product on $V$. Since this is to hold for all $\mathbf{u}$ and $\mathbf{u}'$ we must have

$$T^{\mathrm{T}} J T = J.$$

We can write

$$T^{\mathrm{T}} \begin{pmatrix} \mathbf{q} \\ \mathbf{p} \end{pmatrix} = \begin{pmatrix} A\mathbf{q} + B\mathbf{p} \\ C\mathbf{q} + D\mathbf{p} \end{pmatrix},$$

where $A, B, C$, and $D$ are $n \times n$ matrices; that is,

$$T = \begin{pmatrix} A & B \\ C & D \end{pmatrix}.$$

Then

$$T^{\mathrm{T}} = \begin{pmatrix} A^{\mathrm{T}} & C^{\mathrm{T}} \\ B^{\mathrm{T}} & D^{\mathrm{T}} \end{pmatrix},$$

where $A^{\mathrm{T}}$ denotes the $n$-dimensional transpose of $A$, etc. The condition $T^{\mathrm{T}} J T = J$ becomes the conditions $A^{\mathrm{T}} C = C^{\mathrm{T}} A$, $B^{\mathrm{T}} D = D^{\mathrm{T}} B$, and $A^{\mathrm{T}} D - C^{\mathrm{T}} B = I$. Notice that $T^{-1}$, which is also symplectic, is given by

$$T^{-1} = \begin{pmatrix} D^{\mathrm{T}} & -B^{\mathrm{T}} \\ -C^{\mathrm{T}} & A^{\mathrm{T}} \end{pmatrix},$$

and so we also have

$$DC^{\mathrm{T}} = CD^{\mathrm{T}} \quad \text{and} \quad BA^{\mathrm{T}} = AB^{\mathrm{T}}.$$

We now turn to the problem of justifying the assertion that the group of linear symplectic transformations (in four dimensions) is precisely the collection of all transformations of linear optics. As in the case of Gaussian optics, the argument can be split into two parts. The first is a physical part showing that (in the linear approximation) the matrix $\begin{pmatrix} I & 0 \\ -P & I \end{pmatrix}$, where $P = P^{\mathrm{T}}$ is a symmetric matrix, corresponds to refraction at a surface between two regions of constant index of refraction (and that every $P$ can arise) and that $\begin{pmatrix} I & dI \\ 0 & I \end{pmatrix}$ corresponds to motion in a medium of constant index of refraction, where $d$ is the optical distance along the axis. The second is a mathematical argument showing that every symplectic matrix can be written as a product of matrices of the above types.

We will omit the mathematical part, which is a rather tricky generalization of the arguments of section 9.2. We refer the reader to Guillemin & Sternberg *Symplectic Techniques in Physics* section 4, pp. 27–30. We concentrate on the physical aspects of the problem. As in Gaussian optics, we describe the incoming light ray by its direction $\mathbf{v} = (v_x, v_y, v_z)$ and its intersection with the plane parallel to the $xy$-plane passing through the point $z$ on the optical axis. Here $\|\mathbf{v}\|^2 = v_x^2 + v_y^2 + v_z^2 = 1$. Now

$$v_z = (1 - v_x^2 + v_y^2)^{1/2} = 1 - \tfrac{1}{2}(v_x^2 + v_y^2) + \cdots \doteq 1,$$

since we are ignoring quadratic terms in $v_x$ and $v_y$, which are assumed small. We set

$$p_x = nv_x, \quad p_y = nv_y,$$

where $n$ is the index of refraction. Moving a distance $t$ along the optical axis is the same (up to quadratic terms in $v_x$ and $v_y$) as moving a distance $t$ along the line through $\mathbf{v}$ and hence

$$q_{2x} - q_{1x} = tv_x$$

and

$$q_{2y} - q_{1y} = tv_y$$

or

$$\begin{pmatrix} \mathbf{q}_2 \\ \mathbf{p}_2 \end{pmatrix} = \begin{pmatrix} I & dI \\ 0 & I \end{pmatrix} \begin{pmatrix} \mathbf{q}_1 \\ \mathbf{p}_1 \end{pmatrix}$$

where $d = t/n$ (see figure 9.19).

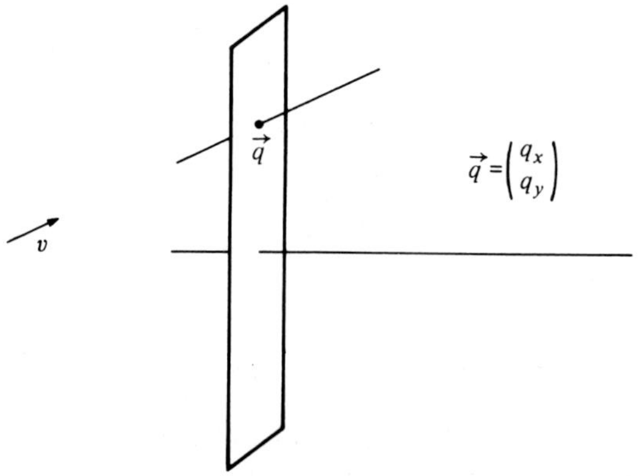


**Figure 9.19**

Now let us turn to refraction. We may assume that our surface is quadratic, and is given by

$$z' - z = \tfrac{1}{2}k\boldsymbol{q}\cdot\boldsymbol{q},$$

where $k$ is a symmetric $2 \times 2$ matrix. The normal to this surface at the point $\mathbf{q}$ is given by

$$\mathbf{u} = (k\boldsymbol{q}, -1).$$

(Up to quadratic terms and higher, $\mathbf{u}$ has length 1.) The projection of a vector $\mathbf{v}$ onto the tangent plane to the surface at $\mathbf{q}$ is given by

$$\mathbf{v} - (\mathbf{v}\cdot\mathbf{u})\mathbf{u}.$$

Writing $\mathbf{v} = (v_x, v_y, 1) = (\boldsymbol{v}, 1)$ we see that $\mathbf{v}\cdot\mathbf{u} = k\boldsymbol{q}\cdot\boldsymbol{v} - 1$ and

$$\mathbf{v} - (\mathbf{v}\cdot\mathbf{u})\mathbf{u} = (\boldsymbol{v}, 1) - (k\boldsymbol{q}\cdot\boldsymbol{v} - 1)(k\boldsymbol{q}, -1).$$

Ignoring the quadratic term $k\boldsymbol{q}\cdot\boldsymbol{v}$ this becomes

$$(\boldsymbol{v} + k\boldsymbol{q}, 0).$$

Snell's law says that $n_1(\mathbf{v}_1 - (\mathbf{v}_1\cdot\mathbf{u})\mathbf{u}) = n_2(\mathbf{v}_2 - (\mathbf{v}_2\cdot\mathbf{u})\mathbf{u})$. In the linear approximation, with $\boldsymbol{p}_1 = n\boldsymbol{v}_1$ and $\boldsymbol{p}_2 = n\boldsymbol{v}_2$, this becomes

$$\boldsymbol{p}_1 - n_1 k\boldsymbol{q} = \boldsymbol{p}_2 - n_2 k\boldsymbol{q}$$

or

$$\boldsymbol{p}_2 = \boldsymbol{p}_1 - P\boldsymbol{q},$$

where

$$P = -(n_1 - n_2)k.$$

We thus get the refraction matrix $\begin{pmatrix} I & 0 \\ -P & I \end{pmatrix}$. This concludes our proof that linear optics is isomorphic to the study of the group $\mathrm{Sp}(4, \mathbb{R})$.

There is one more point relating to Gaussian optics that deserves mention. Suppose that our optical system is rotationally invariant; then at each refracting surface, the power matrix $P$ is of the form $P = mI$, where $m$ is a scalar and $I$ is the $2 \times 2$ identity matrix. It is clear that the collection of matrices that one can get by multiplying such matrices with $\begin{pmatrix} I & eI \\ 0 & I \end{pmatrix}$ will be of the form $\begin{pmatrix} aI & bI \\ cI & dI \end{pmatrix}$. Note that a matrix of the above form, when acting on

$$\begin{pmatrix} q_x \\ q_y \\ p_x \\ p_y \end{pmatrix},$$

is the same as $\begin{pmatrix} a & b \\ c & d \end{pmatrix}$ acting separately on $\begin{pmatrix} q_x \\ p_x \end{pmatrix}$ and $\begin{pmatrix} q_y \\ p_y \end{pmatrix}$. Thus, in our study of Gaussian optics the restriction to paraxial rays was unnecessary. We could have treated skew rays by simply treating the $x$- and $y$-components separately in the same fashion. This is a consequence of the linear approximation.

The basic formula for the optical length

$$L = L_{\text{axis}} + \tfrac{1}{2}(\mathbf{p}_2 \cdot \mathbf{q}_2 - \mathbf{p}_1 \cdot \mathbf{q}_1)$$

(where the **p** and **q** components are now vectors) is proved exactly as it was in the Gaussian case, by looking at what happens at each of our basic components. There is no point in repeating the proof.

Two planes are called nonconjugate if, in the optical matrix relating them, the matrix $B$ is nonsingular. Then we can solve the equations

$$\mathbf{q}_2 = A\mathbf{q}_1 + B\mathbf{p}_1$$

and

$$\mathbf{p}_2 = C\mathbf{q}_1 + D\mathbf{p}_1$$

for $\mathbf{p}_1$ and $\mathbf{p}_2$ as

$$\mathbf{p}_1 = -B^{-1}A\mathbf{q}_1 + B^{-1}\mathbf{q}_2$$

and

$$\mathbf{p}_2 = (C - DB^{-1}A)\mathbf{q}_1 + DB^{-1}\mathbf{q}_2.$$

We can then write

$$L = L_{\text{axis}} + W(\mathbf{q}_1, \mathbf{q}_2),$$

where

$$W(\mathbf{q}_1, \mathbf{q}_2) = \tfrac{1}{2}[DB^{-1}\mathbf{q}_2 \cdot \mathbf{q}_2 + B^{-1}A\mathbf{q}_1 \cdot \mathbf{q}_1 - (2B^{\mathrm{T}})^{-1}\mathbf{q}_1 \cdot \mathbf{q}_2].$$

(In proving this formula we make use of the identity

$$-(B^{\mathrm{T}})^{-1} = C - DB^{-1}A,$$

which follows for nonsingular $B$ from $A^{\mathrm{T}}D - B^{\mathrm{T}}C = I$.) A direct computation (using the above identity) shows that (in the obvious sense)

$$\frac{\partial L}{\partial \mathbf{q}_2} = \mathbf{p}_2$$

and

$$\frac{\partial L}{\partial \mathbf{q}_1} = -\mathbf{p}_1.$$

Thus a knowledge of $L$ allows us to determine $\mathbf{p}_1$ and $\mathbf{p}_2$ in terms of $\mathbf{q}_1$ and $\mathbf{q}_2$.

We can now briefly describe the transition to (nonlinear) geometrical optics. We can put the condition that the matrix $A$ be symplectic in the following way. Consider the two-form

$$\omega = \mathrm{d}q_x \wedge \mathrm{d}p_x + \mathrm{d}q_y \wedge \mathrm{d}p_y$$

on $\mathbb{R}^4$. Then the linear map $A: \mathbb{R}^4 \to \mathbb{R}^4$ is symplectic if and only if

$$A^*\omega = \omega,$$

in other words, the pullback of $\omega$ under $A$ is again $\omega$. We can now call a differentiable map $\phi$ *symplectic* if

$$\phi^*\omega = \omega.$$

We simply drop the condition that $\phi$ be linear. (In the older literature, symplectic maps were called canonical transformations.) Hamilton showed that the maps (from incoming to outgoing maps) in geometrical optics are precisely the symplectic maps. He also showed that under approximate *non-congruency* hypotheses, a symplectic map is determined by the characteristic function $L$ as above, where $L(\mathbf{q}_1, \mathbf{q}_2)$ is the *optical length* of the path joining $\mathbf{q}_1$ to $\mathbf{q}_2$. (Of course, $L$ no longer has the simple formula given above.)

Some ten years after writing his fundamental papers on optics, Hamilton made a startling observation: that the same formalism applies to mechanics of point particles. Let $q_1,\ldots,q_n$ represent the (generalized) position coordinates of a system of particles and $p_1,\ldots,p_n$ the corresponding momenta. Replace the optical axis, $z$, by the time. Then the transformation from initial position and momenta to final position and momenta is always symplectic. This discovery led to remarkable progress in theoretical mechanics in the nineteenth century. In the 1920s – almost a century later – Hamilton's analogy between optics and mechanics served as one of the major clues in the discovery of quantum mechanics.

## Summary

### A Matrix formulation of Gaussian optics

You should understand the use of a two-component vector to represent a ray passing through a reference plane.

You should be able to develop and use the $2 \times 2$ matrices that represent the effect of a translation, a refracting surface, or a thin lens.

### B Lens systems

You should be able to calculate the matrix for a system of refracting surfaces or thin lenses between two given reference planes.

Given such a lens system, you should be able to locate the principal planes and focal planes, use them for ray tracing, and locate the image of a given object.

### C Hamiltonian optics

For a Gaussian optical system, you should know how to write down the Hamiltonian point characteristic between two reference planes and to use it to determine what ray connects a pair of points in the two planes.

## Exercises

9.1. Figure 9.20 shows the focal planes and principal planes for a thick lens. Rays incident from the left which are parallel to the axis are refracted so that they pass through a focal point in the plane $F_2$, while rays emanating from the focal point in the plane $F_1$ are refracted so that they emerge parallel to the axis. Principal planes $H_1$ and $H_2$ are associated with $F_1$ and $F_2$ respectively.

(a) By ray tracing on the diagram, locate the image of the object in the plane $z_1$. Trace the ray $R_1$ plus *two* other rays.
(b) Use Newton's equation to calculate the position of the image which you located in (a). Specify the location of this image with respect to one of the planes in figure 9.20.
(c) Construct the matrix of the system between planes $z_1$ and $z_2$. Use this matrix to determine the position and slope of ray $R_1$ as it emerges from the lens at $z_2$.

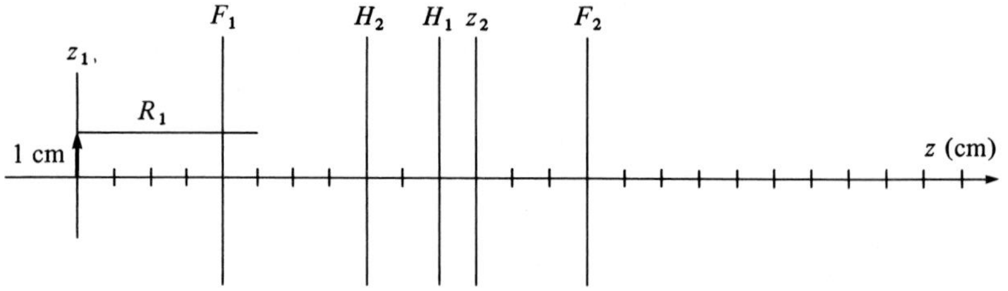


**Figure 9.20**

9.2. The thick lens shown in figure 9.21 is made of glass with $n = \frac{3}{2}$. Construct the matrix between reference planes $z_1$ and $z_2$. Locate the focal planes $F_1$ and $F_2$ and principal planes $H_1$ and $H_2$, and show them on a diagram. By tracing rays on the diagram, locate the image of an object located 1 cm to the left of $z_1$, and check your result by using Newton's equation.

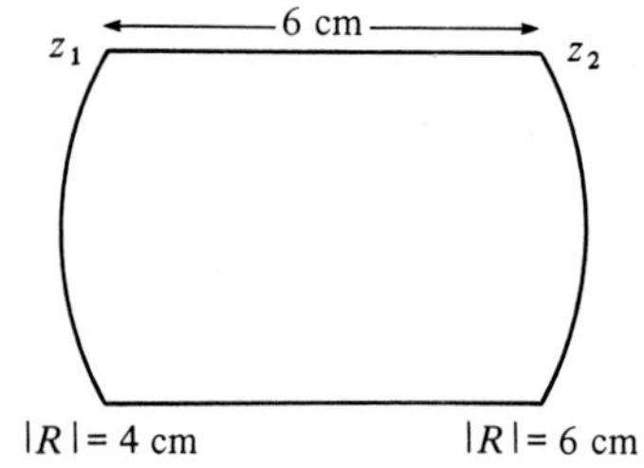


**Figure 9.21**

9.3. Suppose that you take ray tracing as the fundamental characterization of the properties of a thin lens; i.e., you assume that the intersection of a ray through the center of the lens with a ray which is parallel to the axis on the left and is bent through the focal point on the right determines the intersection of all the rays from a given object.
(a) Derive the thin lens equation from this assumption. Consider only the case where $p, q$ and $f$ are all positive.
(b) Prove from the same assumptions that a thin lens can be represented by a $2 \times 2$ matrix, and derive the form of this matrix.

9.4. A crystal ball of radius 6 cm is made of glass with index of refraction $\frac{3}{2}$. For rays which are close to a diameter, this crystal ball behaves like a linear thick lens (i.e., a cylindrical *core*, with a diameter as its axis, is just a thick lens). Construct the matrix for this lens between the reference planes $z_1$ and $z_2$, between the focal planes, and between the principal planes. Draw a diagram showing all these planes.

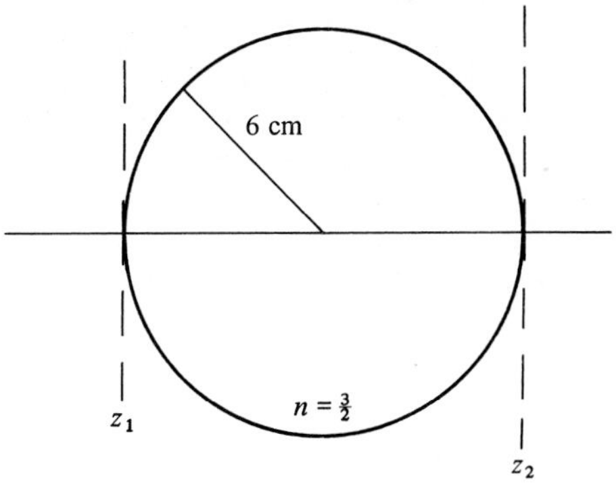


**Figure 9.22**

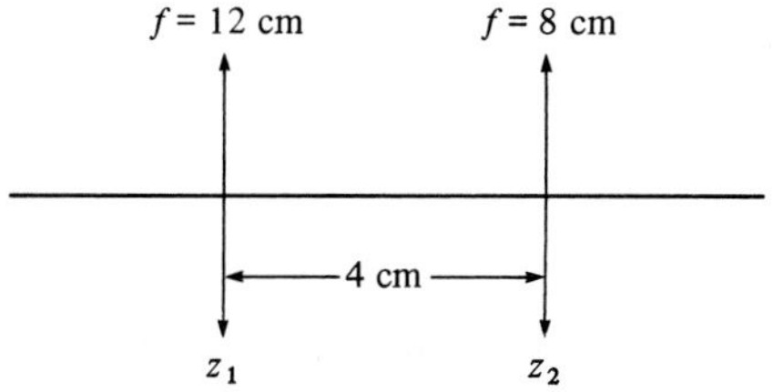


**Figure 9.23**

9.5.(a) For the system of two converging lenses shown in figure 9.23, work out the matrix between planes $z_1$ and $z_2$.

(b) Locate the focal planes $z_1'$ and $z_2'$, both by using the thin lens equation and by using the matrix for the system, as described in the notes. Construct the matrix between the focal planes.

(c) Locate the principal planes $H_1$ and $H_2$ and construct the matrix between them. The easy way to do this is to use the fact that the focal length is $f = 6$ and that each principal plane is therefore 6 cm away from the corresponding focal plane. Notice that both principal planes lie between the two lenses, and that $H_1$ lies to the *right* of $H_2$ in this case.

(d) Make a diagram of this optical system, showing the focal planes and principal planes.

(e) Let $z_3$ be the plane 12 cm to the left of $z_1$. Find the plane conjugate to this plane in four ways: by matrix multiplication, by using the thin lens equation twice, by using Newton's equation $x_1 x_2 = f^2$, and by ray tracing.

9.6. A lens system consists of two thin lenses, whose focal lengths are $f_1$ and $f_2$

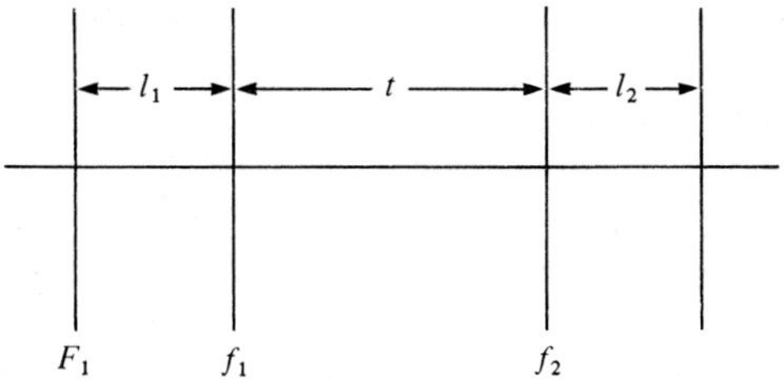


**Figure 9.24**

respectively, mounted a distance $t$ apart. The first focal plane is located at a distance $l_1$ to the left of lens 1, the second focal plane is located a distance $l_2$ to the right of lens 2. Prove that the focal length $f$ of this system satisfies the equation $f^2 - tf - l_1 l_2 = 0$. Bearing in mind that $l_1, l_2, f, f_1$ and $f_2$ all make sense even if they are negative, decide which root of this quadratic equation is physically meaningful.

9.7. Invent a system of thin lenses whose optical matrix is the identity matrix $\begin{pmatrix} 1 & 0 \\ 0 & 1 \end{pmatrix}$. (Note: this takes several lenses. You might wish to start by constructing a system whose matrix $M$ satisfies $M^2 = I$.)

9.8. A ray enters the optical system shown in figure 9.25 at $z_1$ with coordinates $\begin{pmatrix} q_1 \\ p_1 \end{pmatrix} = \begin{pmatrix} 1 \\ 2 \end{pmatrix}$. Find the coordinates of the outgoing ray at $z_2$.

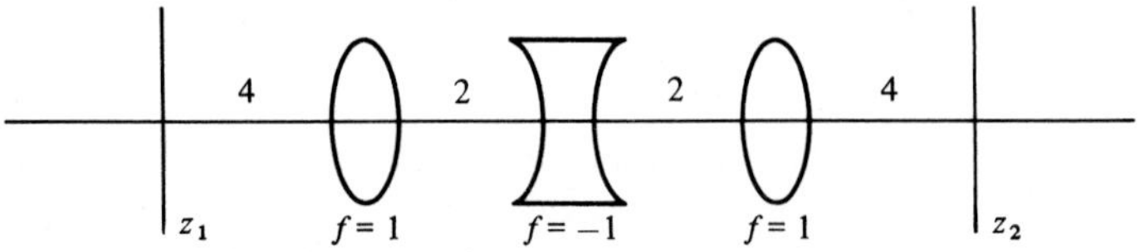


**Figure 9.25**

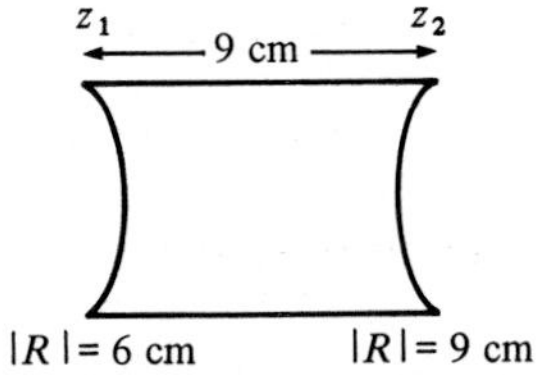


**Figure 9.26**

9.9. The thick lens shown in figure 9.26 is made of glass with index of refraction $n = \frac{3}{2}$.

(a) Construct the matrix between reference planes $z_1$ and $z_2$.

(b) Determine what incoming ray $\begin{pmatrix} q_1 \\ p_1 \end{pmatrix}$ is transformed into the outgoing ray $\begin{pmatrix} q_2 \\ p_2 \end{pmatrix} = \begin{pmatrix} \frac{1}{2} \\ \frac{1}{6} \end{pmatrix}$ at plane $z_2$.

9.10. The converging lens shown in figure 9.27 has $f = 10$ cm. It is made of glass with $n = 1.4$, and its two convex surfaces both have the same radius of curvature $R$.

(a) Calculate $R$, and determine the thickness $b$ of the lens as a function of the distance $q$ from the axis. (Note: $b(2) = 0$.)

(b) A ray from $A$ will follow the path $ACF$. Show that this path requires a minimum time compared with any path which passes through the lens at a different value of $q$.

(c) Show that the path $ACB$ requires greater time than any other path from $A$ to $B$ via the lens.

(d) Write down the function $W(q_A, q_F)$ for the planes of $A$ and $F$, and

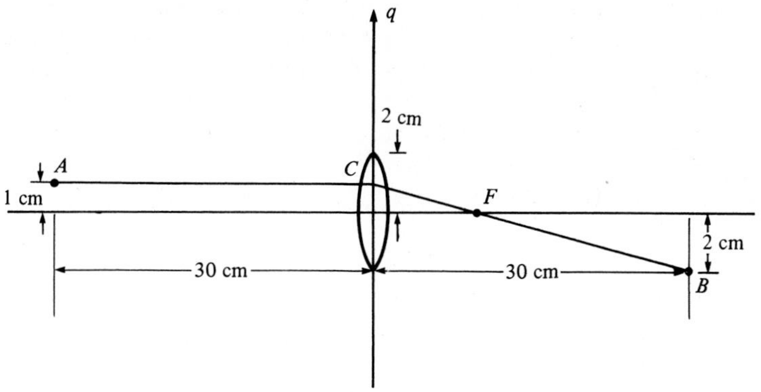


**Figure 9.27** Not to scale.

show that Hamilton's equations give the correct slopes for the ray with $q_A = 1$, $q_F = 0$. Do the same for the planes of $A$ and $B$. Finally, use $W(q_A, q_B)$ to determine what ray passes through the axis in the planes of both $A$ and $B$.

9.11. Let $\mathbf{v}_1 = \begin{pmatrix} q_1 \\ p_1 \end{pmatrix}$ and $\tilde{\mathbf{v}}_1 = \begin{pmatrix} \tilde{q}_1 \\ \tilde{p}_1 \end{pmatrix}$ denote two rays entering an arbitrary Gaussian optical system. The symplectic scalar product of these vectors is defined by $\omega(\mathbf{v}_1, \tilde{\mathbf{v}}_1) = q_1 \tilde{p}_1 - \tilde{q}_1 p_1$.

(a) Show that this scalar product is preserved by the action of the optical system: i.e., $\omega(\mathbf{v}_2, \tilde{\mathbf{v}}_2) = \omega(\mathbf{v}_1, \tilde{\mathbf{v}}_1)$.

(b) Show that $\omega(\mathbf{v}_1, \tilde{\mathbf{v}}_1) = 0$ if $\mathbf{v}_1, \tilde{\mathbf{v}}_1$ denote rays which meet anywhere on the optical axis.

(c) Suppose that two rays pass through the same point $q_1$ in reference plane $z_1$, with an angle $\phi_1$ between them. If these rays meet in the conjugate plane $z_2$ with angle $\phi_2$ between them, what is their distance $q_2$ from the axis? (Assume $n = 1$ at planes $z_1$ and $z_2$.)

# 10

# Vector spaces and linear transformations




In Chapter 10 we go back and prove the basic facts about finite-dimensional vector spaces and their linear transformations. The treatment here is a straightforward generalization, in the main, of the results obtained in the first four chapters in the two-dimensional case. The one new algorithm is that of row reduction. Two important new concepts (somewhat hard to get used to at first) are introduced: those of the dual space and the quotient space. These concepts will prove crucial in what follows.


## Introduction

We have worked extensively with two-dimensional vector spaces, but so far always with one of two specific models in mind. A vector space $V$ was either the set of displacements in an affine plane, or it was $\mathbb{R}^2$, the set of ordered pairs of real numbers. By introducing coordinates, we were able to identify any two-dimensional vector space with $\mathbb{R}^2$ and thereby to represent any linear transformation of the space by a $2 \times 2$ matrix.

We shall now begin to view more general vector spaces from an abstract and axiomatic point of view. The advantage of this approach is that it will permit us to consider vector spaces that are not defined either in geometrical terms or as $n$-tuples of real numbers. It will turn out that any such vector space containing only a finite number of *linearly independent* elements can be identified with $\mathbb{R}^n$ for some integer $n$ so that eventually we shall return to the study of $\mathbb{R}^n$ and the use of matrices to represent linear transformations. In what follows, you should keep in mind the familiar two-dimensional geometrical model of a vector space in order to remind yourself that the definitions and axioms are reasonable. The emphasis in the examples, however, will be on vector spaces that do not arise in a geometrical context. Such vector spaces are part of the natural mathematical language of many branches of physics, notably electromagnetic theory and quantum mechanics.

## 10.1. Properties of vector spaces

We begin by repeating the basic definitions.

A vector space, also known as a linear space, consists of a set of elements called vectors which satisfy certain axioms listed below. We shall denote vector spaces by capital letters, e.g., $V, W^*, C_1$, and elements by lower-case bold letters, e.g., $\mathbf{v}, \mathbf{w}_2, \mathbf{b}^1$.

Part of the characterization of a vector space $V$ is a rule that assigns to any two elements $\mathbf{v}_1$ and $\mathbf{v}_2$ a unique third element $\mathbf{v}$, usually called the *sum* and denoted $\mathbf{v}_1 + \mathbf{v}_2$. This operation satisfies the same axioms as addition of real numbers:

Commutative law: $\mathbf{v}_1 + \mathbf{v}_2 = \mathbf{v}_2 + \mathbf{v}_1$. (10.1)

Associative law: $(\mathbf{v}_1 + \mathbf{v}_2) + \mathbf{v}_3 = \mathbf{v}_1 + (\mathbf{v}_2 + \mathbf{v}_3)$. (10.2)

Existence of zero: There is an element $\mathbf{0}$ such that $\mathbf{v} + \mathbf{0} = \mathbf{v}$ for all $\mathbf{v}$. (10.3)

Existence of negative: For any $\mathbf{v}$ there is an element $-\mathbf{v}$ such that $-\mathbf{v} + \mathbf{v} = \mathbf{0}$. (10.4)

In some cases the operation of addition is defined directly in terms of addition of real numbers, so that it is clear from the definition that these axioms are satisfied. For example, in $\mathbb{R}^2$ we define

$$\begin{pmatrix} a_1 \\ a_2 \end{pmatrix} + \begin{pmatrix} b_1 \\ b_2 \end{pmatrix} = \begin{pmatrix} a_1 + b_1 \\ a_2 + b_2 \end{pmatrix}.$$

Similarly, we might consider the two-dimensional vector space of all functions defined on a two-element set, $\{A, B\}$, with addition defined pointwise, so that $\mathbf{h} = \mathbf{f} + \mathbf{g}$ is the function with the property that

$$\mathbf{h}(A) = \mathbf{f}(A) + \mathbf{g}(A),$$
$$\mathbf{h}(B) = \mathbf{f}(B) + \mathbf{g}(B).$$

As a final example, we might consider the space of all continuous functions on the interval $[0, 1]$, with addition again defined pointwise, so that if $\mathbf{f}$ and $\mathbf{g}$ are elements of the space, their sum is the function $\mathbf{h}$ given by $\mathbf{h}(x) = \mathbf{f}(x) + \mathbf{g}(x)$. In this case it is crucial to notice that, for any $\mathbf{f}$ and $\mathbf{g}$, the sum $\mathbf{h}$ is also a continuous function and so lies in the vector space.

In all these examples it is clear that the zero element and inverse element are unique. In fact this is true in any vector space, but it need not be assumed, since it is easily proved from the axioms. The proofs are left to the reader.

The other operation that must be defined as part of the characterization of a vector space is multiplication of a vector by a scalar. In this chapter the scalar will always be a real number and the resulting vector space a *real vector space*, but we later consider complex vector spaces in which elements may be multiplied by complex numbers. Again the axioms are those of ordinary multiplication, so that, if $c_1$ and $c_2$ are scalars and $\mathbf{v}_1$ and $\mathbf{v}_2$ are vectors, we have

$$\text{Associative law: } c_1(c_2\mathbf{v}) = (c_1c_2)\mathbf{v}. \tag{10.5}$$

$$\text{Distributive laws: } \begin{Bmatrix} (c_1 + c_2)\mathbf{v} = c_1\mathbf{v} + c_2\mathbf{v}, \\ c(\mathbf{v}_1 + \mathbf{v}_2) = c\mathbf{v}_1 + c\mathbf{v}_2 \end{Bmatrix}. \tag{10.6}$$

$$\text{Multiplication by 1 is the identity: } 1\mathbf{v} = \mathbf{v} \text{ for all } \mathbf{v}. \tag{10.7}$$

Because the axioms of addition and scalar multiplication in any vector space are the same as in ordinary arithmetic, almost any property which is true in arithmetic is also true in vector algebra. Here is a list of such properties, all readily provable from the axioms. Think about these, and convince yourself that they really require proof: they are not true just by definition.

(a) $0\mathbf{v} = \mathbf{0}$

(b) $c\mathbf{0} = \mathbf{0}$

(c) $(-c)\mathbf{v} = -(c\mathbf{v}) = c(-\mathbf{v})$

(d) $\mathbf{v} + \mathbf{v} = 2\mathbf{v}, \mathbf{v} + \mathbf{v} + \mathbf{v} = 3\mathbf{v}$, etc.

(e) If $a\mathbf{v} = \mathbf{0}$ then either $a = 0$ or $\mathbf{v} = \mathbf{0}$

(f) $-(\mathbf{v} + \mathbf{w}) = -\mathbf{v} + -\mathbf{w}$

## 10.2. The dual space

Given any vector space $V$, we can consider the set of all *linear functions* from $V$ to $\mathbb{R}$. These form a vector space, called the *dual space* $V^*$, as we shall now show. We shall denote elements of $V^*$ by bold Greek letters and also, introducing a convention which will be useful later on, identify them by superscripts rather than by subscripts. Thus $\mathbf{v}_1, \mathbf{v}_2, \ldots$ are elements of $V$, while $\boldsymbol{\alpha}^1, \boldsymbol{\alpha}^2, \ldots$ are elements of $V^*$. The action of an element $V^*$ on an element of $V$ will be denoted by using square brackets, e.g., $\boldsymbol{\alpha}[\mathbf{v}]$.

We define the sum of two elements of $V^*$ in the usual manner for functions: i.e., for any $\mathbf{v} \in V$, $(\boldsymbol{\alpha}^1 + \boldsymbol{\alpha}^2)[\mathbf{v}] = \boldsymbol{\alpha}^1[\mathbf{v}] + \boldsymbol{\alpha}^2[\mathbf{v}]$. Since the sum of linear functions is is also a linear function, $\boldsymbol{\alpha}^1 + \boldsymbol{\alpha}^2$ is indeed an element of $V^*$, and it is easy to see that all the addition axioms (10.1)–(10.4) are satisfied, with the zero element in $V^*$ being the zero function, which is certainly linear. Similarly, we define scalar multiplication by

$$(c\boldsymbol{\alpha})[\mathbf{v}] = c(\boldsymbol{\alpha}[\mathbf{v}])$$

and thus see immediately that $c\boldsymbol{\alpha}$ is linear and that the axioms for multiplication are satisfied.

While the abstract concept of dual space is straightforward, there are a bewildering variety of ways in which elements of a dual space may be defined. Here are some examples:

1. $V$ is $\mathbb{R}^2$, with a typical element $\mathbf{v} = \begin{pmatrix} x \\ y \end{pmatrix}$. Then $V^*$ may be identified with two-component row vectors, for example $\boldsymbol{\alpha} = (a, b)$, with $\boldsymbol{\alpha}[\mathbf{v}] = (a, b)\begin{pmatrix} x \\ y \end{pmatrix} = ax + by$.

2. $V$ is the space of all functions on the two-element set $\{A, B\}$. Then the rule $\boldsymbol{\alpha}^A: V \to \mathbb{R}$ which assigns to an element $\mathbf{f} \in V$ its value on the element $A$, so that $\boldsymbol{\alpha}^A[\mathbf{f}] = f(A)$, is an element of $V^*$. In this case, in fact, the general element of $V^*$ is of the form

$$\boldsymbol{\alpha}[\mathbf{f}] = af(A) + bf(B)$$

for arbitrary $a$ and $b$. What is interesting about this example is that we have identified $A$ with $\boldsymbol{\alpha}^A$ and similarly can identify $B$ with $\boldsymbol{\alpha}^B$. Although an expression like '$aA + bB$' makes no sense, $a\boldsymbol{\alpha}^A + b\boldsymbol{\alpha}^B$ makes perfect sense as an element of $V^*$. Thus we have a procedure for associating a vector space to any finite set so that the elements of the set become vectors: just take the dual space of the space of functions on the set! This construction will prove useful in the theory of electric networks.

3. $V$ is the space of differentiable functions $f(t)$ on the interval $[0, 1]$. Then all the following are elements of $V^*$:

$$\boldsymbol{\alpha}: \mathbf{f} \mapsto f(0)$$
$$\boldsymbol{\beta}: \mathbf{f} \mapsto f'(0)$$
$$\boldsymbol{\gamma}: \mathbf{f} \mapsto \int_0^1 f(t)\,\mathrm{d}t$$
$$\boldsymbol{\delta}: \mathbf{f} \mapsto \int_0^1 tf(t)\,\mathrm{d}t$$
$$\boldsymbol{\varepsilon}: \mathbf{f} \mapsto f(\tfrac{1}{2}) + f'(\tfrac{3}{4}) + \int_{\frac{1}{3}}^{\frac{2}{3}} tf(t)\,\mathrm{d}t$$

## 10.3. Subspaces

Frequently a vector space $W$ arises as a subspace of a larger vector space $V$ with addition and scalar multiplication defined in $W$ just as in $V$. In such a case, since $V$ is known to satisfy all the vector space axioms, there is no need to check them for $W$. All that must be done to confirm that $W$ is a vector space is to show that it is *closed* under addition and multiplication; i.e., that for any $\mathbf{w}_1, \mathbf{w}_2 \in W$, the sum $\mathbf{w}_1 + \mathbf{w}_2$ is an element of $W$, and, for any real number $c$ and any $\mathbf{w} \in W$, $c\mathbf{w}$ is an element of $W$. In particular, the zero vector must be an element of $W$.

In practice, subspaces are usually defined by one of two methods, either by specifying a set of elements of $V$ or a set of elements of $V^*$.

**Method 1.** Let $\mathbf{w}_1, \mathbf{w}_2, \ldots, \mathbf{w}_k$ be vectors in $V$. Then the set of all linear combinations of the form

$$\mathbf{w} = \sum_{i=1}^{k} c_i \mathbf{w}_i$$

is a subspace of $V$. (It may, of course, be the entire space $V$.)

**Method 2.** Let $\alpha^1, \alpha^2, \ldots, \alpha^k$ be elements of $V^*$. Then the set $W$ of elements $\mathbf{v} \in V$ satisfying

$$\begin{aligned} \alpha^1[\mathbf{v}] &= 0, \\ \alpha^2[\mathbf{v}] &= 0, \\ &\vdots \\ \alpha^k[\mathbf{v}] &= 0 \end{aligned}$$

is a subspace of $V$. The proof is simple. Let $\mathbf{w}_1$ and $\mathbf{w}_2$ be two vectors in this set $W$. Then, because the functions $\alpha^1, \alpha^2, \ldots$ are all *linear*,

$$\alpha^i[\mathbf{w}_1 + \mathbf{w}_2] = \alpha^i[\mathbf{w}_1] + \alpha^i[\mathbf{w}_2] = 0 \quad i = 1, 2, \ldots, k$$

so that $\mathbf{w}_1 + \mathbf{w}_2 \in W$. Similarly,

$$\alpha^i[c\mathbf{w}] = c\alpha^i[\mathbf{w}] = 0$$

so that $c\mathbf{w} \in W$. Thus $W$ is closed under addition and scalar multiplication and is a subspace. (It may, of course, be $\{\mathbf{0}\}$ – the *zero* subspace consisting of $\mathbf{0}$ above.)

A familiar example of these two methods is the construction of a plane (through the origin) in $\mathbb{R}^3$. Method 1 describes the plane in terms of two vectors that span it; e.g.,

$$\begin{pmatrix} 1 \\ 1 \\ 0 \end{pmatrix} \text{ and } \begin{pmatrix} 0 \\ 1 \\ 2 \end{pmatrix}, \text{ or } \begin{pmatrix} 1 \\ 2 \\ 2 \end{pmatrix} \text{ and } \begin{pmatrix} -1 \\ 0 \\ 2 \end{pmatrix}.$$

Method 2 describes the plane by means of a linear equation, e.g.,

$$2x - 2y + z = 0,$$

which is the same as saying that $\alpha[\mathbf{w}] = 0$ where

$$\alpha = (2, -2, 1) \quad \text{and} \quad \mathbf{w} = \begin{pmatrix} x \\ y \\ z \end{pmatrix}.$$

As another example, consider the space $V$ of polynomial functions of degree $\leqslant 2$, with a typical element

$$f(t) = a + bt + ct^2.$$

A one-dimensional subspace $W$ can be described by method 1 as the space of all constant multiples of the function $1 - t^2$. The same subspace can alternatively be described by method 2 in terms of the two conditions

$$f(1) = 0 \quad \text{and} \quad f(-1) = 0$$

or, more obscurely, by

$$f'(0)=0 \quad \text{and} \quad \int_0^{\sqrt{3}} f(t)\mathrm{d}t=0.$$

## 10.4. Dimension and basis

To proceed further with the study of vector spaces, we need the notions of linear dependence and linear independence of a set of vectors. A set of vectors $\{\mathbf{v}_1, \mathbf{v}_2, \ldots, \mathbf{v}_k\}$ is said to be *linearly dependent* if there exist real numbers $\lambda_1, \lambda_2, \ldots, \lambda_k$, not all zero, such that

$$\lambda_1\mathbf{v}_1+\lambda_2\mathbf{v}_2+\cdots+\lambda_k\mathbf{v}_k=\mathbf{0}.$$

If this equation holds *only* for $\lambda_1=\lambda_2=\cdots=\lambda_k=0$, then the set of vectors is said to be *linearly independent.*

Here are some examples of these important concepts:

1. Let $V$ be $\mathbb{R}^3$, and consider

$$\mathbf{v}_1=\begin{pmatrix}1\\1\\0\end{pmatrix}, \quad \mathbf{v}_2=\begin{pmatrix}0\\2\\1\end{pmatrix}, \quad \mathbf{v}_3=\begin{pmatrix}2\\4\\1\end{pmatrix}.$$

The set $\{\mathbf{v}_1, \mathbf{v}_2, \mathbf{v}_3\}$ is linearly dependent because

$$2\mathbf{v}_1+\mathbf{v}_2-\mathbf{v}_3=\begin{pmatrix}2\\2\\0\end{pmatrix}+\begin{pmatrix}0\\2\\1\end{pmatrix}-\begin{pmatrix}2\\4\\1\end{pmatrix}=\mathbf{0}.$$

On the other hand, the set $\{\mathbf{v}_1, \mathbf{v}_2\}$ is linearly independent, because

$$\lambda_1\begin{pmatrix}1\\1\\0\end{pmatrix}+\lambda_2\begin{pmatrix}0\\2\\1\end{pmatrix}=\begin{pmatrix}\lambda_1\\\lambda_1+2\lambda_2\\\lambda_2\end{pmatrix}$$

and it is apparent on inspection that this last vector can only be zero if $\lambda_1=0$ and $\lambda_2=0$.

2. Let $V$ be the space of functions on $[0, 2\pi]$ and consider

$$\mathbf{v}_1=\cos^2 t, \quad \mathbf{v}_2=\sin^2 t, \quad \mathbf{v}_3=\cos 2t.$$

This set of vectors is linearly dependent because $\mathbf{v}_1-\mathbf{v}_2-\mathbf{v}_3=\mathbf{0}$.

3. Let $V$ be the space of functions on the set $\{A, B\}$, and consider

$$\begin{aligned}&\mathbf{f}_1\colon f_1(A)=1, f_1(B)=2,\\&\mathbf{f}_2\colon f_2(A)=2, f_2(B)=-3,\\&\mathbf{f}_3\colon f_3(A)=-3, f_3(B)=1.\end{aligned}$$

This set is linearly dependent, because $\mathbf{f}_1+\mathbf{f}_2+\mathbf{f}_3=\mathbf{0}$ (it is the zero function).

4. Let $V$ be the space of polynomials of degree $\leq 2$: Consider the following

elements of $V^*$:

$$\boldsymbol{\alpha}: f \to \int_{-2}^{2} t f(t)\,\mathrm{d}t,$$

$$\boldsymbol{\beta}: f \to f'(0).$$

Writing $f(t) = A + Bt + Ct^2$ we find

$$\boldsymbol{\alpha}[\mathbf{f}] = \int_{-2}^{2} (At + Bt^2 + Ct^3)\,\mathrm{d}t = \tfrac{16}{3}B$$

and

$$\boldsymbol{\beta}[\mathbf{f}] = B$$

so $(\boldsymbol{\alpha} - \tfrac{16}{3}\boldsymbol{\beta})[\mathbf{f}] = \mathbf{0}$ and the set $\{\boldsymbol{\alpha}, \boldsymbol{\beta}\}$ is linearly dependent.

It is probably clear from these examples that there are situations in which it may not be apparent on inspection whether a set of vectors is linearly dependent or independent. We shall have to develop a systematic procedure for investigating this question.

We say that a set of vectors $\{\mathbf{v}_1, \mathbf{v}_2, \ldots, \mathbf{v}_k\}$ *spans* a vector space $V$ if any vector $\mathbf{v} \in V$ can be written as a linear combination $\sum_{i=1}^{k} \mu_i \mathbf{v}_i$. (The set $\{\mathbf{v}_1, \ldots, \mathbf{v}_k\}$ may be linearly dependent, in which case the coefficients $\mu_1, \ldots, \mu_k$ are not uniquely determined.) Consider the following examples:

1. Let $V$ be $\mathbb{R}^3$. The set

$$\left\{ \begin{pmatrix} 1 \\ 0 \\ 0 \end{pmatrix}, \begin{pmatrix} 0 \\ 1 \\ 0 \end{pmatrix}, \begin{pmatrix} 0 \\ 0 \\ 1 \end{pmatrix} \right\}$$

clearly spans $\mathbb{R}^3$, since any element $\begin{pmatrix} \mu_1 \\ \mu_2 \\ \mu_3 \end{pmatrix}$ can be written

$$\begin{pmatrix} \mu_1 \\ \mu_2 \\ \mu_3 \end{pmatrix} = \mu_1 \begin{pmatrix} 1 \\ 0 \\ 0 \end{pmatrix} + \mu_2 \begin{pmatrix} 0 \\ 1 \\ 0 \end{pmatrix} + \mu_3 \begin{pmatrix} 0 \\ 0 \\ 1 \end{pmatrix}.$$

Less obviously, the set

$$\left\{ \begin{pmatrix} 1 \\ 1 \\ 0 \end{pmatrix}, \begin{pmatrix} 0 \\ 1 \\ 1 \end{pmatrix}, \begin{pmatrix} 1 \\ 0 \\ 1 \end{pmatrix} \right\}$$

also spans $\mathbb{R}^3$, but the set

$$\left\{ \begin{pmatrix} 1 \\ 1 \\ 0 \end{pmatrix}, \begin{pmatrix} 0 \\ -1 \\ 1 \end{pmatrix} \right\}$$

does not.

2. Let $V$ be the space of functions $\mathbf{f}(t)$ on $[0, \infty)$ which satisfy the differential equation

$$\mathbf{f}'' + 3\mathbf{f}' + 2\mathbf{f} = \mathbf{0}.$$

The vectors $e^{-t}$ and $e^{-2t}$ span $V$, because the general solution to the equation is of the form

$$\mathbf{f}(t) = Ae^{-t} + Be^{-2t}.$$

3. Let $V$ be the space of functions of the form $\mathbf{f}(t) = A + Bt^3$. Then the vectors

$$\alpha : \mathbf{f} \to f(0),$$
$$\beta : \mathbf{f} \to \int_{-1}^{1} t f(t)\,\mathrm{d}t$$

span the dual space $V^*$. Clearly

$$\boldsymbol{\alpha}[\mathbf{f}] = A,$$
$$\boldsymbol{\beta}[\mathbf{f}] = \int_{-1}^{1} (2At + Bt^4)\,\mathrm{d}t = \tfrac{2}{5}B.$$

But any element $\gamma \in V^*$ must be of the form $\gamma[\mathbf{f}] = aA + bB$ for some constants $a$ and $b$. Thus $\gamma = a\boldsymbol{\alpha} + \frac{5}{2}b\boldsymbol{\beta}$, and $\boldsymbol{\alpha}$ and $\boldsymbol{\beta}$ span $V^*$.

Let $\mathbf{v}_1, \ldots, \mathbf{v}_n$ be a finite set of linearly independent vectors that spans a vector space $V$. The number $n$ of vectors in such a collection is called the *dimension* of $V$. To establish that dimension is a well-defined integer; i.e., that all such sets for a given space contain the same number of elements, we must prove the following result:

**Theorem.** *Let $\{\mathbf{v}_1, \mathbf{v}_2, \ldots, \mathbf{v}_k\}$ be a set of vectors that span a vector space $V$. Then any set of $k+1$ vectors in $V$ is linearly dependent.*

The proof is by induction: we first establish the result for $k = 1$; then we show that if it is true for a space spanned by $k-1$ vectors, it is true for a space spanned by $k$ vectors. When $k = 1$, the theorem states that, if $V$ is spanned by one vector $\mathbf{v}$, then any two vectors in $V$ are linearly dependent. Indeed, consider two such vectors, $\mathbf{w}_1$ and $\mathbf{w}_2$. Since $\mathbf{v}$ spans $V$, there exist real numbers $\mu_1$ and $\mu_2$ such that $\mathbf{w}_1 = \mu_1\mathbf{v}$ and $\mathbf{w}_2 = \mu_2\mathbf{v}$. Clearly, then,

$$\mu_2\mathbf{w}_1 - \mu_1\mathbf{w}_2 = \mu_2\mu_1\mathbf{v} - \mu_1\mu_2\mathbf{v} = \mathbf{0}$$

so that $\mathbf{w}_1$ and $\mathbf{w}_2$ are linearly dependent.

We now assume that the theorem is true for any set of $k$ vectors in a space spanned by $k-1$ vectors, and we consider a set of $k+1$ vectors, $\{\mathbf{w}_1, \ldots, \mathbf{w}_{k+1}\}$, in a space spanned by $\{\mathbf{v}_1 \ldots \mathbf{v}_k\}$. We can write $\mathbf{w}_1 = a_{11}\mathbf{v}_1 + a_{12}\mathbf{v}_2 + \cdots + a_{1k}\mathbf{v}_k$ because the vectors $\{\mathbf{v}_i\}$ span $V$. If $\mathbf{w}_1 = \mathbf{0}$, then $r\mathbf{w}_1 = 0$ with $r \neq 0$ gives a non-trivial relation among the $\mathbf{w}$s and there is nothing further to prove. So we may assume that $\mathbf{w}_1 \neq \mathbf{0}$ and hence we may as well assume that we have ordered the vectors $\{\mathbf{v}_1, \mathbf{v}_2, \ldots, \mathbf{v}_k\}$ so that $a_{11} \neq 0$. Thus

$$\frac{1}{a_{11}}\mathbf{w}_1 = \mathbf{v}_1 + \frac{a_{12}}{a_{11}}\mathbf{v}_2 + \cdots + \frac{a_{1k}}{a_{11}}\mathbf{v}_k.$$

But

$$\mathbf{w}_2 = a_{21}\mathbf{v}_1 + a_{22}\mathbf{v}_2 + \cdots + a_{2k}\mathbf{v}_k.$$

Thus

$$\mathbf{w}_2 - \frac{a_{21}}{a_{11}}\mathbf{w}_1 = \left(a_{22} - \frac{a_{21}a_{12}}{a_{11}}\right)\mathbf{v}_2 + \cdots + \left(a_{2k} - \frac{a_{k1}a_{1k}}{a_{11}}\right)\mathbf{v}_k$$

and similarly we can express

$$\mathbf{w}_3 - \frac{a_{31}}{a_{11}}\mathbf{w}_1, \quad \text{etc.}$$

in terms of the $k-1$ vectors $\{\mathbf{v}_2, \ldots, \mathbf{v}_k\}$. But we are assuming that the theorem is true for $k-1$ vectors, so that the set of $k$ vectors

$$\left\{\mathbf{w}_2 - \frac{a_{21}}{a_{11}}\mathbf{w}_1, \mathbf{w}_3 - \frac{a_{31}}{a_{11}}\mathbf{w}_1, \ldots, \mathbf{w}_{k+1} - \frac{a_{(k+1)1}}{a_{11}}\mathbf{w}_1\right\}$$

is linearly dependent. Thus there exist constants $\lambda_2, \ldots, \lambda_k$, not all zero, such that

$$\lambda_2\left(\mathbf{w}_2 - \frac{a_{21}}{a_{11}}\mathbf{w}_1\right) + \lambda_3\left(\mathbf{w}_3 - \frac{a_{31}}{a_{11}}\mathbf{w}_1\right) + \cdots + \lambda_k\left(\mathbf{w}_{k+1} - \frac{a_{(k+1)1}}{a_{11}}\mathbf{w}_1\right) = \mathbf{0}.$$

But this means that $\{\mathbf{w}_1, \mathbf{w}_2, \ldots, \mathbf{w}_{k+1}\}$ is a linearly dependent set, as we wished to show.

Now we can easily show that the dimension of a vector space is well-defined. Suppose we have, in a vector space $V$, a collection $\{\mathbf{v}_1, \ldots, \mathbf{v}_k\}$ which is linearly independent and spans, and a second such collection $\{\mathbf{w}_1, \ldots, \mathbf{w}_n\}$. By the theorem just proved, $n \leqslant k$, otherwise the vectors $\{\mathbf{w}_1, \ldots, \mathbf{w}_n\}$ would be linearly dependent. By the same argument $k \leqslant n$, otherwise $\{\mathbf{v}_1, \ldots, \mathbf{v}_k\}$ would be linearly dependent. We conclude that $k = n$, so that any finite collection of linearly independent and spanning vectors in a vector space contains the same number of vectors. Hence we have the right to call this number the dimension of $V$.

In fact, in a vector space $V$ of dimension $n$, *any* set of $n$ independent vectors spans the space. Let $\{\mathbf{v}_1, \ldots, \mathbf{v}_n\}$ be a set of independent vectors in $V$ and let $\mathbf{w}$ be an arbitrary non-zero vector. The set $\{\mathbf{w}, \mathbf{v}_1, \ldots, \mathbf{v}_n\}$, which contains $n+1$ vectors, must be linearly dependent, so there exist constants $\lambda_0, \lambda_1, \ldots, \lambda_n$ such that

$$\lambda_0\mathbf{w} + \lambda_1\mathbf{v}_1 + \cdots + \lambda_n\mathbf{v}_n = \mathbf{0}.$$

Now $\lambda_0$ cannot be zero; otherwise $\{\mathbf{v}_1, \ldots, \mathbf{v}_n\}$ would be dependent, contrary to hypothesis. Hence we can write

$$\mathbf{w} = -\frac{1}{\lambda_0}(\lambda_1\mathbf{v}_1 + \cdots + \lambda_n\mathbf{v}_n)$$

and we have expressed the arbitrary non-zero vector $\mathbf{w}$ as a linear combination of $\{\mathbf{v}_1, \ldots, \mathbf{v}_n\}$, which therefore spans.

A linearly independent and spanning collection of vectors, $\{\mathbf{v}_1, \ldots, \mathbf{v}_n\}$, when *written in a specified order* is called a *basis* of $V$. Thus $\{\mathbf{v}_1, \mathbf{v}_2, \ldots, \mathbf{v}_n\}$ is a different basis from $\{\mathbf{v}_2, \mathbf{v}_1, \mathbf{v}_3, \ldots, \mathbf{v}_n\}$.

Starting with fewer than $n$ independent vectors in an $n$-dimensional space $V$, say $\mathbf{w}_1, \ldots, \mathbf{w}_k$ $(k < n)$, we can always find a vector $\mathbf{v}_{k+1}$ which is not in the subspace spanned by $\{\mathbf{w}_1, \ldots, \mathbf{w}_k\}$. Continuing this process for $n-k$ steps, we eventually arrive at a basis for $V$ which includes the vectors $\mathbf{w}_1, \ldots, \mathbf{w}_k$. In particular, given a vector space $V$ of dimension $n$ with a subspace $W$ of dimension $k$, we can always *construct a basis for V in which the first k vectors form a basis for W*. This process is called *extending* a basis for the subspace $W$ to a basis for the entire space $V$.

Once we have chosen a basis, say $\{\mathbf{e}_1, \ldots, \mathbf{e}_n\}$, for a vector space $V$, we can write any element of $V$ *uniquely* as a linear combination of basis vectors

$$\mathbf{v} = x_1\mathbf{e}_1 + x_2\mathbf{e}_2 + \cdots + x_n\mathbf{e}_n.$$

The numbers $x_1, \ldots, x_n$ are called the *components* of $\mathbf{v}$ with respect to the given basis. To show that they are uniquely determined, we imagine that $\mathbf{v}$ can be expressed alternatively as

$$\mathbf{v} = y_1\mathbf{e}_1 + y_2\mathbf{e}_2 + \cdots + y_n\mathbf{e}_n.$$

Then, subtracting, we have

$$\mathbf{0} = (x_1 - y_1)\mathbf{e}_1 + (x_2 - y_2)\mathbf{e}_2 + \cdots + (x_n - y_n)\mathbf{e}_n.$$

But, since the basis elements are linearly independent, $x_1 - y_1 = x_2 - y_2 = \cdots = x_n - y_n = 0$ which proves the uniqueness of the components.

Thus, a basis determines an isomorphism, $L$, of $V$ with $\mathbb{R}^n$, where

$$L\mathbf{v}_1 = \begin{pmatrix} 1 \\ 0 \\ \vdots \\ 0 \end{pmatrix}, \quad L\mathbf{v}_2 = \begin{pmatrix} 0 \\ 1 \\ 0 \\ \vdots \end{pmatrix}, \quad \text{etc.}$$

Conversely, if $L$ is such an isomorphism, then

$$\mathbf{v}_1 = L^{-1}\begin{pmatrix} 1 \\ 0 \\ \vdots \\ 1 \end{pmatrix}, \quad \mathbf{v}_2 = L^{-1}\begin{pmatrix} 0 \\ 1 \\ 0 \\ \vdots \end{pmatrix}, \quad \text{etc.}$$

is a basis $V$. We may thus *identify* a basis $\{\mathbf{v}_1, \ldots, \mathbf{v}_n\}$ with the corresponding isomorphism $L$, just as we did in Chapter 1 in the two-dimensional case.

Let $L\colon V \to \mathbb{R}^n$ and $L'\colon V \to \mathbb{R}^n$ be two bases of the same $n$-dimensional space, $V$.

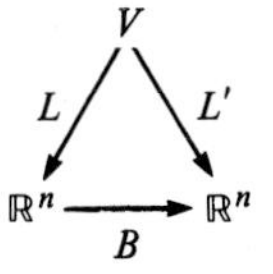


$L' = BL$

Then $B = L' \circ L^{-1}$ is a linear isomorphism of $\mathbb{R}^n \to \mathbb{R}^n$, hence an invertible $n \times n$ matrix. It is called the *change of basis* matrix.

Let $V$ be a vector space of dimension $k$ and $W$ a vector space of dimension $l$. Let $T\colon V \to W$ be a linear transformation. Suppose that we choose bases of $V$ and of $W$. So we have isomorphisms $L\colon V \to \mathbb{R}^k$ and $M\colon W \to \mathbb{R}$ and we can define the map

$$MTL^{-1}\colon \mathbb{R}^k \to \mathbb{R}^l.$$

We can regard $MTL^{-1}$ as a matrix with $l$ rows and $k$ columns. We call $MTL^{-1}$

the matrix of $T$ relative to the bases $L$ and $M$, and denote it by $\mathrm{Mat}_{L,M}(T)$. So

$$\mathrm{Mat}_{L,M}(T) = MTL^{-1}.$$

We can picture the situation by the diagram

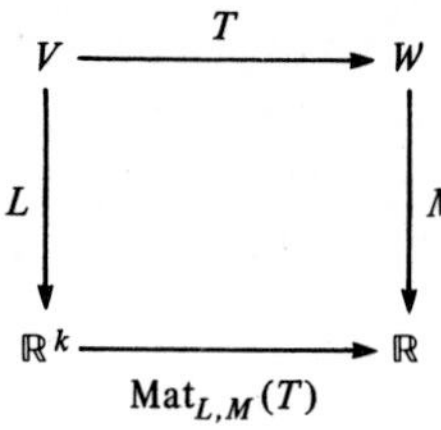


If we make a different choice $L' = PL$ of basis on $V$ and $M' = QM$ of basis on $W$, then

$$L'^{-1} = L^{-1}P^{-1}$$

so

$$M'TL'^{-1} = QMTL^{-1}P^{-1}$$

or

$$\mathrm{Mat}_{L',M'}(T) = Q(\mathrm{Mat}_{L,M}(T))P^{-1}$$

when

$$L' = PL, \; M' = QM$$

is the change of basis formula. It tells us how the matrix representation of a linear transformation changes when we change the basis.

## 10.5. The dual basis

Having constructed a basis for a vector space $V$, we can readily construct *a dual basis* for the dual space $V^*$. Let $\{\mathbf{e}_1, \mathbf{e}_2, \ldots, \mathbf{e}_n\}$ be a basis for $V$. Then any vector $\mathbf{v} \in V$ can be written uniquely in the form

$$\mathbf{v} = x_1\mathbf{e}_1 + x_2\mathbf{e}_2 + \cdots + x_n\mathbf{e}_n.$$

Now let $\boldsymbol{\alpha}$ be an element of $V^*$. Since $\boldsymbol{\alpha}$ is a *linear* function on $V$,

$$\boldsymbol{\alpha}[\mathbf{v}] = x_1\boldsymbol{\alpha}[\mathbf{e}_1] + x_2\boldsymbol{\alpha}[\mathbf{e}_2] + \cdots + x_n\boldsymbol{\alpha}[\mathbf{e}_n].$$

This means that $\boldsymbol{\alpha}$ is determined completely by its values on the basis vectors $\{\mathbf{e}_1, \ldots, \mathbf{e}_n\}$. We therefore introduce vectors $\boldsymbol{\varepsilon}^1, \ldots, \boldsymbol{\varepsilon}^n$ in $V^*$ with the property that

$$\boldsymbol{\varepsilon}^i[\mathbf{e}_j] = \begin{cases} 1 & \text{if} \quad i = j, \\ 0 & \text{if} \quad i \neq j \end{cases}.$$

To prove that the elements $\boldsymbol{\varepsilon}^i$ are linearly independent, we consider $\Sigma\lambda_i\boldsymbol{\varepsilon}^i$. Applying this to an arbitrary basis element $\mathbf{e}_j$, we obtain

$$\sum_{i=1}^{n} \lambda_i \boldsymbol{\varepsilon}^i[\mathbf{e}_j] = \lambda_j.$$

Thus, if $\Sigma\lambda_i\boldsymbol{\varepsilon}^i$ is the zero element in $V^*$, $\lambda_j = 0$ for all $j$. This proves that the set $\{\boldsymbol{\varepsilon}^1, \ldots, \boldsymbol{\varepsilon}^n\}$ is independent.

Now, given any $\alpha \in V^*$, we write

$$\alpha = \alpha[\mathbf{e}_1]\varepsilon^1 + \alpha[\mathbf{e}_2]\varepsilon^2 + \cdots + \alpha[\mathbf{e}_n]\varepsilon^n.$$

Clearly both sides of this expression have the same value on any basis elements $\mathbf{e}_j$ and so are the same element of $V^*$. This proves that the elements $\varepsilon^1, \ldots, \varepsilon^n$ span $V^*$. Since these elements are also independent, we conclude that $V^*$ is also $n$-dimensional and $\{\varepsilon^1, \ldots, \varepsilon^n\}$ form a basis for it.

We can use this basis to identify $V^*$ with $\mathbb{R}^{n*}$. When we express an element $\alpha \in V^*$ in terms of the dual basis:

$$\alpha = \lambda_1 \varepsilon^1 + \lambda_2 \varepsilon^2 + \cdots + \lambda_n \varepsilon^n$$

we find it convenient to identify elements of $\mathbb{R}^{n*}$ as row vectors. So $\alpha$ becomes identified with the *row* vector $(\lambda_1, \lambda_2, \ldots, \lambda_n)$. An advantage of this notation is that the action of $\alpha$ on $\mathbf{v}$ is then described by the usual rule for multiplying matrices:

$$\alpha[\mathbf{v}] = (\lambda_1, \lambda_2, \ldots, \lambda_n)\begin{pmatrix} x_1 \\ x_2 \\ \vdots \\ x_n \end{pmatrix} = \lambda_1 x_1 + \lambda_2 x_2 + \cdots + \lambda_n x_n.$$

It is important to bear in mind that this technique is correct only if the identification of $V$ and $V^*$ has been done consistently: the basis used in identifying $V^*$ with $\mathbb{R}^n$ must be dual to the basis used in identifying $V$ with $\mathbb{R}^n$.

Suppose now that we have an $n$-dimensional space $V$ with a $k$-dimensional subspace $W$. We can choose a basis for $V$ in which the first $k$ vectors form a basis for $W$:

$$\{\mathbf{v}_1, \ldots, \mathbf{v}_k; \mathbf{v}_{k+1}, \ldots, \mathbf{v}_n\}$$

and then construct the dual basis

$$\{\alpha^1, \ldots, \alpha^k; \alpha^{k+1}, \ldots, \alpha^n\}$$

so that

$$\alpha^i[\mathbf{v}_j] = \begin{cases} 1 & \text{if } i = j, \\ 0 & \text{otherwise} \end{cases}.$$

The $(n-k)$-dimensional subspace spanned by $\{\alpha^{k+1}, \ldots, \alpha^n\}$ is called the *annihilator space* of $W$, denoted $W^\perp$. It derives its name from the fact that if $\alpha \in W^\perp$ and $\mathbf{w} \in W$, then $\alpha[\mathbf{w}] = \mathbf{0}$; that is, $W^\perp$ 'annihilates' the subspace $W$. What was earlier called method 2 for describing a subspace was in fact a specification in terms of the annihilator space. For example, the vector $(a, b, c)$ defines a one-dimensional subspace $W^\perp$ of the *dual* of $\mathbb{R}^3$. The subspace $W$ of $\mathbb{R}^3$ annihilated by $W^\perp$ is two-dimensional: it is the plane $ax + by + cz = 0$. If we specify two independent elements of the dual of $\mathbb{R}^3$, $(a_1, b_1, c_1)$ and $(a_2, b_2, c_2)$, then the subspace of $\mathbb{R}^3$ annihilated by these is one-dimensional: it is the line which satisfies the pair of equations

$$a_1 x + b_1 y + c_1 z = 0,$$
$$a_2 x + b_2 y + c_2 z = 0.$$

Notice that the annihilator space $W^{\perp}$ of a subspace $W \subset V$ does not depend on any specific choice of basis for $V$. Introducing a basis was only a convenient device to permit us to calculate the dimension of $W^{\perp}$.

We still lack a systematic procedure for calculating the dimension of a subspace spanned by specified elements of a vector space, or of a subspace annihilated by specified elements of the dual of a vector space. Such a procedure is the *row reduction* algorithm, which will be presented in a later section.

## 10.6. Quotient spaces

We continue to consider an $n$-dimensional vector space $V$ with a subspace $W$ of dimension $k$.

It seems reasonable that there should be a space of dimension $n-k$ which is in some sense the 'difference' between $V$ and $W$. This space is called the *quotient space* $V/W$. Its elements are not elements of $V$, however; they are *sets* of elements of $V$ called *equivalence classes.* Before defining these classes, we should first see why something simpler will not suffice.

For a concrete example of a vector space $V$ with subspace $W$, we can take $V$ to be the plane $\mathbb{R}^2$ and $W$ a line in the plane, as depicted in figure 10.1. One possibility for forming the 'difference' between $V$ and $W$ would be to consider the set of elements of $V$ which are *not* in $W$. Alas, these span the entire space $V$; for example, in figure 10.1, the vectors $\mathbf{v}_1$ and $\mathbf{v}_2$, neither of which is in $W$, clearly span the entire plane.

Another possibility would be to choose a basis of $k$ vectors for $W$, extend it to a basis for $V$, and form the subspace which is spanned by the $n-k$ basis vectors which are *not* in $W$. This gives a subspace of the desired dimension, but one which depends on arbitrary choice of basis elements and so is not well-defined. For example, in

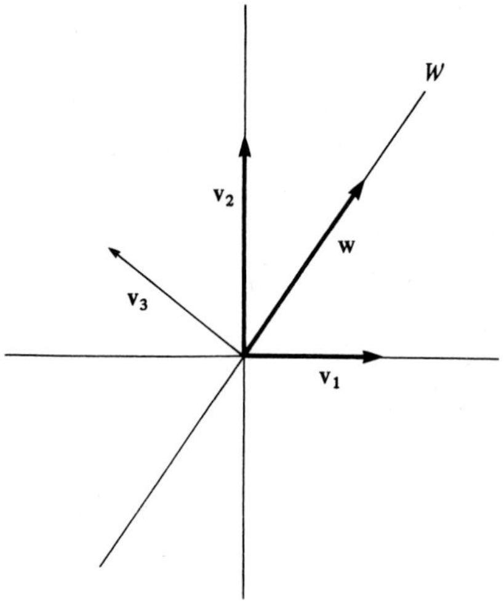


**Figure 10.1**

figure 10.1, we select **w** as the first basis vector, and we could then choose $\mathbf{v}_1$, $\mathbf{v}_2$, or $\mathbf{v}_3$ as a second basis vector, obtaining quite a different subspace with each choice. If there were a scalar product defined on $V$, we could select the subspace orthogonal to $W$, but, lacking a scalar product, there is no way to prescribe a choice of the second basis element.

The construction which works is to define *equivalence classes* (modulo $W$), each consisting of a set of vectors in $V$ whose differences all lie in $W$. We denote the equivalence class of a vector by writing a bar over it; thus, for example, $\bar{\mathbf{v}}$ denotes the set of all vectors of the form $\mathbf{v} + \mathbf{w}$, where $\mathbf{v}$ is a specified element of $V$ and $\mathbf{w}$ is an arbitrary element of $W$.

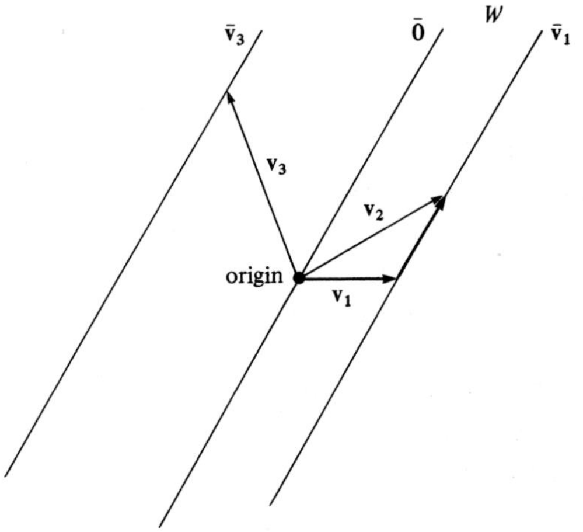


**Figure 10.2**

Referring to figure 10.2, we see, for example, that $\bar{\mathbf{0}}$, the equivalence class of the zero vector, is the subspace $W$, a line through the origin. The vectors $\mathbf{v}_1$ and $\mathbf{v}_2$, which differ by an element of $W$, belong to the same equivalence class, which we may denote $\bar{\mathbf{v}}_1$ or $\bar{\mathbf{v}}_2$. This equivalence class is a line which does *not* pass through the origin. The equivalence class $\bar{\mathbf{v}}_3$ is a different line, again not passing through the origin. In this case the equivalence classes are a family of lines parallel to $W$. More generally, we can view a subspace $W$ as a $k$-dimensional *hyperplane* through the origin of $V$ and the equivalence classes modulo $W$ as a family of hyperplanes parallel to this one.

To introduce the operation of addition of equivalence classes, we look first at the arithmetic of the integers *modulo* 4, with which you are probably familiar. Here there are four equivalence classes:

$$\begin{aligned}
\bar{0} &= \{0, 4, -4, \ 8, -8, \ldots\} = \{4n\},\\
\bar{1} &= \{1, 5, -3, \ 9, -7, \ldots\} = \{4n+1\},\\
\bar{2} &= \{2, 6, -2, 10, -6, \ldots\} = \{4n+2\},\\
\bar{3} &= \{3, 7, -1, 11, -5, \ldots\} = \{4n+3\}.
\end{aligned}$$

To add two equivalence classes, we select any integer from each class, add these together and then find the class to which the sum belongs. For example, to add $\bar{2}$ and $\bar{3}$, we could select 6 from the class $\bar{2}$, 3 from the class $\bar{3}$, and form the sum $6+3=9$, which belongs to the class $\bar{1}$. So $\bar{2}+\bar{3}=\bar{1}$. Since any other choice (say $-2+-1=-3$) would have led to the same conclusion, this operation of addition is well defined.

Addition of equivalence classes of vectors modulo the subspace $W$ is defined similarly. We simply make the definition $\bar{\mathbf{v}}_1+\bar{\mathbf{v}}_2=\overline{(\mathbf{v}_1+\mathbf{v}_2)}$; i.e., add any two vectors from the classes $\bar{\mathbf{v}}_1$ and $\bar{\mathbf{v}}_2$, and find the class to which the sum belongs. Suppose we choose $\mathbf{v}_1+\mathbf{w}_1$ from $\bar{\mathbf{v}}_1$ and $\mathbf{v}_2+\mathbf{w}_2$ from $\bar{\mathbf{v}}_2$, where $\mathbf{w}_1$ and $\mathbf{w}_2$ are arbitrary elements of $W$. Then the sum $\bar{\mathbf{v}}_1+\bar{\mathbf{v}}_2$ is the equivalence class containing $(\mathbf{v}_1+\mathbf{v}_2)+(\mathbf{w}_1+\mathbf{w}_2)$; which is $\overline{(\mathbf{v}_1+\mathbf{v}_2)}$, no matter what choice of $\mathbf{w}_1$ and $\mathbf{w}_2$ may have been made.

This operation of addition is illustrated geometrically in figure 10.3. The point is that $\bar{\mathbf{v}}_1+\bar{\mathbf{v}}_2=\bar{\mathbf{v}}_3$, no matter whether $\mathbf{v}_1$ and $\mathbf{v}_2$ or $\mathbf{u}_1$ and $\mathbf{u}_2$ are chosen as representatives of the classes $\bar{\mathbf{v}}_1$ and $\bar{\mathbf{v}}_2$.

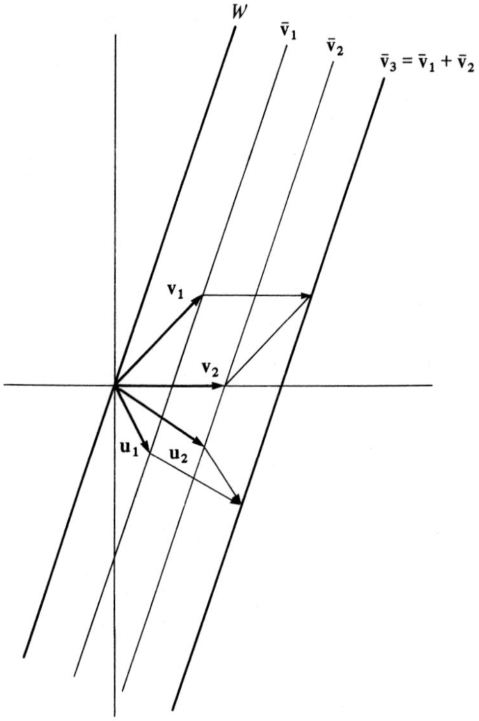


**Figure 10.3**

We define multiplication of an equivalence class by a scalar in a similar way: $c\bar{\mathbf{v}}_1=\overline{(c\mathbf{v}_1)}$. That is, multiply any element of $\bar{\mathbf{v}}_1$ by $c$, and take the equivalence class of the result. Because $W$ is a subspace, the result is unique.

It is now straightforward to check that the equivalence classes of $V$ modulo a subspace $W$ form a vector space. This space is called the *quotient space* $V/W$. To construct a basis for it, simply choose a basis for $W$ and extend it to a basis for all of $V$. The basis elements of $V$ are therefore $\mathbf{w}_1, \mathbf{w}_2, \ldots, \mathbf{w}_k$ (elements of $W$) and $\mathbf{v}_1, \mathbf{v}_2, \ldots, \mathbf{v}_{n-k}$ (not elements of $W$). We claim that the equivalence classes $\bar{\mathbf{v}}_1, \bar{\mathbf{v}}_2, \ldots, \bar{\mathbf{v}}_{n-k}$ form a basis for $V/W$. To prove this, we must show that they are independent and that they span $V/W$.

Let us deal first with the question of independence. Suppose that $\bar{\mathbf{v}}_1, \bar{\mathbf{v}}_2, \ldots, \bar{\mathbf{v}}_{n-k}$ were not independent. Then constants $\lambda_1, \lambda_2, \ldots, \lambda_{n-k}$ exist such that

$$\lambda_1 \bar{\mathbf{v}}_1 + \lambda_2 \bar{\mathbf{v}}_2 + \cdots + \lambda_{n-k} \bar{\mathbf{v}}_{n-k} = \bar{\mathbf{0}}$$

which implies that

$$\lambda_1 \mathbf{v}_1 + \lambda_2 \mathbf{v}_2 + \cdots + \lambda_{n-k} \mathbf{v}_{n-k} \in W$$

contradicting the assumption that the set of vectors $\{\mathbf{w}_1, \ldots, \mathbf{w}_k; \mathbf{v}_1, \ldots, \mathbf{v}_{n-k}\}$ is linearly independent.

We can write a vector $\mathbf{v} \in V$ as a linear combination of basis elements:

$$\mathbf{v} = x_1 \mathbf{v}_1 + x_2 \mathbf{v}_2 + \cdots + x_{n-k} \mathbf{v}_{n-k} + \text{element of } W$$

which implies that

$$\bar{\mathbf{v}} = x_1 \bar{\mathbf{v}}_1 + x_2 \bar{\mathbf{v}}_2 + \cdots + x_{n-k} \bar{\mathbf{v}}_{n-k}.$$

This proves that the equivalence classes $\bar{\mathbf{v}}_1, \bar{\mathbf{v}}_2, \ldots, \bar{\mathbf{v}}_{n-k}$ span the space $V/W$. We conclude that $\bar{\mathbf{v}}_1, \bar{\mathbf{v}}_2, \ldots, \bar{\mathbf{v}}_{n-k}$ form a basis for $V/W$, and that

$$\dim(V/W) = \dim V - \dim W.$$

The time has come for some examples of quotient spaces:

**Example 1.** $V$ is $\mathbb{R}^3$, $W$ is the one-dimensional subspace spanned by $\begin{pmatrix}1\\1\\0\end{pmatrix}$. Since $\begin{pmatrix}1\\1\\0\end{pmatrix}$, $\begin{pmatrix}0\\1\\0\end{pmatrix}$, and $\begin{pmatrix}0\\0\\1\end{pmatrix}$ span $\mathbb{R}^3$, we can choose $\overline{\begin{pmatrix}0\\1\\0\end{pmatrix}}$ and $\overline{\begin{pmatrix}0\\0\\1\end{pmatrix}}$ as a basis for the two-dimensional quotient space $V/W$. Now, for example,

$$\begin{pmatrix}3\\1\\1\end{pmatrix} = 3\begin{pmatrix}1\\1\\0\end{pmatrix} - 2\begin{pmatrix}0\\1\\0\end{pmatrix} + \begin{pmatrix}0\\0\\1\end{pmatrix}$$

so

$$\overline{\begin{pmatrix}3\\1\\1\end{pmatrix}} = -2\overline{\begin{pmatrix}0\\1\\0\end{pmatrix}} + \overline{\begin{pmatrix}0\\0\\1\end{pmatrix}}.$$

In a similar manner we can express the equivalence class of any vector in $\mathbb{R}^3$ as a linear combination of $\overline{\begin{pmatrix}0\\1\\0\end{pmatrix}}$ and $\overline{\begin{pmatrix}0\\0\\1\end{pmatrix}}$.

Notice, incidentally, that $\overline{\begin{pmatrix}1\\0\\0\end{pmatrix}}$ and $\overline{\begin{pmatrix}0\\1\\0\end{pmatrix}}$ would *not* serve as a basis for $V/W$. Because their sum is an element of $W$, they are not linearly independent elements of $V/W$:

$$\begin{pmatrix}1\\0\\0\end{pmatrix}+\begin{pmatrix}0\\1\\0\end{pmatrix}=\begin{pmatrix}1\\1\\0\end{pmatrix}, \quad \text{so} \quad \overline{\begin{pmatrix}1\\0\\0\end{pmatrix}}+\overline{\begin{pmatrix}0\\1\\0\end{pmatrix}}=\bar{\mathbf{0}}.$$

**Example 2.** $V$ is the space of polynomials $f(t)$ of degree $\leqslant 2$; $W$ is the two-dimensional subspace of such polynomials satisfying the additional condition $f(1)=0$. A basis for $W$ is $f_1(t)=1-t$ and $f_2(t)=1-t^2$. A basis for $V/W$ is the equivalence class $\bar{1}$. In this case, the general element of $V$ is

$$f(t)=A+Bt+Ct^2$$

so

$$f(t)=A+B+C-B(1-t)-C(1-t^2).$$

This means that

$$\overline{f(t)}=(A+B+C)\bar{\mathbf{1}}.$$

If you think of elements of $V/W$ as planes, this is obvious: the subspace $W$ is the plane $f(1)=0$, and the equivalence class of any other function $f(t)$ is determined by the value of $f(1)$.

We can now put together the concepts of dual space and quotient space to obtain a powerful result. Earlier, we found that, if $W$ is a subspace of $V$, the annihilator space $W^{\perp}$ is a subspace of $V^*$. Now suppose that $\boldsymbol{\alpha}$ is an element of $W^{\perp}$, and consider its action on an equivalence class $\bar{\mathbf{v}}$. Because $\boldsymbol{\alpha}[\mathbf{w}]=\mathbf{0}$ for all $\mathbf{w}\in W$, and $\boldsymbol{\alpha}$ is linear,

$$\boldsymbol{\alpha}[\mathbf{v}+\mathbf{w}]=\boldsymbol{\alpha}[\mathbf{v}].$$

That is, $\boldsymbol{\alpha}$ has the same value on *any* element in a equivalence class, and it can therefore be regarded as a linear function on the space of equivalence classes. Conversely, any linear function on $V/W$ can be regarded as a linear function on $V$. Simply define $\boldsymbol{\beta}[\mathbf{v}]$ as

$$\boldsymbol{\beta}[\mathbf{v}]=\boldsymbol{\beta}[\bar{\mathbf{v}}]$$

Then $\boldsymbol{\beta}[\mathbf{w}]=\boldsymbol{\beta}[\bar{\mathbf{0}}]=\mathbf{0}$. So $\boldsymbol{\beta}$ is an element of $W^{\perp}$. *We can therefore identify $W^{\perp}$ with the dual space of the quotient space $V/W$.* Recall that both $W^{\perp}$ and $V/W$ have dimension equal to $\dim V-\dim W$.

Similarly, we may consider the quotient space $V^*/W^{\perp}$, whose elements are equivalence classes $\bar{\boldsymbol{\beta}}$ whose elements differ by elements of $W^{\perp}$, i.e.,

$$\bar{\boldsymbol{\beta}}=\{\boldsymbol{\beta}+\boldsymbol{\alpha}:\boldsymbol{\alpha}\in W^{\perp}\}.$$

Define the action of $\bar{\boldsymbol{\beta}}$ on a vector $w\in W$ by $\bar{\boldsymbol{\beta}}[\mathbf{w}]=\boldsymbol{\beta}[\mathbf{w}]$. This is legitimate since, for any $\boldsymbol{\alpha}\in W^{\perp}$, $(\boldsymbol{\beta}+\boldsymbol{\alpha})[\mathbf{w}]=\beta[\mathbf{w}]$. If $\bar{\boldsymbol{\beta}}[\mathbf{w}]=\mathbf{0}$ for all $\mathbf{w}\in W$ then $\boldsymbol{\beta}[\mathbf{w}]=\mathbf{0}$ for all

$\mathbf{w} \in W$. This says that $\boldsymbol{\beta} \in W^{\perp}$ or $\bar{\boldsymbol{\beta}} = \mathbf{0}$. So $\bar{\boldsymbol{\beta}}$ is completely determined by the linear function it defines on $W$.

Conversely, we claim that every $\gamma \in W^*$ is of the form $\gamma = \bar{\boldsymbol{\beta}}$ for some $\boldsymbol{\beta} \in V^*$. Indeed choose a basis $\mathbf{w}_1, \ldots, \mathbf{w}_k$ of $W$ and extend it to a basis $\mathbf{w}_1, \ldots, \mathbf{w}_k, \mathbf{v}_1, \ldots, \mathbf{v}_{n-k}$ of $V_0$.

Let $\boldsymbol{\beta}$ be any linear function with $\boldsymbol{\beta}[\mathbf{w}_i] = \gamma[\mathbf{w}_i]$ for all $i$ and let $\boldsymbol{\beta}$ take any values on the vs. Then $\bar{\boldsymbol{\beta}} = \gamma$. *We can therefore identify the space of these equivalence classes, the quotient space* $V^*/W^{\perp}$, *with the dual space* $W^*$.

The results just proved may be summarized in the following diagram:

$$\begin{array}{ccccc} V^*/W^{\perp} & \leftarrow & V^* & \leftarrow & W^{\perp} \\ W & \rightarrow & V & \rightarrow & V/W \end{array}$$

Here the spaces which are dual to one another are arranged vertically: $V$ and $V^*$ (dimension $n$) are dual, $W$ and $V^*/W^{\perp}$ (dimension $k$) are dual, $V/W$ and $W^{\perp}$ (dimension $n-k$) are dual.

Much of linear algebra and its applications to electric network theory rests on this single theorem, which deserves your most careful consideration.

As an illustration of the theorem, let $V$ be the space of polynomials $f(t)$ of degree $\leqslant 2$, and let $W$ be the two-dimensional subspace of *even* polynomials. Then $V/W$ is one-dimensional, and a basis element, which we shall call $\mathbf{h}_0$, is the equivalence class of the function $f(t) = t$. Thus if $f(t) = A + Bt + Ct^2$, $\overline{f(t)} = B\mathbf{h}_0$.

In this case, the annihilator space $W^{\perp}$ is also one-dimensional. One choice for a basis element is the linear operator

$$\boldsymbol{\alpha}: f(t) \mapsto \int_{-1}^{1} t f(t) \mathrm{d}t.$$

Since

$$\int_{-1}^{1} t(A + Bt + Ct^2) \mathrm{d}t = \tfrac{2}{3} B$$

we see that $\boldsymbol{\alpha}$ does indeed annihilate any even polynomial and assign the value $\tfrac{2}{3}$ to the polynomial $f(t) = t$, which specifies the basis $\mathbf{h}_0$ of $V/W$. That is, $\tfrac{3}{2}\boldsymbol{\alpha}$ is the basis element dual to $\mathbf{h}_0$.

We can extend $\boldsymbol{\alpha}$ to a complete basis for $V^*$ by adjoining the basis elements

$$\boldsymbol{\beta}_1 : f(t) \mapsto f(0)$$

and

$$\boldsymbol{\beta}_2 : f(t) \mapsto \tfrac{1}{2} f''(0)$$

whose effect is to pick out the coefficients $A$ and $C$ respectively. That is,

$$\boldsymbol{\beta}_1[A + Bt + Ct^2] = A,$$
$$\boldsymbol{\beta}_2[A + Bt + Ct^2] = C.$$

The equivalence classes $\bar{\boldsymbol{\beta}}_1$ and $\bar{\boldsymbol{\beta}}_2$, which form a basis for $V^*/W^{\perp}$, clearly also form a basis for $W^*$; indeed, they are the dual basis elements for the basis elements 1 and $t^2$ in the subspace $W$ of even polynomials.

## 10.7. Linear transformations

We consider now a linear transformation

$$A: V \to W$$

where $V$ is a vector space of dimension $m$ and $W$ is a space of dimension $n$. As always, to state that $A$ is linear means that $A(c_1\mathbf{v}_1 + c_2\mathbf{v}_2) = c_1 A\mathbf{v}_1 + c_2 A\mathbf{v}_2$.

Associated with a linear transformation $A: V \to W$ are two subspaces, the *kernel* of $A$ and the *image* of $A$.

The kernel of $A$, denoted $\ker A$, is the set of vectors $\mathbf{v} \in V$ such that $A\mathbf{v} = \mathbf{0}$. To verify that $\ker A$ is a subspace of $V$, we note that, if $\mathbf{v}_1$ and $\mathbf{v}_2$ are in the kernel of $A$, then $A(c_1\mathbf{v}_1 + c_2\mathbf{v}_2) = c_1 A\mathbf{v}_1 + c_2 A\mathbf{v}_2 = \mathbf{0}$ so that $c_1\mathbf{v}_1 + c_2\mathbf{v}_2 \in \ker A$ also. This proves that $\ker A$ is closed, and hence a subspace.

The image of $A$, denoted $\operatorname{im} A$, is the set of vectors $\mathbf{w} \in W$ which are of the form $A\mathbf{v}$ for some $\mathbf{v} \in V$. If $\mathbf{w}_1$ and $\mathbf{w}_2$ are vectors in $\operatorname{im} A$, then $\mathbf{w}_1 = A\mathbf{v}_1$ and $\mathbf{w}_2 = A\mathbf{v}_2$ for some $\mathbf{v}_1, \mathbf{v}_2 \in V$. Because of the linearity of $A$,

$$A(c_1\mathbf{v}_1 + c_2\mathbf{v}_2) = c_1 A\mathbf{v}_1 + c_2 A\mathbf{v}_2 = c_1\mathbf{w}_1 + c_2\mathbf{w}_2$$

so that $c_1\mathbf{w}_1 + c_2\mathbf{w}_2$ is also an element of $\operatorname{im} A$. This proves that $\operatorname{im} A$ is a subspace of $W$.

The dimensions of $\ker A$ and of $\operatorname{im} A$ are related by the equation

$$\dim(\operatorname{im} A) + \dim(\ker A) = \dim V. \tag{10.8}$$

The dimension of the image of $A$ is called the *rank* of $A$, the dimension of the kernel of $A$ is called the *nullity of* $A$, and equation (10.8) is called the *rank-nullity theorem.* You are already familiar with the theorem in the special case of transformations of the plane into itself. Recall that there were three possibilities for $A: \mathbb{R}^2 \to \mathbb{R}^2$:

(1) $A$ has the entire plane as its image and carries no non-zero vector into the origin (rank 2, nullity 0).

(2) $A$ collapses the plane into a line, and carries a line into the origin (rank 1, nullity 1).

(3) $A$ collapses the entire plane into the origin (rank 0, nullity 2).

To prove the rank–nullity theorem we choose a convenient basis for $V$. Suppose that $\dim V = n$, $\dim(\ker A) = k$. We choose a basis $\{\mathbf{u}_1, \mathbf{u}_2, \ldots, \mathbf{u}_k\}$ for $\ker A$, then extend this to a basis for all of $V$ *by choosing* $r = n - k$ vectors $\mathbf{v}_1, \mathbf{v}_2, \ldots, \mathbf{v}_r$. For convenience, we order this basis as

$$\{\mathbf{v}_1, \mathbf{v}_2, \ldots, \mathbf{v}_r; \mathbf{u}_1, \mathbf{u}_2, \ldots, \mathbf{u}_k\}$$

so that the *first* $r$ vectors in the basis do *not* lie in $\ker A$. The problem is now to show that the $r$ vectors $\{A\mathbf{v}_1, A\mathbf{v}_2, \ldots, A\mathbf{v}_r\}$ form a basis for $\operatorname{im} A$.

We first show that the vectors $\{A\mathbf{v}_1, A\mathbf{v}_2, \ldots, A\mathbf{v}_r\}$ are linearly independent. Suppose that

$$\sum_{i=1}^{r} \lambda_i A\mathbf{v}_i = \mathbf{0}.$$

Because $A$ is linear, this is the same as

$$A\left(\sum_{i=1}^{r} \lambda_i \mathbf{v}_i\right) = \mathbf{0}$$

which implies that $\sum_{i=1}^{r} \lambda_i \mathbf{v}_i \in \ker A$. But the vectors $\{\mathbf{v}_i\}$, along with the basis $\{\mathbf{u}_j\}$ for $\ker A$, form a basis for $V$. Therefore

$$\sum_{i=1}^{r} \lambda_i \mathbf{v}_i = \mathbf{u} \quad \text{with} \quad \mathbf{u} \in \ker A$$

implies that all the $\lambda_i$ are zero, and therefore that $\{A\mathbf{v}_1, A\mathbf{v}_2, \ldots, A\mathbf{v}_r\}$ are linearly independent.

To show that the vectors $\{A\mathbf{v}_1, A\mathbf{v}_2, \ldots, A\mathbf{v}_r\}$ span $\operatorname{im} A$, we consider an arbitrary vector $\mathbf{w} \in \operatorname{im} A$. There is some vector $\mathbf{v} \in V$ such that $\mathbf{w} = A\mathbf{v}$. We can write

$$\mathbf{v} = \sum_{i=1}^{r} a_i \mathbf{v}_i + \sum_{j=1}^{k} b_j \mathbf{u}_j.$$

But all the basis vectors $\mathbf{u}_j$ are in $\ker A$, so

$$\mathbf{w} = A\left(\sum_{i=1}^{r} a_i \mathbf{v}_i + \sum_{j=1}^{k} b_j \mathbf{u}_j\right) = \sum_{i=1}^{r} a_i (A\mathbf{v}_i);$$

i.e., any $\mathbf{w}$ can be written as a linear combination of $\{A\mathbf{v}_1, A\mathbf{v}_2, \ldots, A\mathbf{v}_r\}$.

We conclude that the $r$ vectors $\{A\mathbf{v}_1, A\mathbf{v}_2, \ldots, A\mathbf{v}_r\}$ form a basis for $\operatorname{im} A$. It follows that $\dim(\operatorname{im} A) = r$. But $r = n - k$, where $\dim V = n$ and $\dim(\ker A) = k$. Thus $\dim(\operatorname{im} A) = \dim V - \dim(\ker A)$, which is the rank–nullity theorem.

The rank–nullity theorem provides a proof of a result which you have probably already conjectured about the annihilator space of a subspace. Suppose that $\{\boldsymbol{\alpha}^1, \boldsymbol{\alpha}^2, \ldots, \boldsymbol{\alpha}^n\}$ are elements of $V^*$. Then we can define a linear transformation

$$A: V \to \mathbb{R}^n$$

by

$$A\mathbf{v} = \begin{pmatrix} \boldsymbol{\alpha}^1[\mathbf{v}] \\ \boldsymbol{\alpha}^2[\mathbf{v}] \\ \vdots \\ \boldsymbol{\alpha}^n[\mathbf{v}] \end{pmatrix}.$$

The vectors $\{\boldsymbol{\alpha}^1, \boldsymbol{\alpha}^2, \ldots, \boldsymbol{\alpha}^n\}$ span a subspace $U^\perp \subset V^*$ which annihilates the subspace $\ker A$. The rank–nullity theorem says that $\dim(\operatorname{im} A) = \dim V - \dim(\ker A)$. But we saw in section 10.4 that

$$\dim(U^\perp) = \dim V - \dim(\ker A).$$

It follows that

$$\dim(\operatorname{im} A) = \dim U^\perp.$$

In terms of matrices, each element $\boldsymbol{\alpha}^i$ is a *row* of the matrix, and $\dim U^\perp$ is the dimension of the subspace of $V^*$ spanned by the *rows* of the matrix. On the other hand, $\dim(\operatorname{im} A)$ is the dimension of the subspace of $W$ spanned by the *columns* of the matrix. Both of these numbers equal $r$, the rank of the matrix.

This view of the rows of a matrix as elements of the dual space $V^*$ is particularly useful when we are trying to solve systems of linear equations. For example, the system of equations

$$\begin{aligned} x + y + z &= 0, \\ x + 2y + 3z &= 0, \\ 2x + 3y + 4z &= 0 \end{aligned}$$

may be represented as $A\mathbf{v} = \mathbf{0}$ where

$$A = \begin{pmatrix} 1 & 1 & 1 \\ 1 & 2 & 3 \\ 2 & 3 & 4 \end{pmatrix}, \quad \mathbf{v} = \begin{pmatrix} x \\ y \\ z \end{pmatrix}.$$

Here the *rows* of $A$ are associated with individual equations. Because the third equation is the sum of the first two, the three rows span only a two-dimensional subspace of $V^*$, the rank of the matrix $A$ is 2 and its nullity is $3 - 2 = 1$. Therefore $\ker A$ is one-dimensional, and there exists a one-dimensional subspace of non-trivial solutions to the equation $A\mathbf{v} = \mathbf{0}$.

## 10.8. Row reduction

Consider now a linear transformation

$$T: V \to W$$

where $V$ is $m$-dimensional, $W$ is $n$-dimensional, and the rank of $T$ is $r$. We have seen that by a proper choice of basis for $V$ and $W$ we can assure that $T$ has an especially simple matrix representation. We simply choose as a basis for $V$ the vectors

$$\{\mathbf{v}_1, \mathbf{v}_2, \ldots, \mathbf{v}_r; \mathbf{v}_{r+1}, \ldots, \mathbf{v}_m\}$$

where the *last* $m - r$ basis vectors form a basis for $\ker T$, so that $T\mathbf{v}_{r+1} = T\mathbf{v}_{r+2} = \cdots = T\mathbf{v}_m = 0$. Then $T\mathbf{v}_1, T\mathbf{v}_2, \ldots, T\mathbf{v}_r$ form a basis for $\operatorname{im} T$. We choose $\mathbf{w}_1 = T\mathbf{v}_1, \mathbf{w}_2 = T\mathbf{v}_2, \ldots, \mathbf{w}_r = T\mathbf{v}_r$ as a basis for $\operatorname{im} T$, then extend to a basis for all of $W$. Now the matrix representation of $T$ relative to this basis is simply the matrix

$$\underbrace{\begin{pmatrix} I_r & 0 \\ 0 & 0 \end{pmatrix}}_{m \text{ columns}} \Big\} n \text{ rows} \tag{10.9}$$

which has a string of $r$ 1s down the diagonal from the upper left-hand corner and all its other entries zero.

Usually, alas, the transformation $T$ is described by a matrix $A$ which represents it relative to some other, less convenient basis. An important computational problem is then to find the change of basis for $V$ and $W$ which converts the matrix representation relative to the given basis, $A$, to $I_r$. In practice this is most efficiently achieved by the algorithm of *row reduction*, which is in essence just a systematic procedure for solving linear equations by the familiar process of elimination. We first describe the process and illustrate it, then explain why it solves the general problem.

Suppose we are given the matrix

$$M = \begin{pmatrix} 0 & 4 & -4 & 8 \\ 2 & 4 & 0 & 2 \\ 3 & 0 & 6 & -9 \end{pmatrix}.$$

The *index* of any non-zero row of $M$ is the position of the first non-vanishing entry, and this entry is called the *leading entry*. Thus, for the first row of $M$, the index is 2 and the leading entry is 4, while for the third row the index is 1 and the leading entry is 3.

The first step in row reduction is to locate a row of smallest index, to move it to the top position by interchanging it with the top row if necessary, and to divide it by its leading entry. For the given matrix $M$, we interchange the first and second rows to obtain

$$\begin{pmatrix} 2 & 4 & 0 & 2 \\ 0 & 4 & -4 & 8 \\ 3 & 0 & 6 & -9 \end{pmatrix}$$

and we then divide the top row by its leading entry, 2, to obtain

$$\begin{pmatrix} 1 & 2 & 0 & 1 \\ 0 & 4 & -4 & 8 \\ 3 & 0 & 6 & -9 \end{pmatrix}.$$

The second step is to clear the column under the leading entry of the top row. This is achieved by subtracting the appropriate multiple of the top row from each other row in turn. In our example, we subtract 3 times the top row from the third row, obtaining

$$\begin{pmatrix} 1 & 2 & 0 & 1 \\ 0 & 4 & -4 & 8 \\ 0 & -6 & 6 & -12 \end{pmatrix}.$$

The matrix now has a leading entry one in the top row, and all other rows which are not zero have an index greater than the index of the top row. We next move a row of next smallest index to the second position and divide by its leading entry. In the example, the second row already has next smallest index, and we divide it by its leading entry, 4, to obtain

$$\begin{pmatrix} 1 & 2 & 0 & 1 \\ 0 & 1 & -1 & 2 \\ 0 & -6 & 6 & -12 \end{pmatrix}.$$

We now clear the column corresponding to the leading entry in the second row by subtracting a suitable multiple of the second row from all other rows. In the example, we subtract twice the second row from the first and substract $-6$ times the second row from the third, obtaining

$$\begin{pmatrix} 1 & 0 & 2 & -3 \\ 0 & 1 & -1 & 2 \\ 0 & 0 & 0 & 0 \end{pmatrix}. \tag{10.10}$$

Now the first and second columns both contain just a single 1, which is the leading entry of a row.

In the general case, we now again interchange rows, if necessary, to move a row of smallest leading entry to the third position, divide this row by its leading entry, and subtract multiples of it from all other rows to clear the column of the leading entry. Eventually there are no more non-zero rows, and we have a matrix in row-reduced form. In the example this has already happened. Note the following features of a row-reduced matrix such as given in (10.10).

(a) All zero rows, if any, are at the bottom.
(b) The non-zero rows are arranged in order of increasing index.
(c) Every column containing the leading entry of a non-zero row has a one as its leading entry and zeros elsewhere.

Each operation in the row-reduction process can be achieved by *left* multiplication by an invertible $n \times n$ matrix. For example, multiplying on the left by the matrix

$$S_1 = \begin{pmatrix} 0 & 1 & 0 \\ 1 & 0 & 0 \\ 0 & 0 & 1 \end{pmatrix}$$

interchanges the first and second rows:

$$\begin{pmatrix} 0 & 1 & 0 \\ 1 & 0 & 0 \\ 0 & 0 & 1 \end{pmatrix}\begin{pmatrix} 0 & 4 & -4 & 8 \\ 2 & 4 & 0 & 2 \\ 3 & 0 & 6 & -9 \end{pmatrix} = \begin{pmatrix} 2 & 4 & 0 & 2 \\ 0 & 4 & -4 & 8 \\ 3 & 0 & 6 & -9 \end{pmatrix}.$$

Multiplying on the left by the matrix

$$S_2 = \begin{pmatrix} \frac{1}{2} & 0 & 0 \\ 0 & 1 & 0 \\ 0 & 0 & 1 \end{pmatrix}$$

divides the first row by 2:

$$\begin{pmatrix} \frac{1}{2} & 0 & 0 \\ 0 & 1 & 0 \\ 0 & 0 & 1 \end{pmatrix}\begin{pmatrix} 2 & 4 & 0 & 2 \\ 0 & 4 & -4 & 8 \\ 3 & 0 & 6 & -9 \end{pmatrix} = \begin{pmatrix} 1 & 2 & 0 & 1 \\ 0 & 4 & -4 & 8 \\ 3 & 0 & 6 & -9 \end{pmatrix}.$$

Multiplying on the left by the matrix

$$S_3 = \begin{pmatrix} 1 & 0 & 0 \\ 0 & 1 & 0 \\ -3 & 0 & 1 \end{pmatrix}$$

subtracts three times the first row from the third:

$$\begin{pmatrix} 1 & 0 & 0 \\ 0 & 1 & 0 \\ -3 & 0 & 1 \end{pmatrix}\begin{pmatrix} 1 & 2 & 0 & 1 \\ 0 & 4 & -4 & 8 \\ 3 & 0 & 6 & -9 \end{pmatrix}=\begin{pmatrix} 1 & 2 & 0 & 1 \\ 0 & 4 & -4 & 8 \\ 0 & -6 & 6 & -12 \end{pmatrix}.$$

Thus we can write the final row-reduced matrix $B$ as

$$B = S_k S_{k-1} \cdots S_3 S_2 S_1 M$$

or as

$$B = SM$$

where $S$ is an invertible $n \times n$ matrix. Notice that, since $S$ is invertible, dim im $B$ = dim im $M$.

The image and kernel of the row-reduced matrix $B$ are easy to determine. Clearly the image is the $r$-dimensional subspace corresponding to the $r$ non-zero rows of $B$, spanned by the columns

$$\begin{pmatrix} 1 \\ 0 \\ 0 \\ \vdots \\ 0 \end{pmatrix}, \begin{pmatrix} 0 \\ 1 \\ 0 \\ \vdots \\ 0 \end{pmatrix}, \ldots$$

which contain the leading entries of all the non-zero rows. By the rank–nullity theorem, the kernel of $B$ has dimension $m - r$, equal to the number of columns that do not contain leading entries. To find a basis for ker $B$, we consider vectors which have a 1 in one of the $m - r$ positions corresponding to the columns with non-leading entries, and zeros in the remaining $m - r$ positions, then use the rows of the matrix $B$ one at a time to calculate the components in the positions of the leading-entry columns.

For example, with

$$B = \begin{pmatrix} 1 & 0 & 2 & -3 \\ 0 & 1 & -1 & 2 \\ 0 & 0 & 0 & 0 \end{pmatrix}$$

a basis for im $B$ is clearly $\begin{pmatrix} 1 \\ 0 \\ 0 \end{pmatrix}$ and $\begin{pmatrix} 0 \\ 1 \\ 0 \end{pmatrix}$. The columns without leading entries are the third and fourth, so we search for basis vectors of ker $B$ which have the form

$$\mathbf{u}_1 = \begin{pmatrix} x_1 \\ x_2 \\ 1 \\ 0 \end{pmatrix} \quad \text{and} \quad \mathbf{u}_2 = \begin{pmatrix} y_1 \\ y_2 \\ 0 \\ 1 \end{pmatrix}.$$

Setting $B\mathbf{u}_1 = \mathbf{0}$, we find

$$x_1 + 2 = 0, \quad x_2 - 1 = 0$$

so

$$\mathbf{u}_1 = \begin{pmatrix} -2 \\ 1 \\ 1 \\ 0 \end{pmatrix}.$$

Setting $B\mathbf{u}_2 = \mathbf{0}$, we find

$$y_1 - 3 = 0, y_2 + 2 = 0$$

so

$$\mathbf{u}_2 = \begin{pmatrix} 3 \\ -2 \\ 0 \\ 1 \end{pmatrix}.$$

Of course, we were interested in the kernel and image of the original matrix $A$, not of the row-reduced matrix $B$. However, $B = SA$, where $S$ is *invertible*, so

$$A = S^{-1}B.$$

Clearly any vector in the kernel of $B$ is also in the kernel of $A$, so *by finding the kernel of $B$ we have also found the kernel of $A$*. To find the image of $A$ we must invert $S$ and let $S^{-1}$ act on the image of $B$. To summarize, we have $B = SA$, $\ker B = \ker A$, and $\dim \operatorname{im} B = \dim \operatorname{im} A$.

Suppose now that we wish to solve an equation of the form

$$A\mathbf{v} = \mathbf{w}.$$

We apply the operations of row reduction *both* to the matrix $A$ and to the vector $\mathbf{w}$, obtaining

$$SA\mathbf{v} = S\mathbf{w} \quad \text{or} \quad B\mathbf{v} = \mathbf{u}$$

where $B$ is row-reduced and $\mathbf{u} = S\mathbf{w}$. This equation is of a form like

$$\begin{pmatrix} 1 & 0 & 2 & -3 \\ 0 & 1 & -1 & 2 \\ 0 & 0 & 0 & 0 \end{pmatrix} \begin{pmatrix} x_1 \\ x_2 \\ x_3 \\ x_4 \end{pmatrix} = \begin{pmatrix} u_1 \\ u_2 \\ u_3 \end{pmatrix}$$

and it can be solved by inspection, as follows.

(1) If any component of $\mathbf{u}$ corresponding to a zero row of $B$ is different from zero, the equation has no solution.

(2) If the components of $\mathbf{u}$ corresponding to the zero rows of $B$ are all zero, then

$$\mathbf{v}_0 = \begin{pmatrix} u_1 \\ u_2 \\ 0 \\ 0 \end{pmatrix}$$

is one solution to the equation.

(3) The general solution to the equation is of the form $\mathbf{v}_0 + \mathbf{v}$, where $\mathbf{v} \in \ker A$.

In practice, before applying the row-reduction procedure, it is convenient to combine the matrix $A$ and the vector $\mathbf{w}$ into a single array so that row reduction can be applied to both at once. Here is an example of the complete process. We wish to solve

$$A\mathbf{v} = \mathbf{w},$$

where

$$A = \begin{pmatrix} 2 & 4 & 2 & 2 \\ 1 & 3 & 2 & 0 \\ 3 & 1 & -2 & 8 \end{pmatrix} \quad \text{and} \quad \mathbf{w} = \begin{pmatrix} 0 \\ -1 \\ 5 \end{pmatrix}.$$

We combine $A$ and $\mathbf{w}$ into the array

$$\left(\begin{array}{cccc|c} 2 & 4 & 2 & 2 & 0 \\ 1 & 3 & 2 & 0 & -1 \\ 3 & 1 & -2 & 8 & 5 \end{array}\right)$$

and apply row reduction, obtaining successively

$$\left(\begin{array}{cccc|c} 1 & 2 & 1 & 1 & 0 \\ 1 & 3 & 2 & 0 & -1 \\ 3 & 1 & -2 & 8 & 5 \end{array}\right),$$

$$\left(\begin{array}{cccc|c} 1 & 2 & 1 & 1 & 0 \\ 0 & 1 & 1 & -1 & -1 \\ 0 & -5 & -5 & 5 & 5 \end{array}\right),$$

$$\left(\begin{array}{cccc|c} 1 & 0 & -1 & 3 & 2 \\ 0 & 1 & 1 & -1 & -1 \\ 0 & 0 & 0 & 0 & 0 \end{array}\right).$$

One solution to the equation is therefore

$$\begin{pmatrix} 2 \\ -1 \\ 0 \\ 0 \end{pmatrix}.$$

To find the general solution, we must construct a basis for the kernel of $A$. One basis vector, with one in the third position and zero in the fourth, is $\begin{pmatrix} 1 \\ -1 \\ 1 \\ 0 \end{pmatrix}$. The other, with zero in the third position and one in the fourth, is $\begin{pmatrix} -3 \\ 1 \\ 0 \\ 1 \end{pmatrix}$. So the general solution to $A\mathbf{v} = \mathbf{w}$ is

$$\mathbf{v} = \begin{pmatrix} 2 \\ -1 \\ 0 \\ 0 \end{pmatrix} + \lambda_1 \begin{pmatrix} 1 \\ -1 \\ 1 \\ 0 \end{pmatrix} + \lambda_2 \begin{pmatrix} -3 \\ 1 \\ 0 \\ 1 \end{pmatrix}$$

where $\lambda_1$ and $\lambda_2$ are arbitrary real numbers. Note the characteristic form of the solution in relation to the columns of the row reduced matrix which do not contain leading entries of rows (in this case, columns 3 and 4). The particular solution to the equation has zeros in both the third and fourth positions, while the basis vectors for ker $A$ each have zeros in all but one of these positions. There are many ways to write the general solution to the equation, but this is the simplest.

For a non-singular square matrix $A$, the technique just described provides an efficient method of matrix inversion. The transformation $S$ that row-reduces $A$ to the identity matrix is just the matrix $A^{-1}$, and it can be calculated step by step if each individual row-reduction operation is applied to a matrix that begins as the identity matrix. Suppose, for example, that $A = \begin{pmatrix} 1 & 2 & 1 \\ 2 & 3 & 3 \\ -1 & -1 & 0 \end{pmatrix}$. We begin with $A$ and the identity matrix,

$$\left(\begin{array}{ccc|ccc} 1 & 2 & 1 & 1 & 0 & 0 \\ 2 & 3 & 3 & 0 & 1 & 0 \\ -1 & -1 & 0 & 0 & 0 & 1 \end{array}\right),$$

and apply row-reduction operations to both. Substract twice row 1 from row 2; add row 1 to row 3:

$$\left(\begin{array}{ccc|ccc} 1 & 2 & 1 & 1 & 0 & 0 \\ 0 & -1 & 1 & -2 & 1 & 0 \\ 0 & 1 & 1 & 1 & 0 & 1 \end{array}\right).$$

Divide row 2 by $-1$:

$$\left(\begin{array}{ccc|ccc} 1 & 2 & 1 & 1 & 0 & 0 \\ 0 & 1 & -1 & 2 & -1 & 0 \\ 0 & 1 & 1 & 1 & 0 & 1 \end{array}\right).$$

Subtract twice row 2 from row 1; subtract row 2 from row 3:

$$\left(\begin{array}{ccc|ccc} 1 & 0 & 3 & -3 & 2 & 0 \\ 0 & 1 & -1 & 2 & -1 & 0 \\ 0 & 0 & 2 & -1 & 1 & 1 \end{array}\right).$$

Divide row 3 by 2:

$$\left(\begin{array}{ccc|ccc} 1 & 0 & 3 & -3 & 2 & 0 \\ 0 & 1 & -1 & 2 & -1 & 0 \\ 0 & 0 & 1 & -\frac{1}{2} & \frac{1}{2} & \frac{1}{2} \end{array}\right).$$

Subtract 3 times row 3 from row 1; add row 3 to row 2:

$$\left(\begin{array}{ccc|ccc} 1 & 0 & 0 & -\frac{3}{2} & +\frac{1}{2} & -\frac{3}{2} \\ 0 & 1 & 0 & \frac{3}{2} & -\frac{1}{2} & \frac{1}{2} \\ 0 & 0 & 1 & -\frac{1}{2} & \frac{1}{2} & \frac{1}{2} \end{array}\right).$$

It follows that $A^{-1}$ is the matrix

$$\frac{1}{2}\begin{pmatrix} -3 & 1 & -3 \\ 3 & -1 & 1 \\ -1 & 1 & 1 \end{pmatrix}.$$

Let us now return to the case of a general rectangular matrix. Instead of performing row operations, we could perform column operations: just the same operations as in row reduction, but with the word 'row' replaced by 'column'. We would end up with a matrix

$$C = MT$$

where $T$ is an invertible square matrix and $C$ is column reduced. That is:

(a′) All zero columns of $C$, if any, are on the right;

(b′) The non-zero columns of $C$ are arranged in order of increasing index (when the index of a non-zero column is the position of the first non-vanishing entry);

and

(c′) Every row containing the leading entry of a non-zero column has a one as its leading entry and zeros elsewhere.

Notice that now

$$\operatorname{im} C = \operatorname{im} M$$

and the $r$ non-zero columns of $C$ will be linearly independent. Hence they will give a basis of $\operatorname{im} M$. Thus, to summarize:

> To find a basis of $\operatorname{im} M$, apply column reduction. The non-zero columns of the resulting matrix $C$ provide such a basis.
>
> To find a basis of $\ker M$ apply row reduction. The resulting rows of the row reduced matrix $B = SM$ give a set of $r$ equations for $\ker B$ which are in 'solved' from – solved for the positions of the columns containing leading entries in terms of the remaining $m - r$ positions. A basis can be found by successively choosing 1 for one of the remaining positions with the other remaining positions zero and solving.

We can also perform both column and row operations. For example, suppose we perform column operations to the row-reduced matrix

$$B = \begin{pmatrix} 1 & 0 & 2 & 3 \\ 0 & 1 & -1 & 2 \\ 0 & 0 & 0 & 0 \end{pmatrix}.$$

Subtracting multiples of the first column from the third and fourth yields

$$\begin{pmatrix} 1 & 0 & 0 & 0 \\ 0 & 1 & -1 & 2 \\ 0 & 0 & 0 & 0 \end{pmatrix}$$

and subtracting multiples of the second column from the third and fourth yields

$$\begin{pmatrix} 1 & 0 & 0 & 0 \\ 0 & 1 & 0 & 0 \\ 0 & 0 & 0 & 0 \end{pmatrix}.$$

In general, by performing column operations to row-reduced matrix $B$, we can first arrange (by switching columns) that the leading columns are exactly the first $r$ columns. (This step was not needed in the example above). Then, successively subtracting off multiples of each of the first $r$ columns from the remaining $m-r$ columns, we end up with a matrix whose only non-zero entries are $r$ 1s down the principal diagonal, i.e., of the form (10.9). We have thus described an effective algorithm for finding matrices $S$ and $T$ such that

$$SMT \text{ has the form (10.9).}$$

## 10.9. The constant rank theorem*

If we combine the results of the preceding section with those of section 6.3, we obtain some very powerful information about the behavior of differentiable maps. Let $V$ and $W$ be vector spaces of dimension $m$ and $n$ respectively. Let $O$ be some (open) region in $V$ and suppose that

$$f: O \to W$$

is a differentiable map. At each point $\mathbf{p} \in O$ we can compute the differential, $\mathrm{d}f_{\mathbf{p}}$, of $f$ at $\mathbf{p}$. The differential $\mathrm{d}f_{\mathbf{p}}$ is a linear map of $V$ into $W$, and so we may compute its rank. Of course, this rank depends on the point $\mathbf{p}$. Our purpose is to prove the following theorem.

**The constant rank theorem.** *Suppose that there is a constant integer $r$ such that the rank $\mathrm{d}f_{\mathbf{p}} = r$ for all $\mathbf{p} \in O$. Then for any $\mathbf{x} \in O$ we can find a one-to-one differentiable map $\phi$ mapping a neighborhood of $\mathbf{x}$ into $\mathbb{R}^m$ such that $\phi$ has a differentiable inverse, and a one-to-one differentiable map $\psi$ mapping a neighborhood of $f(\mathbf{x})$ in $W$ into $\mathbb{R}^n$, also with differentiable inverse, such that the composite map*

$$\psi \circ f \circ \phi^{-1}: \mathbb{R}^m \to \mathbb{R}^n$$

*is a linear map with matrix* (10.9).

In short, this theorem says that, for differentiable maps of constant rank, the main theorem of row reduction holds: we can 'make changes of variables', i.e., find maps $\phi$ and $\psi$, such that

$$(\psi \circ f \circ \phi^{-1})\left(\begin{pmatrix} x_1 \\ \vdots \\ x_m \end{pmatrix}\right) = \begin{pmatrix} x_1 \\ \vdots \\ x_r \\ 0 \\ \vdots \\ 0 \end{pmatrix}.$$

* This section can be omitted on first reading.

*Proof.* By making a preliminary change of variables consisting of a translation in $V$, we may assume that $\mathbf{x} = \mathbf{0}$ and, by another translation in $W$, we may assume that $f(\mathbf{x}) = \mathbf{0}$. Thus, in order to simplify the notation, we may assume that $f(\mathbf{0}) = \mathbf{0}$, and that we are interested in $f$ near $\mathbf{0}$. By row reduction, we can bring the *linear* map $\mathrm{d}f_0$ to the form (10.9), that is, we can find invertible linear maps $R\colon W \to \mathbb{R}^n$ and $S\colon V \to \mathbb{R}^n$ such that

$$\mathrm{d}(RfS^{-1})_0 = R\mathrm{d}f_0 S^{-1}$$

has the form (10.9). Thus we can write

$$RfS^{-1} = \begin{pmatrix} f_1 \\ \vdots \\ f_r \\ f_{r+1} \\ \vdots \\ f_n \end{pmatrix}$$

where the matrix

$$\left(\frac{\partial f_i}{\partial x_j}\right)_{\substack{i=1,\ldots,r \\ j=1,\ldots,r}}$$

is the identity matrix at $\mathbf{0}$. Hence it is invertible in some neighborhood of $\mathbf{0}$. Consider the map $g$ defined near $\mathbf{0}$ in $\mathbb{R}^m$ by

$$g\left(\begin{pmatrix} x_1 \\ \vdots \\ x_m \end{pmatrix}\right) = \begin{pmatrix} f_1 \\ \vdots \\ f_r \\ x_{r+1} \\ \vdots \\ x_m \end{pmatrix}.$$

That is, the first $r$ components of $g$ are given by $f_1, \ldots, f_r$, while the last $m - r$ components are just the last $m - r$ coordinates. Notice that the matrix

$$\left(\frac{\partial g_i}{\partial x_j}\right) = \left(\begin{array}{c|c} \dfrac{\partial f_i}{\partial x_j} & \cdots \\ \hline 0 & \begin{matrix} 1 & & 0 \\ & \ddots & \\ 0 & & 1 \end{matrix} \end{array}\right) \begin{array}{l} r \text{ rows} \\ \\ m-r \text{ rows} \end{array}$$

$$\begin{array}{cc} r & m-r \\ \text{columns} & \text{columns} \end{array}$$

is invertible at $\mathbf{0}$, and hence is invertible near $\mathbf{0}$. We now apply the inverse function theorem of section 6.3 to the map $g$. Review the proof there to see that it was valid for arbitrary finite-dimensional vector spaces. (Observe also that the implicit function theorem and its proof are valid, where $x$ and $y$ are taken as vector variables and the condition there on $\partial g/\partial y$ (which is now a matrix) is that it be invertible.) Getting back to our current problem, we can find an inverse for $g$. Let

$$h = (RfS^{-1}) \circ g^{-1}.$$

$g : \mathbb{R}^m \to \mathbb{R}^m$
$h = (RfS^{-1}) \circ g^{-1}$

Then

$$(RfS^{-1})\left(\begin{pmatrix} x_1 \\ \vdots \\ x_m \end{pmatrix}\right) = \begin{pmatrix} f_1 \\ \vdots \\ f_n \end{pmatrix}$$

and

$$g\left(\begin{pmatrix} x_1 \\ \vdots \\ x_m \end{pmatrix}\right) = \begin{pmatrix} f_1 \\ \vdots \\ f_r \\ x_{r+1} \\ \vdots \\ x_m \end{pmatrix}$$

so

$$h\left(\begin{pmatrix} y_1 \\ \vdots \\ y_m \end{pmatrix}\right) = \begin{pmatrix} y_1 \\ \vdots \\ y_r \\ g_{r+1}(y) \\ \vdots \\ g_n(y) \end{pmatrix}$$

for suitable functions $g_{r+1}, \ldots, g_n$ defined on $\mathbb{R}^m$ near $\mathbf{0}$. Now the chain rule says that

$$\mathrm{d}h_{\mathbf{q}} = \mathrm{d}(Rf_{\mathbf{p}}S^{-1}) \circ \mathrm{d}g_{\mathbf{q}}^{-1} \text{ where } \quad \mathbf{q} = g(S\mathbf{p}).$$

Hence

$$\operatorname{rank} \mathrm{d}h_{\mathbf{q}} \equiv r.$$

Looking at the matrix

$$\left(\frac{\partial h_i}{\partial y_j}\right) = \left(\begin{array}{c|c} I_r & 0 \\ \hline \vdots & \dfrac{\partial g_i}{\partial y_j} \end{array}\right) \begin{array}{l} r \text{ rows} \\ n-r \text{ rows} \end{array}$$

$$\begin{array}{cc} r & n-r \\ \text{columns} & \text{columns} \end{array}$$

the only way that this can happen is if all the partial derivatives occuring in the lower right-hand corner vanish identically. Thus the last $n-r$ $h$s must depend only on the *first* $r$ coordinates, $y$. That is,

$$g_{r+1} = g_{r+1}(y_1, \ldots, y_r, 0, \ldots, 0), \text{ etc.}$$

Now introduce the transformation $H$ on $\mathbb{R}^n$ given by

$$H\left(\begin{pmatrix} z_1 \\ \vdots \\ z_r \\ z_{r+1} \\ \vdots \\ z_n \end{pmatrix}\right) = \begin{pmatrix} z_1 \\ \vdots \\ z_r \\ z_{r+1} - g_{r+1}(z_1, \ldots, z_r, 0, \ldots, 0) \\ \vdots \\ z_n - g_n(z_1, \ldots, z_r, 0, \ldots, 0) \end{pmatrix}.$$

This is a smooth map defined near $z=0$ in $\mathbb{R}^n$. It is clearly invertible. Indeed, its inverse is given by

$$H^{-1}\left(\begin{pmatrix} w_1 \\ \vdots \\ w_n \end{pmatrix}\right) = \begin{pmatrix} w_1 \\ \vdots \\ w_r \\ w_{r+1} + g_{r+1}(w_1,\ldots,w_r,0,\ldots,0) \\ \vdots \\ w_n + g_n(w_1,\ldots,w_r,0,\ldots,0) \end{pmatrix}.$$

Then

$$(H\circ h)\left(\begin{pmatrix} y_1 \\ \vdots \\ y_m \end{pmatrix}\right) = \begin{pmatrix} y_1 \\ \vdots \\ y_r \\ 0 \\ \vdots \\ 0 \end{pmatrix}.$$

Substituting the definition $h=(RfS^{-1})\circ g^{-1}$ into $H\circ h$ gives

$$H\circ h=(HR)\circ f\circ(S^{-1}g^{-1}).$$

Defining

$$\psi = H\circ R$$

and

$$\phi = g\circ S$$

shows that $\psi\circ f\circ\phi^{-1}$ has the desired form:

$$\psi\circ f\circ\phi^{-1}\left(\begin{pmatrix} y_1 \\ \vdots \\ y_m \end{pmatrix}\right) = \begin{pmatrix} y_1 \\ \vdots \\ y_r \\ 0 \\ \vdots \\ 0 \end{pmatrix}$$

Q.E.D.

The most important application of this theorem is to the case where $r=n$, the dimension of the image. If $\mathrm{d}f_{\mathbf{p}}$ is continuous and has rank $n$ at some point $\mathbf{p}$, then we claim that it has rank $n$ at all points sufficiently close to $\mathbf{p}$. Indeed, by row reduction, we can find $R$ and $S$ such that $R\mathrm{d}f_{\mathbf{p}}S^{-1}$ has the form (10.9) with $r=n$. Now, for $\mathbf{q}$ close to $\mathbf{p}$, the upper left-hand block of $R\mathrm{d}f_{\mathbf{q}}S^{-1}$ will be close to the identity matrix. Hence the dimension of the image of $\mathrm{d}f_{\mathbf{q}}$ is at least $n$. Since the dimension of this image cannot exceed $\dim W=n$, we conclude that $\operatorname{rank}\mathrm{d}f_{\mathbf{q}}\equiv n$ for all $\mathbf{q}$ in some neighborhood of $\mathbf{p}$. From the constant rank theorem, we conclude:

**The solution set theorem.** *Suppose that $f\colon O\to W$ is a continuously differentiable map and that $\mathrm{d}f_{\mathbf{p}}$ is surjective at $\mathbf{p}$, that is, that $\operatorname{rank}\mathrm{d}f_{\mathbf{p}}=\dim W$. Then we can find differentiable maps $\phi$ mapping a neighborhood of $\mathbf{p}$ into $\mathbb{R}^m$ and $\psi$ mapping a* *neighborhood of $f(\mathbf{p})$ into $\mathbb{R}^n$, both with continuously differentiable inverses such that $\phi(\mathbf{p}) = \mathbf{0}$ in $\mathbb{R}^m$ and $\psi(f(\mathbf{q})) = \mathbf{0}$ in $\mathbb{R}^n$ and*

There's some confusion here: for rank $df_p$ = dim W = n we require m ≥ n. So what we have from the theorem in the m's and n's have clearly gotten switched about in mid-stream.

$$\psi f \phi^{-1}\left(\begin{pmatrix} x_1 \\ \vdots \\ x_m \end{pmatrix}\right) = \begin{pmatrix} x_1 \\ \vdots \\ x_m \\ 0 \\ \vdots \\ 0 \end{pmatrix}.$$

$\psi f \phi^{-1}\begin{pmatrix} y_1 \\ \vdots \\ y_m \end{pmatrix} = \begin{pmatrix} x_1 \\ \vdots \\ x_n \end{pmatrix}$

*In particular, a point $\mathbf{x}$ is the solution to the equation*

$$f(\mathbf{x}) = f(\mathbf{p}) \quad \mathbf{x} \text{ near } \mathbf{p}$$

*if and only if*

$$\mathbf{x} = \phi^{-1}\left(\begin{pmatrix} 0 \\ \vdots \\ 0 \\ y_{m+1} \\ \vdots \\ y_m \end{pmatrix}\right)$$

*with $y_{m+1}, y_{m+2}, \ldots, y_n$ all near 0.*

Introducing a little terminology will make the statement of the solution set theorem more succinct. Let $H_{m-n}$ denote the subspace of $\mathbb{R}^m$ determined by the equations

$$x_1 = 0, \ldots, x_n = 0.$$

(Here $n$ is assumed to be $\leqslant m$.) A subset $M$ of an $m$-dimensional vector space $V$ is called a *submanifold* of *codimension* $m - n$ if it has the following property: about each point $\mathbf{x} \in M$ we can find a neighborhood $O$ in $V$ and a differentiable map $\phi: O \to \mathbb{R}^m$ with $\phi(\mathbf{x}) = \mathbf{0}$, such that $\phi$ maps $O$ in a one-to-one fashion into a neighborhood, $U$, of $\mathbf{0}$, and $\phi^{-1}$ is differentiable, and such that

$$\phi(M \cap O) = U \cap H_{m-n}.$$

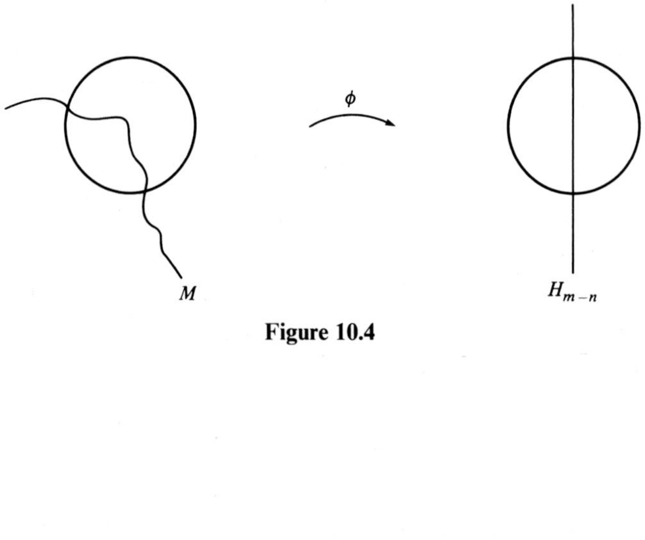


**Figure 10.4**

In other words, the condition on $M$ says that, near each of the points, we can find a smooth distortion of the full space (the distortion being given by the map $\phi$) so that $M$ gets flattened out and looks like a piece of hyperplane. For example, the circle $x^2 + y^2 = 1$ is a submanifold of codimension 1 in $\mathbb{R}^2$, because we can introduce the map

$$\phi(x, y) = \begin{pmatrix} r^2 - 1 \\ \arctan(y/x) \end{pmatrix}$$

at all points with $x \neq 0$ (followed by an appropriate shift in the vertical direction to center the image at the origin). At the points $x = 0$, we can use $\arctan(x/y)$. The perimeter of a square is *not* a submanifold, because there is no smooth way of straightening out the corners.

Let $f: O \to W$ be a continuously differentiable map. A point $\mathbf{p} \in O$ is called a *regular point* of $f$ if $\mathrm{d}f_{\mathbf{p}}$ is surjective; in other words, if rank $\mathrm{d}f_{\mathbf{p}} = \dim W$. A point which is not a regular point is called a critical point. If $W = \mathbb{R}$, then $\mathbf{p}$ is a critical point if $\mathrm{d}f_{\mathbf{p}} = 0$. This agrees with our earlier notation.

A point $\mathbf{q} \in W$ is called a *regular value* if all points $\mathbf{p}$ in $f^{-1}(\mathbf{q})$ are regular points. Then we can formulate our theorem as

| |
|---|
| *If* $\mathbf{q}$ *is a regular value of* $f$, *then* $f^{-1}(\mathbf{q})$ *is a submanifold.* |

Here are some examples.

(a) Take $n = 1$ and $f: \mathbb{R}^m \to \mathbb{R}^1 = \mathbb{R}$ given by $f\left(\begin{pmatrix} x_1 \\ \vdots \\ x_m \end{pmatrix}\right) = x_1^2 + \cdots + x_m^2$. Then $\mathrm{d}f_{\mathbf{p}} \neq 0$ if $\mathbf{p} \neq \mathbf{0}$. Thus any non-zero value of $f$ is a regular value. For $c > 0$, $f^{-1}(c)$ is the sphere of radius $\sqrt{c}$. Thus spheres are submanifolds.

(b) Let $M(k)$ denote the vector space of all $k \times k$ matrices, so $m = \dim V = k^2$. Let $W = S(R)$ denote the vector space of all $k \times k$ symmetric matrices – those matrices $D$ which satisfy $D^{\mathrm{T}} = D$. Then $n = \dim W = \frac{1}{2}k(k+1)$. Consider the map

$$f: V \to W \quad f(A) = AA^{\mathrm{T}}.$$

Then

$$\mathrm{d}f_A(B) = BA^{\mathrm{T}} + AB^{\mathrm{T}}$$

as you can easily check. We claim that the identity matrix, $I$, is a regular value of $f$. We must show that $\mathrm{d}f_A$ is surjective if $AA^{\mathrm{T}} = I$. That is, we must be able to solve

$$BA^{\mathrm{T}} + AB^{\mathrm{T}} = C$$

for $B$ given any symmetric matrix $C$. Indeed, take $B = \frac{1}{2}CA$. Then

$$\begin{aligned} BA^{\mathrm{T}} + AB^{\mathrm{T}} &= \tfrac{1}{2}CAA^{\mathrm{T}} + \tfrac{1}{2}AA^{\mathrm{T}}C^{\mathrm{T}} \\ &= \tfrac{1}{2}C + \tfrac{1}{2}C^{\mathrm{T}} && \text{since } AA^{\mathrm{T}} = A^{\mathrm{T}}A = I \\ &= C && \text{since } C = C^{\mathrm{T}}. \end{aligned}$$

Thus the set of all orthogonal matrices – those satisfying $AA^{\mathrm{T}} = I$ – is a submanifold of the space of all matrices.

## 10.10. The adjoint transformation

When studying transformations from one affine plane to another, we made use of the concept of *pullback* of a function. Recall that a transformation $\phi$ from an affine plane $A$ to another affine plane $B$ gave rise, in a natural way, to a *linear* transformation from the functions on $B$ to the functions on $A$, as depicted in figure 10.5.

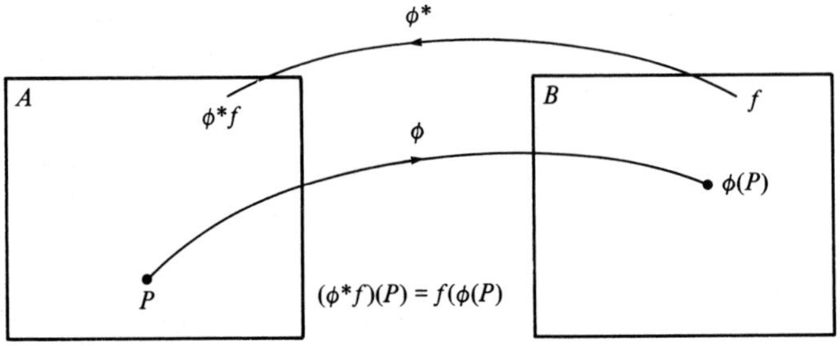


**Figure 10.5**

The pullback of a function $f$ on plane $B$, denoted by $\phi^* f$, is a function on plane $A$ defined by

$$(\phi^* f)(P) = f(\phi(P)).$$

This concept of pullback can be extended immediately to the case of a transformation from any vector space $V$ to any other vector space $W$. We will take special interest in the pullback of *linear* functions on $W$ (i.e., of elements of the dual space $W^*$) which arises as a consequence of a *linear* transformation $A$ from $V$ to $W$. In this case the pullback transformation from $W^*$ to $V^*$ is called the *adjoint* of $A$. We denote the adjoint by $A^*$, and define its action on an element $\boldsymbol{\beta} \in W^*$ by

$$(A^*\boldsymbol{\beta})[\mathbf{v}] = \boldsymbol{\beta}[A\mathbf{v}].$$

The proof that $A^*$, thus defined, is linear in $\boldsymbol{\beta}$ is the same as the proof that pullback in general is linear: if $\boldsymbol{\beta}^1$ and $\boldsymbol{\beta}^2$ are elements of $W^*$, then

$$\begin{aligned}(A^*(c_1\boldsymbol{\beta}^1 + c_2\boldsymbol{\beta}^2))[\mathbf{v}] &= (c_1\boldsymbol{\beta}^1 + c_2\boldsymbol{\beta}^2)[A\mathbf{v}] \\ &= c_1\boldsymbol{\beta}^1[A\mathbf{v}] + c_2\boldsymbol{\beta}^2[A\mathbf{v}] \\ &= c_1 A^*\boldsymbol{\beta}^1[\mathbf{v}] + c_2 A^*\boldsymbol{\beta}^2[\mathbf{v}]\end{aligned}$$

so that

$$A^*(c_1\boldsymbol{\beta}^1 + c_2\boldsymbol{\beta}^2) = c_1 A^*\boldsymbol{\beta}^1 + c_2 A^*\boldsymbol{\beta}^2$$

and $A^*$ is linear. Notice that the linearity of $\boldsymbol{\beta}$ and $A$ imply that $A^*\boldsymbol{\beta}$ is a linear function of $\mathbf{v}$, so that $A^*\boldsymbol{\beta}$ does indeed lie in $V^*$.

It is crucial to observe that the adjoint $A^*$ acts 'in the opposite direction' to $A$. Note carefully: if $A$ transforms a vector $\mathbf{v}\in V$ into a vector $A\mathbf{v}\in W$, the adjoint $A^*$ transforms a vector $\boldsymbol{\beta}\in W^*$ into a vector $A^*\boldsymbol{\beta}\in V^*$. This can be summarized in the diagram

$$V^* \xleftarrow{A^*} W^*$$

$$V \xrightarrow{A} W$$

As an example of the adjoint transformation, let $V$ be the three-dimensional space of even polynomials of degree $\leqslant 4$, and let $W$ be the two-dimensional space of odd polynomials of degree $\leqslant 3$. Then the operation of differentiation defines a linear transformation D from $V$ to $W$:

$$\mathrm{D}: f(t) \mapsto f'(t).$$

A typical element of $W^*$ is

$$\boldsymbol{\beta}: \mathbf{g} \mapsto g(1).$$

For example, $\boldsymbol{\beta}(t+2t^3) = 1+2 = 3$. To calculate $\mathrm{D}^*\boldsymbol{\beta}$ we use the definition

$$\mathrm{D}^*\boldsymbol{\beta}[\mathbf{f}] = \boldsymbol{\beta}[\mathrm{D}\mathbf{f}].$$

In the case at hand, $\mathrm{D}\mathbf{f} = f'(t)$ and $\boldsymbol{\beta}[\mathrm{D}\mathbf{f}] = f'(1)$. We conclude that $\mathrm{D}^*\boldsymbol{\beta}$ is the linear function on $V$ (element of $V^*$):

$$\mathrm{D}^*\boldsymbol{\beta}: f(t) \to f'(1).$$

Another element of $W^*$ is

$$\boldsymbol{\alpha}: \mathbf{g} \mapsto \int_0^1 g(t)\,\mathrm{d}t.$$

Now

$$\mathrm{D}^*\boldsymbol{\alpha}[\mathbf{f}] = \boldsymbol{\alpha}[\mathrm{D}\mathbf{f}] = \int_0^1 f'(t)\,\mathrm{d}t = f(1) - f(0)$$

so $\mathrm{D}^*\boldsymbol{\alpha}$ is the linear function

$$\mathrm{D}^*\boldsymbol{\alpha}: f(t) \mapsto f(1) - f(0).$$

When we introduce bases for $V$, $W$, $V^*$, and $W^*$, the description of the adjoint becomes particularly simple. Suppose that $V$ is $m$-dimensional, with basis $\{\mathbf{v}_1, \mathbf{v}_2, \ldots, \mathbf{v}_m\}$, and that we have introduced a dual basis $\{\boldsymbol{\alpha}^1, \boldsymbol{\alpha}^2, \ldots, \boldsymbol{\alpha}^m\}$ in the dual space $V^*$. Similarly, let $\{\mathbf{w}_1, \mathbf{w}_2, \ldots, \mathbf{w}_n\}$ and $\{\boldsymbol{\beta}^1, \boldsymbol{\beta}^2, \ldots, \boldsymbol{\beta}^n\}$ be dual bases in $W$ and $W^*$ respectively. If $A$ is a linear transformation from $V$ to $W$, then

$$A\mathbf{v}_i = \sum_{j=1}^{n} a_{ji}\mathbf{w}_j,$$

where the quantities $\{a_{1i}, a_{2i}, \ldots, a_{ni}\}$ form the $i$th *column* of the matrix which represents $A$. Now we can calculate how $A^*$ acts on a basis element of $W^*$:

$$(A^*\boldsymbol{\beta}^k)[\mathbf{v}_i] = \boldsymbol{\beta}^k[A\mathbf{v}_i] = \boldsymbol{\beta}^k\left[\sum_{j=1}^{n} a_{ji}\mathbf{w}_j\right] = \sum_{j=1}^{n} a_{ji}\boldsymbol{\beta}^k[\mathbf{w}_j].$$

But since the basis $\{\boldsymbol{\beta}^1, \ldots, \boldsymbol{\beta}^n\}$ is dual to $\{\mathbf{w}_1, \ldots, \mathbf{w}_n\}$, we have

$$\boldsymbol{\beta}^k[\mathbf{w}_j] = \begin{cases} 1 & j = k, \\ 0 & j \neq k \end{cases}$$

so that

$$(A^*\boldsymbol{\beta}^k)[\mathbf{v}_i] = a_{ki}.$$

Thus we may express $A^*\boldsymbol{\beta}^k$ in terms of the basis elements $\{\boldsymbol{\alpha}^1, \ldots, \boldsymbol{\alpha}^m\}$ of $V^*$ as

$$A^*\boldsymbol{\beta}^k = \sum_{l=1}^{n} a_{kl}\boldsymbol{\alpha}^l.$$

The quantities $\{a_{kl}\} = \{a_{k1}, a_{k2}, \ldots, a_{km}\}$ form the $k$th *column* of the matrix which represents $A^*$, but they are also the $k$th row of the matrix which represents $A$. This means that the matrix which represents $A^*$ is just the *transpose* of the matrix which represents $A$. Thus, for example, if $V$ is three-dimensional, $W$ two-dimensional and $A: V \to W$ is represented by

$$A = \begin{pmatrix} a_{11} & a_{12} & a_{13} \\ a_{21} & a_{22} & a_{23} \end{pmatrix}$$

the adjoint transformation $A^*: W^* \to V^*$ is represented by

$$A^* = \begin{pmatrix} a_{11} & a_{21} \\ a_{12} & a_{22} \\ a_{13} & a_{23} \end{pmatrix}.$$

An easy way to see that the matrices representing $A$ and $A^*$ are transposes of one another is to recall that elements of the dual space may be represented as *row* vectors. In the present example, an element of $W^*$ may be thought of as a two-component *row* vector:

$$\boldsymbol{\beta} = (\lambda_1, \lambda_2)$$

while an element of $V$ is a three-component column vector:

$$\mathbf{v} = \begin{pmatrix} x_1 \\ x_2 \\ x_3 \end{pmatrix}.$$

Now $(A^*\boldsymbol{\beta})[\mathbf{v}] = \boldsymbol{\beta}[A\mathbf{v}]$ is written as

$$(\lambda_1, \lambda_2)\begin{pmatrix} a_{11} & a_{12} & a_{13} \\ a_{21} & a_{22} & a_{23} \end{pmatrix}\begin{pmatrix} v_1 \\ v_2 \\ v_3 \end{pmatrix}.$$

It is most natural to think of the matrix $A$ as acting first on the column vector to its right to yield $A\mathbf{v}$, which is then acted upon by $\boldsymbol{\beta}$. Alternatively, though, we can think of it as acting first on the row vector to its *left*:

$$(\lambda_1, \lambda_2)\begin{pmatrix} a_{11} & a_{12} & a_{13} \\ a_{21} & a_{22} & a_{23} \end{pmatrix} = (\mu_1, \mu_2, \mu_3)$$

where $(\mu_1, \mu_2, \mu_3)$ represents the vector $(A^*\boldsymbol{\beta}) \in V^*$, which then acts on $\mathbf{v} \in V$. Thus, if we reverse the usual conventions of matrix multiplication, letting a matrix act on a *row* vector to its *left*, *the same* matrix represents both $A$ and $A^*$. If we want to represent $A^*$ more conventionally, by a matrix which acts on a column vector to its right, we must write

$$\begin{pmatrix} \mu_1 \\ \mu_2 \\ \mu_3 \end{pmatrix} = \begin{pmatrix} a_{11} & a_{21} \\ a_{12} & a_{22} \\ a_{13} & a_{23} \end{pmatrix} \begin{pmatrix} \lambda_1 \\ \lambda_2 \end{pmatrix}.$$

Now, of course, the matrix representing $A^*$ is the transpose of the one which represents $A$.

We turn now to an investigation of the kernel and image of the adjoint $A^*$. If $\boldsymbol{\beta} \in W^*$ is in the kernel of $A^*$, then $A^*\boldsymbol{\beta}[\mathbf{v}] = \mathbf{0}$ for all $\mathbf{v} \in V$, so that $\boldsymbol{\beta}[A\mathbf{v}] = \mathbf{0}$. This means that $\boldsymbol{\beta}$ annihilates all vectors of the form $A\mathbf{v}$. We conclude that *the kernel of $A^*$ annihilates the image of $A$.*

Now suppose that $\boldsymbol{\alpha} \in V^*$ is in the image of $A^*$, so that

$$\boldsymbol{\alpha} = A^*\boldsymbol{\beta} \quad \text{for some} \quad \boldsymbol{\beta} \in W^*.$$

Suppose that $\mathbf{v}$ is an element of $\ker A$. Then

$$\boldsymbol{\alpha}[\mathbf{v}] = A^*\boldsymbol{\beta}[\mathbf{v}] = \boldsymbol{\beta}[A\mathbf{v}] = \mathbf{0}.$$

We conclude that *the image of $A^*$ annihilates the kernel of $A$.*

Putting these results together with the general results about dual spaces and quotient spaces proved at the end of section 10.5, we can construct two diagrams which summarize our general picture of vector spaces and linear transformations. Looking at subspaces of $V$ and $V^*$, we have

$$\begin{array}{c} V^*/\text{im}\, A^* \leftarrow V^* \leftarrow \text{im}\, A^* \\ \ker A \rightarrow V \rightarrow V/\ker A. \end{array}$$

This diagram reflects the fact that the quotient space $V^*/\text{im}\, A^*$ may be identified with the dual of $\ker A$, while the image of $A^*$ is dual to $V/\ker A$.

Looking at subspaces of $W$ and $W^*$, we have the diagram

$$\begin{array}{c} W^*/\ker A^* \leftarrow W^* \leftarrow \ker A^* \\ \text{im}\, A \rightarrow W \rightarrow W/\text{im}\, A. \end{array}$$

Here $W^*/\ker A^*$ may be identified with the dual of $\text{im}\, A$, while $\ker A^*$ is dual to $W/\text{im}\, A$. Numerous examples of these relationships will appear as we study electric network theory.

---

## Summary

### A Vector spaces

You should know the axioms for a vector space and be able to apply them.

Given a basis for a vector space, you should be able to recognize or construct a dual basis for its dual space.

Given a subspace $U$ of a vector space $V$, you should be able to construct and use a basis for the annihilator space $U^{\perp}$ and the quotient space $V/U$.

B Linear transformations

You should be able to write down the matrix that represents a linear transformation $A: V \to W$ between given bases.

You should be able to state, prove and apply the rank–nullity theorem.

Given the matrix of a linear transformation $A$, you should know how to use row reduction to determine bases for $\ker A$ and $\operatorname{im} A$ and to find the general solution to $A\mathbf{v} = \mathbf{w}$.

You should know the definition of the adjoint $A^*$ of a linear transformation $A$ and be able to state, prove, and apply relations between the kernel and image of $A$ and the kernel and image of $A^*$.

---

## Exercises

10.1. Consider the five-dimensional vector space $V$ of polynomials $f(t)$ of degree $\leqslant 4$. Determine whether each of the following is a subspace. If not, explain why. If so, find a basis.
(a) Elements of $V$ satisfying $f(t) = f(-t)$.
(b) Elements satisfying $f(0) = 1$.
(c) Elements satisfying $f(1) = f(-1)$.
(d) Elements satisfying $\int_{-1}^{1} t f(t)\,dt = 0$.

10.2.(a) Find a basis for the subspace of $\mathbb{R}^3$ defined by $\boldsymbol{\alpha}[\mathbf{v}] = 0$, where

$$\boldsymbol{\alpha} = (2, -3, 1)$$

(b) Find a basis for the subspace of $\mathbb{R}^3$ defined by $\boldsymbol{\alpha}[\mathbf{v}] = 0$ and $\boldsymbol{\beta}[\mathbf{v}] = 0$, where $\boldsymbol{\alpha} = (2, -3, 1)$ and $\boldsymbol{\beta} = (2, 1, -1)$.
(c) Find a basis for the annihilator space of the subspace $W \in \mathbb{R}^3$ spanned by

$$\begin{pmatrix} 2 \\ -3 \\ 1 \end{pmatrix} \text{ and } \begin{pmatrix} 2 \\ 1 \\ -1 \end{pmatrix}.$$

10.3.(a) Show that the set of functions $f(t)$ satisfying $f'' + 5f' + 6f = 0$ is a vector space $V$.
(b) Three elements of the space $V^*$ dual to this space are

$$\boldsymbol{\alpha}^1 = f \mapsto f(0),$$
$$\boldsymbol{\alpha}^2 = f \mapsto f'(0),$$
$$\boldsymbol{\alpha}^3 = f \mapsto \int_0^{\infty} e^t f(t)\,dt.$$

Find a relation among $\boldsymbol{\alpha}^1$, $\boldsymbol{\alpha}^2$, $\boldsymbol{\alpha}^3$ which shows that they are linearly dependent.
(c) As a basis for $V$, choose

$$\mathbf{v}_1 = e^{-2t} + e^{-3t},$$
$$\mathbf{v}_2 = 2e^{-2t} - e^{-3t}.$$

Express the dual basis elements $\boldsymbol{\beta}^1$ and $\boldsymbol{\beta}^2$ in terms of $\boldsymbol{\alpha}^1$ and $\boldsymbol{\alpha}^2$ above.

10.4.(a) Show that, if $S$ and $T$ are subspaces of a vector space $V$, then their intersection $S \cap T$ (the set of elements common to $S$ and $T$) is also a subspace of $V$.

(b) Show that $S + T$ (the set of vectors that are linear combinations of vectors in $S$ and $T$) is a subspace of $V$.

(c) Show that $\dim(S + T) = \dim(S) + \dim(T) - \dim(S \cap T)$.
Hint: Start with a basis for $S \cap T$ and extend it to a basis for $S + T$.

(d) Suppose $V$ is $\mathbb{R}^4$, $S$ is spanned by $\begin{pmatrix} 1 \\ 0 \\ 1 \\ 0 \end{pmatrix}$ and $\begin{pmatrix} 2 \\ 1 \\ 2 \\ 1 \end{pmatrix}$, and $T$ is spanned by $\begin{pmatrix} 1 \\ 2 \\ 1 \\ 1 \end{pmatrix}$ and $\begin{pmatrix} 0 \\ 1 \\ 0 \\ 0 \end{pmatrix}$. Construct a basis for $S \cap T$, for $S + T$, and for the annihilator space of $(S + T)$.

10.5. Let $W$ be the subspace of $\mathbb{R}^3$ spanned by $\begin{pmatrix} 1 \\ 2 \\ 4 \end{pmatrix}$.

(a) Show that $\mathbf{e}_1 = \overline{\begin{pmatrix} 1 \\ 1 \\ 0 \end{pmatrix}}$ and $\mathbf{e}_2 = \overline{\begin{pmatrix} 1 \\ 1 \\ 1 \end{pmatrix}}$ form a basis for the quotient space $\mathbb{R}^3/W$. Express $\overline{\begin{pmatrix} 1 \\ 0 \\ 0 \end{pmatrix}}$, $\overline{\begin{pmatrix} 0 \\ 1 \\ 0 \end{pmatrix}}$ and $\overline{\begin{pmatrix} 0 \\ 0 \\ 1 \end{pmatrix}}$ as linear combinations of these basis elements.

(b) Show that $\boldsymbol{\alpha} = (2, -1, 0)$ and $\boldsymbol{\beta} = (4, 0, -1)$, both elements of the dual of $\mathbb{R}^3$, are a basis for the annihilator space $W^\perp$. In terms of $\boldsymbol{\alpha}$ and $\boldsymbol{\beta}$, construct a basis $\{\boldsymbol{\varepsilon}^1, \boldsymbol{\varepsilon}^2\}$ for $W^\perp$ which is dual to the basis $\{\mathbf{e}_1, \mathbf{e}_2\}$ for $\mathbb{R}^3/W$.

10.6.(a) Let $W$ be a subspace of a vector space $V$; let $U$ be a subspace of $W$. Prove that $W/U$ is a subspace of $V/U$, and show that there is a natural identification of $\dfrac{V/U}{W/U}$ with $V/W$.

(b) For the case where $V$ is $\mathbb{R}^3$, $W$ is the plane $x + y + z = 0$, and $U$ is the line spanned by $\begin{pmatrix} 1 \\ -2 \\ 1 \end{pmatrix}$, construct explicit bases for these spaces.

(c) Figure out what is happening in the dual space, that is, construct subspaces of $V^*$ which are the dual spaces or annihilator spaces for the various spaces in (a). Do this first in general, then for the explicit case described in (b).

10.7. Consider the space $V$ of functions on $[1,2]$ of the form $f(t) = At + B/t$. Choose a basis $\mathbf{v}_1 = t$, $\mathbf{v}_2 = 1/t$.

Two elements of the dual space $V^*$ are

$$\boldsymbol{\beta}^1: f(t) \to f(1),$$
$$\boldsymbol{\beta}^2: f(t) \to f(2).$$

Express the basis elements $\{\boldsymbol{\alpha}^1, \boldsymbol{\alpha}^2\}$ which are *dual* to $\{\mathbf{v}_1, \mathbf{v}_2\}$ as linear combinations of $\boldsymbol{\beta}^1$ and $\boldsymbol{\beta}^2$. Before working exercises 10.8–10.12 reread section 4.2 on the Gram–Schmidt process.

10.8.(a) Find the dimension of the space of trigonometric polynomials spanned by $\{1, \sin^2 x, \cos^2 x, \sin^4 x, \sin^2 x \cos^2 x, \cos^4 x\}$.

(b) Define a scalar product on this space by

$$(f, g) = \frac{1}{\pi}\int_0^{\pi} f(x)g(x)\mathrm{d}x.$$

With respect to this scalar product, construct an orthonormal basis.

10.9(a) Consider the vector space of odd polynomials of degree $\leqslant 3$ with basis $\{t, t^3\}$ and scalar product $\int_0^1 f(t)g(t)\mathrm{d}t$. Construct an orthogonal basis (you need not normalize).

(b) Given the linear operator $f: V \to V$ defined by $f(p(t)) = tp'(t)$, construct the matrix $A$ which represents $f$ with respect to the $\{t, t^3\}$ basis and the matrix $\hat{A}$ which represents $f$ with respect to the orthogonal basis which you constructed.

(c) Three elements of $V^*$, the dual space to $V$, may be defined as follows: $\boldsymbol{\alpha}^1[p] = p(1)$, $\boldsymbol{\alpha}^2[p] = p^1(0)$, $\boldsymbol{\alpha}^3[p] = \int_0^1 tp(t)\mathrm{d}t$.
Find a relationship among $\boldsymbol{\alpha}^1$, $\boldsymbol{\alpha}^2$, $\boldsymbol{\alpha}^3$ which shows explicitly that they are dependent.

10.10. Let $V$ be the two-dimensional vector space of functions that are linear combinations of $f_1 = 1$ and $f_2 = \cos^2 t$. Define a scalar product on this space by $(f, g) = (2/\pi)\int_0^{\pi/2} f(t)g(t)\mathrm{d}t$.

(a) Construct an orthonormal basis for $V$.
Note:

$$\frac{2}{\pi}\int_0^{\pi/2} \cos^2 t\,\mathrm{d}t = \frac{1}{2};\ \frac{2}{\pi}\int_0^{\pi/2} \cos^4 t\,\mathrm{d}t = 3/8$$

(b) Three elements of the dual space $V^*$ are:

$$\alpha^1: f \to f(\pi/2),\quad \alpha^2: f \to \int_0^{\pi/2} f(t)\cos t\,\mathrm{d}t,\quad \alpha^3: f \to \int_0^{\pi/2} f(t)\sin t\,\mathrm{d}t.$$

Show *explicitly* that these are linearly dependent.

10.11. Consider the three-dimensional vector space $V$ whose elements are the polynomials of degree $\leqslant 2$ multiplied by $\mathrm{e}^{-2t}$. A basis for this space is

$$\mathbf{v}_1 = \mathrm{e}^{-2t},\quad \mathbf{v}_2 = t\mathrm{e}^{-2t},\quad \mathbf{v}_3 = t^2\mathrm{e}^{-2t}.$$

(a) With respect to this basis, write down the matrix D which represents the operation of differentiation; i.e.,

$$\mathrm{D}f(t) = f'(t).$$

(b) Construct the matrix $\mathrm{D}^2 + 4\mathrm{D} + 4$. (There are a lot of zeros in this matrix!)

(c) The *general* solution to the equation $\ddot{x} + 4\dot{x} + 4x = \mathrm{e}^{-2t}$ lies in the vector space $V$. Find it by using the matrix that you constructed in (b).

10.12. Consider the vector space $V$ of solutions to the differential equation

$$\ddot{x} + 3\dot{x} + 2x = 0.$$

Define a scalar product on this space by

$$(\mathbf{f}, \mathbf{g}) = \int_0^\infty \mathbf{f}(t)\mathbf{g}(t)\,\mathrm{d}t$$

(Remember that $\int_0^\infty \mathrm{e}^{-\alpha t} = 1/\alpha$.)

(a) Take $\mathbf{f}_1 = \mathrm{e}^{-2t}$ as the first basis vector for $V$. Construct a second vector $\mathbf{f}_2$ that is orthogonal to $\mathbf{f}_1$.

(b) With respect to the basis $\{\mathbf{f}_1, \mathbf{f}_2\}$, construct the matrix D that represents the operation of differentiation with respect to $t$. Verify that $\mathrm{D}^2 + 3\mathrm{D} + 2I = 0$.

Three elements of the dual space $V^*$ are the following:

$$\boldsymbol{\alpha}_1[\mathbf{f}] = \mathbf{f}(0),$$
$$\boldsymbol{\alpha}_2[\mathbf{f}] = \dot{\mathbf{f}}(0),$$
$$\boldsymbol{\alpha}_3[\mathbf{f}] = \int_0^\infty \mathbf{f}(t)\,\mathrm{d}t.$$

(c) State on what grounds you know that there exist numbers $\lambda_1, \lambda_2$ and $\lambda_3$ (not all zero) such that

$$\lambda_1\boldsymbol{\alpha}_1 + \lambda_2\boldsymbol{\alpha}_2 + \lambda_3\boldsymbol{\alpha}_3 = \mathbf{0}.$$

Then determine $\lambda_1, \lambda_2$ and $\lambda_3$.

(d) In terms of $\boldsymbol{\alpha}_1$ and $\boldsymbol{\alpha}_2$, construct a basis $\{\boldsymbol{\beta}_1, \boldsymbol{\beta}_2\}$ for $V^*$ that is *dual* to the basis $\{\mathbf{f}_1, \mathbf{f}_2\}$ that you constructed in part (a).

10.13. Let $V$ be the four-dimensional vector space of polynomials of degree $\leqslant 3$, with basis elements $1, t, t^2, t^3$. Let D be the differentiation operator on this space: $\mathrm{D}f(t) = f'(t)$, and let $T_a$ be the translation operator: $T_a f(t) = f(t+a)$.

(a) Construct the matrices which represent D and $T_a$ relative to the given basis, and show that

$$T_a = I + \mathrm{D}a + \tfrac{1}{2}\mathrm{D}^2a^2 + \tfrac{1}{6}\mathrm{D}^3a^3.$$

(b) Prove, that if $V$ is the space of polynomials of degree $\leqslant n$,

$$T_a = \mathrm{e}^{\mathrm{D}a}.$$

(Hint: Think of Taylor's theorem, and you need not construct any matrices.)

10.14. Let $V$ be the space of one-forms on the plane for which the coefficients of $\mathrm{d}x$ and $\mathrm{d}y$ are quadratic functions. A basis for $V$ is $x^2\mathrm{d}x$, $xy\mathrm{d}x$, $y^2\mathrm{d}x$, $x^2\mathrm{d}y$, $xy\mathrm{d}y$, $y^2\mathrm{d}y$. Any curve $\Gamma$ in the plane defines an element $\boldsymbol{\alpha}_\Gamma$ of the dual space $V^*$ by the rule

$$\boldsymbol{\alpha}_\Gamma[\omega] = \int_\Gamma \omega.$$

(a) Invent a non-trivial curve $\Gamma$ for which $\boldsymbol{\alpha}_\Gamma$ is the zero element of $V^*$.

(b) Find a basis for the subspace of $V$ which is annihilated by $\boldsymbol{\alpha}_\Gamma$, where $\Gamma$ is *any* closed curve.

(c) Let $\Gamma_1, \Gamma_2$ and $\Gamma_3$ be any three *closed* curves in the plane. Prove that $\boldsymbol{\alpha}_{\Gamma_1}, \boldsymbol{\alpha}_{\Gamma_2}$, and $\boldsymbol{\alpha}_{\Gamma_3}$ are linearly dependent. (Hint: Use Green's theorem to convert the integrals to double integrals.)

(d) Find curves $\Gamma_1, \Gamma_2, \ldots, \Gamma_6$ (straight line segments will do the job) so that the elements $\alpha_{\Gamma_1}, \alpha_{\Gamma_2}, \ldots, \alpha_{\Gamma_6}$ form a basis for $V^*$ which is *dual* to the basis listed above. For example

$$\int_{\Gamma_1} x^2 dx = 1 \quad \text{while} \quad \int_{\Gamma_j} x^2 dx = 0 \quad j = 2,3,4,5,6$$

$$\int_{\Gamma_2} xy dx = 1 \quad \text{while} \quad \int_{\Gamma_j} xy dx = 0 \quad j = 1,3,4,5,6.$$

10.15. Consider the linear transformation $f$ from $\mathbb{R}^4$ to $\mathbb{R}^3$ whose matrix is

$$\begin{pmatrix} 1 & 2 & 0 & 1 \\ 1 & 0 & 2 & -3 \\ 0 & 1 & -1 & 2 \end{pmatrix}.$$

(a) Find a basis for the kernel (null space) of $f$, and construct the general solution to the equation

$$f(\mathbf{v}) = \begin{pmatrix} 1 \\ 3 \\ -1 \end{pmatrix}.$$

(b) Let $N$ denote the kernel of $f$; let $G$ be the quotient $\mathbb{R}^4/N$. Construct a basis for $G$, and explain how you are sure that your basis vectors are independent elements of $G$.

10.16. Consider the linear transformation $f$ from $\mathbb{R}^4$ to $\mathbb{R}^3$ whose matrix is

$$\begin{pmatrix} 2 & 4 & 2 & 2 \\ 1 & 3 & 2 & 0 \\ 3 & 1 & -2 & 8 \end{pmatrix}.$$

(a) Find a basis for the kernel of $f$, and construct the general solution to the equation

$$f(\mathbf{v}) = \begin{pmatrix} 0 \\ -1 \\ 5 \end{pmatrix}.$$

(b) Write down a basis for the image of $f$.

(c) Two elements of the quotient space $H$ are

$$\mathbf{h}_1 = \overline{\begin{pmatrix} 0 \\ 1 \\ 0 \end{pmatrix}} \quad \text{and} \quad \mathbf{h}_2 = \overline{\begin{pmatrix} 0 \\ 0 \\ 1 \end{pmatrix}}.$$

Show explicitly that $\mathbf{h}_1$ and $\mathbf{h}_2$ are linearly *dependent*. (Hint: Look back at part (a).)

10.17. Let $A$ be the matrix of a linear transformation $f: \mathbb{R}^4 \to \mathbb{R}^4$ given by

$$A = \begin{pmatrix} 1 & 0 & -1 & 1 \\ 0 & 2 & 4 & -8 \\ 2 & 1 & -4 & 6 \\ 1 & 1 & -3 & 5 \end{pmatrix}$$

(a) By row reduction, find the rank of $A$ and find a basis for the kernel. Label this basis $\mathbf{v}_1, \ldots, \mathbf{v}_k$ ($k = \dim \ker A$).

(b) Find the solutions $\mathbf{w}_1, \ldots, \mathbf{w}_k$ to the equations $A\mathbf{w}_i = \mathbf{v}_i$.

(c) Do the vectors $\mathbf{w}_1, \ldots, \mathbf{w}_k, \mathbf{v}_1, \ldots, \mathbf{v}_k$ form a basis for $\mathbb{R}^4$? What does your answer imply about the transformation $f$?

(d) Find a basis for im$A$ and express this basis in terms of the $\mathbf{w}_i$ and the $\mathbf{v}_i$. How does this answer relate to your answer to (c)? Hint: Something strange is going on!)

10.18. Let $A$ denote the $4 \times 4$ matrix

$$A = \begin{pmatrix} 2 & 4 & 2 & 2 \\ 1 & 2 & 0 & 2 \\ 3 & 6 & 5 & 1 \\ 0 & 0 & 3 & -3 \end{pmatrix}.$$

(a) Using row reduction, construct a basis for the kernel of $A$ and a basis for the image of $A$, and construct the general solution to the equation

$$A\mathbf{v} = \begin{pmatrix} 4 \\ 3 \\ 4 \\ -3 \end{pmatrix}.$$

(b) Construct basis vectors $\mathbf{u}_1$ and $\mathbf{u}_2$ for the quotient space $U = \mathbb{R}^4/\ker A$.

Express the vector $\begin{pmatrix} 4 \\ 3 \\ 4 \\ -3 \end{pmatrix}$ in terms of $\mathbf{u}_1$ and $\mathbf{u}_2$.

10.19. Let $W$ be the subspace of $\mathbb{R}^4$ spanned by the vectors

$$\mathbf{w}_1 = \begin{pmatrix} 1 \\ 1 \\ 1 \\ -1 \end{pmatrix}, \quad \mathbf{w}_2 = \begin{pmatrix} 3 \\ 3 \\ 5 \\ 3 \end{pmatrix}, \quad \mathbf{w}_3 = \begin{pmatrix} 1 \\ 1 \\ 3 \\ 5 \end{pmatrix}.$$

The scalar product in $\mathbb{R}^4$ is the ordinary Euclidean one.

(a) Construct an orthonormal basis for $W$.

(b) Write the vector $\mathbf{v} = \begin{pmatrix} 4 \\ 0 \\ -1 \\ 7 \end{pmatrix}$ as the sum of a vector in $W$ and a vector orthogonal to $W$.

(c) Let $f: \mathbb{R}^4 \to \mathbb{R}^3$ be defined by

$$f(\mathbf{v}) = \begin{pmatrix} (\mathbf{w}_1, \mathbf{v}) \\ (\mathbf{w}_2, \mathbf{v}) \\ (\mathbf{w}_3, \mathbf{v}) \end{pmatrix}.$$

Write down the matrix representing $f$. By row-reduction, construct the *general* solution to the equation

$$f(\mathbf{v}) = \begin{pmatrix} 4 \\ 6 \\ -2 \end{pmatrix}.$$

10.20. Using row reduction, find a basis for the subspace $U \subset \mathbb{R}^5$ whose annihilator space $U^\perp$ is spanned by the row vectors

$$\begin{aligned}\boldsymbol{\alpha}^1 &= (1, 2, 0, -1, 2),\\ \boldsymbol{\alpha}^2 &= (2, 4, 3, 4, 4),\\ \boldsymbol{\alpha}^3 &= (0, 0, 1, 2, 1),\\ \boldsymbol{\alpha}^4 &= (3, 6, 5, 7, 8).\end{aligned}$$

Then construct the general solution to the simultaneous linear equations

$$\boldsymbol{\alpha}^1[\mathbf{v}] = 5;\quad \boldsymbol{\alpha}^2[\mathbf{v}] = 16;\quad \boldsymbol{\alpha}^3[\mathbf{v}] = 3;\quad \boldsymbol{\alpha}^4[\mathbf{v}] = 27.$$

10.21. Use row reduction to calculate the inverse of the matrix

$$A = \begin{pmatrix} 1 & -2 & 1 \\ -2 & 5 & -4 \\ 1 & -4 & 6 \end{pmatrix}.$$

10.22. Let $V$ be the space of functions $f$ on $\mathbb{R}^2$ with the property that $f(\lambda x, \lambda y) = \lambda^3 f(x, y)$. A basis for $V$ is

$$\{\mathbf{v}_1 = x^3,\quad \mathbf{v}_2 = x^2 y,\quad \mathbf{v}_3 = xy^2,\quad \mathbf{v}_4 = y^3\}.$$

Let $W$ be the space of one-forms on the plane which are quadratic functions of $x$ and $y$, with basis

$$\{\mathbf{w}_1 = x^2\mathrm{d}x,\ \mathbf{w}_2 = xy\mathrm{d}x,\ \mathbf{w}_3 = y^2\mathrm{d}x,\ \mathbf{w}_4 = x^2\mathrm{d}y,\ \mathbf{w}_5 = xy\mathrm{d}y,\ \mathbf{w}_6 = y^2\mathrm{d}y\}.$$

The operator d is then a linear transformation from $V$ to $W$.

(a) Write down the matrix which represents d relative to the given bases.

(b) Construct a basis for the image of d and for the quotient space $G = W/(\mathrm{im}\,\mathrm{d})$.

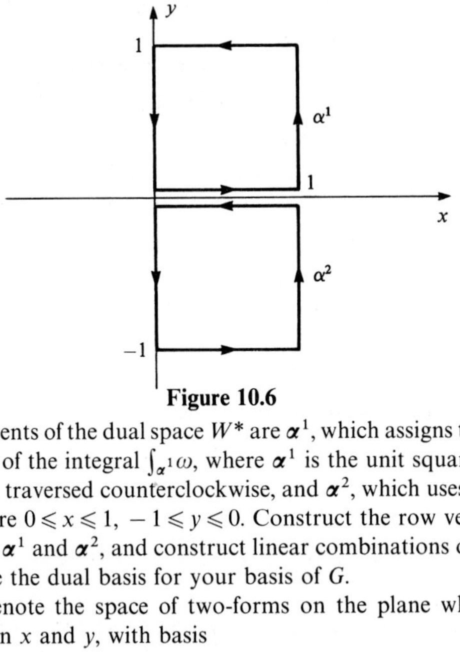


**Figure 10.6**

(c) Two elements of the dual space $W^*$ are $\boldsymbol{\alpha}^1$, which assigns to any $\omega \in W$ the value of the integral $\int_{\alpha^1}\omega$, where $\boldsymbol{\alpha}^1$ is the unit square $0 \leqslant x \leqslant 1$, $0 \leqslant y \leqslant 1$, traversed counterclockwise, and $\boldsymbol{\alpha}^2$, which uses instead the unit square $0 \leqslant x \leqslant 1$, $-1 \leqslant y \leqslant 0$. Construct the row vectors which represent $\boldsymbol{\alpha}^1$ and $\boldsymbol{\alpha}^2$, and construct linear combinations of $\boldsymbol{\alpha}^1$ and $\boldsymbol{\alpha}^2$ which are the dual basis for your basis of $G$.

(d) Let $U$ denote the space of two-forms on the plane which depend *linearly* on $x$ and $y$, with basis

$$\{\mathbf{u}_1 = x\mathrm{d}x \wedge \mathrm{d}y,\ \mathbf{u}_2 = y\mathrm{d}x \wedge \mathrm{d}y\}.$$

Construct the matrix which represents the operator d from $W$ to $U$. What is the kernel of this operator?

10.23. Let $A$ be a linear transformation from $V$ to $W$, $A^*$ the adjoint transformation from $W^*$ to $V^*$. Suppose that we have *not* chosen dual bases in $V$ and $V^*$. Instead, we have a basis $\{\mathbf{v}_1, \mathbf{v}_2, \ldots, \mathbf{v}_m\}$ for $V$, a basis $\{\boldsymbol{\alpha}^1, \boldsymbol{\alpha}^2, \ldots, \boldsymbol{\alpha}^m\}$ for $V^*$, with $\boldsymbol{\alpha}^i[\mathbf{v}_j] = S_{ij}$. The numbers $S_{ij}$ form an $m \times m$ matrix $S$. Similarly, we have bases $\{\mathbf{w}_1, \mathbf{w}_2, \ldots, \mathbf{w}_n\}$ and $\{\boldsymbol{\beta}^1, \boldsymbol{\beta}^2, \ldots, \boldsymbol{\beta}^n\}$ in $W$ and $W^*$ respectively, with $\boldsymbol{\beta}^k[\mathbf{w}_l] = T_{kl}$. The numbers $T_{kl}$ form an $n \times n$ matrix $T$. (If we had chosen dual bases, $S$ and $T$ would both be identity matrices.)

If $A$ is the *matrix* of the transformation relative to bases $\{\mathbf{v}_i\}$ and $\{\mathbf{w}_i\}$, what is the matrix of $A^*$ relative to bases $\{\boldsymbol{\beta}^i\}$ and $\{\boldsymbol{\alpha}^i\}$? Express your answer in terms of $A^{\mathrm{T}}$, the transpose of the matrix $A$, and the matrices $S$ and $T$.

10.24. For a vector space $V$ with a scalar product $(\mathbf{v}_1\ \mathbf{v}_2)$, the adjoint $A^*$ of a linear transformation $A: V \to V$ is another linear transformtion $A^*: V \to V$ defined by

$$(A^*\mathbf{v}_1, \mathbf{v}_2) = (\mathbf{v}_1, A\mathbf{v}_2).$$

(a) Show that this definition follows from the definition of $A^*$ as a transformation from $V^*$ to $V^*$, combined with the usual identification of $V^*$ with $V$ which arises from the scalar product.

(b) Show that, relative to an *orthonormal* basis $\{\mathbf{e}_1, \ldots, \mathbf{e}_n\}$, the matrix representing $A^*$ is the transpose of the matrix representing $A$.

(c) Let $\pi$ denote the linear operation of orthogonal projection from $V$ onto a subspace $W$; i.e., for any $\mathbf{v} \in V$, $\pi\mathbf{v}$ lies in $W$, and $\mathbf{v} - \pi\mathbf{v}$ is orthogonal to $\pi\mathbf{v}$. (Note: $\pi$ is a transformation from $V$ to $V$, with $\operatorname{im} \pi = W$.) Show that $\pi^* = \pi$.

(d) Let $V$ be the space of polynomials of degree $\leqslant 2$, with scalar product $(\mathbf{f}, \mathbf{g}) = \int_0^1 f(t) g(t) \mathrm{d}t$. Choose a basis $\mathbf{v}_1 = 1$, $\mathbf{v}_2 = t$, $\mathbf{v}_3 = t^2$ (which is *not* orthonormal). Let $A$ be the linear transformation defined by

$$Af(t) = \dot{f}\,(t+1).$$

Construct the matrix which represents $A^*$ relative to the given basis.

10.25. Consider the linear transformation from a four-dimensional vector space $V$ to a three-dimensional vector space $W$, which is represented by the matrix

$$A = \begin{pmatrix} 1 & 1 & 0 & 3 \\ 2 & 1 & 2 & 2 \\ 2 & 3 & -2 & 10 \end{pmatrix}.$$

(a) Let $M$ denote the image of $A$. Show that the vectors $\mathbf{m}_1 = \begin{pmatrix} 1 \\ 2 \\ 2 \end{pmatrix}$ and $\mathbf{m}_2 = \begin{pmatrix} 1 \\ 1 \\ 3 \end{pmatrix}$ form a basis for $M$.

(b) Let $H = W/M$. Show that a basis for this quotient space consists of the single element $\mathbf{h}_1 = \overline{\begin{pmatrix} 1 \\ 0 \\ 0 \end{pmatrix}}$, which is the equivalence class containing all vectors $\begin{pmatrix} 1 \\ 0 \\ 0 \end{pmatrix} + \mathbf{m}$, where $\mathbf{m}$ is an element of $M$. Show that $\overline{\begin{pmatrix} 0 \\ 1 \\ 0 \end{pmatrix}} =$

$-\frac{1}{4}\mathbf{h}_1$ $\left(\text{i.e., find an element } \mathbf{m} \text{ of } M \text{ such that} \begin{pmatrix}0\\1\\0\end{pmatrix} = -\frac{1}{4}\begin{pmatrix}1\\0\\0\end{pmatrix} + \mathbf{m}\right)$.

Then express $\overline{\begin{pmatrix}0\\0\\1\end{pmatrix}}$ in terms of $\mathbf{h}_1$.

(c) Let $N$ denote the kernel of $A$. One element of $N$ is $\mathbf{n}_1 = \begin{pmatrix}1\\-4\\0\\1\end{pmatrix}$ Find a second vector $\mathbf{n}_2$ such that $\mathbf{n}_1$ and $\mathbf{n}_2$ form a basis for $N$. $\Bigg($Hint: Apply row reduction to $A$, then look for a vector of the form $\begin{pmatrix}a\\b\\1\\0\end{pmatrix}$.$\Bigg)$

(d) Let $G$ denote the quotient space $V/N$. Define basis vectors in this space by $\mathbf{g}_1 = \overline{\begin{pmatrix}1\\0\\0\\0\end{pmatrix}}$, $\mathbf{g}_2 = \overline{\begin{pmatrix}0\\1\\0\\0\end{pmatrix}}$. Then show that $\overline{\begin{pmatrix}0\\0\\1\\0\end{pmatrix}} = 2\mathbf{g}_1 - 2\mathbf{g}_2$ and express $\overline{\begin{pmatrix}0\\0\\0\\1\end{pmatrix}}$ in terms of $\mathbf{g}_1$ and $\mathbf{g}_2$.

(e) Construct the non-singular $2 \times 2$ matrix $C$ which represents $A$ as a transformation from $G$ (basis $\mathbf{g}_1, \mathbf{g}_2$) to $M$ (basis $\mathbf{m}_1, \mathbf{m}_2$).

(f) Express $\begin{pmatrix}2\\1\\7\end{pmatrix}$ in terms of $\mathbf{m}_1$ and $\mathbf{m}_2$. (Hint: Apply the same operations that you used to row-reduce the matrix.) Then apply $C^{-1}$, and thereby solve the equation $A\mathbf{v} = \begin{pmatrix}2\\1\\7\end{pmatrix}$, obtaining the answer in the form $\mathbf{v} = a\mathbf{g}_1 + b\mathbf{g}_2 = \begin{pmatrix}a\\b\\0\\0\end{pmatrix} + \mathbf{n}$, where $\mathbf{n}$ is an arbitrary element of $N$.

10.26. Let $f^*$ denote the adjoint of the transformation $f$ in the preceding problem. The adjoint $f^*$ is a transformation from the dual space $W^*$ to the dual space $V^*$, defined by $f^*\boldsymbol{\beta}(\mathbf{v}) = \boldsymbol{\beta}(f\mathbf{v})$, where $\boldsymbol{\beta}$ and $\mathbf{v}$ are arbitrary elements of $W^*$ and $V$. With respect to the bases which are dual to the original bases in $V$ and $W$, it is represented by the transpose of $A$:

$$A^{\mathrm{T}} = \begin{pmatrix} 1 & 2 & 2 \\ 1 & 1 & 3 \\ 0 & 2 & -2 \\ 3 & 2 & 10 \end{pmatrix}.$$

(a) Show that the *image* of $f^*$ is dual to the space $G$. Let $\mathbf{g}_1^*$ and $\mathbf{g}_2^*$ be elements of $V^*$ which are dual to $g_1$ and $g_2$ and express each column of $A^{\mathrm{T}}$ in terms of $\mathbf{g}_1^*$ and $\mathbf{g}_2^*$.

(b) Show that the *kernel* of $f^*$ is dual to the space $H$. Let $\gamma'$ be dual to $\mathbf{h}_1$:i.e., $\gamma'(\mathbf{h}_1) = 1$. Let $\boldsymbol{\beta}^1$ be the element of $W^*$ which picks out the first component of a vector; i.e., $\mathbf{w}_1^*\begin{pmatrix} a \\ b \\ c \end{pmatrix} = a$. Express $\boldsymbol{\beta}^1$ in terms of $\gamma'$. Do the same for $\boldsymbol{\beta}^2$ and $\boldsymbol{\beta}^3$, which pick out the second and third components respectively.

(c) Find vectors $\boldsymbol{\alpha}^1$ and $\boldsymbol{\alpha}^2$ in $W^*$ such that $f^*\boldsymbol{\alpha}^1 = \boldsymbol{\beta}^1, f^*\boldsymbol{\alpha}^2 = \boldsymbol{\beta}$. Show that $\boldsymbol{\alpha}^1, \boldsymbol{\alpha}^2, \gamma'$ form a basis for $W^*$.

# 11

# Determinants



Chapter 11 is devoted to proving the central facts about determinants of $n \times n$ matrices. The subject is developed axiomatically, and the basic computational algorithms are presented.

## Introduction

In this chapter we discuss properties of the determinants of $n \times n$ matrices. Let $A$ be an $n \times n$ matrix. We will let $A_1, \ldots, A_n$ denote the columns of $A$. Thus, if $I$ is the $n \times n$ identity matrix,

$$I_1 = \begin{pmatrix} 1 \\ 0 \\ \vdots \\ 0 \end{pmatrix}, \quad I_2 = \begin{pmatrix} 0 \\ 1 \\ 0 \\ \vdots \\ 0 \end{pmatrix} \ldots \text{etc.}$$

For any matrix $A$, then

$$A_1 = AI_1, \ldots, A_n = AI_n;$$

in other words, $A_i$ is just the image of $I_i$, the $i$th element of the standard basis, under $A$.

We expect to be able to define $\operatorname{Det} A$ as the *oriented volume of the parallelepiped spanned by* $A_1, \ldots, A_n$. Our experiences in Chapters 1 and 9 suggest that this oriented volume may be multi-linear – that is, linear in $A_1$, when $A_2, \ldots, A_n$ are held fixed, linear in $A_2$ when $A_1, A_3, \ldots, A_n$ are held fixed, and so on. Also (due to the orientation), we expect that $\operatorname{Det} A$ should be antisymmetric in the columns; that is, interchanging the columns of a matrix changes the sign of its determinant. We must define the determinant, and prove that it has the requisite properties. In fact, we shall follow the classical treatment of Artin and characterize the determinant axiomatically. That is, we shall write down a simple list of properties we expect the determinant function to have, and shall show that these properties uniquely characterize the determinant. In other words, there is only at most *one* such function. Also we shall get some rules for computing it (if it exists). We will then show that there does exist a function satisfying the axioms. By showing that various other definitions also satisfy the axioms, we will be able to conclude that all these definitions must give the *same* function.

In what follows, we will be interested in a function, $D$, of matrices. When it is evaluated on a matrix $A$ we write $D(A)$ or $D(A_1,\ldots,A_n)$ when we want to emphasize that it is a function of the $n$ column vectors $A_1,\ldots,A_n$. If we keep all the columns but the $k$th constant, we obtain a function of a single column. We shall write this function as $D_k$. (It is understood that all the other columns are held fixed with given values.) For example, we shall write

$$D_2\left(\begin{pmatrix} x \\ y \\ z \end{pmatrix}\right) \quad \text{for} \quad D\left(\begin{pmatrix} 1 & x & 7 \\ -2 & y & 9 \\ 3 & z & -1 \end{pmatrix}\right).$$

(Strictly speaking, we should specify the vectors $A_1 = \begin{pmatrix} 1 \\ -2 \\ 3 \end{pmatrix}$ and $A_3 = \begin{pmatrix} 7 \\ 9 \\ -1 \end{pmatrix}$ in the notation for $D_k$, and write

$$D_{2;\begin{pmatrix} 1 \\ -2 \\ 3 \end{pmatrix},\begin{pmatrix} 7 \\ 9 \\ -1 \end{pmatrix}}\left(\begin{pmatrix} x \\ y \\ z \end{pmatrix}\right)$$

but the notation would be overly cumbersome.)

## 11.1. Axioms for determinants

A function $D$ of matrices is called a *determinant* if it satisfies the following conditions:

Each of the functions $D_k$ is linear:

$$D_k(A_k + A_k') = D_k(A_k) + D_k(A_k'),$$
$$D_k(cA_k) = cD_k(A_k). \tag{11.1}$$

In other words, $D$ is linear in each column when all the other columns are kept fixed.

If two adjacent columns of a matrix $A$ are equal, then $D(A) = 0$. (11.2)

And

$$D(I) = 1. \tag{11.3}$$

We will now draw some consequences from (11.1) and (11.2) – assuming that some function $D$ exists satisfying (11.1) and (11.2).

*Adding any multiple of one column to an adjacent column does not change the value of D.*

*Proof*

$$D(A_1,\ldots,A_k,cA_k+A_{k+1},\ldots,A_n)$$
$$=D(A_1,\ldots,A_k,A_{k+1},\ldots,A_n)+cD(A_1,\ldots,A_k,A_k,\ldots,A_n)$$

by (11.1). But $D(A_1,\ldots,A_k,A_k,\ldots,A_n)=0$ by (1.2). So

$$D(A_1,\ldots,cA_k+A_{k+1},\ldots,A_n)=D(A_1,\ldots,A_n). \tag{11.4}$$

For example,

$$D\left(\begin{pmatrix}1&4&7\\2&5&8\\3&6&9\end{pmatrix}\right)=D\left(\begin{pmatrix}1&4-4\cdot1&7\\2&5-4\cdot2&8\\3&6-4\cdot3&9\end{pmatrix}\right)$$
$$=D\left(\begin{pmatrix}1&0&7\\2&-3&8\\3&-6&9\end{pmatrix}\right)$$

Now add the $k$th column to the $(k+1)$st, then subtract the resulting $(k+1)$st column from the $k$th, then add the $k$th to the $(k+1)$st again – so

$$\begin{aligned}D(A)&=D(A_1,\ldots,A_k,A_k+A_{k+1},A_{k+2},\ldots,A_n)\\&=D(A_1,\ldots,A_k-(A_k+A_{k+1}),A_k+A_{k+1},\ldots,A_n)\\&=D(A_1,\ldots,-A_{k+1},A_k+A_{k+1},\ldots,A_n)\\&=D(A_1,\ldots,-A_{k+1},A_k+A_{k+1}-A_{k+1},\ldots,A_n)\\&=D(A_1,\ldots,-A_{k+1},A_k,A_{k+2},\ldots,A_n)\\&=-D(A_1,\ldots,A_{k+1},A_k,A_{k+1},\ldots,A_n)\quad\text{by (11.1).}\end{aligned} \tag{11.5}$$

Thus

*Interchanging two adjacent columns changes the sign of $D(A)$.*

Now this implies that

*If any two columns of $A$ are equal, then $D(A)=0$.* (11.6)

Indeed, if two columns are equal, we can keep interchanging adjacent columns until the two columns are adjacent then apply (11.5) to conclude (11.6). We can now apply the argument proving (11.4) to conclude:

*Adding any multiple of one column to another does not change the value of $D(A)$.* (11.6)

Thus, continuing our example,

$$D\left(\begin{pmatrix}1&4&7\\2&5&8\\3&6&9\end{pmatrix}\right)=D\left(\begin{pmatrix}1&0&7\\2&-3&8\\3&-6&9\end{pmatrix}\right)\quad\text{subtract }7\times\text{ first column from last}$$
$$=D\left(\begin{pmatrix}1&0&0\\2&-3&-6\\3&-6&-12\end{pmatrix}\right)\quad\text{now add }-2\times\text{ second column to last}$$

$$= D\begin{pmatrix} 1 & 0 & 0 \\ 2 & -3 & 0 \\ 3 & -6 & 0 \end{pmatrix} = 0 \text{ by (11.1).}$$

Let us do another example.

$$D\begin{pmatrix} 2 & 1 & 4 \\ 3 & 2 & 1 \\ 4 & 2 & -1 \end{pmatrix} \quad \text{subtract } 2 \times \text{second column from first}$$

$$= D\begin{pmatrix} 0 & 1 & 4 \\ -1 & 2 & 1 \\ 0 & 2 & -1 \end{pmatrix} \quad \text{add } 2 \times \text{first column to second}$$

$$= D\begin{pmatrix} 0 & 2 & 4 \\ -1 & 0 & 1 \\ 0 & 2 & -1 \end{pmatrix} \quad \text{add first column to third}$$

$$= D\begin{pmatrix} 0 & 2 & 4 \\ -1 & 0 & 0 \\ 0 & 2 & -1 \end{pmatrix} \quad \text{subtract } 4 \times \text{second column from third}$$

$$= D\begin{pmatrix} 0 & 1 & 0 \\ -1 & 0 & 0 \\ 0 & 2 & -9 \end{pmatrix} \quad \text{by (11.1)}$$

$$= -9D\begin{pmatrix} 0 & 1 & 0 \\ -1 & 0 & 0 \\ 0 & 2 & 1 \end{pmatrix} \quad \text{subtract } 2 \times \text{third column from second}$$

$$= -9D\begin{pmatrix} 0 & 1 & 0 \\ -1 & 0 & 0 \\ 0 & 0 & 1 \end{pmatrix} \quad \text{interchange first and second columns}$$

$$= +9D\begin{pmatrix} 1 & 0 & 0 \\ 0 & -1 & 0 \\ 0 & 0 & 1 \end{pmatrix} \quad \text{by (11.1)}$$

$$= -9D\begin{pmatrix} 1 & 0 & 0 \\ 0 & 1 & 0 \\ 0 & 0 & 1 \end{pmatrix}.$$

If we now apply (11.3) we conclude from the axioms that

$$D\begin{pmatrix} 2 & 1 & 4 \\ 1 & 2 & 1 \\ 4 & 2 & -1 \end{pmatrix} = -9.$$

We also observe that it follows from (11.7) (as in the proof of (11.5)) that

*Interchanging any two columns changes the sign of $D(A)$.* (11.8)

We can also conclude:

*If the columns of $A$ are linearly dependent, then $D(A) = 0$.* (11.9)

Indeed, if $\lambda_1 A_1 + \cdots + \lambda_n A_n = 0$ and some $\lambda_i \neq 0$, then we can solve for the $i$th column in terms of the others

$$A_i - c_1 A_1 - \cdots - c_n A_n = 0, \quad c_i = 0.$$

So subtracting $c_1 A_1$, etc. from the $i$th column does not change $D(A)$ (by (11.6)) and yields a matrix whose $i$th column vanishes. Then by (11.1) $D(A) = 0$.

In particular, we know that any $n$ vectors of the form

$$\left(\!\!\left(\begin{matrix} 0 \\ a_2 \\ \vdots \\ a_n \end{matrix}\right)\!\!\right)$$

(with a zero in the first position) are linearly dependent. Therefore

If all the entries of the top row of $A$ are zero, then $D(A) = 0$. (11.10)

Suppose that at least one entry in the top row of $A$ does not vanish. By an interchange of columns, if necessary, we can arrange that the *first* column has a non-zero entry in the top row. So

$$D(A) = \pm D(B) \quad \text{where} \quad b_{11} \neq 0.$$

Now

$$D(B) = b_{11} D(B')$$

where the first column, $B'_1$ of $B'$, is $B'_1 = (1/b_{11}) B_1$.

By subtracting off suitable multiples of the first column from each of the remaining columns we can arrange that all the other entries in the top row vanish, i.e.,

$$D(B') = D(B'')$$

where

$$B'' = \left(\begin{array}{c|c} 1 & 0 \ldots 0 \\ \hline B''_{21} & \\ \vdots & C \\ B''_{n1} & \end{array}\right).$$

Now consider $D(B'')$ as a function of the columns of $C$. Clearly it satisfies conditions (11.1) and (11.2). Also, if $C$ were the $(n-1) \times (n-1)$ identity matrix, we could without changing the value of $D(B'')$ make all the entries $b''_{21}, b''_{31}$, etc., in the first column vanish just by subtracting off multiples of the second, third, etc., columns from the first. For example,

$$D\left(\!\!\left(\begin{matrix} 1 & 0 & 0 & 0 \\ 2 & 1 & 0 & 0 \\ 3 & 0 & 1 & 0 \\ 4 & 0 & 0 & 1 \end{matrix}\right)\!\!\right) = \begin{array}{l} \text{subtracting} \\ 2 \times \text{the second} \\ \text{column from the first} \end{array}$$

$$D\left(\!\!\left(\begin{matrix} 1 & 0 & 0 & 0 \\ 0 & 1 & 0 & 0 \\ 3 & 0 & 1 & 0 \\ 4 & 0 & 0 & 1 \end{matrix}\right)\!\!\right) = \begin{array}{l} \text{subtracting } 3 \times \text{the} \\ \text{third column from the} \\ \text{second} \end{array}$$

$$D\left(\begin{pmatrix}1&0&0&0\\0&1&0&0\\0&0&1&0\\4&0&0&0\end{pmatrix}\right) = \begin{matrix}\text{subtracting } 4\times \text{the}\\ \text{fourth column from}\\ \text{the third}\end{matrix}$$

$$D\left(\begin{pmatrix}1&0&0&0\\0&1&0&0\\0&0&1&0\\0&0&0&1\end{pmatrix}\right).$$

In other words,

$$D\left(\left(\begin{array}{c|c}1 & 0\ldots0\\ \hline B''_{21} & \\ B''_{n1} & C\end{array}\right)\right)$$

as a function of $C$, satisfies all the axioms for a determinant for $(n-1)\times(n-1)$ matrices. Therefore

$$D(B'') = D(C)$$

where, on the right, we mean the $D$-function (if it exists) for $(n-1)\times(n-1)$ matrices.

Applying the same argument over again, we conclude that either $D(C)=0$ (if all the entries in *its* top row are zero) or we can express $D(C)$ in terms of a $D$-function for $(n-2)\times(n-2)$ matrices. Eventually we get down to a $1\times 1$ 'matrix' where the axioms (11.1) and (11.3) imply that

$$D(a) = a(D(1)) = a.$$

This proves that the $D$ function, if it exists, is *unique* and gives a definite recipe for computing it.

For example, suppose

$$A = \begin{pmatrix}0&3&-1&4\\2&4&1&5\\-4&7&2&8\\2&9&3&2\end{pmatrix}.$$

$D(A) = -D(B)$ where

$$B = \begin{pmatrix}-1&3&0&4\\1&4&2&5\\2&7&-4&8\\3&9&2&2\end{pmatrix},$$

$D(B) = D(B'')$ where

$$B'' = \begin{pmatrix}1&0&0&0\\-1&7&2&9\\-2&13&-4&16\\-3&18&2&14\end{pmatrix}$$

and $D(B'') = D(C)$ where

$$C = \begin{pmatrix} 7 & 2 & 9 \\ 13 & -4 & 16 \\ 18 & 2 & 14 \end{pmatrix}.$$

Now

$$\begin{aligned} D(C) &= 2D\left(\begin{pmatrix} 7 & 1 & 9 \\ 13 & -2 & 16 \\ 18 & 1 & 14 \end{pmatrix}\right) \\ &= -2D\left(\begin{pmatrix} 1 & 0 & 0 \\ -2 & 13 & 16 \\ 1 & 1 & 14 \end{pmatrix}\right) \\ &= -2D\left(\begin{pmatrix} 1 & 0 & 0 \\ -2 & 27 & 34 \\ 1 & -6 & 5 \end{pmatrix}\right) \\ &= -2D\left(\begin{pmatrix} 27 & 34 \\ -6 & 5 \end{pmatrix}\right) \\ &= -2{\cdot}27D\left(\begin{pmatrix} 1 & 34 \\ -\frac{6}{27} & 5 \end{pmatrix}\right) \\ &= -2{\cdot}27D\left(\begin{pmatrix} 1 & 0 \\ -\frac{6}{27} & 5 + 34{\cdot}\frac{6}{27} \end{pmatrix}\right) \\ &= -2{\cdot}27(5 + 34{\cdot}\tfrac{6}{27}). \end{aligned}$$

In the next section we shall give a different proof of the uniqueness of $D(A)$, and a different recipe. The uniqueness implies that these two recipes must give the same answer. But we must still show that a $D(A)$ satisfying (11.1), (11.2) and (11.3) exists.

We can, however, derive an important consequence from our current algorithm procedure for computing $D(A)$. Suppose the matrix $A$ is of the form

$$A = \begin{pmatrix} L & M \\ 0 & N \end{pmatrix}$$

where $L$ is a $k \times k$ matrix, $N$ an $(n-k) \times (n-k)$ matrix and $M$ a matrix with $k$ rows and $n-k$ columns. In other words, suppose that the first $k$ columns of $A$ all have zeros in their last $n-k$ positions. Then the first $k$ columns of $A$ are linearly dependent or independent if and only if the columns of $L$ are. That is, the last $n-k$ zero positions in these vectors do not affect the dependence or independence of these columns. If these columns are linearly dependent, then

$$D(A) = 0 \quad \text{and} \quad D(L) = 0.$$

If these columns are linearly independent, then in applying our algorithm, we can use the first $k$ columns in the first $k$ steps and thus replace $M$ by the zero matrix but not affect the entries of $N$ at all, until we get to the $(k+1)$st step. Thus

$$D\begin{pmatrix} L & M \\ 0 & N \end{pmatrix} = D(L)D(N). \tag{11.11}$$

This is the generalization to $n$ dimensions of the 'base times height' formula for the area of a parallelogram. In two dimensions it says that

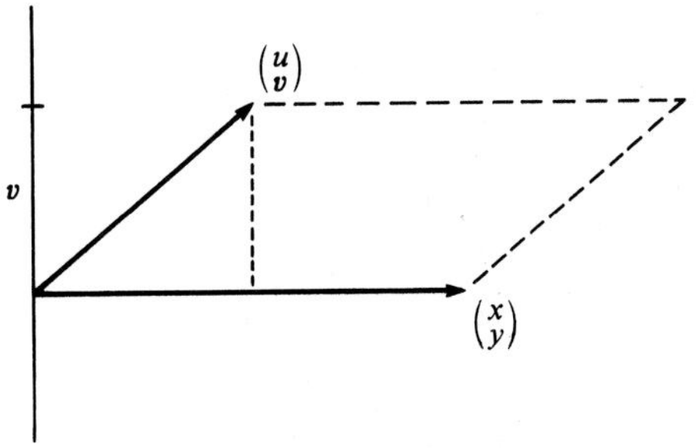


**Figure 11.1**

$$D\left(\begin{pmatrix} x & u \\ 0 & v \end{pmatrix}\right) = xv$$

i.e., that the oriented area of the parallelogram spanned by $\begin{pmatrix} x \\ 0 \end{pmatrix}$ and $\begin{pmatrix} u \\ v \end{pmatrix}$ is the same as the oriented area of the parallelogram spanned by $\begin{pmatrix} x \\ 0 \end{pmatrix}$ and $\begin{pmatrix} 0 \\ v \end{pmatrix}$. In three dimensions it says that

$$D\left(\begin{pmatrix} x & u & p \\ y & v & q \\ 0 & 0 & r \end{pmatrix}\right) = D\left(\begin{pmatrix} x & u \\ y & v \end{pmatrix}\right) D((r))$$

with a similar interpretation.

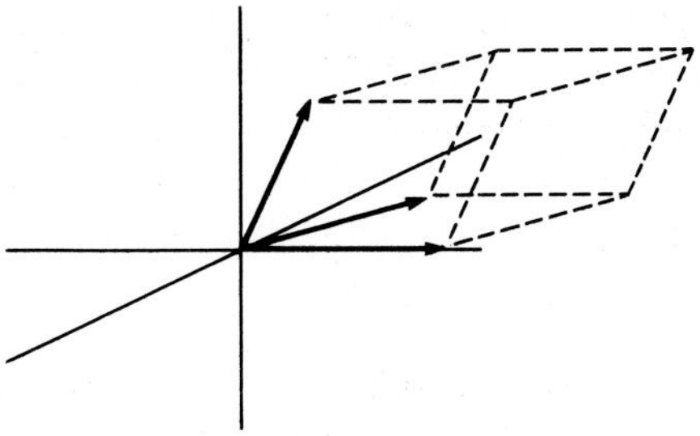

**Figure 11.2**

## 11.2. The multiplication law and other consequences of the axioms

Let us draw some further consequences from (11.8). Let $(v_1,\ldots,v_n)$ be any permutation of $(1,\ldots,n)$. We can rearrange the columns, one at a time in the matrix

$$D((A_{v_1},\ldots,A_{v_2},\ldots,A_{v_n}))$$

until they are back in their original order. At each stage we apply (11.8), and conclude that

$$D((A_{v_1},\ldots,A_{v_n}))=\pm D((A_1,\ldots,A_n)) \tag{11.12}$$

where the $\pm$ does not depend on the particular entries of $A$. Applying (11.12) to the identity matrix, we see that

$$D((I_{v_1},\ldots,I_{v_n}))=\pm 1$$

and hence that

$$D((A_{v_1},\ldots,A_{v_n}))=D((I_{v_1},\ldots,I_{v_n}))D(A). \tag{11.13}$$

Now let $B=(b_{ij})$ be a second $n\times n$ matrix and let

$$C=AB.$$

The columns of $C$ are given by

$$C_k=b_{1k}A_1+b_{2k}A_2+\cdots+b_{nk}A_n.$$

Now in computing $D(C)$ we may first apply (11.1) to the first column of $C$ getting a sum; then to each summand we apply (11.1) to the second column and so on. For example, in the $3\times 3$ case

$$\begin{aligned}C_1&=b_{11}A_1+b_{21}A_2+b_{31}A_3,\\ C_2&=b_{12}A_1+b_{22}A_2+b_{32}A_3,\\ C_3&=b_{13}A_1+b_{23}A_2+b_{33}A_3\end{aligned}$$

so

$$\begin{aligned}D(C_1,C_2,C_3)&=b_{11}D(A_1,C_2,C_3)+b_{21}D(A_2,C_2,C_3)+b_{31}D(A_3,C_2,C_3)\\ &=b_{11}\{b_{12}D(A_1,A_1,C_3)+b_{22}D(A_1,A_2,C_3)+b_{32}D(A_1,A_3,C_3)\}\\ &\quad+b_{21}\{b_{12}D(A_2,A_1,C_3)+b_{22}D(A_2,A_2,C_3)+b_{32}D(A_2,A_3,C_3)\}\\ &\quad+b_{31}\{b_{12}D(A_3,A_1,C_3)+b_{22}D(A_3,A_2,C_3)+b_{32}D(A_3,A_3,C_3)\}.\end{aligned}$$

Before proceeding to the next step, we can eliminate all repeated columns. It is clear that at the end only expressions of the form $D((A_{v_1},A_{v_2},A_{v_3}))$ will be left and these, by (11.12), are equal to $\pm D((A))$. Thus, in general, we see that

$$D(C)=D(A)\sum\pm b_{v_1 1}b_{v_2 2}\ldots b_{v_n n} \tag{11.14}$$

where the sum is taken over all permutations and the $\pm$ sign is the one given by (11.12).

Suppose we use (11.14) and take $A=I$. Then $C=B$ and we conclude that

$$D(B)=\sum\pm b_{v_1 1}\cdots b_{v_n n}. \tag{11.15}$$

This gives an explicit formula for $D(B)$ and again proves that it is unique – if it exists.

If we substitute (11.15) into (11.14), we get

$$D(AB) = D(A)D(B) \tag{11.16}$$

the multiplication law for determinants.

Each term in (11.15) contains exactly one entry from each row of $B$. Therefore

*$D(B)$ is a linear function of each row, when all other rows are held fixed.* (11.1)′

Now in (11.14) take $i$ to be some number $1 \leqslant i \leqslant n$ and take $A$ to be the matrix with

$$A_k = I_k \quad k \neq i, i+1,$$
$$A_i = I_i + I_{i+1},$$
$$A_{i+1} = 0.$$

For example, with $n = 3$ and $i = 2$ we would have the matrix

$$\begin{pmatrix} 1 & 0 & 0 \\ 0 & 1 & 0 \\ 0 & 1 & 0 \end{pmatrix}.$$

Notice that $AB$ has all its rows except the $(i+1)$st the same as the rows of $B$. The $(i+1)$st row has been replaced by the $i$th. In the above example

$$\begin{pmatrix} 1 & 0 & 0 \\ 0 & 1 & 0 \\ 0 & 1 & 0 \end{pmatrix}\begin{pmatrix} 1 & 5 & 9 \\ 2 & 6 & 8 \\ 3 & 7 & 4 \end{pmatrix} = \begin{pmatrix} 1 & 5 & 9 \\ 2 & 6 & 8 \\ 2 & 6 & 8 \end{pmatrix}.$$

Also $D(A) = 0$ since it has one whole column zero. For this we conclude that

*If $B$ has two adjacent rows equal, then $D(B) = 0$.* (11.2)′

This means that

$$D(B^{\mathrm{T}}) \text{ satisfies axioms (11.1) and (11.2)}$$

because replacing $B$ by its transpose, $B^{\mathrm{T}}$, interchanges the role of rows and columns. But $I^{\mathrm{T}} = I$, so (11.3) is also satisfied. Thus $D(B^{\mathrm{T}})$ satisfies axioms (11.1)–(11.3), hence by uniqueness must coincide with $D(B)$. In other words

$$D(B^{\mathrm{T}}) = D(B). \tag{11.17}$$

## 11.3. The existence of determinants

We shall now prove the existence of determinants. That is, we shall construct a function of $n \times n$ matrices that clearly satisfies (11.1), (11.2) and (11.3).

For $n = 1$

$$\text{define } D((a)) = a.$$

For $n = 2$

$$\text{define } D\left(\begin{pmatrix} a & b \\ c & d \end{pmatrix}\right) = ad - bc.$$

It is easy to check directly that (11.1)–(11.3) are satisfied. We now proceed inductively. Suppose that we assume the existence of $(n-1)\times(n-1)$ determinants. Let

$$A = (a_{ik})$$

be an $n \times n$ matrix. Consider some definite position, say the position at the $i$th row and $k$th column. Let us cancel the $i$th row and $k$th column in $A$ and take the determinant of the remaining $(n-1)$-rowed matrix. This determinant multiplied by $(-1)^{i+k}$ will be called the *cofactor* of $a_{ik}$ and be denoted by $A_{ik}$. The distribution of the sign $(-1)^{i+k}$ follows the chessboard pattern, namely

$$\begin{matrix} + & - & + & - & \cdot & \cdot & \cdot \\ - & + & - & + & \cdot & \cdot & \cdot \\ + & - & + & - & \cdot & \cdot & \cdot \\ - & + & - & + & \cdot & \cdot & \cdot \\ \cdot & \cdot & \cdot & \cdot & \cdot & \cdot & \cdot \end{matrix}$$

Let $i$ be any number from 1 to $n$. We consider the following function $D$ of the matrix $(A)$:

$$D = a_{i1}A_{i1} + a_{i2}A_{i2} + \cdots + a_{in}A_{in}. \tag{11.18}$$

It is the sum of the products of the $i$th row and their cofactors.

Consider this $D$ and its dependence on a given column, say $A_k$. For $v \neq k$, $A_{iv}$ depends linearly on $A_k$ and $a_{iv}$ does not depend on it; for $v = k$, $A_{ik}$ does not depend on $A_k$ but $a_{ik}$ is one element of this column. Thus (11.1) is satisfied. Assume next that two adjacent columns $A_k$ and $A_{k+1}$ are equal. For $v \neq k, k+1$, we have then two equal columns in $A_{iv}$ so that $A_{iv} = 0$. The determinants used in the computation of $A_{i,k}$ and $A_{i,k+1}$ are the same but the signs are opposite; hence $A_{i,k} = -A_{i,k+1}$ whereas $a_{i,k} = a_{i,k+1}$. Thus $D = 0$ and (11.2) holds. For the special case $A_v = I_v$, $v = 1, 2, \ldots, n$, we have $a_{iv} = 0$ for $v \neq i$, while $a_{ii} = 1$, $A_{ii} = 1$. Hence $D = 1$ and this is (11.3). This proves the existence of an $n$-rowed determinant as well as the truth of formula (11.18), the so-called development of a determinant according to its $i$th row. Equation (11.18) may be generalized as follows. In our determinant replace the $i$th row by the $j$th row and develop according to this new row. For $i \neq j$ that determinant is 0 and for $i = j$ it is $D$:

$$a_{j1}A_{i1} + a_{j2}A_{i2} + \cdots + a_{jn}A_{in} = \left\{\begin{matrix} D & \text{for} & j = i, \\ 0 & \text{for} & j \neq i \end{matrix}\right\}. \tag{11.19}$$

If we interchange the rows and the columns, we get the following formula:

$$a_{1h}A_{1k} + a_{2h}A_{2k} + \cdots + a_{nh}A_{nh} = \left\{\begin{matrix} D & \text{for} & h = k, \\ 0 & \text{for} & h \neq k \end{matrix}\right\}. \tag{11.20}$$

Equation (11.20) says that, if we form the matrix

$$B = (A_{ij})$$

called the *cofactor* matrix of $A$, then

$$B^{\mathrm{T}}A = D(A)I.$$

Notice that we have already proved that if $A$ is singular (so that the columns of $A$ are linearly dependent), then $D(A)=0$. The preceding equation gives a formula for $A^{-1}$ if $D(A)\neq 0$. Thus we have proved

*A matrix A is invertible if and only if $D(A)\neq 0$. If $D(A)\neq 0$, then*

$$A^{-1}=\frac{1}{D(A)}B^{\mathrm{T}} \tag{11.21}$$

*where B is the cofactor matrix of A.*

This formula for $A^{-1}$ is known as *Cramer's rule*. It is not an effective way of computing $A^{-1}$ if $n>2$. (For $n=2$, it coincides with the prescription in Chapter 1.) For $n>2$, it is better to use the algorithm described in Chapter 10. However, Cramer's rule does have theoretical importance. For instance, it shows that the entries of $A^{-1}$ are all quotients of a polynomial in the entries of $A$ by the determinant.

## Summary

A Determinants

You should know the axioms for determinants and be able to use them directly for evaluation of determinants.

You should be able to state and apply the rule for evaluating a determinant by use of cofactors.

You should be able to state and apply Cramer's rule for the inverse of a nonsingular square matrix.

## Exercises

You should write down several $3\times 3$ and $4\times 4$ matrices and evaluate their determinants. You will find that already in the $4\times 4$ case the expression (11.15) or (11.19) becomes quite unpleasant (involving 4! multiplications on 4! − 1 additions) while the algorithm described in section 11.1 is quite manageable. Here are several against which you can check your arithmetic.

(a) $\mathrm{Det}\begin{pmatrix}1 & 2 & 3 & 4\\ 5 & 8 & 11 & 12\\ 6 & 9 & 13 & 15\\ 7 & 10 & 14 & 16\end{pmatrix}=-2.$

(b) $\mathrm{Det}\begin{pmatrix}0 & 2 & 2 & 2\\ 3 & 1 & 2 & 2\\ 3 & 3 & 4 & 2\\ 3 & 3 & 3 & 5\end{pmatrix}=12.$

(c) $\mathrm{Det}\begin{pmatrix}1 & 1 & 1 & 1\\ 2 & 3 & 3 & 3\\ 2 & 2 & 3 & 3\\ 2 & 2 & 2 & 3\end{pmatrix}=1.$

(d) $\operatorname{Det}\begin{pmatrix} 1 & 1 & 1 & 1 \\ 2 & 3 & 5 & 6 \\ 4 & 9 & 25 & 36 \\ 8 & 27 & 125 & 216 \end{pmatrix} = 72 = (3-2)(5-2)(6-2)(5-3) \times (6-3)(6-5).$

11.1. In generalization of example (b), show that

$$\operatorname{Det}\begin{pmatrix} r_1 & a & a & a \\ b & r_2 & a & a \\ b & b & r_3 & a \\ b & b & b & r_4 \end{pmatrix} = \frac{bf(a) - af(b)}{b-a} \quad \text{if} \quad a \neq b$$

where $f(x) = (r_1 - x)(r_2 - x)(r_3 - x)(r_4 - x)$.
Hint: The determinant of the matrix below is a function $F(x)$.

$$\begin{pmatrix} r_1 - x & a-x & a-x & a-x \\ b-x & r_2 - x & a-x & a-x \\ b-x & b-x & r_3 - x & a-x \\ b-x & b-x & b-x & r_4 - x \end{pmatrix}$$

But it is a *linear* function of $x$ since we may subtract the first row from all the remaining rows to eliminate the $x$ from all but the first row. Hence

$$F(x) = A + Bx$$

for some constants $A$ and $B$. But $F(a) = f(a)$ and $F(b) = f(b)$. So solve for $A$ and set $x = 0$. What does the formula become when $a = b$?

11.2. In generalization of example (c), show that

$$\operatorname{Det}\begin{pmatrix} 1 & 1 & 1 & 1 \\ x & a & a & a \\ x & y & b & b \\ x & y & z & c \end{pmatrix} = (a-x)(b-y)(c-z).$$

11.3. In generalization of (d), show that

$$\begin{pmatrix} 1 & 1 & 1 & 1 \\ x & y & z & w \\ x^2 & y^2 & z^2 & w^2 \\ x^3 & y^3 & z^3 & w^3 \end{pmatrix} = (y-x)(z-x)(w-x)(z-y)(w-y)(w-z).$$

State and prove the corresponding fact with 4 replaced by $n$.

11.4. Show that

$$|\operatorname{Det}(A_1, \ldots, A_n)| \leqslant \|A_1\| \cdots \|A_n\|.$$

When does equality hold?
(Hint: Use the interpretation of $|\operatorname{Det}|$ in terms of volume.)

11.5. Show that if $O$ is an orthogonal matrix (so $OO^{\mathrm{T}} = I$), then $\operatorname{Det} O = \pm 1$.

11.6. A matrix $R$ is a *rotation* if $RR^{\mathrm{T}} = I$ and $\operatorname{Det} R = +1$. Show that a rotation in an *odd*-dimensional space always leaves at least one non-zero factor fixed; i.e., $R$ has 1 as an eigenvalue.
(Hint: Consider $\operatorname{Det}(R - I)$.)

# SUGGESTED READING

The short list of books that we give at the end of this section is not meant as bibliography. Rather, it consists of books that students of the course have found helpful in supplementing and extending the material covered in this volume. The book by **Loomis** and **Sternberg** can be considered as a companion text. The presentation there is more abstract and formal, with more of an emphasis on mathematical proof. The actual mathematical prerequisites are the same as for this book, but the demands on mathematical sophistication and on tolerance for formal definitions and argumentation are greater. On one or two occasions in this book and in Volume 2 we have referred to Loomis and Sternberg for the detailed proof of some key theorems. The **Feynman Lectures** form another general reference giving an elegant presentation of physics at the level of this book.

One of our main subjects is linear algebra. The text by **Halmos** is a classic, with a tilt towards extension of the finite dimensional theory in the direction of Hilbert space. The text by **Lang** discusses the subject from the viewpoint of abstract algebra. The text by **Strang** emphasizes computational techniques and applications that we barely touch upon here. A good strategy is to read all three to get a balanced view of the subject.

Our discussion in Chapter 1 started with the geometry of lines. The natural place to go from there is to the study of projective geometry, and we gave some indications in this direction in the appendix to Chapter 1 and in Exercises 1.16–1.20. The text by **Hartshorne** gives a coherent introduction to the subject.

At the end of Chapter 2 we make a brief mention of probability theory, and it is one of our major gaps that we don't give a serious discussion of this important topic. A good all-round introduction to probability which does not make heavy mathematical demands are the three volumes by **Hoel, Port**, and **Stone**. Probability theory can easily lead into rather imposing mathematical machinery such as measure theory and intricate questions in Fourier analysis. These books have the advantage of illustrating the important ideas without getting into the subject deeply enough to be entangled in heavy mathematics. The book by **Moran** is harder reading, but worth the effort. The book by **Kemeny** and **Snell** gives a self-contained discussion of finite Markov chains and can be read as a continuation of Chapter 2. The book by **Doyle** and **Snell** is a delightful short introduction to both Markov chains and to networks, which we shall study in Volume 2.

Chapter 3 gives an introduction to differential equations. The text by **Hirsch** and **Smale** would be a natural next book from a point of view close to the one we espouse here. The book by **Braun** is organized much more along standard lines. (The exponential of a matrix does not make its appearance until p. 321!) However, the attention paid to the details of many and varied applications makes this book worthwhile. The book by **Simmons** is also fairly standard but has interesting historical information. The two books by **Arnol'd** are classics and are delightful reading. The interplay between geometry and analysis displays the sure hand of one of the masters of the subject.

In Chapter 4 we spend two sections on relativity theory. We have listed three books on the subject. The book by **Misner** *et al.* is big and heavy, but full of ideas. The book by **Taylor** and **Wheeler** is short and inspirational and involves a minimum of mathematics. The book *Spacetime, Geometry, Cosmology* by **Burke** develops many of the mathematical ideas we try to explain in our book and gives a very well thought out discussion of the physics of relativity. It is 'user friendly' and we recommend it strongly.

The other book by **Burke**, together with the books by **Flanders** and by **Spivak**, can be regarded as parallel with ours, and are recommended as general supplementary and collateral reading.

In Chapter 6 and again in section 10.9 we touch on topics which naturally belong in a course on differential topology. The book by **Guillemin** and **Pollack** is written in a discursive style with many pictures and intuitive guides along with the formal presentation of the theory. The book by **Bröcker** and **Lander** is just the opposite. It is very terse, with concise statements of the theorems and proofs and a minimum of discussion. But of course this can be an advantage. The books cover somewhat different topics and we recommend them both.

## References


Arnol'd, V.I. *Oridinary Differential Equations* MIT Press 1973

Arnol'd, V.I. *Geometric Methods in the Theory of Ordinary Differential Equations* Springer-Verlag 1983

Braun, M. *Differential Equations and their Applications* Springer-Verlag 1975

Bröcker, Th. and Lander, L. *Differentiable Germs and Catastrophes* Cambridge Univ. Press London Mathematical Society Lecture Note Series 17

Burke, W.L. *Spacetime, Geometry, Cosmology* University Science Books 1980

Burke, W.L. *Applied Differential Geometry* Cambridge Univ. Press 1985

Doyle, P.G. and Snell, J.L. *Random Walks and Electric Networks* Carus Mathematical Monographs Math. Ass. of Amer. 1985

Feynman, R.P., Leighton, R.B. and Sands, M. *The Feynman Lectures on Physics* Addison-Wesley 1963

Flanders, H. *Differential Forms with Applications to the Physical Sciences* Academic Press 1963
Guillemin, V.W. and Pollack, A. *Differential Topology* Prentice-Hall 1974
Guillemin, V.W. and Sternberg, S. *Symplectic Techniques in Physics* Cambridge University Press 1984
Halmos, P.F. *Finite Dimensional Vector Spaces* Springer-Verlag 1972
Hartshorne, R. *Foundations of Projective Geometry* Benjamin 1967
Hirsch, M.W. and Smale, S. *Differential Equations, Dynamical Systems and Linear Algebra*
Hoel, P.G., Port, S.C. and Stone, C.J. *Introduction to Probability, Introduction to Stochastic Processes, Introduction to Statistical Theory*, Houghton Mifflin 1971–2
Kemeny, J.G. and Snell, L. *Finite Markov Chains* van Nostrand Reinhold 1960
Lang, S. *Linear Algebra* Springer-Verlag
Loomis, L.H. and Sternberg, S. *Advanced Calculus* Addison-Wesley 1968
Misner, C.W., Thorne, K.S. and Wheeler, J.A. *Gravitation* W.H. Freeman 1977
Moran, P.A.P. *An Introduction to Probability Theory* Clarendon Press 1968
Simmons *Differential Equations with Applications and Historical Notes* McGraw-Hill 1972
Spivak, M. *Calculus on Manifolds* Benjamin 1965
Strang, G. *Linear Algebra and its Applications* Academic Press 1980
Taylor, E.F. and Wheeler, J.A. *Spacetime Physics* W.H. Freeman 1966

# INDEX

Harrogate, North Yorkshire
September 1990